Microorganisms to Combat Pollution

edited by

E. Rosenberg

Department of Molecular Microbiology & Biotechnology,
Tel Aviv University, Israel

Reprinted from *Biodegradation* 3: 2/3

Springer Science+Business Media, B.V.

Library of Congress Cataloging-in-Publication Data

Microorganisms to combat pollution / edited by E. Rosenberg.
p. cm.
Includes bibliographical references.
ISBN 978-94-010-4730-2 ISBN 978-94-011-1672-5 (eBook)
DOI 10.1007/978-94-011-1672-5
1. Bioremediation--Congresses. I. Rosenberg, Eugene.
TD192.5.M33 1993
628.5'2--dc20 93-3223

ISBN 978-94-010-4730-2

Printed on acid-free paper

Contents

Preface v

A. Genetically Engineered Microorganisms

Daubaras D. & Chakrabarty A. M.: The environment, microbes and bioremediation: microbial activities modulated by the environment [1]

Atlas R. M.: Molecular methods for environmental monitoring and containment of genetically engineered microorganisms [13]

B. Metals

Barkay T., Turner R., Saouter E. & Horn J.: Mercury biotransformations and their potential for remediation of mercury contamination [23]

Ron E. Z., Minz D., Finkelstein N. P. & Rosenberg E.: Interactions of bacteria with cadmium [37]

C. Lignocellulose

Bayer E. A. & Lamed R.: The cellulose paradox: pollutant *par excellence* and/or a reclaimable natural resource? [47]

Hadar Y., Kerem Z., Gorodecki B. & Ardon O.: Utilization of lignocellulosic waste by the edible mushroom, *Pleurotus* [65]

Shoham Y., Schwartz Z., Khasin A., Gat O., Zosim Z. & Rosenberg E.: Delignification of wood pulp by a thermostable xylanase from *Bacillus stearothermophilus* strain T-6 [83]

Broda P.: Biotechnology in the degradation and utilization of lignocellulose [95]

D. Nitrogen and Sulphur

Kuenen J. G. & Robertson L. A.: The use of natural bacterial populations for the treatment of sulphur-containing wastewater [115]

Abeliovich A.: Transformations of ammonia and the environmental impact of nitrifying bacteria [131]

E. Halogenated Compounds

van der Meer J. R., Bosma T. N. P., de Bruin W. P., Harms H., Holliger C., Rijnaarts H. H. M., Tros M. E., Schraa G. & Zehnder A. J. B.: Versatility of soil column experiments to study biodegradation of halogenated compounds under environmental conditions [141]

Boyle A. W., Silvin C. J., Hassett J. P., Nakas J. P. & Tanenbaum S. W.: Bacterial PCB biodegradation [161]

Belkin S.: Biodegradation of haloalkanes [175]

F. Petroleum

Pritchard P. H., Mueller J. G., Rogers J. C., Kremer F. V. & Glaser J. A.: Oil spill bioremediation: experiences, lessons and results from the *Exxon Valdez* oil spill in Alaska [191]

Rosenberg E., Legmann R., Kushmaro A., Taube R., Adler E. & Ron E. Z.: Petroleum bioremediation – a multiphase problem [213]

Cerniglia C. E.: Biodegradation of polycyclic aromatic hydrocarbons [227]

G. Military / Industrial Pollution

Walker J. E. & Kaplan D. L.: Biological degradation of explosives and chemical agents [245]

Oren A., Gurevich P., Azachi M. & Henis Y.: Microbial degradation of pollutants at high salt concentrations [263]

List of contributors 275

Index 277

Preface

This volume contains material first presented at an international workshop on the 'Use of Microorganisms to Combat Pollution', held in Israel, May 10–18, 1992. The workshop was sponsored by the Bat-Sheva de Rothschild Foundation for the Advancement of Science and included microbiologists, biochemists and geneticists from universities, environmental agencies and the military. Each of the contributors to this volume is an acknowledged expert on the treatment of one or more types of pollution using microorganisms or their enzymes.

This book differs from most published symposia proceedings in the breadth of coverage of each subject. Most of the chapters are divided into three parts: (a) A general presentation of the source and toxicity of the pollutant, (b) a review of the current state-of-the-science on the biodegradation of that pollutant and (c) the authors' unique research experiences on the problem. In several examples, the authors have presented data from both laboratory studies and field trials. Thus, the book contains not only the theoretical background on the biodegradation of pollutants, but also practical experiences in applying this knowledge to solving significant pollution problems.

The opening chapters deal with the use of genetically engineered microorganisms (GEM's) in bioremediation research and application. The major use of GEM's is as a tool for an improved understanding of the biodegradation pathways. The actual use of GEM's to treat or avoid pollution appears to be limited at present to totally recalcitrant pollutants, such as certain aromatic halides, and situations where high levels of a specific enzyme are required, e.g., biobleaching of wood pulp with xylanase (Shoham et al.). The presentation of Ron Atlas demonstrates how new molecular biological techniques, such as the use of PCRs, can be applied to detection and containment of GEM's.

The remaining chapters are divided into the type of pollution potentially treatable by microorganisms: metals, lignocellulose, inorganic nitrogen and sulfur molecules, halogenated compounds, hydrocarbons and products produced by the military which must now be destroyed in an environmentally safe manner. In some cases the pollutants are mixtures of compounds that are difficult to define, e.g., petroleum and PCBs. This, of course, makes the quantitative determination of biodegradation more difficult. Nevertheless, the data presented here document and explain the generally successful use of microorganisms to treat a wide variety of pollution problems.

It is hoped that the wide scope of pollution problems presented and the broad approach to discussing them will make this book useful to students and scientists interested in applied microbiology and the environmental sciences.

E. Rosenberg

Biodegradation **3**: 125–135, 1992.

The environment, microbes and bioremediation: microbial activities modulated by the environment

Dayna Daubaras & A.M. Chakrabarty*
*Department of Microbiology & Immunology (M/C 790), University of Illinois College of Medicine, 835 South Wolcott Avenue, Chicago, IL 60612, USA (*requests for offprints)*

Key words: natural evolution, directed evolution, biodegradation, environmental pollutants, environmental signal transduction, gene expression

Abstract

Microorganisms in nature are largely responsible for the biodegradation and removal of toxic and non-toxic chemicals. Many organisms are also known to have specific ecological niches for proliferation and colonization. The nature of the environment dictates to a large extent the biodegradability of synthetic compounds by modulating the evolutionary processes in microorganisms for new degradative genes. Similarly, environmental factors often determine the extent of microbial gene expression by activating or repressing specific gene or sets of genes through a sensory signal transduction process. Understanding how the environment modulates microbial activity is critical for successful bioremediative applications.

Introduction

The environment exerts profound influence on microbial activities. Many microbes are fully active in certain environments but are rendered inactive in others. Bioremediation, which employs biological agents for the detoxification and removal of environmental pollutants, is often carried out in the fields where microbial activities or their products are essential for bioremediation to be effective. If the pollutants are synthetic and new to the environment, the microorganisms may not have the capability to degrade the pollutants since the appropriate biodegradative genes may not have evolved. Even if the genes are present, functional expression of the genes is essential for the appropriate degradative enzymes to be produced. Various factors present in the environment are known to activate or repress gene expression and thereby modulate microbial activities. If bioremediation is to succeed, it is imperative that we learn how the environment controls evolution of new genes in bacteria or the expression of microbial genes. This article deals with an example in each of these two areas.

Environmental modulation of microbial gene evolution

A prominent example of bioremediation often cited in newspapers and scientific magazines is the Exxon Valdez oil spill. The indigenous microorganisms in Prince William Sound were supplied with an oleophilic fertilizer allowing them to proliferate and consume the spilled oil (Pritchard & Costa 1991). In addition, numerous bioremediation studies are being conducted using natural (often indigenous) or adapted microorganisms to degrade and remove a variety of pollutants from the environment (Keeler 1991). Emphasis so far, however, has been on the use of natural microorganisms, rather than genetically manipulated ones, due to adverse public reaction to, as well as various regulatory constraints on, the release of genetically engineered microorganisms in the environment (Bakst 1991). However, natural mixed cultures are known to degrade pollutants, including crude oil, at slow rates because of interactions among themselves (Friello et al. 1976). Also, since many contaminated sites contain more than one pollutant, it is often difficult for natural microorganisms to degrade, efficiently and simultaneously, a mixture of pollutants. It has been shown (Haugland et al. 1990a) that a single culture with appropriate genetic competence to degrade a mixture of chemicals is much more efficient than a mixture of cultures with equal genetic capability. In addition, indigenous microorganisms, when exposed to a mixture of pollutants, may produce toxic intermediates that do not allow an overall reduction of the toxicity or teratogenicity of the biotreated sample to any significant extent (Mueller et al. 1991). Thus, genetically improved single cultures are considered more efficient than mixed cultures in degrading single or mixed toxic chemicals.

Another important reason for using genetic selection for decontamination of polluted environments is that in many cases natural microorganisms have not evolved the genetic competence to utilize a synthetic compound. For example, the half-life of chlorinated dioxins or a number of polychlorinated compounds is of the order of years, which means that natural microorganisms do not have efficient enzyme systems to break down these compounds. To degrade a newly-made synthetic compound, a microorganism must evolve the appropriate genes encoding enzymes that have high affinities for the target chemical or intermediate products as substrates. This sort of natural evolution may take years, depending on the chemical structure and the solubility of the compound, as well as the nature of the environment. Therefore, genetic selection resulting in an enhancement of the evolutionary process may play a critical role in facilitating the evolution of new biodegradative genes in natural microorganisms for the utilization of such compounds.

How do we enhance the process of natural evolution for biodegradative purposes? To address this question, it is necessary to understand how new genes encoding new types of catabolic enzymes evolve in nature. An interesting example of natural evolution of new catabolic enzymes is the microbial degradation of chlorinated aromatic compounds. Large quantities of these compounds have been synthesized and released into the environment in the form of herbicides and pesticides, or as industrially useful compounds such as PCBs (polychlorinated biphenyls). Because these compounds have many chlorine substituents, natural microorganisms that rapidly degrade the non-chlorinated parent compounds are unable to mineralize the chlorinated ones to any significant extent. However, because of exposure to natural microflora for many years, a number of chlorinated aromatic compounds with fewer chlorine substituents can be degraded by natural microorganisms through the use of newly evolved degradative genes.

An example of such an evolutionary process is that of chlorocatechol degradative genes. Catechol is a central intermediate of aromatic degradation (Wheelis & Ornston 1972) and its mode of degradation is illustrated in Fig. 1. Catechol is utilized by a large number of bacteria, including the genus *Pseudomonas*. In *P. putida*, catechol is degraded by a set of chromosomal genes *cat*A, *cat*B, *cat*C etc., leading to the formation of β-ketoadipate. This is finally converted to succinate and other members of the tricarboxylic acid cycle for oxidative metabolism. The enzymes such as pyrocatechase I and cycloisomerase I (muconate lactonizing enzyme I) have high affinity for their substrates, catechol and *cis*, *cis*-muconate, but have little affinity for the chlorinated catechols (Schlomann et al. 1990). Consequently, catechol degrading microorganisms are unable to utilize chlorinated catechols. However, it is possible to isolate from nature various *Pseudomonas* species and *Alcaligenes eutrophus* which are capable of degrading chlorinated benzoic acids, phenoxyacetic acids, and benzenes such as 3-chlorobenzoic acid (3-Cba; Chatterjee et al. 1981), 2,4-dichlorophenoxyacetic acid (2,4-D; Don & Pemberton 1981) and 1,2,4-trichlorobenzene (1,2,4-Tcb; van der Meer et al. 1991a). These compounds are metabolized to their corresponding

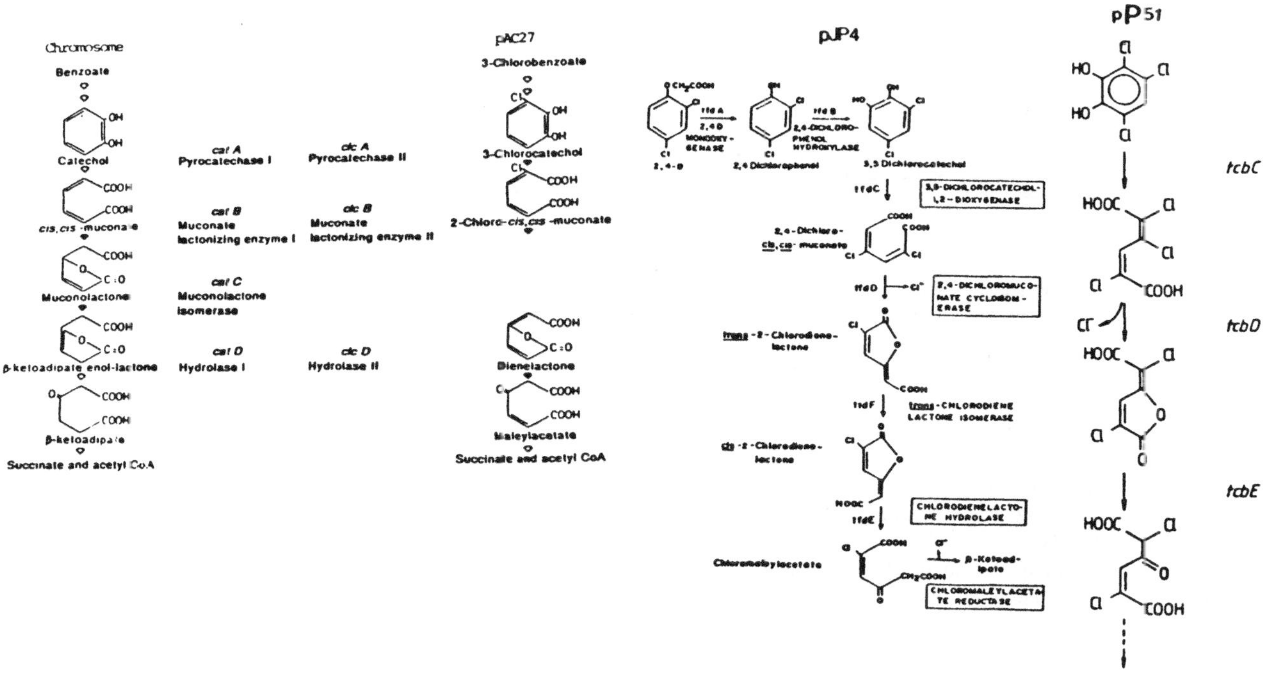

Fig. 1. The degradation pathways of catechol, 3-chlorocatechol, 3,5-dichlorocatechol and 3,4,6-trichlorocatechol encoded by chromosomal genes and the plasmids pAC27, pJP4 and pP51. The genes for each step of the pathway are indicated, as well as the enzymes of the first three pathways. Type II enzymes involved in 3-chlorocatechol and 3,5-dichlorocatechol utilization are also involved in 3,4,6-trichlorocatechol dissimilation.

chlorocatechols, *viz.* 3-chlorocatechol, 3,5-dichlorocatechol, and 3,4,6-trichlorocatechol, which are further degraded via chloromuconates and dienelactones (Reineke & Knackmuss 1988). Thus, the pathways of catechol and chlorocatechol degradation are quite similar, as shown in Fig. 1, where *cat*A, *clc*A, *tfd*C, and *tcb*C encode type I or type II 1,2-dioxygenases acting on catechol or chlorinated catechols, *cat*B, *clc*B, *tfd*D, and *tcb*D encode cycloisomerases acting on *cis*, *cis*-muconate or its chlorinated derivatives, and *cat*D, *clc*D, *tfd*E, and *tcb*E encode hydrolases having affinity for enol-lactone or dienelactones. Although the pathways are similar, the evolved enzymes active on chlorinated catechols or their intermediate metabolites have altered substrate specificities. For example, the *cat*A gene product (type I dioxygenase) has little activity towards chlorinated catechols, while the *clc*A gene product (type II dioxygenase) has high activity towards 3-chlorocatechol, with lower activity towards 3,5-dichlorocatechol. In contrast, the *tcb*C gene product preferentially acts on 3,4-dichlorocatechol, whereas the *tfd*C product has a higher affinity for 3,5-dichlorocatechol (Schlomann et al. 1990; van der Meer et al. 1991a). In addition, similar substrate specificity differences among the enzymes acting on chlorinated and non-chlorinated compounds exist in subsequent steps of the pathway.

How do microorganisms evolve new degradative genes that encode enzymes with new substrate specificities? It is interesting to note that the catechol degradative (*cat*) genes are chromosomal while the chlorocatechol degradative (*clc*, *tfd*, and *tcb*) genes are all plasmid-borne. Recruitment of genes in nature is facilitated by the transmission of plasmids to other natural microorganisms to allow degradation of chlorinated compounds, as shown by mesocosm studies in the presence of chlorobenzoates (Fulthorpe & Wyndham 1991). Since chlorobenzoate, chlorophenoxyacetate, and chlorobenzene degradation initially requires oxygenases to generate

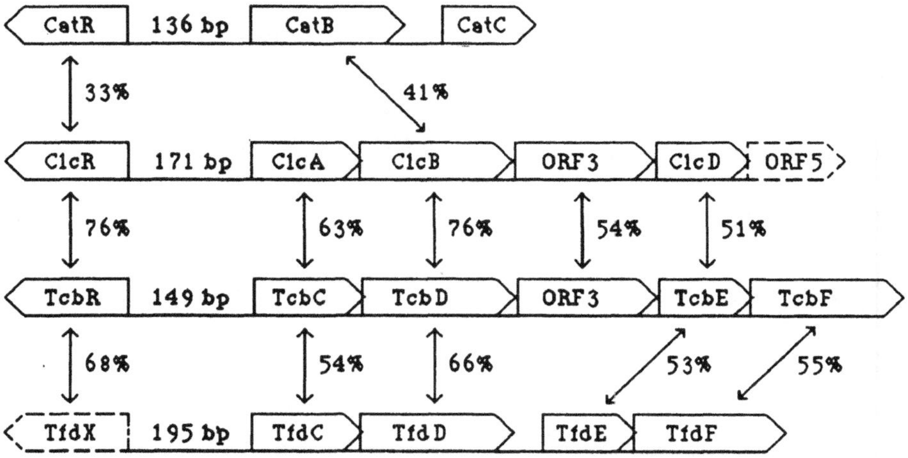

Fig. 2. The organizational similarity between the catechol and chlorocatechol degradative operons as shown by their gene products. Each of the operons contains a divergently transcribed regulatory gene with 100 to 200 base pairs between the ATG of the first gene in the operon and the ATG of the divergently transcribed regulatory gene. TfdX is the translated gene product of part of an open reading frame with amino acid identity to TcbR and other regulatory proteins, but its exact function has not been delineated. The extent of amino acid identities among the gene products is shown by arrows (the amino acid identity of TcbR to ClcR is only over a portion of ClcR, not the whole protein). In addition, other identities between the gene products are: CatR and TcbR, 31%; ClcA and TfdC, 59%; CatB and TcbD, 44%; CatB and TfdD, 42%; ClcB and TdfD, 63%; ClcD and TfdE, 54%. Also, ORF5 whose function is unknown, shows about 47% identity to TfdF in its N-terminal region.

chlorocatechol, it is interesting to note that the genes for oxygenases are often recruited by transposable elements on a plasmid either as separate gene clusters flanked by IS elements or as a composite transposon (van der Meer et al. 1991b; Nakatsu et al. 1991). A dehalogenase gene involved in dechlorinating chlorinated aliphatic compounds has been reported to be present on a mobile genetic element capable of recombining at high frequencies with plasmid and chromosomal DNA (Thomas et al. 1992). Molecular cloning, DNA sequencing, as well as purification and amino acid sequence determination of some of the catechol and chlorocatechol degradative enzymes, have provided considerable insight into the mode of evolution of the chlorocatechol degradative genes.

It should be noted that the host strains harboring the three plasmids pAC27, pJP4, and pP51 containing the *clc*ABD, *tfd*CDEF and *tcb*CDEF gene clusters (Fig. 1) were isolated in different continents *viz.* the United States, Australia, and Europe. These plasmids appear to be different with respect to their host range, *inc* property, and substrate profiles. If these three plasmids evolved independently in the natural microflora in response to the release of chlorinated substrates into the environment, then we can ask if the genes of the pathway will show any similarity to one another as well as to any presumptive ancestral genes, such as the *cat* genes. Indeed, the organization of the *cat*BC operon which is positively regulated by *cat*R regulatory gene, shows similarity to that of the *clc*ABD operon and its regulatory gene *clc*R (Rothmel et al. 1991a). In addition, both operons are similar in their organization (Fig. 2) to the *tcb*CDEF operon and its regulator gene *tcb*R (van der Meer et al. 1991c). In the above cases, the positive regulatory genes are transcribed divergently from the operons they control in such a way that the promoters overlap (Rothmel et al. 1991a; van der Meer et al. 1991c; Rothmel et al. 1991b). In contrast, the *tfd*CDEF operon is controlled by a negative regulatory gene *tfd*R, which acts as an activator in the presence of inducers such as 2,4-D, 2,4-dichlorophenol, and 4-chlorocatechol; also *tfd*R maps several kilobases upstream of the *tfd*CDEF operon (Kaphammer et al. 1990). However, just upstream of *tfd*C, there is a divergently transcribed open reading frame (*tfd*X, Fig. 2) which shows similarity to the other regulatory genes (Perkins et al. 1988). In addition, CatR shows significant amino acid sequence homology with the other two regulatory proteins ClcR and TfdX (Rothmel et al. 1991a), while TcbR shows significant homology with parts of CatR, ClcR, and TfdX (van der Meer et al. 1991c). Similarly, the structural genes of the four operons *cat*BC, *clc*ABD, *tfd*CDEF, and *tcb*CDEF exhibit extensive homology with one another (Fig. 2). The chromosomally encoded CatA shows a good deal of homology to ClcA (Neidle et al. 1988; Frantz et al. 1987) while ClcA shows 50 to 60% homology (Fig. 2) with TcbC and TfdC (van der Meer et al. 1991a; Perkins et al. 1990; Ghosal & You 1988). Similarly CatB shows 40 to 50% homology with ClcB, TfdD, and TcbD (Aldrich et al. 1987; Frantz & Chakrabarty 1987; van der Meer et al. 1991a; Perkins et al. 1990; Ghosal & You 1988), while ClcD shows about 50% homology with TfdE and TcbE (van der Meer et al. 1991a; Perkins et al. 1990). Finally, about 50% homology exists between TfdF and TcbF and between TfdF and the N-terminal portion of a putative open reading frame at the end of the *clc*ABD operon.

In addition to sequence identity among the evolved genes, there is striking organizational similarity in the *clc*, *tfd*, and *tcb* operons (Fig. 2). For example, not only are the genes organized in the same order as the steps of the pathway but there is a 4 base pair overlap (ATGA) between the stop codon of *clc*A and the start codon of *clc*B (Frantz & Chakrabarty 1987). This 4 base pair overlap also exists between the stop codons of *tfd*C and *tcb*C and the start codons of *tfd*D and *tcb*D (Perkins et al. 1990; van der Meer et al. 1991c). Similarly, this 4 base pair overlap can be found between the stuctural genes of *clc*D and ORF5, *tcb*E and *tcb*F and *tfd*E and *tfd*F. Even more interesting is the presence of an additional open reading frame, called ORF3, between the *clc*B and *clc*D genes and also between the *tcb*D and *tcb*E genes. These ORF3 regions have no known function in chlorocatechol degradation but are present in both operons with more than 50% homology between the two (van der Meer et al. 1991c). Thus, the evolved gene clusters exhibit a great deal of organizational and sequence similarity to each other and to the chromosomal *cat* genes that allow degradation of non-chlorinated catechol.

We have so far discussed the evolution of genes that took place over a long period of time in response to the release of chlorinated compounds in the environment. In nature, these compounds are often present with other biodegradable lignocellulosic materials, and therefore, the urgency to recruit and evolve new biodegradative genes is absent. We have described another system (Rothmel et al. 1991a) in which microorganisms from dump sites were subjected to strong selection in a chemostat for the utilization of 2,4,5-trichlorophenoxyacetic acid (2,4,5-T). This recalcitrant compound was supplied to the microorganisms as the only major source of carbon and energy. Thus, the evolution of a degradative pathway for the utilization of 2,4,5-T was necessary for the survival of the microorganisms in the chemostat. It is interesting to note that 2,4,5-T degradative (*tft*) genes which evolved in a strain of *Pseudomonas cepacia* AC1100, showed little DNA homology to other members of the genus *Pseudomonas* (Tomasek et al. 1989; Haugland et al. 1991). A transposable element termed IS931 (Haugland et al. 1990b) present near the *tft* genes with a large number of copies on the genome of *P. cepacia* AC1100 also did not show any hybridization with genomic digests from a large number of pseudomonads (Tomasek et al. 1989). Recently, a cluster of *tft* genes has been sequenced and DNA (and protein) databank searches revealed homologies with genes encoding glutathione reductases, glutathione transferases, and (chloro) catechol 1,2-dioxygenases (Fig. 3). Glutathione transferases are often associated with xenobiotic detoxification in a variety of organisms including eukaryotes and have not been reported in pseudomonads. Thus, it appears that a stressed environment such as the chemostat with a single recalcitrant carbon source favors gene recruitment from any available source. Therefore, environmental conditions to a large extent dictate the mode of evolution of new degradative genes.

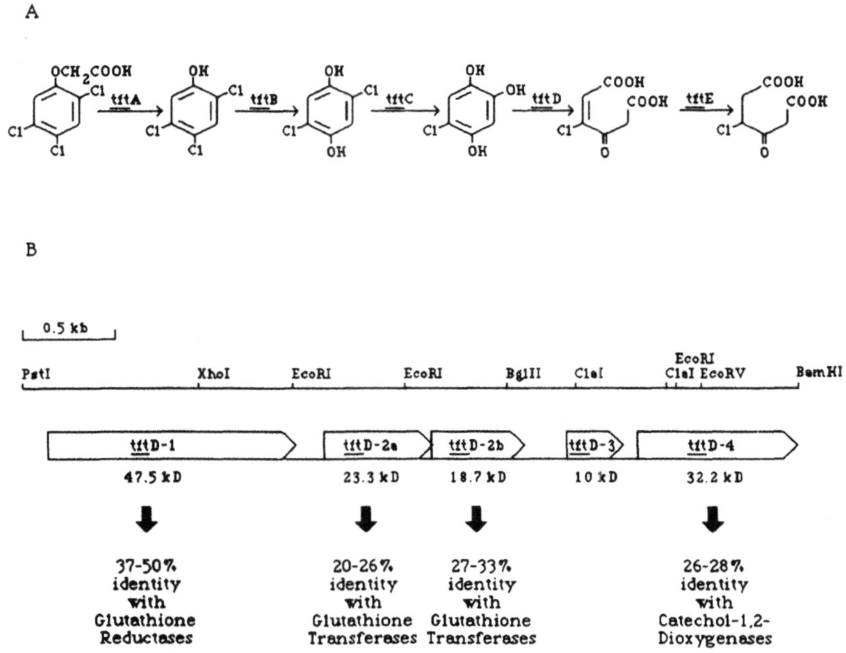

Fig. 3. (A) A tentative pathway of 2,4,5-T degradation and (B) DNA sequence analysis of a 4.2 kb *Pst*I-*Bam*H1 fragment which complements a 2,4,5-T negative mutant. The *tft*D gene cluster is believed to encode a chlorohydroquinone oxygenase complex. This cluster revealed the presence of five potential open reading frames, *tft*D-1, *tft*D-2a, *tft*D-2b, *tft*D-3 and *tft*D-4. The molecular mass of each polypeptide encoded by these genes is indicated. Four of the polypeptides, excluding the gene product of *tft*D-2b, were confirmed by expression of these genes from the T7-promoter in *E. coli*. In addition, amino acid comparison searches of the protein databanks found significant homologies with the gene products of *tft*D-1, *tft*D-2a/b, and *tft*D-4 and enzymes such as glutathione reductase, glutathione transferase, and catechol 1,2-dioxygenase respectively.

Environmental modulation of microbial gene expression

In the preceding section, we discussed how the environment modulates the recruitment and evolution of degradative pathway genes in bacteria. The environment also plays a major role in regulating the expression of specific genes in microorganisms, such that the habitat of some microorganisms is dictated by the environment they live in. An example of how the ecological niche of a bacterium is influenced by the environment is the proliferation of mucoid cells of *P. aeruginosa* in the lungs of cystic fibrosis (CF) patients. Normally, *P. aeruginosa* cells are nonmucoid and do not produce appreciable amounts of extracellular or capsular polysaccharide. The *P. aeruginosa* cells are, however, known to infect the lungs of CF patients, where they become heavily mucoid on continued proliferation in this environment due to production of an exopolysaccharide termed alginic acid (May et al. 1991). Alginate is a polymer of D-mannuronic and L-guluronic acids (Fig. 4A) which is a strong gelling agent. Alginate is believed to encapsulate *P. aeruginosa* cells to help the infecting bacteria adhere to the epithelial cells of the CF lung tissues, thereby giving rise to microcolonies on the tissue surfaces. The strong gel formation by the alginate capsule is also beneficial to the infecting *P. aeruginosa* cells since the alginate layer is believed to protect the cells from antibiotic therapy and the body's immune system. Therefore, mucoid, alginate-producing *P. aeruginosa* cells proliferate extensively in the CF lung, causing serious problems for the patients due to production of other virulence factors such as elastase, various proteases, phospholipase C, exotoxin A, etc (May et al. 1991) that cause extensive damage to the lung.

It should be pointed out that the CF lung environment is different from a normal lung environment in that the CF disease is characterized by a defective chloride channel which leads to an accumulation of abnormal fluids in the CF lung. Such a genetic defect allows accumulation of high levels of electrolytes such as NaCl and a highly dehydrated, sticky mucus in the respiratory tissues. The CF lung is, therefore, another example of a stressed environment similar to the chemostat with 2,4,5-T as the only major carbon source, as described previously. When mucoid *P. aeruginosa* cells isolated from the CF lung are cultured in the laboratory in a rich medium, they tend to revert back to non-mucoidy (i.e. no longer produce alginate). This suggests that there are specific genetic mechanisms that allow the *P. aeruginosa* cells to switch on the alginate genes in the stressed CF lung (to help survive in a dehydrated, salty environment), but to turn the switch off when the environment is favorable. In addition, although *P. aeruginosa* is known to cause other infections in burn and eye or urinary-tract, they are seldom mucoid. This suggests that mucoidy due to alginate production is primarily a characteristic of growth in the CF lung. Only mucoid *P. aeruginosa* cells appear to use the CF lung as their primary habitat.

How do *P. aeruginosa* cells sense the CF lung environment when they happen to end up in the lung, and how do they transduce the signals to trigger the activation of the alginate genes? In nature, microorganisms are known to scan the environment, resulting in the expression of a specific gene or sets of genes. Such environmental sensory signal transduction is well known for gene expression involving chemotaxis, nitrogen assimilation, medium osmolarity, virulence, etc (Bourret et al. 1991; Stock et al. 1990). Sensory signal transduction often employs a so-called two-component system; a sensor, usually a trans-membrane protein, scans the environment and in the presence of appropriate signals, undergoes a conformational change, which triggers autophosphorylation. The phosphate is then transferred from the sensor kinase to a cytoplasmic regulator protein, which in its phosphorylated form activates the promoter of a gene or set of genes by functionally binding the promoter region (Stock et al. 1990; Bourett et al. 1991). If the kinase is a soluble protein, it often needs additional transducer proteins that can sense external signals by a membrane spanning domain and interact with the kinase to trigger phosphorylation.

To understand how the alginate genes, which are normally silent in *P. aeruginosa*, are specifically expressed in the CF lung environment, we have delineated the alginate biosynthetic pathway (Fig. 4B) and the organization of the alginate biosynthetic (*alg*) genes (Fig. 4C). The mannuronate units of alginate are synthesized from fructose 6-phosphate *via* formation of mannose 6-phosphate, mannose 1-phosphate, GDP-mannose and GDP-mannuronic acid (Fig. 4B). Four enzymatic activities *viz*. PMI (Phosphomannose isomerase), PMM (Phosphomannomutase), GMP (GDP-mannose pyrophosphorylase) and GMD (GDP-mannose dehydrogenase) are involved in this conversion. Two of the enzymes, PMI and GMP, are encoded by a single gene *alg*A, while PMM is encoded by *alg*C and GMD by *alg*D (Fig. 4B). Another gene, *alg*E, believed to be involved in alginate polymerization (Chu et al. 1991) is present downstream of *alg*D. All these genes which are clustered at 34-min region of the chromosome (except *alg*C), have been completely sequenced and their gene products hyperproduced from the *tac* promoter (Shinabarger et al. 1991; Zielinski et al. 1991; Roychoudhury et al. 1989; Roychoudhury et al. 1992a). The functions of other genes present downstream of the *alg*D gene such as *alg*8, *alg*44, *alg*60 etc, are not known although they are believed to be involved in alginate polymerization or excretion. The gene *alg*G, downstream of *alg*E (Fig. 4C), is known to be required for insertion of guluronate residues in alginate (Chitnis & Ohman 1990). Very little is known about the detailed steps involving epimerization, acetylation or excretion of alginate.

The gene *alg*C, encoding PMM, is not part of the cluster at 34-min but has been shown to map (Shortridge et al. 1991) on a 60-kb *Dra*I fragment at 10-min region near three *alg* regulatory genes *alg*R1, *alg*R2 and *alg*R3 (Fig. 4C). The nucleotide sequence of the *alg*R1 gene demonstrates that it is a member of the two-component response regulator

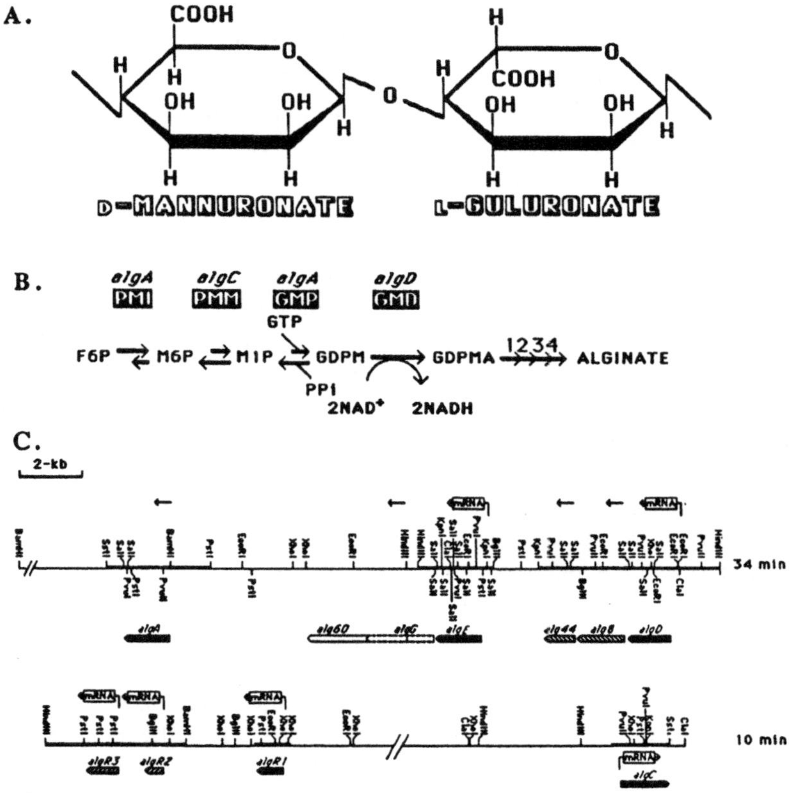

Fig. 4. (A). Chemical structure of alginate. (B). Alginate biosynthesis pathway. Arrows (1, 2, 3, 4) indicate the undefined steps of polymerization, acetylation, export, and epimerization. The gene encoding each enzyme is indicated above the enzyme name. Equilibria for the alginate reactions are known for PMI, GMP, and GMD and are indicated by the relative sizes of the arrows for each direction of the reaction. F6P, fructose 6-phosphate; M6P, mannose 6-phosphate; M1P, mannose 1-phosphate; GDPM, GDP-mannose; GDPMA, GDP-mannuronic acid. (C). Organization of alginate genes. The locations of the two gene clusters on the *P. aeruginosa* chromosome are indicated in minutes. Restriction enzyme sites are indicated by vertical lines: (——) Sequenced DNA); (——) unsequenced regions; (——>) direction of transcription; (←mRNA⌉) denotes the direction of transcription and transcriptional start; (◀■) protein encoded by the indicated gene and confirmed by N-terminal amino acid sequencing; (◁□) protein-coding region predicted from DNA sequence analysis of the indicated gene; (◁□) protein encoded by the indicated gene for which the exact coding region has not been determined.

proteins which in their phosphorylated form, are involved in binding and activating or repressing specific promoters. Indeed, AlgR1 has been shown to be required for the activation of both the *alg*D (May et al. 1991) and the *alg*C (Zielinski et al. 1991) promoters. AlgR1 has been purified and shown to bind two far upstream regions centered at -382 and -458 of the *alg*D promoter (Kato & Chakrabarty 1991). The binding of AlgR1 at such far upstream sites is believed to lead to looping out of the intervening region (Fig. 5) so that AlgR1 molecules bound at -382 and -458 regions can make contact with the RNA polymerase bound at the GG-N10-GC promoter region (at -21 to -33 of the *alg*D transcription initiation site). The bending of the DNA is believed to be facilitated by proteins such as AlgR3. AlgR3 is a basic histone-type of protein homologous to sea-urchin histone H1 (Kato et al. 1990). The looping of the *alg*D promoter region is also facilitated by a putative analogue of the *Escherichia coli* cyclic AMP receptor protein (CRP), which is known to be a DNA-bending protein involved in gene regulation (DeVault et al. 1991). It was previously demonstrated that AlgR2 is required for alginate synthesis and for maximal activation of the *alg*D promoter by AlgR1 (Kato et al.

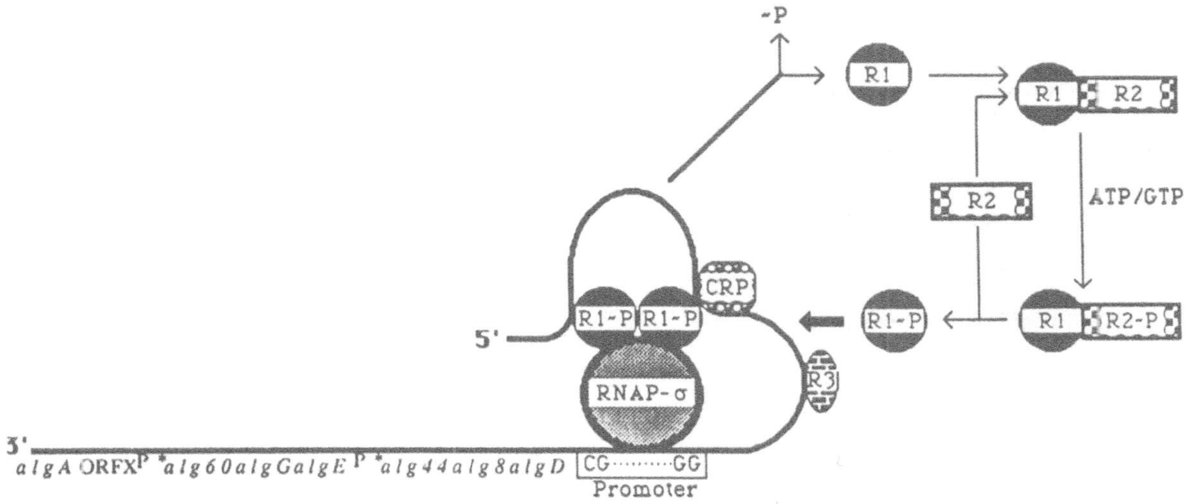

Fig. 5. The pathway for AlgR1 phosphorylation and the DNA looping model for *alg*D promoter activation. AlgR1 and AlgR2 have been shown to occur in a 80 kDa complex where AlgR2 undergoes autophosphorylation in presence of either ATP or GTP and subsequently transfers the phosphate to AlgR1. AlgR1-P functionally binds at the two binding sites centered at -382 and -458 of the *alg*D upstream region and allows bending of the intervening DNA (aided by CRP and AlgR3) to interact with the RNA polymerase at the promoter region. While this allows transcription to be initiated, it is not clear if the entire 16 kb segment down-stream of the *alg*D gene is transcribed as a unit. The location of AlgR3 is arbitrary while CRP site is known to be centered at -362 near one of the AlgR1 binding sites.

1989). It has been subsequently shown that AlgR2 is a protein kinase capable of autophosphorylation in the presence of either ATP or GTP (Roychoudhury et al. 1992b). AlgR2 can then transfer the phosphate to AlgR1 and allow binding of the phosphorylated AlgR1 to the upstream region of the *alg*D promoter for functional initiation of transcription (Roychoudhury et al. 1992c). Since the *alg*D promoter is activated under conditions of high osmolarity due to the presence of high NaCl concentrations in the growth medium (Berry et al. 1989) or under conditions of membrane perturbation induced by ethanol, a known dehydrating agent (DeVault et al. 1990) and since such conditions are reminiscent of the environment present in the CF lung, it appears that specific environmental factors unique to this diseased tissue trigger autophosphorylation of AlgR2. AlgR2 allows phosphorylation of AlgR1 which can then initiate efficient transcription from the *alg*D and the *alg*C promoters, resulting in alginate synthesis. Since AlgR2 is a soluble protein without a membrane spanning region, it is not clear how the osmolarity or membrane perturbation induced signals are transmitted to AlgR2, nor is it completely clear as to how many genes downstream of the *alg*D promoter (other than *alg*D) are under the control of this environmentally regulated promoter. It is interesting to note that expression of other pathogenesis-related genes such as neuraminidase, which appears to be involved in the adherence of *P. aeruginosa* to eukaryotic cell surfaces, might be modulated by high osmolarity and require AlgR1 as an activator (Cacalano et al. 1992).

In conclusion, environmental signal transduction is an important feature in enabling microorganisms to scan their environment and take appropriate measures to establish themselves in a particular ecological niche. Since the CF lung is a highly stressed environment, it is interesting to note that *P. aeruginosa* uses an efficient signal transduction system to turn on the alginate genes for production of the alginate capsule. This capsule not only protects the infecting cells from dehydration or high osmolarity in the CF lung, but also from the body's defense mechanisms or antibiotic treatments, allowing them to colonize the host tissue. If bioremediation is to succeed, the remediating microorganisms must not only evolve and express the degradative genes, but must also establish themselves in the environment where they are to be used. Understanding how microorganisms establish themselves in certain ecological niches as in biofilms on plastic or metal surfaces and human or animal tissues, as well as in soil or sediments (Costerton et al. 1987), is an important aspect of developing successful bioremediation processes.

Acknowledgements

The environmental modulation of microbial gene evolution part of the work is funded by grants from the National Science Foundation (DMB 91-00015) and the National Institutes of Environmental Health Sciences (ES-04050-07). The environmental modulation of microbial gene expression part of the work is supported by Public Health Service grant AI 16790-13 from the National Institutes of Health and in part by grant P-455 from the Cystic Fibrosis Foundation.

References

Aldrich TL, Frantz B, Gill JF, Kilbane JJ & Chakrabarty AM (1987). Gene 52: 185–195
Bakst JS (1991). J. Indust. Microbiol. 8: 13–22
Berry A, DeVault JD & Chakrabarty AM (1989). J. Bacteriol. 171: 2312–2317
Bourret RB, Borkovich KA & Simon MI (1991). Annu. Rev. Biochem. 60: 401–441
Cacalano G, Kays M, Saiman L & Prince A (1992). J. Clin. Invest. 89: 1866–1874
Chatterjee DK, Kellogg ST, Hamada S & Chakrabarty AM (1981). J. Bacteriol. 146: 639–646
Chitnis C & Ohman DE (1990). J. Bacteriol. 172: 2894–2900
Chu L, May TB, Chakrabarty AM & Misra TK (1991). Gene 107: 1–10
Costerton JW, Cheng KJ, Geesey GG, Ladd TI, Nickel JC, Dasgupta M & Marrie TJ (1987). Annu. Rev. Microbiol. 41: 435–464
DeVault JD, Kimbara K & Chakrabarty AM (1990). Molec. Microbiol. 4: 737–745
DeVault JD, Hendrickson W, Kato J & Chakrabarty AM (1991). Molec. Microbiol. 5:2503–2509
Don RH & Pemberton JM (1981). J. Bacteriol. 145: 681–686
Frantz B & Chakrabarty AM (1987). Proc. Natl. Acad. Sci. USA. 84: 4460–4464
Frantz B, Aldrich TL & Chakrabarty AM (1987). Biotech. Adv. 5: 85–99
Friello DA, Mylroie JR & Chakrabarty AM (1976). In: Sharpley JM (Ed) Proceedings of Third International Biodegradation Symposium (pp 205–214). Applied Science Publication, Essex.
Fulthorpe RR & Wyndham RC (1991). Appl. Environ. Microbiol. 57: 1546–1553
Ghosal D & You IS (1988). Mol. Gen. Genet. 211: 113–120
Haugland RA, Schlem DJ, Lyons RP, Sferra PR & Chakrabarty AM (1990a). Appl. Environ. Microbiol. 56: 1357–1362
Haugland RA, Sangodkar UMX & Chakrabarty AM (1990b). Mol. Gen. Genet. 220: 222–228
Haugland RA, Sangodkar UMX, Sferra PR & Chakrabarty AM (1991). Gene 100: 65–73
Kaphammer B, Kukor JJ & Olsen RH (1990). J. Bacteriol. 172: 2280–2286
Kato J & Chakrabarty AM (1991). Proc. Natl. Acad. Sci. USA 88: 1760–1764
Kato J, Chu L, Kitano K, DeVault JD, Kimbara K, Chakrabarty AM & Misra TK (1989). Gene 84: 31–38
Kato J, Misra TK & Chakrabarty AM (1990). Proc. Natl. Acad. Sci. USA 87: 2887–2891
Keeler R (1991). R & D Mag. 33: 34–40
May TB, Shinabarger D, Maharaj R, Kato J, Chu L, DeVault JD, Roychoudhury S, Zielinski NA, Berry A, Rothmel RK, Misra TK & Chakrabarty AM (1991). Clin. Microbiol. Rev. 4: 191–206
Mueller JG, Middaugh DP, Lantz SE & Chapman PJ (1991). Appl. Environ. Microbiol. 57: 1277–1285
Nakatsu C, Ng J, Singh R, Straus N & Wyndham C (1991). Proc. Natl. Acad. Sci. USA 88: 8312–8316
Neidle EL, Hartnett C, Bonitz S & Ornston LN (1988). J. Bacteriol. 170: 4874–4880
Perkins EJ, Bolton GW, Gordon MP & Lurquin PF (1988). Nucleic Acids Res.16: 7200
Perkins EJ, Gordon MP, Caceres O & Lurquin PF (1990). J. Bacteriol. 172: 2351–2359
Pritchard PH & Costa CF (1991). Environ. Sci. Technol. 25: 372–383
Reineke W & Knackmuss HJ (1988). Annu. Rev. Microbiol. 42: 263–287
Rothmel RK, Haugland RA, Sangodkar UMX, Coco WM & Chakrabarty AM (1991a). In: Rossmore HW (Ed) Biodeterioration and Biodegradation 8 (pp 276–291). Elsevier Applied Science, London,
Rothmel RK, Shinabarger DL, Parsek MR, Aldrich TL & Chakrabarty AM (1991b). J. Bacteriol. 173: 4717–4724
Roychoudhury S, May TB, Gill JF, Singh SK, Feingold DS & Chakrabarty AM (1989). J. Biol. Chem. 264: 9380–9385
Roychoudhury S, Chakrabarty K, Ho Y-K & Chakrabarty AM (1992a). J. Biol. Chem. 267: 990–996.

Roychoudhury S, Sakai K & Chakrabarty AM (1992b). Proc. Natl. Acad. Sci. USA 89: 2659–2663

Roychoudhury S, Sakai K, Schlictman D. & Chakrabarty AM (1992c). Gene 112: 45–51

Schlomann M, Pieper DH & Knackmuss H-J (1990). In: Silver S, Chakrabarty AM, Iglewski B, Kaplan S (Eds) *Pseudomonas*: Biotransformation, Pathogenesis, and Evolving Biotechnology (pp 185–196). American Society for Microbiology, Washington DC

Shinabarger D, Berry A, May TB, Rothmel R, Fialho A & Chakrabarty AM (1991). J. Biol. Chem. 266: 2080–2088

Shortridge VD, Pato ML, Vasil AI & Vasil ML (1991). Infect. Immun. 59: 3596–3603

Stock JB, Stock AM & Mottonen JM (1990), Nature 334: 395–400

Thomas AW, Topping AW, Slater JH & Weightman AJ (1992). J. Bacteriol. 174: 1941–1947

Tomasek PH, Frantz B, Sangodkar UMX, Haugland RA & Chakrabarty AM (1989). Gene 76: 227–238

van der Meer JR, Eggen RIL, Zehnder AJB & de Vos WM (1991a). J. Bacteriol. 173: 2425–2434

van der Meer JR, Zehnder AJB & de Vos WM (1991b). J. Bacteriol. 173: 7077–7083

van der Meer JR, Frijters ACJ, Levearu JHJ, Eggen RIL, Zehnder AJB & de Vos WM (1991c). J. Bacteriol. 173: 3700–3708

Wheelis ML & Ornston LN (1972). J. Bacteriol. 109: 790–795

Zielinski NA, Chakrabarty AM & Berry A (1991). J. Biol. Chem. 266: 9754–9763

Biodegradation 3: 137–146, 1992.

Molecular methods for environmental monitoring and containment of genetically engineered microorganisms

Ronald M. Atlas
Department of Biology, University of Louisville, Louisville, KY 40292, USA

Key words: genetically engineered microorganisms, gene probes, polymerase chain reaction, suicide vectors

Abstract

Plans to introduce genetically engineered microorganisms into the environment has led to concerns over safety and has raised questions about how to detect and to contain such microorganisms. Specific gene sequences, such as *lac*Z, have been inserted into genetically engineered microorganisms to permit their phenotypic detection. Molecular methods have been developed based upon recovery of DNA from environmental samples and gene probe hybridization to specific diagnostic gene sequences for the specific detection of genetically engineered microorganisms. DNA amplification using the polymerase chain reaction has been applied to enhance detection sensitivity so that single gene targets can be detected. Detection of messenger RNA has permitted the monitoring of gene expression in the environment. The use of reporter genes, such as the *lux* gene for bioluminescence, likewise has permitted the observation of gene expression. Conditional lethal constructs have been developed as models for containment of genetically engineered microorganisms. Suicide vectors, based upon the *hok* gene have been developed as model containment systems.

Introduction

When molecular methods developed the capacity for genetically engineering organisms it was clear that a scientific revolution would ensue that would impact human health, agricultural productivity, and environmental quality. The power of recombinant DNA technology to create organisms with novel genetic characteristics offered great promise for solving environmental problems but as an unknown technology also was met with caution within the scientific community and fear among the public (Curtiss 1976; Sharples 1983; Brown et al. 1984). Scientists sought to reassure the public of the safety of recombinant DNA technology and declared a moratorium in 1975 at the Asilomar Conference until host strains could be developed for recombinant research that could not escape from the laboratory. By assuring the containment of genetically engineered microorganisms within the laboratory, molecular biologists reduced the risk of the organisms they created causing any untoward effects on human health or the environment. They, perhaps unwittingly, furthered public fears by fostering an image of harmful 'Andromeda strains' that would cause great harm if they escaped the containment of the laboratory.

Given the atmosphere of peril surrounding the power of genetic engineering, proposals for the deliberate release of genetically engineered organisms to enhance agricultural productivity and environmental quality have been met with great caution. Regulatory constraints have been placed on such releases of genetically engineered organisms to a greater extent than upon nonrecombinant ones (Federal Register 1986). Several symposia have been held to examine what we know about releases of organisms into the environment, survival of organisms in various environments, rates and processes of natural recombination in the environment, and other factors relating to risks and potential applications of genetically engineered organisms in the environment (Halvorson et al. 1985; Sussman et al. 1988).

The scientific community, having examined the organisms that can be created by recombinant DNA technology and having gained a sense of safety through several years of experience, now feels that there are no greater risks associated with recombinant organisms than with other naturally occurring and genetically modified organisms. Scientists have proposed (a) that risk analysis be used as the basis for establishing the relative safety of deliberate releases and (b) that all genetically modified organisms be evaluated in terms of their phenotypic properties and our familiarity with releases of organisms with identical or similar phenotypes.

Cultural methods for detecting genetically engineered microorganisms

Selective markers

Most genetically engineered microorganisms contain marker genes that can be used for their detection. These marker genes often are essential in the construction of the genetically engineered microorganism as they allow for the selection of recombinants that are likely to contain the desired combination of genes. Frequently, antibiotic resistance markers are employed so that by plating onto media containing the antibiotic, only resistant microorganisms will grow. The problem of using antibiotic resistance markers for microorganisms that are deliberately released into the environment is twofold. First, there is a background of resistant microorganisms in environmental samples, especially in soils, so that the selection of antibiotic resistant microorganisms will include both target genetically engineered microorganisms and nontarget background microbial populations. Second, it is not desirable to increase the prevalence of antibiotic resistance genes that may become disseminated into populations of pathogenic microorganisms. To reduce the risk of creating medical problems with drug-resistant pathogens, only antibiotic resistance markers associated with antibiotics of limited therapeutic use should be employed.

Additional markers have been developed as reporter genes that can be introduced into microorganisms so that they subsequently can be detected. Adding the genes for lactose metabolism permits the use of conventional selective and differential media for the detection of organisms containing this marker gene. The *lac*Z system, which encodes lactose utilization (β-galactosidase production) in *E. coli*, has been used as a marker or reporter of recombinant *Pseudomonas* in studies to determine whether genetically engineered microorganisms will survive and disperse if released at an environmental site (Drahos et al. 1986). This marker system has proved very effective in tracking the movement of genetically engineered pseudomonads in soils and especially within the rhizosphere. *Pseudomonas* strains that incorporate the *lac*Z genes are capable of cleaving the X-gal substrate to produce a characteristic blue color. This trait makes them very distinguishable from the nonrecombinant *Pseudomonas* species that are indigenous to the site.

The *xyl*E gene of the TOL plasmid has been cloned for use as a transcriptional fusion reporter gene. This gene encodes catechol 2,3 oxygenase; a simple aerosol spray technique has been used to find recombinant colonies (Zukowski et al. 1983; Ingram et al. 1989). A similar assay system has recently been described by King et al. (1991) using the 2,4-dichlorophenoxyacetate monooxygenase gene, which codes for the conversion of phenoxyacetate to phenol that then can be reacted with a dye to form a colored compound. The *lux* genes that encode bioluminescence are useful as reporter genes that permit not only detection of microorganisms but also monitoring of their activities (Boivin et al. 1988; Engebrecht et al. 1982). Burlage et al. (1990) described a fusion between a promoter (nah) for the degradation of naphthalene in *Pseu-*

domonas and the *lux* genes. Using this plasmid construction, they were able to demonstrate gene expression on a continuous basis, revealing an unexpected pattern. This work suggests the biotechnological applications that may arise from further use of the *lux* genes. King et al. (1990) used another nah-lux construction to demonstrate naphthalene degradation in soil slurries.

Colony hybridization

Colony hybridization is the simplest molecular approach for the detection of genetically engineered microorganisms that can be integrated with conventional environmental microbiological sampling and analysis (Sayler & Layton 1990). In the colony hybridization procedure bacteria are grown on solidified agar media to form colonies. Gene probes and nucleic acid hybridization are then used to detect colonies with specific target nucleic acid sequences. Bacterial colonies are transferred from primary environmental cultivation media to hybridization filters. The colonies are lysed by alkaline or enzymatic treatment and hybridization is then conducted. These methods are dependent on the ability of the target microorganisms to grow on the primary isolation medium and not be totally overgrown by nontarget populations. Growth on the isolation medium increases the number of copies of the target gene to a level that can be detected by a gene probe.

The original colony hybridization protocol developed by Grunstein & Hogness (1975) is suitable for high density plate screening for pure cultures (Hanahan & Meselson 1980). In cultures containing both nontarget *E. coli* and target *P. putida*, one *P. putida* colony containing a toluene (TOL) catabolic plasmid in a background of approximately one million *E. coli* colonies was detected using a whole TOL plasmid probe (Sayler et al. 1985).

Gene probes can be hybridized with primary isolations from environmental samples (Echeverria et al. 1982; Fitts et al. 1983; Hill et al. 1983; Miliotis et al. 1989). The rationale for direct colony hybridization on primary cultivation includes: (a) avoiding a cultivation bias encountered by selective media which may underestimate total abundance of a given genotype; (b) assuring that a given genotype is represented in the population sampled even if the genes are poorly expressed or are poorly selected; (c) providing optimal permissive growth conditions for stressed organisms that may be unculturable on selective media; and (d) reducing the analysis time for cultivation, presumptive quantification and confirmation of a genotype/phenotype (Atlas et al. 1992).

The major uses for colony hybridization in environmental studies have been for the detection, enumeration, and isolation of bacteria with specific genotypes and/or phenotypes, and for development of gene probes (Ford & Olson 1988; Holben & Tiedje 1988). Colony hybridization has been used to detect *Salmonella* (Fitts et al. 1983), *Shigella* (Sethabutr et al. 1985), *Escherichia coli* (Echeverria et al. 1982; Bialkonska-Hobrzanska 1987; Seriwantana et al. 1987; Nishibuchi et al. 1989), *Listeria* (Datta et al. 1987, 1988; Flamm et al. 1989; Nortermans et al. 1989), *Yersinia* (Hill et al. 1983; Miliotis et al. 1989), *Rhizobium* (Hodgson & Roberts 1983), *Erwinia amylovora* (Falkenstein et al. 1988), *Bacillus subtilis* (Kraus et al. 1986), *Pseudomonas fluorescens* (Festl et al. 1986), 4-chlorobiphenyl degrader (Pettigrew & Sayler 1986), toluene degraders (Jain et al. 1987), naphthalene degraders (Blackburn et al. 1987), and mercury resistant bacteria (Barkay 1985, 1987; Barkay & Olson 1986). Jain et al. (1987) used colony hybridization and gene probes to study the maintenance of catabolic and antibiotic resistance plasmids in groundwater aquifer microcosms. They found that introduced catabolic plasmids or organisms can be maintained in groundwater aquifers without selective pressure. In another study, colony hybridization with a naphthalene gene probe was used to correlate gene frequency with naphthalene degradation by naphthalene degrading bacteria in activated sludge; the gene probe analysis for the catabolic genotype was nearly two orders of magnitude more sensitive than the standard plate assay for naphthalene degradation (Blackburn et al. 1987). Colony hybridization has been used with environmental samples to enumerate mercury resistant bacteria in contaminated environments (Barkay et al. 1985; Barkay & Olson 1986) and naphthalene degrading bacteria in aromatic hydrocarbon contaminated soils (Jain et al. 1987).

In general, for environmental assays the organism of interest must be relatively abundant in the population so that at least one positive colony on a agar plate of 100 to 1000 colonies can be found. Additional sensitivity can be achieved by plating the isolated bacteria onto selective agar before colony hybridization.

Direct gene probe detection

Molecular analyses can also be performed without culturing microorganisms to detect the presence of genetically engineered microorganisms in environmental samples. DNA can be recovered from water samples by collecting cells using filtration and then lysing the cells to release their nucleic acids (Fuhrman et al. 1988; Sommerville et al. 1989; Bej et al. 1990; Giovannoni et al. 1990). Sommerville et al. (1989) isolated DNA from aquatic samples by filtering large volumes of water through a filter cartridge. Paul et al. (1990) reported that purification by cesium chloride-ethidium bromide (CsCl-EtBr) ultracentrifugation was required for isolation of purified DNA. In other studies multiple phenol or phenol:chloroform extractions have produced sufficiently pure DNA from cyanobacteria (Zehr & McReynolds 1989).

For soils, differential centrifugation can be used to recover cells; the cells can then be lysed by chemical and/or physical description methods (Balkwill et al. 1975; Faegri et al. 1977; Torsvik & Goksoyr 1978; Torsvik 1980; Bakken 1985; Holben et al. 1988; Steffan et al. 1988; Jansson et al. 1989; Steffan & Atlas 1990; Torsvik et al. 1990). The inclusion of a polyvinylpolypyrrolidone (PVPP) decreases humic content prior to cell lysis (Holben et al. 1988; Steffan et al. 1988), simplifying DNA purification.

DNA can also be extracted directly from environmental samples without first recovering cells (Ogram et al. 1987). In this method cells are lysed while still within the soil matrix for example, by incubation with SDS followed by physical disruption with a bead beater. The DNA is then extracted with alkaline phosphate buffer. This method, with some modifications, has been successfully used to recover DNA from soils and sediments (Ogram et al. 1987; Steffan et al. 1988; Pillai et al. 1991; Tsai & Olson 1991; Tsai & Olson 1992). Purification procedures can be any combination of, CsCl-EtBr ultracentrifugation (Holben et al. 1988; Steffan et al. 1988; Paul et al. 1990), hydroxylapatite or affinity chromatography (Torsvik & Goksoyr, 1978; Ogram et al. 1987; Steffan et al 1988; Paul et al. 1990), phenol/chloroform extractions, ethanol precipitations (Fuhrman et al. 1988), dialysis, or repeated PVPP treatments (Holben et al. 1988; Steffan et al. 1988; Weller & Ward 1989; Paul et al. 1990). Significantly higher yields of DNA are recovered with the direct extraction method than with the cell recovery procedure, but the DNA may contain impurities that can inhibit enzymatic manipulation (Steffan et al. 1988).

Gene probes and nucleic acid hybridization permit the detection of DNA sequences recovered by extracting DNA from environmental samples that are homologous to the probe sequence (Holben et al. 1988; Barkay et al. 1989; Sayler & Layton 1990). The stringency of the hybridization is determined by temperature, salt concentration, and the length and concentration of the target and probe sequences (Berent et al. 1985; Britton & Davidson 1985). In most hybridization protocols the target DNA probe is immobilized on a nitrocellulose or nylon filter. Labelled probe DNA is added to the filters so that the gene probe can hybridize to complementary target DNA sequences. The probe often is labelled ^{32}P which is detected by autoradiography. Using gene probes and DNA extraction procedures, it is possible to detect introduced genetically engineered microorganisms from soils (Holben et al. 1988; Steffan et al. 1988; Jansson et al. 1989).

Specific gene sequences, diagnostic of specific genetically engineered microorganisms, can be amplified using the polymerase chain reaction (PCR). PCR is a method for the *in vitro* replication of defined sequences of DNA whereby gene segments can be amplified exponentially (Mullis & Faloona

1987; Mullis 1990; Ehrlich 1989, Ehrlich et al. 1991; Innis et al. 1990; Saiki et al. 1988). One application of this technique is to enhance gene probe detection of specific gene sequences. By exponentially amplifying a target sequence, PCR significantly enhances the probability of detecting rare sequences in complex mixtures of DNA such as those recovered from environmental samples (Bej & Mahbubani 1992; Bej et al. 1991; Steffan & Atlas 1992; Tsai & Olson 1992).

Steffan & Atlas (1988) used PCR to amplify a 1.0-kilobase (kb) probe-specific region of DNA from the genetically engineered, herbicide-degrading bacterium *Pseudomonas cepacia* AC1100 in order to increase the sensitivity of detecting the organism by dot-blot analysis. The 1.0-kb region was an integral portion of a larger 1.3-kb repeat sequence which is present as 15 to 20 copies on the *P. cepacia* AC1100 genome. PCR was performed by melting the target DNA, annealing 24-base oligonucleotide primers to unique sequences flanking the 1.0-kb region, and performing extension reactions with DNA polymerase. After extension, the DNA was again melted, and the procedure was repeated for a total of 25 to 30 cycles. After amplification, the reaction mixture was transferred to nylon filters and hybridized against radiolabelled 1.0-kb fragment probe DNA. Amplified target DNA was detectable in samples initially containing as little as 0.3 pg of target. The addition of 20 μg of nonspecific DNA isolated from sediment samples did not hinder amplification or detection of the target DNA. The detection of 0.3 pg of target DNA was at least a 103-fold increase in the sensitivity of detecting gene sequences compared with dot-blot analysis of nonamplified samples. PCR performed after bacterial DNA was isolated from sediment samples permitted the detection of as few as 100 cells of *P. cepacia* ACl100 per 100 g of sediment sample against a background of 1011 diverse nontarget organisms; that is, *P. cepacia* ACl100 was positively detected at a concentration of 1 cell per g of sediment. This represented a 103-fold increase in sensitivity compared with nonamplified samples.

Chaudhry et al. (1989) also used PCR for detecting genetically engineered microorganisms. They cloned 0.3-kilobase napier grass (*Pennisetum purpureum*) genomic DNA that did not hybridize to DNAs isolated from various microorganisms, soil sediments, and aquatic environments, into a derivative of a 2,4-dichlorophenoxyacetic acid degradative plasmid, pRC10, and transferred the construct into *Escherichia coli*. This genetically altered microorganism was seeded into filter-sterilized lake and sewage water samples (104/ml). The PCR method amplified and detected the 0.3-kilobase DNA marker of the genetically engineered microorganism even after 10 to 14 days of incubation. The PCR method required only picogram amounts of DNA and had an advantage over the plate count technique, which can detect only culturable microorganisms. They concluded that the method may be useful for monitoring genetically engineered microorganisms in complex environments, where discrimination between genetically engineered microorganisms and indigenous microorganisms is either difficult or requires time consuming tests.

Containment of genetically engineered microorganisms

Debilitated strains and vectors

The initial containment strategy for genetically engineered microorganisms was to employ debilitated host bacteria that could not survive outside of a laboratory environment (Curtiss et al. 1977). An *Escherichia coli* strain that obligately requires exogenous diaminopimelic acid and thymine and is very sensitive to bile, detergents, and antibiotics was created as a fail safe containment host; accidental releases of genetically engineered DNA in this and similar microorganisms would be contained because the bacterial host was extremely fastidious. An alternate means of containment would be to create host strains with reduced ATP generating capacities so that they could not compete with wild types in the environment (Curtiss 1988). Such strains would not likely be useful, however, for environmental applications because of their limited survival capacity in the environment.

Vector systems have been proposed and developed to prevent transfer of genes to other organisms, that is, to contain the recombinant genes to the host organism. One strategy for containing recombinant genes is to construct cloning vectors that will not survive or transfer genes to indigenous microbial populations outside of the laboratory. Vectors carrying mutations that decrease the host range reduce the possibility of transferring genes (Blattner et al. 1977). Plasmid-based cloning vectors that contain only an origin of replication and the desired structural genes lack transfer functions are not readily mobilizable, even by triparental matings, and thus are relatively contained (Levine et al. 1983). Another strategy is to use an impaired transposon delivery system that will insert genetically engineered DNA into the chromosome and is then unable to undergo further transposition (Obukowicz et al. 1986, 1987).

Vectors have also been considered that are suicidal, such that they could not survive transfer to another organism (Curtiss 1988). A plasmid vector with a nonsense mutation in a gene indispensable for plasmid replication and/or maintenance and a suppressor mutation in the chromosome that would allow translational read-through of the message of the gene would be self contained in the host strain (Curtiss et al. 1976, 1977). An alternate strategy would be to endow a plasmid vector with a gene for a restriction enzyme where the ability to synthesize the modification methylase would be specified by a chromosomal gene so that transfer of the plasmid to a recipient cell would lead to the synthesis of the restriction enzyme that would digest the DNA of the recipient and cause death.

Conditional-lethal ('suicide') vectors

One approach for containing genetically engineered microorganisms is to create genetically engineered microorganisms that contain 'suicide vectors' (plasmids with conditionally lethal genes) which could act to prevent movement of genes from the genetically engineered microorganism to other organisms and to contain such organisms within specified bounds (Cuskey & Bourquin 1987; Molin et al. 1987). Organisms could either be designed to remain viable only under the specific environmental conditions of their intended use, for example, only in situations where a particular pesticide or hazardous chemical was present, and to die if they escaped that location or if the environmental condition requirement for their survival changed; alternatively, they could be designed to survive until a specific environmental condition occurred or was introduced that would trigger their death. For industries marketing genetically engineered microorganisms to treat hazardous waste sites, this would be desirable not only because of the reduced potential of environmental risk if the genetically engineered microorganism persisted in the environment, but also because the organisms would have a built-in self destruction mechanism that would lead to new sales for future site bioremediation actions. Thus, creation of suicide vectors may simultaneously meet the economic interest of industry and the public interest for environmental protection.

Model suicide vectors have been constructed using the lethal *hok* gene, which codes for a small polypeptide (Hok) that causes loss of cell membrane potential and rapid death of cells in which it is overproduced (Molin et al. 1987; Bej et al. 1988). We have constructed suicide plasmids containing the *hok* gene under the control of the *lac* promoter. The *lac*Iq gene, which codes for an excess of Lac repressor protein adequately suppresses Hok polypeptide production in strains carrying the suicide vector, permitting the *via*bility of the bacterial cells carrying the conditional suicide vector in the absence of inducer.

IPTG was found to rapidly induce the hok gene in *E. coli* JM101 carrying the suicide plasmid growing *in vitro*. Compared to controls of *E. coli* JM101 lacking the suicide vector, which continued exponential growth after IPTG addition *in vitro*, cultures containing the suicide vector initially ceased to grow and showed evidence of death, both by optical density determinations and *via*ble cell counts between 1 and 2 h after IPTG addition. After 2 h, though, cultures inoculated with *E. coli* JM101 containing the suicide vector resumed exponential growth at a rate comparable to the control cultures. However, by 4 h after IPTG addition no colonies were observed on selective media containing carbenicillin. Moreover, no *hok* genes or suicide vector plasmids could be detected by colony hybridization or Southern blot analyses from cells growing on nonselective media. These results indicate that induction of the *hok* gene had killed all cells containing the suicide vector and that the cells growing after IPTG addition had lost this plasmid. The percentage of *hok*-carrying cells decreased from 97% to 0% within 4 h of IPTG induction of the *hok* gene.

In contrast to these results, in separate *in vitro* experiments in which carbenicillin was added as a selective pressure against loss of the suicide vector plasmid, Hok-resistant cells developed before all *hok*-carrying cells were eliminated from the cultures; this resulted in the development of a significant population of Hok-resistant-*hok*-carrying cells. After an initial period of significant decline during which the proportion of cells carrying hok decreased from 97% to 14%, the proportion of cells carrying the hok gene increased to 87%; thus, after 8 h the proportion of cells with the suicide vector was only 10% lower than prior to addition of IPTG. Transformation of fresh *E. coli* JM101 cells with the suicide vector from the *hok*-resistant cells resulted in *hok*-sensitive cells that died, indicating that the hok gene was still functional.

Apparently, the presence of antibiotic represented a sufficient selective pressure against loss of the suicide vector such that Hok-resistant mutants, still containing the suicide vector, were selected for. Addition of IPTG to soil microcosms containing *E. coli* JM101 with the suicide vector led to a 90% decline in colonies detected on the selective media, whereas numbers of cells did not decline in soil microcosms where IPTG was not added. The surviving cells were debilitated and not capable of normal growth or prolonged plasmid maintenance even in liquid broth culture. It is likely that the persistence of suicide vector-containing cells in soil was the result of the development of Hok-resistance, which protected the cells against Hok-induced death, but also rendered them unable to generate sufficient energy for long-term survival in the environment or long-term maintenance of the suicide plasmid.

Thus, the model *hok* suicide vector, while not yet an adequate fail safe system for containing genetically engineered microorganisms in environmental samples, appears to have real potential for limiting the spread of genetically engineered microorganisms. Conditional lethal strains of pseudomonads, designed as models for containment of genetically engineered microbes, were also constructed by cloning the *hok* gene into plasmids under the control of the *tac* promoter. *Pseudomonas putida* mt-2 was transformed with this conditional lethal plasmid and growth and survival of the host organism and retention of the suicide vector were examined. *Hok* induction caused elimination of greater than 60% of the cells carrying the suicide vector within 5 h of induction compared to 100% retention in uninduced controls. Resistance to *hok*-induced killing developed in *P. putida*, which may have been due to a mutation or physiological adaptation that rendered the membrane 'resistant' to Hok. Constructs were also designed using a *hok* homolog, *gef*, also under control of the *tac* promoter. These constructs were highly effective in *P. putida*, when inserted into the chromosome or on a stable RK2-derived plasmid.

Knudsen & Karlstrom (1991) investigated factors in *E. coli* which reduced survival after induction of the *E. coli relF* gene, which is homologous to *hok* and *gef*. They found that the use of more tightly-controlled promoters to repress the suicide gene improved the efficiency of their system by preventing selection of cells mutated in the killing function. An additional improvement was observed when the conditional lethal system was introduced into the test organism in multiple copies; in such cases mutation rates were reduced. Our results suggest that an additional factor in containment using such conditional lethal constructs may be the stability of the vector for introducing and maintaining these constructs in the host bacteria. Improved stability (via stably-inherited plasmids or via chromosomal insertion) combined with tightly-controlled promoters and duplicated systems provide a measure of containment for genetically-engineered microbes which could be deliberately released.

References

Atlas RM, Sayler G, Burlage RS & Bej AK (1992) Molecular approaches for environmental monitoring of microorganisms. Bio/Techniques: 12: 706–711

Bakken LR (1985) Separation and purification of bacteria from soil. Appl. Environ. Microbiol. 49: 1482–1487

Balkwill DL, Labeda DP & Casida LE, Jr. (1975) Simplified procedure for releasing and concentrating microorganisms from soil for transmission electron microscopy viewing as thin-section and frozen-etched preparations. Can. J. Microbiol. 21: 252–262

Barkay T (1987) Adaptation of aquatic microbial communities to Hg^{2+} stress. Appl. Environ. Microbiol. 53: 2725–2732

Barkay T & Olson BH (1986) Phenotypic and genotypic adaptation of aerobic heterotrophic sediment bacterial communities to mercury stress. Appl. Environ. Microbiol. 52: 403–406

Barkay T, Fouts DL & Olson BH (1985) The preparation of a DNA gene probe for the detection of mercury resistance genes in gram-negative communities. Appl. Environ. Microbiol. 49: 686–692

Barkay T, Liebert C & Gillman M (1989) Hybridization of DNA probes with whole community genome for detection of genes that encode microbial responses to pollutants: mer genes and Hg^{2+} resistance. Appl. Environ. Microbiol. 55: 1574–1577

Bej AK & Mahbubani MM (1992) Applications of the polymerase chain reaction in environmental microbiology. PCR Methods and Appl. 1: 151–159

Bej AK, Perlin MH & Atlas RM (1988) Model suicide vector for containment of genetically engineered microorganisms. Appl. Environ. Microbiol. 54: 2472–2477

Bej AK, Steffan RJ, DiCesare J, Haff L & Atlas RM (1990) Detection of coliform bacteria in water by polymerase chain reaction and gene probes. Appl. Environ. Microbiol. 56: 307–314

Bej AK, Mahbubani MM & Atlas RM (1991) Amplification of nucleic acids by polymerase chain reaction (PCR) and other methods and their applications. CRC Crit. Rev. Mol. Biol. 26: 301–334

Berent SL, Mahmoudi M, Torczynski RM, Bragg PW & Bollon AP (1985) Comparison of oligonucleotide and long DNA fragments as probes in DNA and RNA dot, Southern, northern, colony and plaque hybridizations. Bio/Techniques 3: 208–220

Bialkonska-Hobrzanska H (1987) Detection of enterotoxigenic *Escherichia coli* by dot blot hybridization with biotinylated DNA probes. J. Clinical Microbiol. 25: 338–343

Blackburn JW, Jain RL & Sayler GS (1987) Molecular microbial ecology of a naphthalene-degrading genotype in activated sludge. Environ. Sci. Tech. 21: 884–890

Blattner FL, Williams BG, Blechl AE, Dennistron-Thompson K, Baber HE, Furlong LA, Grunwald DJ, Kiefer DO, Moore DD, Schumm JW, Sheldon ET & Smithies O (1977) Charon phages: safer derivatives of bacteriophage lambda for DNA cloning. Science 196: 161–169

Boivin R, Chalifour F & Dion P (1988) Construction of a Tn5 derivative encoding bioluminescence and its introduction in *Pseudomonas*, *Agrobacterium*, and *Rhizobium*. Mol. Gen. Genet. 213: 50–55

Britton RJ & Davidson EH (1985) Hybridization strategy. In: Hames BP & Higgins SJ (Eds) Nucleic Acid Hybridization: A Practical Approach (pp 3–15). IRL Press, Oxford

Brown TH, Colwell RK, Lenski RE, Levin BR, Lloyd M, Regal PJ & Simberloff D (1984) Report on workshop on possible ecological and evolutionary impacts of bioengineered organisms released into the environment. Bull. Ecol. Soc. Am. 65: 436

Burlage RS, Sayler GS & Larimer FW (1990) Monitoring of naphthalene catabolism by bioluminescence with nah-lux transcriptional fusions. J. Bacteriol. 172: 4749–4757

Chaudhry GR, Toranzos GA & Bhatti AR (1989) Novel method for monitoring genetically engineered microorganisms in the environment. Appl. Environ. Microbiol. 55: 1301–1304

Curtiss R III (1976) Genetic manipulation of microorganisms: potential benefits. Annu. Rev. Microbiol. 30: 507–533

Curtiss R III (1988) Engineering organisms for safety: what is necessary? In: Sussman M, Collins CH, Skinner FA & Stewart-Tull DE (Eds) The Release of Genetically-Engineered Microorganisms. (pp 7-20). Academic Press, New York

Curtiss R III, Szybalski W, Helsinki DR & Falkow S (1976) Workshop on design and testing of safer prokaryotic vehicles and bacterial hosts for research on recombinant DNA molecules. ASM News 42: 134–138

Curtiss R III, Inoue M, Pereira D, Hsu JC, Alexander L & Rock L (1977) Construction and use of safer bacterial host strains for recombinant DNA research. In: Scott WA & Werner R (Eds) Molecular Cloning of Recombinant DNA (pp 99–114). Academic Press, Inc., New York

Cuskey S & Bourquin A (1987) Construction of bacteria with conditional lethal genetic determinants for control of released microorganisms in the environment. Abstr. Annu. Meet. Am. Soc. Microbiol. Q127, p 3

Datta AR, Wentz BA & Hill WE (1987) Detection of hemolytic *Listeria monocytogenes* by using DNA colony hybridization. Appl. Environ. Microbiol. 53: 2256–2259

Datta AR, Wentz BA, Shook D & Trucksess MW (1988) Synthetic oligodeoxyribonucleotide probes for detection of *Listeria monocytogenes*. Appl. Environ. Microbiol. 54: 2933–2937

Drahos DJ, Hemming BC & McPherson S (1986) Tracking recombinant organisms in the environment: β-galactosidase as a selectable nonantibiotic marker for fluorescent Pseudomonads. Bio/Techniques 4: 439–444

Echeverria P, Seriwatana J, Chityothin 0, Chaicumpa W & Tirapat C (1982) Detection of enterotoxigenic *Escherichia coli* in water by filter hybridization with three enterotoxin gene probes. J. Clin. Microbiol. 16: 1086–1090

Engebrecht J, Simon & M. Silverman M (1985) Measuring gene expression with light. Science 227: 1345–1347
Ehrlich HA (Ed) (1989) PCR Technology: Principles and Applications for DNA Amplification. Stockton Press, New York
Ehrlich HA, Gelfand D & Sninsky JJ (1991) Recent advances in the polymerase chain reaction. Science 252: 1643–1651
Faegri A, Torsvik VL & Goksoyr J (1977) Bacterial and fungal activities in soil: separation of bacteria by a rapid fractionated centrifugation technique. Soil Biol. Biochem. 9: 105–112
Falkenstein H, Bellemann P, Walter S, Zeller W & Geider K (1988) Identification of *Erwina amylovora*, the fireblight pathogen, by colony hybridization with DNA from plasmid pEPA29. Appl. Environ. Microbiol. 54: 2798–2802
Federal Register. June 26, 1986. Office of Science and Technology Policy: Coordinated framework for regulation of biotechnology
Festl H, Ludwig W & Schleifer KH (1986) DNA hybridization probe for the *Pseudomonas fluorescens* group. Appl. Environ. Microbiol. 52: 1190–1194
Fitts R, Diamond M, Hamilton C & Neri M (1983) DNA-DNA hybridization assay for the detection of *Salmonella* spp. in foods. Appl. Environ. Microbiol. 46: 1146–1151
Flamm RK, Hinrichs DJ & Thomashow MF (1989) Cloning of a gene encoding a major secreted polypeptide of *Listeria monocytogenes* and its potential use as a species-specific probe. Appl. Environ. Microbiol. 55: 2251–2256
Ford SF & Olson B (1988) Methods for detecting genetically engineered microorganisms in the environment. Adv. Microbial Ecol. 10: 45–79
Fuhrman JA, Comeau DE, Hagstrom A & Cham AM (1988) Extraction from natural planktonic microorganisms of DNA suitable for molecular biological studies. Appl. Environ. Microbiol. 54: 1426–1429
Giovannoni SJ, Britschgi TB, Moyer CL & Field KG (1990) Genetic diversity in Sargasso Sea bacterioplankton. Nature 345: 60–62
Grunstein M & Hodgness DS (1975) Colony hybridization: a method for the isolation of cloned DNAs that contain a specific gene. Proc. Natl. Acad. Sci. USA 72: 3961–3965
Halvorson HO, Pramer D & Rogul M (1985) Engineered Organisms in the Environment: Scientific Issues. American Society for Microbiology, Washington, DC
Hanahan D & Meselson M (1980) Plasmid screening at high colony density. Gene 10: 63–67
Hill WE, Payne WL & Aulisio CCG (1983) Detection and enumeration of virulent *Yersinia enterocolitica* in food by DNA colony hybridization. Appl. Environ. Microbiol. 46: 636–641
Hodgson ALM & Roberts WP (1983) DNA colony hybridization to identify *Rhizobium* strains. J. Gen. Microbiol. 129: 207–212
Holben WE & Tiedje JM (1988) Applications of nucleic acid hybridization in microbial ecology. Ecology 69: 561–568
Holben WE, Jansson JK, Chelm BK & Tiedje JM (1988) DNA probe method for the detection of specific microorganisms in the soil bacterial community. Appl. Environ. Microbiol. 54: 703–711
Ingram C, Brawner M, Youngman P & Westpheling J (1989) xylE functions as an efficient reporter gene in *Streptomyces* spp.: Use for the study of galP1, a catabolite-controlled promoter. J. Bacteriol. 171: 6617–6624
Innis M, Gelfand D, Snisky D & White T (Eds) (1990) PCR Protocols: A Guide to Methods and Applications. Academic Press, New York
Jain RK, Sayler GS, Wilson JT Houston L & Pacia D (1987) Maintenance and stability of introduced genotypes in groundwater aquifer material. Appl. Environ. Microbiol. 53: 996–1002
Jansson JK, Holben WE & Tiedje JM (1989) Detection in soil of a deletion in an engineered DNA sequence by using gene probes. Appl. Environ. Microbiol. 55: 3022–3025
King JMH, DiGrazia PM, Applegate B, Burlage R, Sanseverino J, Dunbar P, Larimer F & Sayler GS (1990) Bioluminescent reporter plasmid for naphthalene exposure and biodegradation. Science 249: 778–791
King RJ, Short KA & Seidler RJ (1991) Assay for detection and enumeration of genetically engineered microorganisms which is based on the activity of a deregulated 2,4-dichlorophenoxyacetate monooxygenase. Appl. Environ. Microbiol. 57: 1790–1792
Knudsen SM & Karlstrom (1991) Development of efficient suicide mechanisms for biological containment of bacteria. Appl. Environ. Microbiol. 57: 85–92
Kraus J, Ludwig W & Schliefer KH (1986) A cloned 23S rRNA gene fragment of *Bacillus subtilis* and its use as a hybridization probe of conserved character. FEMS Microbiol. Lett. 33: 89–93
Levine MM, Kaper JB, Lockman H, Black RE, Clements ML & Falkow S (1983) Recombinant DNA risk assessment studies in man: efficacy of poorly mobilizable plasmids in biologic containment. Recomb. DNA Tech. Bull. 6: 89–97
Miliotis MD, Galen JE, Kaper JB & Morris JG Jr. (1989) Development and testing of a synthetic oligonucleotide probe for the detection of pathogenic *Yersinia* strains. J. Clin. Microbiol. 27: 1667–1670
Molin S, Klemm P, Poulsen LK, Biehl H, Gerdes K & Andersson P (1987) Conditional suicide system for containment of bacteria and plasmids. Bio/Techniques 5: 1315–1318
Mullis KB (1990) The unusual origin of the polymerase chain reaction. Sci. Amer. 262(4): 56–65
Mullis KB & Faloona FA (1987) Specific synthesis of DNA in vitro *via* a polymerase-catalyzed chain reaction. Methods Enzymol. 155: 335–351
Nortermans S, Chakrobarty T, Leimeister-Wachter M, Dufrenne J, Heuvelman KJ et al. (1989) Specific gene probe for detection of biotyped and serotype *Listeria* strains. Appl. Environ. Microbiol. 55: 902–906
Obukowicz MG, Perlak FJ, Kusano-Dretzmer K, Mayer EJ, Bolten SL & Watrud LS (1986) Tn5-mediatedintegration of the delta-endotoxin gene from *Bacillus thuringiensis* into the chromosome of root-colonizing pseudomonads. J. Bacteriol. 168: 982–989
Obukowicz MG, Perlak FJ, Bolten SL, Kusano-Kretzmer K,

Mayer EJ & Watrud LS (1987) IS50L as a nonself transposable vector used to integrate the *Bacillus thuringiensis* delta-endotoxin gene into the chromosome of root-colonizing pseudomonads. Gene 51: 91–96

Ogram A, Sayler GS & Barkay T (1987) The extraction and purification of microbial DNA from sediments. J. Microbiol. Methods 7: 57–66

Paul JH, Cazares L & Thurmond J (1990) Amplification of the rbcL gene from dissolved and particulate DNA from aquatic environments. Appl. Environ. Microbiol. 56: 1963–1966

Pettigrew CA & Sayler GS (1986) The use of DNA:DNA colony hybridization in the rapid isolation of 4-chlorobiphenyl degradative bacterial phenotypes. J. Microbiol. Methods 5: 205–213

Pillai SD, Josephson KL, Bailey RL, Gerba CP & Pepper IL (1991) Rapid method for processing soil samples for polymerase chain reaction amplification of specific gene sequences. Appl. Environ. Microbiol. 57: 2283–2286

Saiki RK, Glenfand DH, Stoffel S, Scharf SJ, Higuchi R, Horn GT, Mullis KB & Erlich HA (1988) Primer-directed enzymatic amplification of DNA with a thermostable DNA polymerase. Science 239: 487–494

Sayler GS & Layton AC (1990) Environmental application of nucleic acid hybridization. Annual Review of Microbiology 44: 625–648

Sayler GS, Shields MS, Tedford ET, Breen A, Hooper SW et al. (1985) Application of DNA-DNA colony hybridization to the detection of catabolic genotypes in environmental samples. Appl. Environ. Microbiol. 49: 1295–1303

Seriwantana J, Echeverria P, Taylor DN, Sakuldaipeara T, Changchawalit C & Chivoratanond O (1987) Identification of enterotoxigenic *Escherichia coli* with synthetic alkaline phosphatase-conjugated oligonucleotide DNA probes. J. Clin. Microbiol. 25: 1438–1441

Sethabutr O, Hanchalay S, Echeverria P, Taylor DN & Leksomboon U (1985) A nonradioactive DNA probe to identify *Shigella* and enteroinvasive *Escherichia coli* in stools of children with diarrhoea. Lancet 2: 1095–1097

Sharples FE (1983) Spread of organisms with novel genotypes: Thoughts from an ecological perspective. Recomb. DNA Tech. Bull. 6: 43–56

Sommerville CC, Knight IT, Straub WL & Colwell RR (1989) Simple, rapid method for direct isolation of nucleic acids from aquatic environments. Appl. Environ. Microbiol. 55: 548–554

Steffan RJ & Atlas RM (1988) DNA amplification to enhance detection of genetically engineered bacteria in environmental samples. Appl. Environ. Microbiol. 54: 2185–2191

Steffan RJ & Atlas RM (1990) Solution hybridization assay for detecting genetically engineered microorganisms in environmental samples. Appl. Environ. Microbiol. 54: 2185–2191

Steffan RJ & Atlas RM (1991) Polymerase chain reaction: applications in environmental microbiology. Annu. Rev. Microbiol. 45: 137–161

Steffan RJ, Goksoyr J, Bej AK & Atlas RM (1988) Recovery of DNA from soils and sediments. Appl. Environ. Microbiol. 54: 2908–2915

Sussman M, Collins CH, Skinner FA & Stewart-Tull DE (1988) The Release of Genetically-Engineered Micro-Organisms. Academic Press London

Torsvik VL (1980) Isolation of bacterial DNA from soil. Soil Biol. Biochem. 12: 15–21

Torsvik VL & Goksoyr J (1978) Determination of bacterial DNA in soil. Soil Biol. Biochem. 10: 7–12

Torsvik VL, Goksoyr J & Daae FL (1990) High diversity in DNA of soil bacteria. Appl. Environ. Microbiol. 56: 782–787

Tsai YL & Olson BH (1991) Rapid method for direct extraction of DNA from soil and sediments. Appl. Environ. Microbiol. 57: 1070–1074

Tsai YL & Olson BH (1992) Detection of low numbers of bacterial cells in soils and sediments by polymerase chain reaction. Appl. Environ. Microbiol. 58: 754–757

Ward DW, Weller R & Bateson MM (1990) 16S rRNA sequences reveal numerous uncultured microorganisms in a natural community. Nature 345: 63–65

Weller R & Ward DM (1989) Selective recovery of 16S rRNA sequences from natural microbial communities in the form of cDNA. Appl. Environ. Microbiol. 55: 1818–1822

Zehr JP & McReynolds LA (1989) Use of degenerate oligonucleotides for the amplification of the nifH gene from the marine cyanobacterium *Trichodesmium thiebautii*. Appl. Environ. Microbiol. 55: 2522–2525

Zukowski MM, Gaffney DF, Speck D, Kauffmann M, Findeli A, Wisecup A & Lecocq J-P (1983) Chromogenic identification of genetic regulatory signals in *Bacillus subtilis* based on expression of a cloned *Pseudomonas* gene. Proc. Natl. Acad. Sci. USA 80: 1101–1105

Biodegradation **3**: 147–159, 1992.

Mercury biotransformations and their potential for remediation of mercury contamination

Tamar Barkay[1,*], Ralph Turner[2], Erwan Saouter[3] & Joanne Horn[3]
[1] *Microbial Ecology and Biotechnology Branch, U.S. Environmental Research Laboratory, Gulf Breeze, FL 32561, USA;* [2] *Environmental Science Division, Oak Ridge National Laboratory, Oak Ridge, TN 37831, USA;* [3] *Center for Environmental Diagnostics and Bioremediation, University of West Florida, Pensacola, FL 32514, USA (* Address for correspondence: US EPA, 1 Sabine Island Drive, Gulf Breeze, FL 32561, USA)*

Key words: bioremediation, inorganic mercury, organomercury

Abstract

Bacterially mediated ionic mercury reduction to volatile Hg^0 was shown to play an important role in the geochemical cycling of mercury in a contaminated freshwater pond. This process, and the degradation of methylmercury, could be stimulated to reduce the concentration of methylmercury that is available for accumulation by biota. A study testing the utility of this approach is described.

Abbreviations: Hg^R – inorganic mercury resistance; Org-Hg – organomercury; Org-Hg^R – organomercury resistance; SRB – sulfate reducing bacteria; Methyl-B12 – methylcobalamine

Introduction

I. Metals and metaloids – general

i. Metal resistance

Metals are crustal elements in whose presence living organisms have evolved. Many metals are essential for the metabolism of living organisms: as prosthetic groups in metalloproteins (e.g. coenzymes), regulators of ionic environments, and in maintenance of membrane potentials. Yet, all metals are toxic at elevated concentrations, and mechanisms enabling life at high metal concentrations exist in all organisms. Eukaryotes and prokaryotes have inducible metallothioniens. These are small proteins with a high cysteine content that sequester metals and thus control metal toxicity and homeostasis (Hamer 1988; Olafson 1984). Phytochelatins are small peptides (5 to 17 residues) consisting of repetitive τ-glutamylcysteine residues with a carboxyl-terminal glycine that are produced by plants in addition to metallothioniens, and serve an analogous function (Rauser 1990).

A variety of mechanisms to combat toxic metals exist in bacteria (Table 1, Silver 1991). For a detailed description of metal resistance mechanisms see the recent review by Silver & Walderhaug (1992). Most, but not all mechanisms, are plasmid encoded, and the majority involve transport phenomena. Transport-related metal resistance mechanisms include efflux, the export of metal ions, following their 'erroneous' uptake by essential ion transport systems (e.g. Cd^{2+} in *Staphylococcus aureus* by the Mn^+ transport system; Cd^{2+}, Zn^{2+}, Co^{2+}, Ni^{2+} and Mn^{2+} in *Alcaligenes eutrophus* by a transport system that has a broad specificity range; AsO_4^{3-} (arsenate); AsO_2^- (arsenite), Sb^{3+}, and TeO_3^{2-} (Turner et al. 1992) via a PO_4 transport system in both gram positive and negative bacte-

Table 1. Genetically defined, metal resistance mechanisms in bacteria.[1]

Metal	Resistance mechanism	Genetic system/regulation
Cadmium	Gram positive (*S. aureus*):	
	i. ATPase mediated efflux of Cd^{2+} that enters the cell *via* the Mn^{2+} transport system. Also confers resistance to Zn^{2+}	Plasmid encoded *cadA* (specifying the ATPase) and *cadC* (encoding a soluble protein). Induced by Cd^{2+}, Bi^{3+} and Pb^{2+}
	ii. Enhanced Cd^{2+} binding. Also confers resistance to Zn^{2+}	Plasmid encoded *cadB*
	iii. Efflux system. No additional metal resistances	Chromosomally encoded. Not related to *cadA*
	Gram negative (*A. eutrophus*): Efflux system of Cd^{2+} (and Co^{2+} and Zn^{2+}) that are taken-up by a nonspecific cation transport system	*czcABCD*. Induced by Cd^{2+}, Zn^{2+} and Co^{2+}. *czcR*, *czcS* and possibly *czcD* have been identified as regulatory elements
Arsenate, Arsenite, Antimony, Tellurite	i. Arsenite and Sb^{3+} stimulated ATPase. Specific to AsO_4^{3-}, AsO_2^- and Sb^{3+}, in both Gram positive and negative bacteria	*arsABCD*. *arsA* – ATPase; *arsB* – trans-membrane protein (anchors the ATPase, anion channel (?)[2]); *arsC* – specifies arsenate resistance (involved in arsenate reduction). *ars* is negatively regulated by the *arsR* gene product. AsO_4^{3-}, AsO_2^-, Sb^{3+} and Bi^{3+} are inducers
	ii. Oxidation of AsO_2^- to the less toxic AsO_4^{3-} in Gram negative bacteria	
Chromium	Reduced accumulation of Cr^{6+} (by either reduced uptake or an efflux system) by resistant cells of *P. aeruginosa* and *A. eutrophus*	*chrA* specifying a hydrophobic polypeptide. *chrB* (found only in *A. eutrophus*) – role in regulation (?). Inducible by Cr^{6+} in *A. eutrophus*. Constitutive in *P. aeruginosa*
Copper	i. *P. syringae*: Periplasmic binding and extracellular sequestration	Plasmid encoded *copA* and *copC* – periplasmic proteins (Cu^{2+} accumulation (?)); *copB* – an outer membrane protein (Cu^{2+} sequestration (?)). Repression by the *copR* (chromosomal location) gene product and induced by Cu^{2+} and the *copI* (plasmid location) gene product
	ii. *E. coli*: Efflux of Cu^{2+} that is taken up by a Cu^{2+} specific transport system. Resistance requires interaction between components of the transport system and an efflux system (?)	Chromosomally located *cut* system encodes transport, intracellular binding and efflux proteins. Plasmid located *pco* system specifies intracellular binding and efflux
Mercury	In a wide variety of Gram positive and negative bacteria, intracellular reduction of Hg^{2+} to the volatile Hg^0 and the degradation of organomercury to Hg^0 and an organic moiety.	Specified by the *mer* system (plasmid or chromosomal location). *merTP(C)* – transport of Hg^{2+}; *merA* – subunit of mercuric reductase; *merB* – organomercurial lyase. Positively/negatively regulated by the *merR* gene product, Hg^{2+} and organomercury compounds, act as inducers

[1] Prepared after Silver (1991) and Silver & Walderhaug (1992).
[2] (?) – Hypothesized by considering available data.

ria). In contrast, sequestration and periplasmic accumulation of Cu^{2+} was found to be the resistance mechanism in plant-pathogenic *Pseudomonas syringae*. Chemical transformation and detoxification of metals as the basis for resistance are well characterized for Hg^{2+} and organomercury compounds (see below). In addition, AsO_4^{3-} resistance involves reduction prior to its efflux as AsO_2^-. Microbes mediate numerous metal biotransformation reactions that are unrelated to metal tolerance. They reduce Cr^{6+} to Cr^{3+} (Bopp & Ehrlich 1988), methylate inorganic- (Hallas et al. 1982) and organo- (Guard et al. 1981) tin, and reduce (Steinberg & Ormland 1990) and methylate (Karlson & Frankenberger 1989) inorganic selenium.

ii. Metal pollution and remediation

The unique physical-chemical and toxic properties of metals have promoted their wide use in industrial processes and as biocides. As a result, higher concentrations than those naturally present in the biosphere have accumulated in the environment, causing public health hazards and ecological perturbations. The removal of metals is, therefore, a challenge to environmental management. Common methods of metal removal are based on concentration of metals mostly by ion exchange and sorption to a variety of resins, and precipitation as metal sulfides. Recovery and recycling are sometimes possible (Patterson & Passino 1987).

Biodegradation, the microbial-mediated transformation of hazardous material, is not applicable to metals, because, unlike organic pollutants, metals as elements cannot be mineralized to ubiquitous nontoxic compounds, such as H_2O and CO_2. However, biomobilization is a valid concept in the management of metal pollution. Biomobilization contributes to metal remediation by:

- transforming metals to less toxic chemical forms
- compartmentalizing metals in those parts of the environment where they are of less public health concern, and
- concentrating metals to facilitate disposal and/or recycling.

The reduction of uranium by the activity of anaerobic bacteria that use U^{6+} as an electron acceptor (Lovley & Phillips 1992) is an example of metal concentration, because the resulting U^{4+} precipitates and can be easily removed from polluted waters. This process is currently being tested for its efficacy in remediating contaminated groundwater (Lovley et al. 1992). Detoxification and compartmentilization of toxic selenium has been achieved (Frankenburger Jr. 1992) by stimulating soil fungi that methylate Se^{4+} to volatile and less toxic dimethylselenide (Karlson & Frankenberger Jr. 1989). Methylation was stimulated by amendments with nutrients and maintenance of oxic and moisture conditions (Frankenburger Jr. 1992). Microbial transformations of mercurial compounds to volatile Hg^0 are well understood (Silver & Walderhaug 1992). These transformations and their potential for the remediation of mercury contaminated environments are the topic of this publication.

II. Mercury

i. Mercury as an environmental contaminant

The sources of mercury in the environment are both natural and anthropogenic. Natural sources result from the geochemical cycle of mercury (Fig. 1). Anthropogenic sources contribute up to two-thirds of the current input of mercury into the environment. Due to its unique chemical and phys-

Table 2. Physical-chemical properties of mercury and their exploitation in industrial products.[1]

Property	Industrial use
Liquid at ordinary temperature, uniform volume expansion and high surface tension	Measuring devices
Low electrical resistance	Conductors
Ability to form amalgams with other metals	Recovery of metals, dentistry
Toxicity to biological systems	Biocides
Brilliant colors	Paints

[1] Prepared after Nriagu 1979.

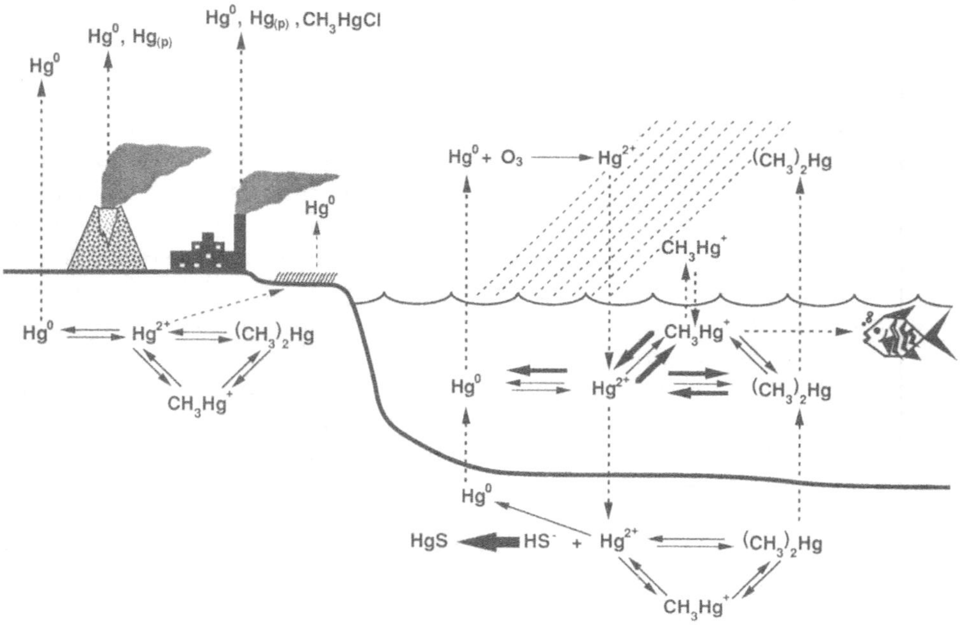

Fig. 1. The mercury cycle in the biosphere. Bold arrows – microbially mediated processes; full line arrows – transformations; broken line arrows – transport processes; a very bold arrow depicts the magnitude of the precipitation reaction between HS^- and Hg^{2+}. (Reproduced with permission from Barkay 1992.)

ical properties, mercury has been an attractive resource for several industries (Table 2). Historically, mercury was used as a pigment in paint production and in medicine (a use that is curiously being continued to this day, mostly as dental amalgams). Applications in metal recovery and felt production were introduced around the turn of the century. Mercury is currently used as an electrode in the chloralkali process, it is an important resource in electrical industries (in fluorescence lights, light switches, thermostats and batteries) (Nriagu 1979). Today, since regulatory restrictions on the disposal of mercury have been enacted, the burning of fossil fuels, the use of mercury in gold mining (mostly in third world countries), and possibly the incineration of solid wastes, remain the major anthropogenic sources of mercury to the environment.

The most common route for mercury toxicosis is by the consumption of tainted food stuff, particularly fish and shellfish (D'Itri & D'Itri 1978). Mercury is a neurotoxin interfering with the normal development and functioning of nerve and muscle systems. Toxicity is manifested as various neuromuscular malfunctions ranging from loss of sensation in mild cases to blindness, paralysis and death in severe cases (Nriagu 1979).

In recent years hazardous mercury concentrations have been discovered in freshwater fish from a broad range of geographic areas. Essentially all tissue-associated mercury in fish is CH_3Hg^+, and the high concentrations observed are the result of biomagnification through trophic level interactions (Watras & Bloom, in press). The amount of CH_3Hg^+ that is available for accumulation by aquatic biota is directly dependent on the various reactions that constitute the geochemical cycle of mercury in aquatic environments.

ii. The role of microorganisms in the geochemical cycling of mercury

a. Degradation of organomercury and reduction of Hg^{2+}

Bioremediation can be viewed as the recruitment of pollutants by geochemical cycles. Thus, mercury biomobilization makes use of the principles that govern its geochemistry (Fig. 1). Microorganisms

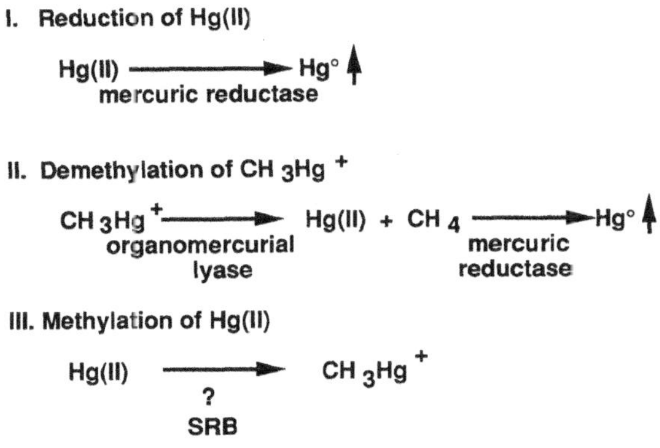

Fig. 2. Microbial transformations of mercury.

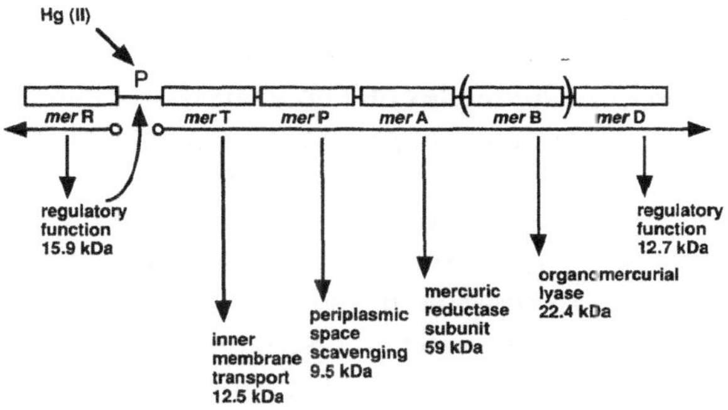

Fig. 3. The *mer* operon. Horizontal arrows – direction of transcription; vertical arrows – polypeptides translated from the corresponding genes. *merTPAD* are from Tn501; *merD* is from the broad spectrum mercury resistance plasmid R331. (Reproduced with permission from Baraky 1992.)

carry out reactions that transform mercury between its three major chemical forms, Hg^{2+}, Hg^0, and organomercury (Fig. 2). They reduce Hg^{2+} to Hg^0, they degrade organomercury (Org-Hg), and they methylate Hg^{2+}. These biological transformations, together with chemical reactions, constitute the mercury cycle in the biosphere (Fig. 1; for a detailed discussion see Barkay 1992). The relative contribution of biological *vs* chemical transformations of mercurial compounds is unknown and probably depends on the concentration of mercury and specific environmental conditions.

The reduction of Hg^{2+} and degradation of Org-Hg are detoxification mechanisms allowing bacterial growth in the presence of Hg^{2+} and Org-Hg, respectively. Bacteria resistant to Hg^{2+} produce mercuric reductase (MR), a flavin-containing disulfide oxidoreductase that catalyzes the reaction:

$$Hg(SR)_2 + NADPH + H^+ \rightarrow Hg^0 + NADP^+ + 2RSH$$

The resulting Hg^0 is subsequently volatilized due to its low aqueous solubility (6×10^{-6} g/100 ml water at 25 °C) and high vapor pressure [Henry's constant (H) = 0.3], leaving the immediate environment of the organism, which can then resume growth. A small proportion of Hg^{2+}-resistant (Hg^R) bacteria are also resistant to Org-Hg due to the activity of an additional enzyme, the organomercurial lyase (OL). The OL cleaves the C-Hg bond by a protonolytic attack, as follows:

$$R'HgSR \xrightarrow[\text{Excess RSH}]{} Hg\,(SR)_2 + R'H$$

The product of this reaction, $Hg(SR)_2$, is then reduced by the MR to Hg^0 (Walsh et al. 1988). Thus, resistance to organomercury is mediated by the sequential activity of OL and MR, and all org-Hg^R bacteria are therefore resistant to Hg^{2+} (broad-spectrum mercury resistance). On the other hand, the majority of Hg^R bacteria do not produce OL and are therefore sensitive to org-Hg (narrow-spectrum mercury resistance).

The genes specifying the various functions needed for reduction of Hg^{2+} and degradation of org-Hg are organized in the mercury resistance (*mer*) operon (Fig. 3). The *mer* operons from several organisms have now been analyzed (Silver & Walderhaug 1992), and were all found to specify functions that transport Hg^{2+} through the cell wall (*merT*, *merP*, and *merC*), the MR (*merA*), the OL (*merB*; in broad-spectrum resistance) and regulatory functions (*merR* and *merD*). Some *mer* operons have only one of the regulatory, or two of the transport, genes. Expression of the *mer* operon is negatively regulated (in the absence of Hg^{2+}), and positively regulated (in presence of Hg^{2+} and some organomercury compounds) by the interaction between a dimmer of MerR (the gene product of *merR*), the DNA at the operator/promoter region, and Hg^{2+}. *In vitro* studies employing DNA fingerprinting and restriction site protection analysis have shown that MerR and RNA polymerase are bound to the

DNA in the absence and the presence of Hg^{2+} (Summers 1992). In the absence of Hg^{2+}, this interaction forms a complex that prevents transcription. A conformational change that occurs upon binding of Hg^{2+} (through a tri-coordinate bond with three cysteine residues of the MerR dimmer; O'Halloran et al. 1989), allows subsequent transcription of the *mer*TPA(B)D operon. In addition, MerR is a negative regulator of *merR* transcription both in the absence or the presence of Hg^{2+} (Summers 1992).

When considering the contribution of *mer*-encoded reactions to the geochemical cycling of mercury, it should be noted that chemical reactions are known to reduce Hg^{2+} to Hg^0 in the absence of biological activities (but they may be indirectly mediated by living organisms). In contrast, the purely chemical degradation of org-Hg proceeds so slowly that it is considered negligible (Walsh et al. 1988). Chemical reactions resulting in Hg^0 volatilization, include reduction of Hg^{2+} to $2Hg^+$ followed by disproportionation to Hg^0 and Hg^{2+} (Baltisburger et al. 1979), and reduction by the interaction of Hg^{2+} with free radical electrons of humic acids (Alberts et al. 1974). Observations indicating evolution of Hg^0 by algae grown in pure cultures (Ben-Bassat & Mayer 1978), and possibly by marine phytoplankton (Kim & Fitzgerald 1986), may be explained as an indirect reduction of Hg^{2+} to $2Hg^+$ by organic acids that are produced during photosynthesis.

b. Methylation of Hg^{2+}

The first report demonstrating that microbes in anaerobic freshwater sediments methylated Hg^{2+} was published in 1969 (Jensen & Jerenelöv 1969). It has since been shown to occur in marine and estuarine sediments (Blum & Bartha 1980), in aerobic and anaerobic waters, and in the fish intestine (see review by Winfrey & Rudd 1990). Whereas many microbes can methylate Hg^{2+} in laboratory cultures (Vonk & Sifpesteijn 1973), the organisms responsible for this activity *in situ* have not been clearly identified. Early suggestions that methanogens could methylate Hg^{2+} (Wood 1974) were rejected when experiments were performed employing specific inhibitors of methanogenesis and sulfate reduction. The results of these experiments

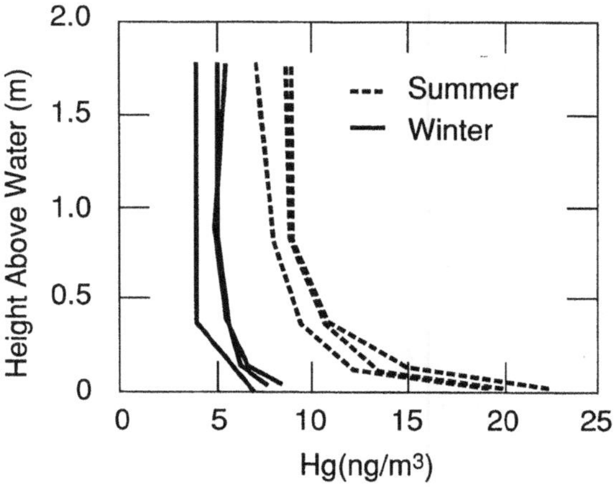

Fig. 4. Elemental mercury gradients formed by the evasion of Hg^0 from RL to the air.

implicated sulfate reducing bacteria (SRB) as the principal Hg^{2+} methylators in estuarine (Compeau & Bartha 1985) and possibly also in freshwater (Winfrey & Rudd 1990) sediments.

A proposed mechanism of methylation involves the non-enzymatic transfer of a negatively charged methyl group, CH_3^-, from methylcorrinoids, such as methylcobalamine (methyl-B12), to Hg^{2+}. Microbial metabolism facilitates this reaction by the enzymatic transfer of CH_3^- from methyltetrahydrofolate to B12. This proposal is supported by the isolation of mutants that concomitantly lost the ability to synthesize methyl-B12 and CH_3Hg^+ (Pan-Hou & Imura 1982). A novel methylcobalamine-like corrinoid (Berman et al. 1990; Choi et al. 1991) was isolated from a methylating SRB, *Desulfovibrio desulfuricans*. Although information regarding the exact mechanism of bacterial Hg^{2+} methylation is scarce, it is generally thought to involve methyl transfer reactions.

Abiotic sources of methylmercury include Hg^{2+} interactions with humic and fulvic substances (Nagase 1984), and methyl transfer from dissolved organic matter to Hg^{2+} may be mediated photochemically in the presence of sulfur (Akagi et al. 1974). Volatile $(CH_3)_2Hg$ may be formed by a disproportionation reaction between $2CH_3Hg^+$ and H_2S (Deacon 1978).

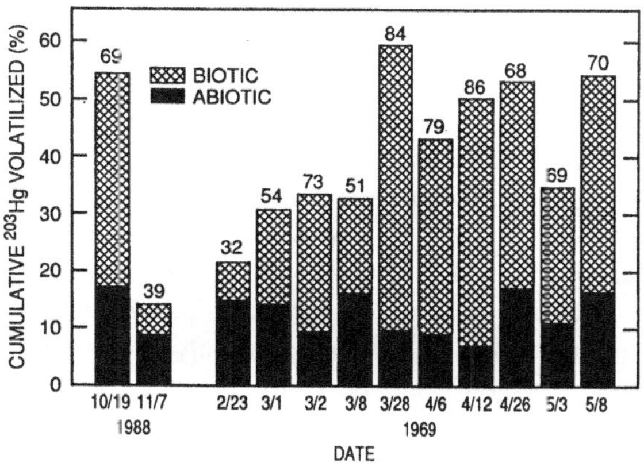

Fig. 5. Biotic and abiotic volatilization of Hg^0 by RL water samples. Numbers above bars indicate percent of volatilization due to biotic activities.

Microbially mediated removal of mercury from a contaminated freshwater stream

East Fork Poplar Creek is a freshwater stream in the vicinity of Oak Ridge, TN. It is a low-order stream-pond ecosystem, contaminated by inorganic mercury released from a nuclear weapons plant located at the headwaters of the stream. Drainage water from the plant is routed through a plastic-lined pond, Reality Lake (RL), with an 8-h retention time prior to discharge into the creek. In recent years, we have been studying the role of microorganisms in the cycling of mercury in RL (Turner et al. 1989; Barkay et al. 1991; Liebert et al. 1991). Control (uncontaminated) samples were collected at a high order river-reservoir system with no known mercury contamination. The results of these studies have encouraged us to propose bioremediation for the management of mercury in aquatic systems.

The role of microbial reactions in the evolution of Hg^0 from RL

Field observations, laboratory experiments with samples taken from RL, and analyses of the indigenous microbial community, have shown that microbes remove mercury, as Hg^0, from RL. A gradient in Hg^0 concentration was detected in air above the pond, surface (Fig. 4), suggesting evasion from the water to the atmosphere. Gradients were measured by pumping air through iodated activated charcoal tubes (to strip Hg^0) that were placed at variable heights above the pond, using a buoy to which a scaffold carrying five air sampling devices was attached. Aerial concentrations of Hg^0 decreased rapidly with height above the water, probably by dilution. Air in close proximity to the surface of the pond contained at least twice as much Hg^0 during summer months as compared to winter months, and as a result stronger gradients were obtained during the summer as compared with winter. The larger flux of Hg^0 during the warmer part

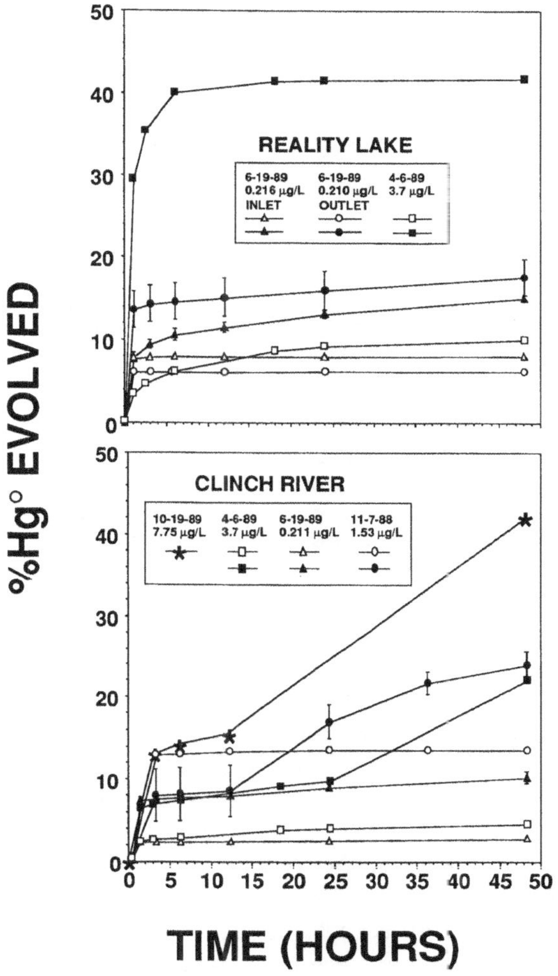

Fig. 6. Evolution of Hg^0 from water samples. Data are presented for each sampling site. Assay dates and total substrate concentrations (indigenous and $^{203}Hg(II)$) are indicated. Full and empty symbols indicate live and heat-killed samples, respectively. (Reproduced with permission from Barkay et al. 1991.)

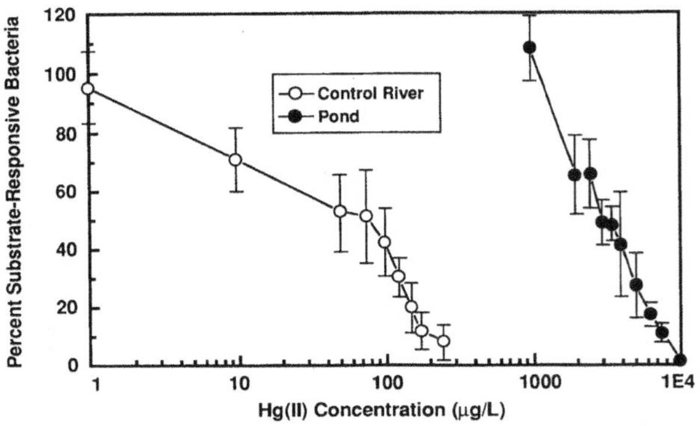

Fig. 7. Tolerance to Hg^{2+} of RL and river microbial communities. Dose response curves are based on direct counts. Percent cells that utilize yeast extract in the presence of increasing concentrations of Hg^{2+}, of cells utilizing yeast extract in the absence of Hg^{2+}, were plotted against test Hg^{2+} concentrations. (Reproduced with permission from Liebert et al. 1991.)

of the year may be due to biological activities and to the effect of temperature on partition of Hg^0 between water and air.

Comparisons of biotic with abiotic Hg^0 reduction/volatilization from RL water samples, measured using a $^{203}Hg^{2+}$, showed that biotic processes were relatively more significant during spring as compared to winter months (Fig. 5). Abiotic Hg^0 evolution was measured in autoclaved samples. Less than 40% of Hg^0 was generated by biotic processes in Nov. 88 and Feb. 89, and this proportion gradually increased to 70–80% through March–May 89. These results suggested that an active microbial community in the pond reduced Hg^{2+} to Hg^0. This conclusion was further supported by studies that compared the microbial communities of the pond with those of communities from the control river site. The pond community was more tolerant to Hg^{2+} (Liebert et al. 1991), which correlated with an enrichment of DNA sequences with homology to *mer* genes, and the biological reduction of Hg^{2+} to Hg^0 (Barkay et al. 1991).

Different kinetics of Hg^0 evolution by active pond and control samples were observed (Fig. 6). Elemental mercury was immediately evolved from pond samples after spiking with $^{203}Hg^{2+}$. On the other hand, Hg^0 from biologically active river samples was evolved in two distinct phases. An initial burst of Hg^0, also present in abiotic samples, was followed by a second increase in activity 12 to 30 h after spiking. In all cases, we did not observe further increase in Hg^0 in autoclaved samples following initial Hg^0 evolution. Thus, we hypothesized that initial activity in river samples was due to abiological processes that reduce Hg^{2+}, whereas reduction of Hg^{2+} by biological activities began after a lag period of 12 to 24 h. A lag period prior to the initiation of Hg^{2+} reduction is considered an acclimation period, and is commonly observed in samples that have not been previously exposed to Hg^{2+} (Barkay 1987). According to this hypothesis, the instantaneous $^{203}Hg^0$ evolution observed in biologically active pond samples strongly suggests that the microbial community of the pond was acclimated to Hg^{2+} and that a substantial proportion of the Hg^0 formed *in situ* was due to microbial activities.

Enrichment of active populations in response to the toxicity of Hg^{2+} is the most likely mechanism of acclimation (Barkay 1987), although induction of enzymatic activities and gene transfer between populations (Spain et al. 1980) cannot be ruled out. Indeed, the pond community had a greater number of organisms resistant to Hg^{2+} than the control river community (Fig. 7). Resistance was quantitated by a direct counting method (Liebert & Barkay 1988) that is based on a modification of methods for enumeration of viable bacteria in aquatic samples (Kogure et al. 1979). Water samples were incubated with yeast extract as a growth substrate, nalidixic acid to inhibit cell division, and variable Hg^{2+} concentrations. Under these conditions, only Hg^{2+}-resistant cells can utilize the available substrate and form large elongated cells. The percent of large cells formed in the presence of Hg^{2+}, of those formed in control samples without Hg^{2+}, is a function of the tolerance of viable bacteria to Hg^{2+}. A dose response curve obtained when these percents are plotted against test Hg^{2+} concentrations (Fig. 7) is used to calculate IC_{50} (concentration that inhibits 50% of elongated cells). Thus, IC_{50} values for the pond and river communities were 3,582 ± 160 and 69.97 ± 4.66 μg Hg^{2+}/L, respectively (see Liebert et al. 1991 for a description of calculations and statistical analyses). The 50-fold difference in IC_{50} values indicates that Hg^{2+}-tolerant popula-

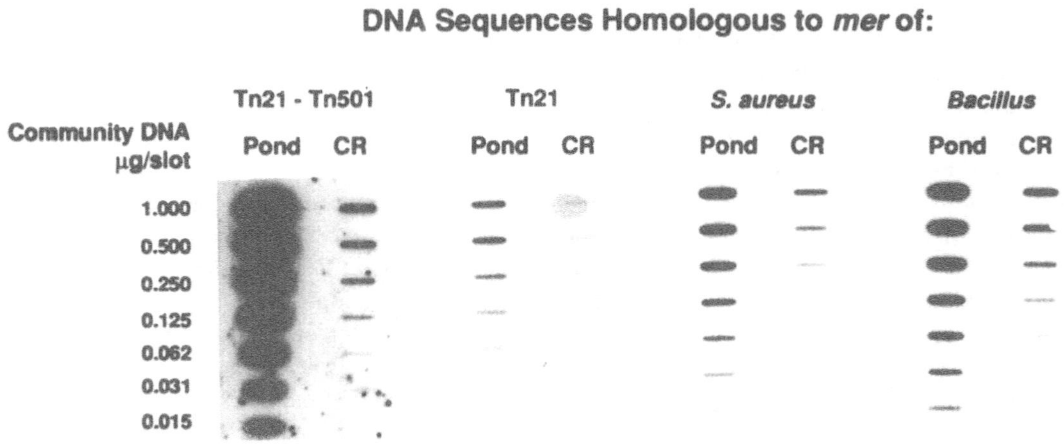

Fig. 8. Abundance of *mer* genes in RL and river microbial communities. Autoradiogram of results obtained by hybridizing DNA that was extracted from pond (Pond) and river (CR) water samples, with *mer* probes. Intensity of hybridization signals is directly related to the abundance of *mer* in DNA extracts. (Reproduced with permission from Barkay et al. 1991.)

tions are enriched in the microbial community of the contaminated pond.

Enriched Hg^R bacteria in the pond most likely carried *mer* genes because DNA-DNA hybridizations with four *mer* probes (Barkay et al. 1989) showed a higher abundance of these genes in DNA extracted from pond, as compared to river, biomass (Fig. 8). Autoradiographic signals were quantitated by densitometry using a calibration curve prepared with known amounts of the various *mer* operons. The abundance of all *mer* systems (per µg community DNA) was higher in the pond community as compared to the control river community. Increase in abundance was as high as 72-fold for DNA homologous to *mer* of transposons 501 (Tn501) and 21 (Tn21) (18.121 ± 4.059 and 0.251 ± 0.023 ng target DNA/µg community DNA for pond and river communities, respectively).

Compilation of the data from RL strongly suggests that *mer*-specified bacterial activities reduce Hg^{2+} to Hg^0, thereby contributing to the volatilization of Hg^{2+} from a mercury-contaminated freshwater pond. This study therefore demonstrates the role of a specific bacterial reaction in the geochemical cycling of mercury in the environment. Could this reaction be stimulated to further remove Hg^{2+} from the pond? A decrease in Hg^{2+} concentration would indirectly affect bioavailable CH_3Hg^+ concentration by substrate limitation, according to the following equation:

$$Hg^0 \longleftarrow Hg^{2+} \rightleftharpoons CH_3Hg^+$$

The feasibility of this approach to the remediation of mercury is currently being tested in our laboratory.

An experimental approach to test the feasibility of mer*-mediated reactions for the remediation of contaminated sites*

The effect of manipulations on the rate of Hg^0 volatilization and CH_3Hg^+ degradation is tested using an experimental approach that progresses from a simple laboratory test system to one incorporating more of the known elements involved in total ecosystem mercury cycling. Analysis is based on quantitation and speciation of mercury by the methods of Bloom & Fitzgerald (1988), and Bloom (1989) in the various compartments of each test system.

At present we do not know whether mercury transformations in the pond are limited by the availability of nutrients for microbial activities, the size of the Hg^{2+} reducing and CH_3Hg^+ demethylating populations, or the availability of substrates for transformation reactions. The following treatments will therefore, be attempted:

(i) stimulation of general microbial activities, among them mercury-specific reactions, by the addition of limiting growth substrates,
(ii) stimulating only Hg^{2+} reduction and CH_3Hg^+ demethylation by the addition of bacterial strains capable of these activities, and
(iii) increasing availability of the substrates, Hg^{2+} and CH_3Hg^+.

Rationale and procedures for treatments (i) and (ii) are described below.

Addition of growth substrates to stimulate activities of indigenous microorganisms has been demonstrated as a successful approach to remediation of contaminated sites (Thomas & Ward 1989). This treatment may stimulate general microbial activities in RL because concentrations of essential nutrients suggest that carbon and phosphorous may be limiting (Table 3). Typical concentrations of available carbon (as DOC), nitrogen (as $NH_4 + NO_3$), and phosphorous (as total PO_4) in RL are 3.67, 4.83, and 0.28 mg/L, respectively. Not only that DOC is low, approaching values found in oligotrophic waters (Morita 1982), but these values constitute an imbalanced C:N:P ratio of 13:17:1. The bacterial cell composition has C:N:P of 17:5:1 (%) (Stanier et al. 1971), and natural planktonic bacterial assemblages have a C:N ratio of 4:1 (on a weight/volume basis) (Nagata 1986). Assuming that approximately 50% of the carbon consumed by microbes is assimilated into cellular material, a C:N:P ratio of available nutrients that deviates from 34:5:1 indicates that bacterial growth and activities are limited. Thus, microbial activities in RL could be stimulated by the addition of carbon (added as glutamate, glucose or as a mixture of amino acids) and/or phosphorous (added as PO_4).

Table 3. Concentrations of essential microbial growth substrates in Reality Lake.

Element	Mg/L	
	Range	Mean ± SD
Carbon (DOC)	1 – 4	3.67 ± 0.82
Nitrogen	3.5 – 6.0	4.83 ± 1.25
Phosphorous	0.2 – 0.4	0.278 ± 0.067

Values are typical to RL, measurements of organic and inorganic nutrient concentrations in RL are performed routinely with samples collected weekly.

Increasing the population size of active microorganisms by application of genetically engineered or native microbes has been a much discussed prospect in environmental management (Halvorson et al. 1985). We have demonstrated Hg^{2+} reduction (see above) and CH_3Hg^+ degradation (Temple et al. 1989) by biologically active water samples; and reducing and demethylating strains were isolated from the pond (Rochelle et al. 1991). The possibility that the abundance of active populations limits mercury transformations will be investigated by introducing active bacterial strains to water samples collected in RL and following their effect on rates of Hg^{2+} reduction/volatilization and CH_3Hg^+ demethylation. Strains with high levels of MR and OL will be selected for testing, from a collection of Hg^R and org-Hg^R bacteria, that was isolated by enrichment from RL water. In addition, genetically engineered *Pseudomonas putida* strains harboring a constitutively expressed broad-spectrum *mer* operon will be employed. The broad-spectrum pDU1358-derived *mer* operon (Griffin et al. 1987) deleted of regulatory elements (*merR* and *merD*) was transposed into *P. putida* using the Tn5-based delivery system of Herrero et al. (1990), and Hg^{2+} and phenylmercuric acetate (PMA) resistant derivatives were selected. Some of these strains, constitutively expressing *mer* functions from *P. putida* promoters, were subsequently shown to be highly tolerant to Hg^{2+} and PMA and to hyper-produce MR (Horn et al. submitted for publication). Furthermore, they effectively removed Hg^{2+} from wastewaters in an experimental, closed, treatment system (Horn, in preparation). Because of concerns regarding release of genetically engineered microorganisms, testing of these engineered *P. putida* strains will be performed in contained systems, shake flasks, and microcosms; field applications (see below) will only employ native non-engineered organisms.

Evaluation of the success of remedial treatments

Our approach to test the effectiveness of remedial treatments is based on evaluating how the geochemical cycle of mercury is affected by accelerat-

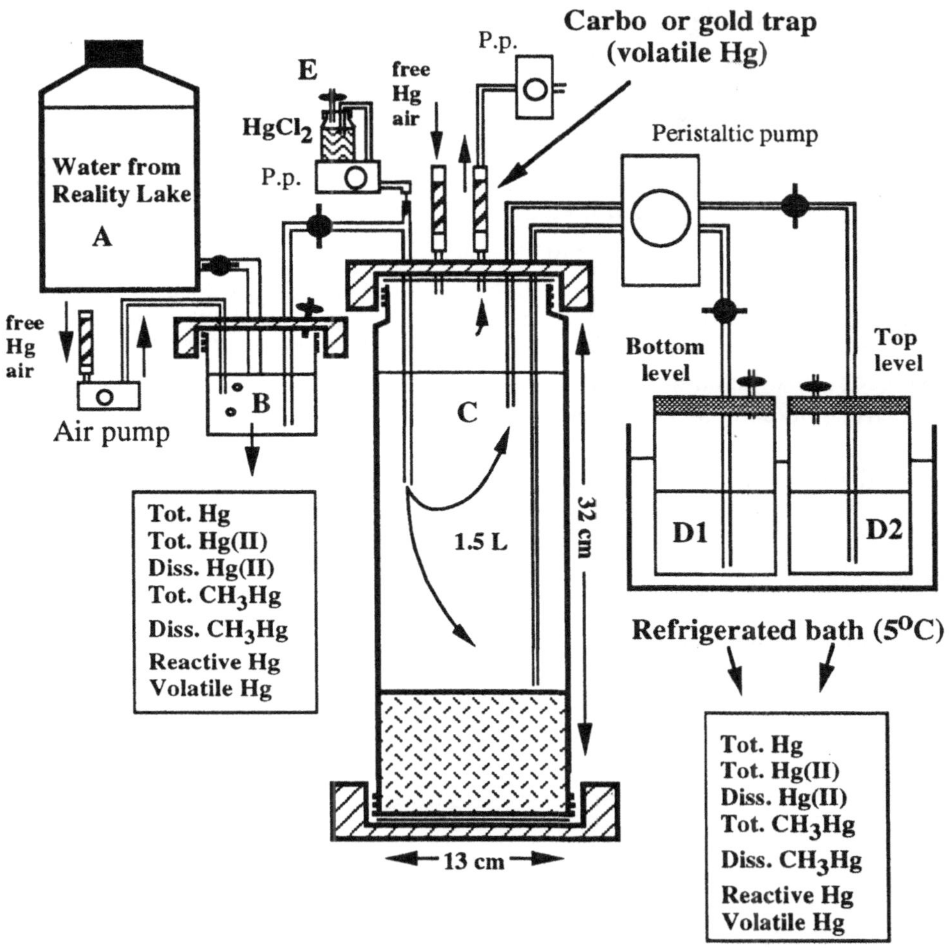

Fig. 9. A microcosm system for the study of geochemical cycling of mercury in aquatic environments. Various species shown to play a role in the geochemical cycling of mercury, are monitored in inlet water, outlet water and air leaving the microcosm. In addition, mercury in sediment and microbiota is quantitated and speciated when microcosm experiments are terminated.

ing specific mercury biotransformations. The approach integrates analyses of mercury pools and fluxes into a mathematical model that describes the mass balance of mercury in aquatic ecosystems. This mercury cycling model (MCM) can be used to predict the pool size of different mercury species using known rates of transformations (Hudson et al. 1991). Thus, once the MCM is shown to describe the cycling of mercury in RL, it can predict the magnitude of increase in Hg^{2+} reduction and CH_3Hg^+ degradation rates that is needed to effectively reduce the concentration of CH_3Hg^+ available for accumulation by fish.

Effect of remedial treatments on rates of Hg^{2+} reduction and CH_3Hg^+ degradation will initially be measured in flasks (Turner et al. 1989) by following evolution of Hg^0. Effective treatments (i.e. with a potential to significantly decrease concentration of CH_3Hg^+) will then be tested in microcosms. A mercury geochemical cycling microcosm has been constructed to allow speciation and quantitation of the major chemical species of mercury that are present in RL (Fig. 9). Inventories of the various species of mercury in air, and inlet and outlet waters, enable determinations of pool size, fluxes between pools, and rates of transformations. This microcosm system utilizes state-of-the-art methods for mercury analyses (Bloom 1989; Bloom & Fitzgerald 1988), therefore allowing evaluation of promising remedial treatments in terms of their effect on the geochemical cycle of mercury in RL. In addition, various aspects of the geochemical cycling of mercury in aquatic environments can be studied using this microcosm because sensitivity of analyses allows for accurate measurements of mercury at background concentrations (10^{-12}–10^{-15} M).

Treatments shown to be effective in microcosms will then be tested in enclosures placed in RL. Field enclosures have been constructed and their utility tested. Our objective is to demonstrate a decrease in CH_3Hg^+ concentration in the water and in fish that are placed in these enclosures.

Acknowledgements

Thanks are extended to Mary Anna Bogle, Cynthia Liebert and Mark Gillman for their assistance. The work reported here was partially supported by Electric Power Research Institute agreement RP3015–03.

References

Akagi H, Takabatake E & Fujita Y (1974) Photochemical methylation of inorganic mercury in the presence of solid sulfur. Chem. Let. 761–764

Alberts JJ, Schindler JE, Miller RW & Nutter Jr. DE (1974) Elemental mercury evolution mediated by humic acid. Science 184: 895–897

Baltisberger RJ, Hildebrand DA, Grieble D & Ballintine TA (1979) A study of the disproportionation of mercury (I) induced by gas sparging in acidic aqueous solutions for cold-vapor atomic absorption spectrometry. Anal. Chim. Acta 111: 111–122

Barkay T (1987) Adaptation of aquatic microbial communities to Hg^{2+} stress. Appl. Environ. Microbiol. 53: 2725–2732

Barkay T (1992) The mercury cycle. Encyclopedia of Microbiology, Vol 3. Academic Press Inc. San Diego

Barkay T, Liebert C & Gillman M (1989) Hybridization of DNA probes with whole-community genome for detection of genes that encode microbial responses to pollutants: *mer* genes and Hg^{2+} resistance. Appl. Environ. Microbiol. 55: 1574–1577

Barkay T, Turner RR, VandenBrook A & Liebert C (1991) The relationships of Hg(II) volatilization from a freshwater pond to the abundance of *mer* genes in the gene pool of the indigenous microbial community. Microb. Ecol. 21: 151–161

Ben-Bassat D & Mayer AM (1978) Light-induced Hg volatilization and O_2 evolution in *Chlorella* and the effect of DCMU and methylamine. Physiol. Plant. 42: 33–38

Berman MT, Chase Jr T & Bartha R (1990) Carbon flow in mercury biomethylation by *Desulfovibrio desulfuricans*. Appl. Environ. Microbiol. 56: 298–300

Bloom N (1989) Determination of picogram levels of methylmercury by aqueous phase ethylation, followed by cryogenic gas chromatography with cold vapour atomic fluorescence. Can. J. Fish. Aquat. Sci. 46: 1131–1140

Bloom NS & Fitzgerald (1988) Determination of volatile mercury species at the picogram level by low temperature gas chromatography with cold-vapor atomic fluorescence detection. Anal. Chim. Acta 209: 151–161

Blum JE & Bartha R (1980) Effect of salinity on methylation of mercury. Bull. Environm. Contam. Toxicol. 25: 404–408

Bopp LH & Ehrlich HL (1988) Chromate resistance and reduction in *Pseudomonas fluorescens* strain LB300. Arch. Microbiol. 150: 426–431

Choi S-C, Berman M & Bartha R (1991) Evidence for a novel Hg^{2+}-methylating corrinoid in *Desulfovibrio* desulfuricans. Abst. 91th Annu. Meet. Am. Soc. Microbiol., Q-261, (p 320)

Compeau GC & Bartha R (1985) Sulfate-reducing bacteria: principle methylators of mercury in anoxic estuarine sediment. Appl. Environ. Microbiol. 50: 498–502

Deacon GB (1978) Volatilization of methyl-mercuric chloride by hydrogen sulphide. Nature (London) 275: 344

D'Itri PA & D'Itri FM (1978) Mercury contamination: a human tragedy. Environ. Manag. 2: 3–16

Frankenberger Jr. WT (1992) Dissipation of soil selenium by microbial volatilization at Kesterson reservoir. Abst. 203rd Am. Soc. Chem. Nat. Meet. San Francisco, CA April 5–10, 1992

Griffin HG, Foster TJ, Silver S & Misra TK (1987) Cloning and DNA sequence of the mercuric- and organomercurial-resistance determinants of plasmid pDU1358. Proc. Natl. Acad. Sci. U.S.A. 84: 3112–3116

Guard HE, Cobet AB & Coleman III WM (1981) Methylation of trimethyltin compounds by estuarine sediments. Science 213: 770–771

Hallas LE, Means JC & Cooney JJ (1982) Methylation of tin by estuarine microorganisms. Science 215: 1505–1507

Halvorson HO, Pramer D & Rogul M (1985) Engineered Organisms in the Environment: Scientific Issues. American Society for Microbiology, Washington DC

Hamer DH (1988) Metallothionein. Annu. Rev. Biochem. 55: 913–951

Herrero M, de Lorenzo V & Timmis KN (1990) Transposon vectors containing non-antibiotic resistance selection markers for cloning and stable chromosomal insertion of foreign genes in gram-negative bacteria. J. Bacteriol. 172: 6557–6567

Horn J, Brunke M, Deckwer W-D & Timmis KN (1993) *Pseudomonas putida* strains which constitutively hyperexpress

mercury resistance for biodetoxification of organomercurial pollutants. Appl. Environ. Microbiol. (in press)
Hudson RJM, Gherini SA & Munson RK (1991) The MTL mercury model: a description of the model, discussion of scientific issues, and presentation of preliminary results. In: Mercury in Temperate Lakes – 1990 Annual Report
Jensen S & Jernelöv A (1969) Biological methylation of mercury in aquatic organisms. Nature 223: 753–754
Karlson U & Frankenberger Jr. WT (1989) Accelerated rates of selenium volatilization from California soils. Soil Sci. Soc. Am. J. 53: 749–753
Kim JP & Fitzgerald WF (1986) Sea-air partitioning of mercury in the equatorial Pacific ocean. Science 231: 1131–1133
Kogure K, Simidu U & Taga N (1979) A tentative direct microscopic method for counting living marine bacteria. Can. J. Microbiol. 25: 415–420
Liebert C & Barkay T (1988) A direct viable counting method for measuring tolerance of aquatic microbial communities to Hg^{2+}. Can. J. Microbiol. 34: 1090–1095
Liebert C, Barkay T & Turner RR (1991) Acclimation of aquatic microbial communities to Hg(II) and CH_3HgCl in polluted freshwater ponds. Microb. Ecol. 21: 139–149
Lovley DR & Phillips EJP (1992) Reduction of uranium by *Desulfovibrio desulfuricans*. Appl. Environ. Microbiol. 58: 850–856
Lovley DR, Landa ER, Phillips EJP & Woodward JC (1992) Remediation of uranium-contaminated soils using uranium extractants and microbial uranium reduction. Abst. 203rd Am. Chem. Soc. Nat. Meeting, San Francisco April 5–10, 1992
Morita RY (1984) Starvation-survival of heterotrophs in the marine environment. Adv. Microb. Ecol. 6: 171–198
Nagase H (1984) Mercury methylation by compounds in humic material. Sci. Tot. Environ. 32: 147–156
Nagata T (1986) Carbon and nitrogen content of natural planktonic bacteria. Appl. Environ. Microbiol. 52: 28–32
Nriagu JO (1979) The Biogeochemistry of Mercury in the Environment. Elsevier/North-Holland Biomedical Press, Amsterdam
Pan-Hou HS & Imura N (1982) Involvement of mercury methylation in microbial mercury detoxication. Arch. Microbiol. 131: 176–177
Patterson JM & Passino R (1987) Metals Speciation, Separation, and Recovery. Lewis Publishers, Inc., Chelsea, MI
O'Halloran TV, Frantz B, Myung KS, Ralston DM & Wright JG (1989) The MerR heavy metal receptor mediates positive activation in a topologically novel transcription complex. Cell 56: 119–129
Olafson RW (1984) Prokaryotic metallothionein. Internat. J. Peptide Protein Res. 24: 303–308
Rauser WE (1990) Phytochelatins. Annu. Rev. Biochem. 59: 61–86
Rochelle PA, Wetherbee MK & Olson BH (1991) Distribution of DNA sequences encoding narrow- and broad-spectrum mercury resistance. Appl. Environ. Microbiol. 57: 1581–1589
Silver S (1991) Resistance systems and detoxification of toxic heavy metals. In: Rossmoore H (Ed) Proceedings of the Eighth International Biodeterioration Symposium (pp 308–339). Elsevier Applied Science, London
Silver S & Walderhaug M (1992) Gene regulation of plasmid- and chromosome-determined inorganic ion transport in bacteria. Microbiol. Rev. 56: 195–228
Spain JC, Pritchard PH & Bourquin AW (1980) Effects of adaptation on biodegradation rates in sediment/water cores from estuarine and freshwater environments. Appl. Environ. Microbiol. 40: 726–734
Stanier RY, Doudoroff M & Adelberg EF (1971) General Microbiology. The Macmillan Press Ltd., London
Steinberg NA & Ormland RS (1990) Dissimilatory selenate reduction potentials in a diversity of sediment types. Appl. Environ. Microbiol. 56: 3550–3557
Summers AO (1992) Untwist and shout: a heavy metal-responsive transcriptional regulator. J. Bacteriol. 174: 3097–3101
Temple KA, Turner RR & Barkay T (1989) Volatilization of divalent mercury and demethylation of methylmercury: microbial adaptation in a mercury-contaminated site. Abst. Q-186, 89th Annu. Meet. Am. Soc. Microbiol. (p 361)
Thomas JM & Ward CH (1989) In situ biorestoration of organic contaminants in the subsurface. Environ. Sci. Technol. 23: 760–766
Turner RJ, Hou Y, Weiner JH & Taylor DE (1992) The arsenical ATPase efflux pump mediates tellurite resistance. J. Bacteriol. 174: 3092–3094
Turner RR, VandenBrook AJ, Barkay T & Elwood JW (1989) Volatilization, methylation, and demethylation of mercury in a mercury-contaminated stream. In: Vernet JP (Ed) Proc. Int. Conf. Heavy Metals in the Environment (pp 353–356). CEP Consultants Ltd., Edinburgh, UK
Vonk JW & Sijpesteijn AK (1973) Studies on the methylation of mercuric chloride by pure cultures of bacteria and fungi. Antonie van Leeuwenhoek 39: 505–513
Walsh CT, Distefano MD, Moore MJ, Shewchuk LM & Verdine GL (1988) Molecular basis of bacterial resistance to organomercurial and inorganic mercuric salts FASEB J. 2: 124–130
Watras CJ & Bloom SN (1992) Mercury and methylmercury in individual zooplankton: implications for bioaccumulation. Limnol. Oceanogr. 37: 1313–1318
Winfrey MR & Rudd JWM (1990) Environmental factors affecting the formation of methylmercury in low pH lakes. Environ. Toxicol. Chem. 9: 853–869
Wood J (1974) Biological cycles for toxic elements in the environment. Science 183: 1049–1052

Biodegradation 3: 161–170, 1992.

Interactions of bacteria with cadmium

Eliora Z. Ron, Dror Minz, N. P. Finkelstein[1] & Eugene Rosenberg
Department of Molecular Microbiology and Biotechnology, Tel-Aviv University, Tel-Aviv, Israel; [1] I.M.I., Haifa, Israel

Key words: bioremediation, cadmium, heavy metals

Abstract

Cadmium pollution arises mainly from contamination of minerals used in agriculture and from industrial processes. The usual situation is of large volumes of soil and water that are contaminated with low – but significant – concentrations of cadmium. Therefore, detoxification of the polluted water and soil involves the concentration of the metal, or binding it in a way that makes it biologically inert.

Cadmium is one of the more toxic metals, that is also carcinogenic and teratogenic. Its effects are short term, even acute (diseases like Itai-itai), or long term. The long term effects are intensified due to the fact that cadmium accumulates in the body.

This paper describes a study involving several hundred cadmium-resistant bacterial isolates. These bacteria could be divided into three groups – the largest group consisted of bacteria resistant to cadmium by effluxing it from the cells. The bacteria of the other two groups were capable of binding cadmium or of detoxifying it. We concentrated on one strain that could bind cadmium very efficiently, depending on the bacterial biomass and on the pH. This strain could effectively remove cadmium from contaminated water and soil.

Introduction

Sources of cadmium pollution

Cadmium occurs as a minor component in many rocks and minerals. When these are mined, processed and later used, some of the cadmium is released into the environment where it can constitute a serious hazard.

Cadmium is also released into the environment as a consequence of the use of large-tonnage commodities such as coal, oil and phosphate rock, in which it occurs as a trace element. Cadmium can be volatilized when coal and oil are burned, and dispersed over the surroundings where it accumulates in the soil, ground and surface waters. Much of the cadmium content of phosphate rock is dispersed over agricultural land in the form of phosphate fertilizers.

The net result of these processes is the pollution of massive volumes of soil and water with low, but significant (1–10 ppm) concentrations of the element. The contaminant is taken up by plants, thereby entering the food chain and becoming a threat to humans. Although the concentrations involved all along the line may be low, the hazard is a serious one because cadmium accumulates in the body.

Another source of cadmium pollution involves industrial wastes. Cadmium is produced as a by-product of the extraction of zinc from its ores. Its principal applications are in rechargable batteries, in pigments and stabilizers for plastics. It is also used heavily for electroplating. The use of cadmium in industry adds up to many thousands of tons yearly. Effluents from factories, as well as discarded cadmium-containing objects are a prominent source of pollution.

Cadmium toxicity

The contaminating cadmium that originates from fertilizers or waste water gets to humans through the food chain. An additional route of exposure to cadmium involves uptake of high concentrations of cadmium – usually from industry – by inhalation or eating. This kind of exposure results in acute cadmium poisoning.

The high toxicity of cadmium consists of short term and long term effects. The short term effects are probably due to the fact that it binds to thiol (-SH) groups and denatures proteins. The long term effects result from the fact that cadmium induces DNA damage and has been proven to be mutagenic and carcinogenic. The long term effects are intensified due to the fact that cadmium accumulates in the body.

When entering the human body, cadmium is bound by polythiols (metallothioneins) and is transported to the liver. From there most of it is secreted in the urine but a fraction accumulates in the kidneys, liver and pancreas. Inhaled cadmium accumulates in the lungs as well. The symptoms of cadmium poisoning include edema and decreased function of the organs (kidneys, lungs). Inhalation is also followed by coughing. The acute disease (Itai-itai disease) involves pain in the bones and fracture.

Long term effect result from the fact that cadmium is mutagenic, teratogenic and carcinogenic. The carcinogenic effect involves a significant incidence of lung cancer following inhalation. The problem is intensified by the fact that cadmium has a very long biological half-life – the half life determined in the kidneys was 30 years.

The importance of cadmium in disturbing cellular metabolism is also demonstrated by the fact that when it enters the cells, cadmium induces the 'stress response'. This cellular response is also called 'heat shock response', as it was first discovered in connection with a shift to higher temperatures. This response involves the synthesis of a large number of new proteins, in response to environmental factors such as high temperature, ethanol and cadmium. It has been observed in eukaryotic (Levinson et al. 1980) and bacterial cells (Van Bogelen et al. 1987). The molecular and biochemical function of the heat shock proteins is not fully understood. However, it is known that several of the inducing factors are agents that denature proteins, and several of the induced proteins are known to act as 'chaperons' that protect other proteins from denaturation.

The other group of proteins that are induced in eukaryotic cells by the presence of cadmium ions are the polythiols and the poly-glutamyl peptides that are common in plant tissues (Steffens et al. 1986). These bind the cadmium with high affinity thereby reducing its binding to sensitive targets, and secreting it out of the tissues. As an example, the metallothioneins are unusual small proteins of less than 10,000 in molecular weight, with cysteine content as high as 30%, that can bind cadmium with dissociation constants as low as 10^{-16} M (Hamer 1986). Metallothioneins have been shown to bind zinc, mercury, copper, silver and bismuth as well. In plant cells exposed to Cd^{2+}, polyglutathione is the major intracellular thiol form (Steffens et al. 1986).

Microbial resistance to cadmium

While eukaryotic organisms detoxify cadmium, as well as other heavy metals, mainly by binding to polythiols, bacteria have developed several different and very efficient mechanisms for tolerating heavy metals. These mechanisms can be so effective that high levels of the relevant toxic metals have no discernible effect on cell growth of resistant strains.

In many but not all organisms, the genes controlling metal resistance are carried on plasmids, which often contain resistance to several metals (Cd, Pb, Cr, Mo, and U). These plasmids provide the bacteria with a competitive advantage over other organisms when heavy metals are present. Many bacterial strains isolated from soil under selective pressure of heavy metals were resistant not only to a large number of heavy metals, but to antibiotics such as ampicillin, cephalosporin, chloramphenicol, furadantin, kanamycin, nalidixic acid, streptomycin, and tetracycline (Marques et al. 1979;

Misra et al. 1985). This finding suggests that metal resistance can be associated with multiple antibiotic resistance on R (resistance) plasmids. Indeed, R plasmids are frequently found in clinical isolates of human pathogens, such as *Pseudomonas aeruginosa, Klebsiella pneumonia, Staphylococcus aureus* and others, that also confer resistance to Hg, Cd, arsenate, Pb, and Zn (Kondo et al. 1974; Mergeay et al. 1985; Nakahara et al. 1977).

Resistance to heavy metals can be due to one or more of several mechanisms. These include:
1. mechanisms and enzymes that make the bacterial cell wall impermeable to the metal(s);
2. efflux mechanisms;
3. binding of the metal ions and
4. enzymes which catalyze the transformation of metals to non-toxic forms (Trevos et al. 1985).

The first three mechanisms have been shown to apply in the case of cadmium. Examples of cadmium resistant bacteria are shown in Table 1.

Resistance to cadmium may be associated with resistance to other (one or more) heavy metals, as a result of the fact that the gene conferring cadmium resistance also confers resistance to the other metal. For example, the *cadA* gene of *Staphylococcus aureus* codes for an ATPase-dependent efflux system that confers resistance to cadmium and zinc (Nucifora et al. 1989). A similar case is the resistance to divalent cations (Cd^{++}, Zn^{++} and Co^{++}) in *Alcaligenes eutrophus*, which is associated with a 9.1 kb plasmid DNA fragment that simultaneously confers resistance to all three metals (Nies et al. 1987). On the other hand, the gene conferring cadmium resistance could be different from the gene (s) conferring resistance to other metals, but located on the same plasmid. For example, we have studied (Minz & Ron, unpublished) a virulent strain of *Escherichia coli* O78 in which cadmium resistance is located on a large, conjugative plasmid. This plasmid also contains, at a different site, the *mer* operon that codes for resistance to mercury due to a specific reductase (Komura & Izaki 1971; Misra et al. 1985; Schottel et al. 1974).

Mechanisms that make cells impermeable to cadmium

Cd^{2+} enters the bacterial cells as a toxic alternative substrate for the cellular Mn^{2+} transport system in gram-positive bacteria (Tynecka et al. 1981), or for the Zn^{2+} transport system in gram-negative bacteria (Ladagga & Silver 1985). Both these systems are chromosomally-coded, nutritionally required cation transport systems.

Mechanisms exist that make the cells impermeable to cadmium. One such mechanism has been studied in the gram-positive bacterium *Bacillus subtilis* where it is associated with a chromosomal mutation. This mutation results in a change in the membrane manganese transport system so that

Table 1. Examples of cadmium-resistant bacteria.

Microorganism	Location of resistance	Suggested mechanism	Other metals
Staphylococcus aureus	R plasmid	Efflux (*cadA*)	Zn
	Chromosome	Efflux	Hg
	R plasmid	Binding (*cadB*)	
Bacillus subtilis	Chromosome	Permeability?	
Pseudomonas aeruginosa	R plasmid	Efflux?	Hg
Pseudomonas putida	Plasmid	Binding	
Pseudomonas cruciviae			
Klebsiella pneumonia	R plasmid	Efflux?	Hg
Klebsiella aerogenes	Chromosome	Binding (capsule)	
Alcaligenes autrophus	Plasmid	Binding?	Zn, Co
Citrobacter	Chromosome?	Precipitation as $CdHPO_4$	
Proteus mirabilis	Chromosome	Binding (envelop)	
Arthrobacter viscosus	Chromosome?	Binding (EPS)	
Rhodococcus fascians	Plasmid		As

Cd^{2+} is no longer taken up (Laddaga et al. 1985; Surowitz et al. 1984).

Cadmium resistance by efflux mechanisms
Several efflux mechanisms have been described (for review see Silver & Walderhaug 1992). The best studied systems are these of *Staphylococcus aureus*. In this organism there are several systems that confer resistance to cadmium. The *cadA* system confers resistance to Cd^{2+} and Zn^{2+}. It codes for an energy-dependent efflux mechanism and is well understood at the genetic, molecular and biochemical level. The *cadA* gene is located on plasmid pI258 from which a DNA fragment was isolated that contains two open reading frames. The larger one, corresponding to a predicted polypeptide of 727 amino acid residues, is necessary and sufficient for expression of cadmium resistance. Comparison of the CadA amino acid sequence with known protein sequences suggested that CadA is a member of the E_1-E_2 cation-translocating ATPases, similar to the K^+-uptake ATPases of Gram-positive and Gram-negative bacteria (Nucifora et al. 1989).

Another plasmid also harbors the *cadB* gene whose product confers resistance by a different mechanism, probably by a change in binding (Perry & Silver 1982). An additional cadmium resistance system in *S. aureus* is chromosomal. Like *cadA* it involves energy dependent Cd^{2+} efflux, but confers Cd^{2+} resistance alone, while *cadA* confers resistance for both Cd^{2+} and Zn^{2+} (Witte et al. 1986). The amino-terminal sequence of this gene product is homologous with the Hg^{2+}-binding protein of the *mer* operon and with MerP protein, the periplasmic Hg^{2+} binding polypeptide and does not hybridize to a *cadA*-specific probe.

Binding of cadmium
It has been shown that some of the cadmium-tolerating cells have developed mechanisms for binding of cadmium by surface factors or intracellular binding factors. This resistance mechanism will be discussed later as it is important for removal of cadmium from the environment. However, several examples of Cd^{2+}-binding systems are summarized here briefly.

1. Precipitation on the cell surface. The cell surface is the most important location for precipitation of heavy metals in general. Cadmium has been shown to bind to capsular material in *Arthrobacter viscosus* and in *Klebsiella aerogenes* (Scott & Palmer 1988, 1990). In several other cases binding at the surface was determined by electron microscopy.
2. Precipitation as $CdHPO_4$. A *Citrobacter* mutant isolated from metal-polluted soil, accumulates Cd^{2+} as insoluble cell-bound $CdHPO_4$ during growth in the presence of Cd^{2+} and glycerol 2-phosphate (Macaskie et al. 1987a, b).
3. Binding of Cd^{2+} by thiols. As mentioned earlier, this is probably the most important mechanism of cadmium resistance in eukaryotic cells. In bacteria, there is also the possibility of producing a polythiol Cd^{2+}-binding peptide analogous to metallothionein of animal cells (Hamer 1986; Sequin & Hamer 1987).

Cadmium transformation
Biological transformation of certain heavy metals is an important detoxification mechanism that can occur in many habitats and can be carried out by a wide variety of microorganisms, bacteria and fungi. As a result of biological action, metals undergo changes in valency and /or conversion into organometallic compounds (for review see Silver 1991).

Bioconversions involving changes in valency and resulting in production of volatile or less toxic compounds have been shown in several cases, reviewed in the preceding paper (Barkay 1992). These include reactions such as the reduction of mercury ions to metallic mercury or the oxidation of arsenite [As(III)] to arsenate [As(V)]. As yet, it is not known if Cd^{2+} can be reduced to Cd^0 biologically.

Another important detoxification mechanism is the transformation of metals into organometallic compounds by methylation. Metals that have been shown to undergo methylation are mercury (Sprangler et al. 1973), and lead (Wong et al. 1975). Although the products of methylation may be more toxic than the free metal, they are often volatile and can be released into the atmosphere. Such is the case with mercury, and its methylated derivatives, methyl mercury (which is water- and lipid-soluble, and more toxic than mercury) and dimethyl mercury (which is volatile). Organometallic compounds can also undergo microbiological and chemical degradation which may result in the metal being liberated in a volatile form (Schottel et al. 1974; Sprangler et al. 1973; Tonomura et al. 1968).

Several organocadmium compounds have been synthesized, and diorganocadmium compounds, analogous to dimethylmercury compounds, have been shown to be light-sensitive and thermolabile. There has been one report indicating that there could be a biological methylation of cadmium and tin (Huey et al. 1975), but it has not been confirmed. Therefore it is safe to say that as yet, there is no conclusive evidence for a microbial transformation of cadmium.

Use of microorganisms to detoxify water and soil contaminated with cadmium

The use of microbial biomass to extract heavy metals from effluent solutions is an area of extensive research and development activity. There are several approaches which vary in the nature and purity of the adsorbing component and in the way of applying it. All these approaches aim at designing biological reactors for the effective treatment of waste waters and soil. Such treatments involve one or more concentration steps, in the course of which the pollutant metal is isolated in a small volume of biomass.

Microbial biomass – attempts at identifying the active component

The simplest approach to biological detoxification involves the utilization of crude biomass in the treatment of heavy-metal-contaminated waters. The crude biomass can be fungal, bacterial or other biological substrates, including waste materials, which are inexpensive and readily available locally. In one such experiment the uptake of cadmium by sludge was studied. At concentrations below 30 mg/l cadmium in solution, large fractions of the cadmium were adsorbed and about 95% of the total cadmium uptake was achieved within a 5-min contact time (Gourdon et al. 1990). It was calculated that some natural inexpensive materials, such as fungal biomass, have a higher cadmium-adsorption capacity than that of activated charcoals, or ion-exchange resins, which are the alternative adsorbents (Salah Azab & Peterson 1989).

However, most researchers feel that in order to achieve efficient and reproducible removal of heavy metals, further microbiological studies are needed. These include identification of the active microorganisms and the cellular component responsible for the binding of the metal. Many bacteria have been shown to bind heavy metals and include strains of *Aeromonas, Flavobacterium, Pseudomonas, Spirillium, Zoogloea, Arthrobacter*, and *Alcaligenes*.

As already mentioned, in most cases, the accumulated cadmium could be located in the cell envelope. Strains of *Proteus mirabilis* were able to grow in the presence of Cd^{2+} concentrations up to 300 mg/l. Eighty percent of the metal was associated with the cell envelope and only 20% accumulated in the cytoplasm (Andreoni et al. 1991). *Alcaligenes eutrophus* CH34, in which cadmium resistance is associated with the presence of a plasmid, immobilized more cadmium than the Cd-sensitive strain that does not have the plasmid. The cadmium was mainly located in the cell envelopes. Several cellular modifications, associated with the resistance (thickening of the cell envelopes and the proportion of peptidoglycan in the envelopes) may contribute to increase the cadmium sequestration capacity of the envelopes (Hambuckers-Berhin & Remacle 1990).

In several cases, the adsorption of cadmium – as well as of other metals – is probably associated with the secretion of exopolysaccharide or capsular material. For example: Exopolysaccharide from *Arthrobacter viscosus* had a 2.3 times greater accumulation capacity for cadmium than the equivalent weight of intact cells and is 13.7 times more effective than the cells of *Arthrobacter globiformis*, an organism that does not produce exopolysaccharide (Scott & Palmer 1988). The soil micro-organisms of the *Arthrobacter* genera are good producers of capsular polysaccharides. The abilities of various strains of *Arthrobacter spp.* (*A. fluorescens, A. giacomelloi, A. globiformis* and *A. viscosus*) to adsorb cadmium from liquid streams was shown to be highest when grown on media containing 1% mannitol, when the highest production of capsular polysaccharides was obtained (Grappelli et al. 1989). Capsular *Klebsiella aerogenes* strains showed minimal intracellular uptake but high metal removal levels due to exocellular adsorption at cadmium concentrations of 5–100 ppm (Scott & Palmer 1990).

On the other hand, *Arthrobacter* and *Pseudomonas* species appear to have detoxification systems that precipitate cadmium internally irrespective of whether or not they excrete polymers (Scott & Palmer 1990). The polysaccharide excreting *Pseudomonas putida* was only 20% more efficient in cadmium removal as compared to *Pseudomonas cruciviae*, a non-capsular organism (Scott et al. 1986).

In a few cases it has been shown that the binding of the metal is mediated by a specific protein, or cellular activity. For example, a protein of 43,000 daltons from marine organisms has been found very efficient in adsorbing cadmium (Kurek et al. 1989). Intensive research is aimed at finding proteins of low molecular weight which resemble metallothionein.

An interesting protein is the one responsible for the cellular activity of a strain of the *Citrobacter* sp. mentioned earlier. Having been pre-grown in cadmium-free continuous culture, this strain accumulated cadmium extensively when resuspended in a buffer that contained Cd^{2+} and glycerol 2-phosphate. The accumulated compound was identified as cell-bound cadmium phosphate, probably $CdHPO_4$ (Macaskie 1987b). The metal uptake mechanism is mediated by the activity of a cell-bound phosphatase that precipitates liberates inorganic phosphate which precipitates with heavy metals at the cell surface (Macaskie et al. 1987a, b).

Application of microbial products for detoxification of water contaminated with cadmium

Several innovative engineering technologies have been applied for detoxification of cadmium-contaminated water using bacteria that bind cadmium. Macaskie & Dean (1989) and Macaskei et al. (1987) developed a system for detoxification of cadmium and other metals in liquid wastes using columns packed with immobilized cells of *Citrobacter* sp. The cells were grown as a biofilm on solid glass bead support, or immobilized by incorporation into polyacrylamide gels that were then shredded. The accumulated cadmium or lead was precipitated on the cell surface after the cell-bound acid phosphatase released inorganic phosphate. In a similar way uranium phosphate or lead phosphate precipitates could be formed. This system was capable of handling up to 10^7 liters a day.

Cells of *Zoogloea ramigera* 115 were immobilized into beads of calcium alginate and used in air-bubbled column reactors to remove cadmium, zinc, manganese, lead, copper and strontium from dilute and concentrated solutions. By placing three bubbled columns in sequence it was possible to achieve cadmium adsorption efficiencies of 99% or greater. During ten applications of approximately 100 μg/ml of cadmium to three reactors in sequence, immobilized cells of *Z. ramigera* adsorbed 99.9% of the metal. The efficiency of the first column decreased from 92.2% on the first day to 53.8% on the tenth day, but the overall efficiency remained high because of the other two reactors. Exposure of bubbled columns to mixed metal solutions yielded similar results (Kuhn & Pfister 1989).

Granulated non-living cellular mass from mixed microbial cultures or algal cultures have been used in a similar way. (For details see Silver 1991.)

Several attempts have been made to use purified proteins bound to columns. Thiol-rich peptides or proteins, like metallothionein, are potentially more efficient in bioaccumulation than whole cells. In one well documented case the microbial polythiol used was from cyanobacteria. It is a methallothionein-like inducible peptide that contains cysteine at the ratio of 11/80 residues (Olafson et al. 1988).

In a different strategy, cells of *Alcaligenes eutrophus* were immobilized on a Flat Sheet Reactor made of composite membranes of polysulfone with inorganic fillers, through which a nutrient solution was passed. In 72 hours of incubation about 90% of the cadmium was removed from a solution containing 320 ppm. It was calculated that the cadmium was bound to an exopolymer, with a binding ratio

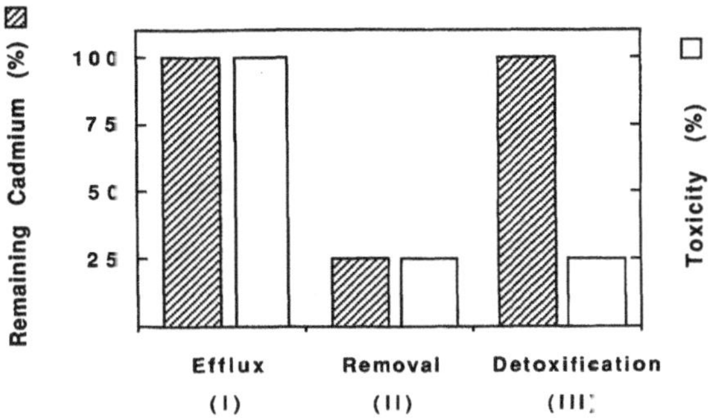

Fig. 1. Various classes of cadmium-resistant bacteria. Bacterial isolates were grown to the stationary phase in tris-buffered minimal salt medium containing 0.2% glucose, and 30 μM of cadmium chloride. The level of cadmium remaining in the medium was determined after the cells were removed by centrifugation. The supernatant was then autoclaved and divided to two parts. One part was used for determining the concentration of cadmium by atomic absorption. The other part was used to determine the toxicity of the cadmium in a bioassay using a sensitive strain.

of 500 μg cadmium per 1 mg of polymer (Diels 1990). *Alcaligenes eutrophus* was also used for the biotreatment of soil in a slurry reactor. The soil contained about 20 ppm of cadmium 50% of which was removed in a two-step reaction.

Experimental results

Overall goal of research

The research carried out in our laboratory was aimed at identifying bacteria capable of detoxifying water and soil contaminated with cadmium, either by removing it from solution or by transforming it into a non-toxic form. We screened for these among a collection of bacteria that were selected on the basis of cadmium resistance.

Table 2. Classes of cadmium-resistant mutants.

Class	Effect on Cd		Mechanism
	Concentration	Toxicity	
I	Unchanged	Unchanged	Efflux
II	Reduced	Reduced	Removal (binding)
III	Unchanged	Reduced	Detoxification

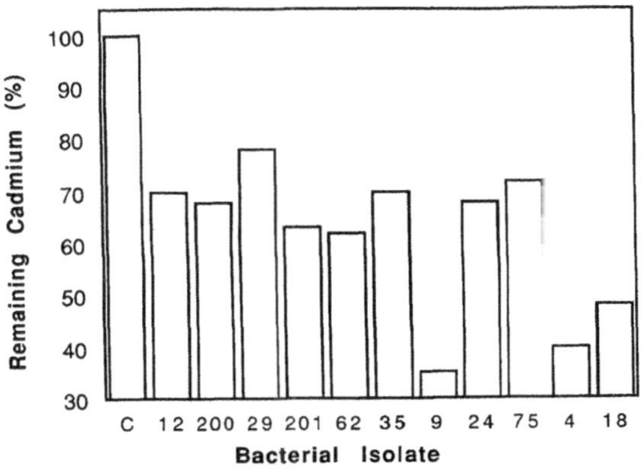

Fig. 2. Removal of cadmium from solutions by various bacterial isolates. Cultures of bacterial isolates (c = control) were grown in minimal medium as described in Fig. 1. After removal of the cells by centrifugation, the concentration of the cadmium remaining in the medium was determined by atomic absorption.

Enrichment of cadmium resistant bacteria

The rationale was to enrich for natural bacterial strains that are resistant to cadmium by ways other than efflux, as the latter would not reduce the concentration of cadmium in the environment. Resistant strains were enriched from polluted soil in medium containing 30 μM cadmium. The medium used was tris-buffered salt glucose minimal medium, pH = 7.5. Several hundred different strains were obtained. Each was screened for its ability to detoxify cadmium. As shown in Fig. 1, these bacteria could be divided into three classes, as summarized in Table 2.

The majority of the bacteria did not change the growth medium (class I) and are probably resistant due to efflux. Only about 10% of the strains detoxified the medium, and in about 75% of these the cadmium was removed in the bacterial pellet (binding, class II). These bacteria were chosen for further examinations, the results of which are summarized in the following paragraphs. The results shown in Fig. 2, represent a typical experiment in which several strains were examined for their ability to remove cadmium from solution. The bacteria were grown to stationary phase in medium containing 30 μM cadmium chloride. They were then centrifuged and the concentration of cadmium remaining in the supernatant was determined. The results

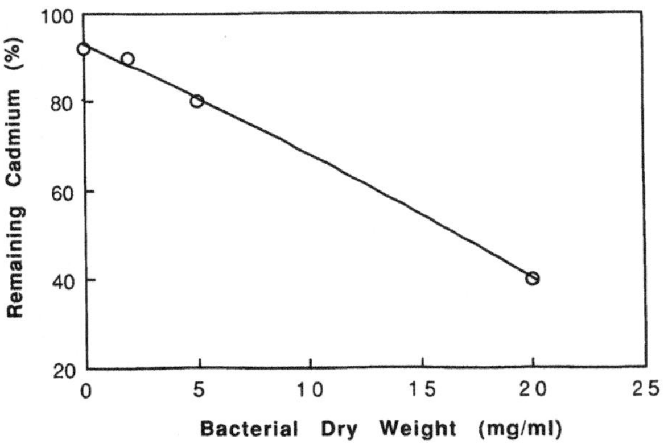

Fig. 3. Removal of cadmium from solutions as a function of bacterial dry weight. Cultures of bacterial isolate number 9 were grown to the stationary phase in tris-buffered minimal salt medium containing 0.2% glucose. The cells were then concentrated by centrifugation and resuspended at increasing cell concentration in fresh medium containing 30 μM of cadmium chloride. The cells were then centrifuged and the level of cadmium remaining in the medium was determined as described in Fig. 1.

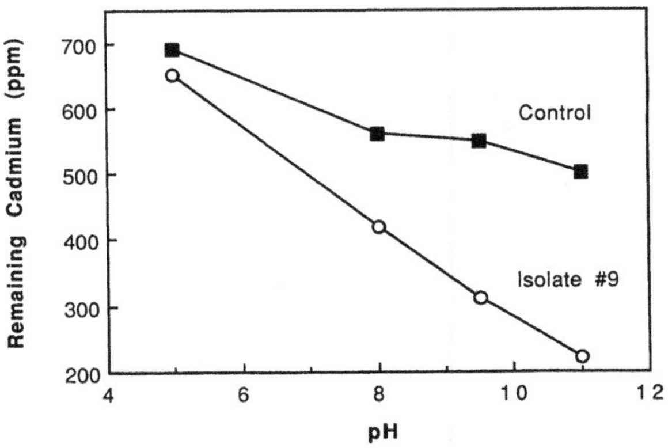

Fig. 4. Removal of cadmium from solutions as a function of pH. Experiment was performed as in Fig. 1, except that the concentration of cadmium was 700 ppm.

indicate that out of 11 strains that were pre-selected by their ability to bind cadmium, 8 reduced the cadmium concentration in the supernatant by 30–40%, and 3 strains reduced the cadmium concentration by more than 50%.

One strain, isolate number 9 was chosen for further studies on the basis of its high efficiency in removing cadmium from solutions. In this strain the removal of cadmium was dependent on the concentration of bacteria, as shown in Fig. 3, and on the pH of the solution (Fig. 4). The kinetics of cadmium adsorbance was fast – most of the cadmium was removed from the solution within two minutes, the time it took to separate the cells from the supernatant fluid by centrifugation.

The efficiency of cadmium removal, the pH profile, and the fast kinetics suggested that cadmium was bound to the surface of the cells, rather than internalized. Additional support for this assumption was the finding that cadmium could easily be desorbed from the cells by lowering the pH. Thus, at pH values lower than 5 the cadmium was released into the solution within a half hour of incubation (Fig. 5).

Isolate number 9 was also shown to be capable of removing cadmium from soil – when incubated with earth contaminated by 28 ppm of cadmium phosphate it adsorbed a large fraction of the cadmium (Fig. 6).

The results presented so far indicate that isolate #9 is potentially useful for removing cadmium from the environment – from contaminated soil, as well as from solutions. It has the added advantage that the bound cadmium can be released and separated from the bacteria at low pH.

Study is now in progress to classify isolate #9. So far it appears to be a strain of *Pseudomonas*. Its

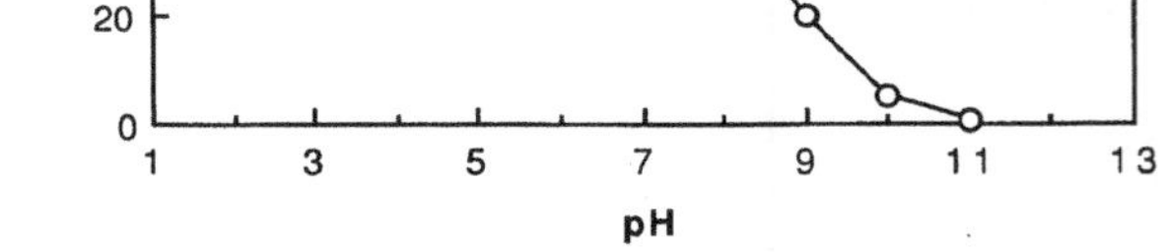

Fig. 5. Release of cadmium from bacteria as a function of pH. Bacteria of isolate number 1 were incubated with 30 μM of cadmium chloride at pH = 11, to adsorbe 94% of the cadmium. They were then centrifuged and resuspended in equal volumes of solutions of various pH values (from pH = 1 to pH = 11). Following 30 minutes of incubation the cells were removed by centrifugation and the concentration of cadmium in the supernatant was determined.

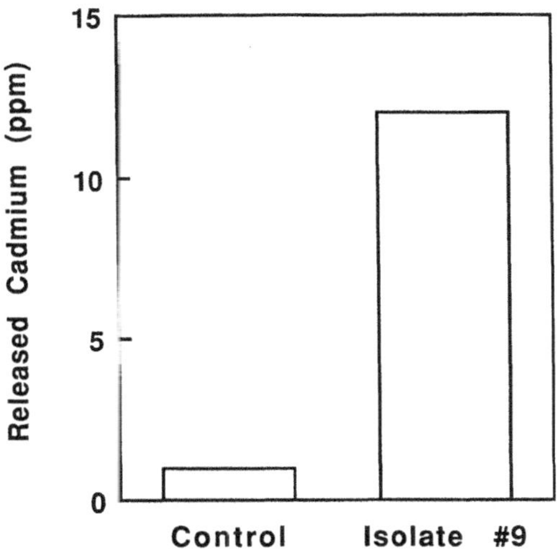

Fig. 6. Release of cadmium from soil. Cells of isolate number 9, grown for 24 hours in minimal salt medium (as in Fig. 1) were incubated with soil containing 30 ppm cadmium, for 12 hours in water. The mixture was then sonicated and the concentration of the cadmium released into the water suspension of bacteria was determined. The control was treated in the same way, except that incubation was with the same medium but without bacteria.

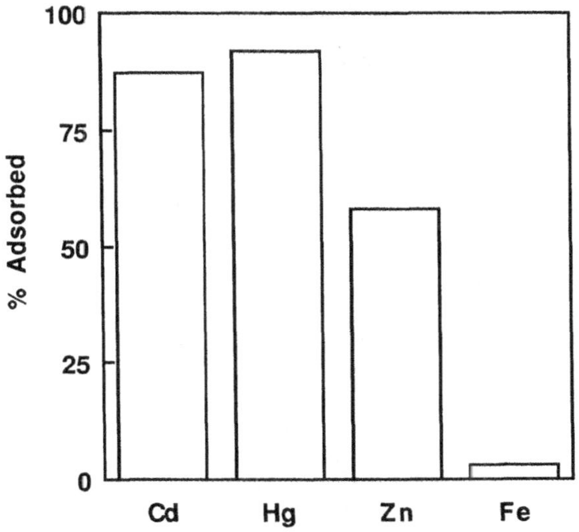

Fig. 7. Adsorbance of various metals by isolate number 9. Bacteria of isolate number 9 grown as described in Fig. 1 were incubated for 1 hour with 30 μM of each metal. The concentration of the metal in the solution was determined by atomic absorption following removal of the cells by centrifugation.

unusual characteristic is in the secretion of a massive amount of extracellular material that binds the cells together to form very large aggregates. This extracellular material appears to also bind cadmium. Studies are in progress to determine whether it is the only cellular component responsible for the binding of the cadmium.

Acknowledgements

We wish to thank Isaac Hendler and his team for their help in the determination of cadmium concentrations. We thank Israel Chemicals Ltd. for supporting this research and for encouraging this publication.

References

Andreoni V, Finoli C, Manfrin P, Pelosi M & Vecchio A (1991) Studies on the accumulation of cadmium by a strain of *Proteus mirabilis*. FEMS Microbiol. Ecol. 85: 183–192

Diels L (1990) Accumulation and precipitation of Cd and Zn ions by *A. eutrophus* strains. Biohydrometallurgy: 369–377

Dunnick JK & Fowler BA (1988) Cadmium. In: Seiler HG & Sigel H (Eds) Handbook on Toxicity of Inorganic Compounds (pp 155–169). Marcel Dekker, Inc., NY, Basel

Grappelli A, Hard JS, Pietrosanti W, Tomati U, Campanella L, Cardarelli E & Cordatore M (1989) Cadmium decontamination of liquid streams by *Arthrobacter* species. In: Lijklema L et al. (Eds) Water Pollution Research and Control, Brighton, Part 5, Vol 21 (pp 1759–1762). Water Sci. Technol.

Gourdon R, Rus E, Bhende S & Sofer SS (1990) Mechanism of cadmium uptake by activated sludge. Appl. Microbiol. Biotechnol. 34: 274–278

Hambuckers-Berhin F & Remacle J (1990) Cadmium sequestration in cells of two strains of *Alcaligenes eutrophus*. FEMS Microbiol. Ecol. 73: 309–316

Hamer DH (1986) Metallothioneins. Ann. Rev. Biochem. 55: 913–951

Huey CW, Brinckman FE, Iverson WP & Grim SO (1975) Bacterial volatilization of cadmium. International Conference on heavy metals in the environment. Toronto, Ontario, Canada, C214–C216

Komura I & Izaki K (1971) Mechanism of mercuric chloride resistance in microorganisms. I. Vaporization of a mercury compound from mercuric chloride by multiple drug resistant strains of *Escherichia coli*. J. Biochem. 70: 885–893

Kondo I, Ishikawa T & Nakahara H (1974) Mercury and cadmium resistances mediated by the penicillinase plasmid in Staphylococcus aureus. J. Bacteriol. 117: 1–7

Kuhn SP & Pfister RM (1989) Adsorption of mixed metals and cadmium by calcium-alginate immobilized *Zoogloea ramigera*. Appl. Microbiol. Biotechnol. 31: 613–618

Kurek E, Francis AJ & Bollag J-M (1989) Immobilization of

cadmium by microbial extracellular products. Arch. Environ. Contam. Toxicol. 21: 106–111

Ladagga R & Silver S (1985) Cadmium uptake in *Escherichia coli* K-12. J. Bacterol. 162: 1100–1105

Laddaga AR, Bessen R & Silver S (1985) Cadmium resistant mutant of *Bacillus subtilis* 168 with reduced cadmium transport. J. Bacteriol. 162: 1106–1101

Levinson W, Oppermann H & Jackson J (1980) Transition series metals and sulfhydryl reagents induce the synthesis of four proteins in eukaryotic cells. Biochim. Biophys. Acta 606: 170–180

Macaskie LE, Dean ACR, Cheetham AK, Jakeman RJB & Skarnulis AJ (1987) Cadmium accumulation by a *Citrobacter* sp.: the chemical nature of the accumulated metal precipitate and its location on the bacterial cells. J. Gen. Microbiol. 133: 539–544

Macaskie LE, Wates JM & Dean ACR (1987) Cadmium accumulation by a *Citrobacter* sp. immobilized on gel and solid supports: applicability to the treatment of liquid wastes containing heavy metal cations. Biotechnol. Bioeng. 38: 66–73

Marques AM, Congregado F & Simon-Pujol DM (1979) Antibiotic and heavy metal resistance of *Pseudomonas aeruginosa* isolated from soils. J. Appl. Bacteriol. 47: 347–350

Mergeay M, Nies D, Schlegel HG, Gerits J, Charles P & van Gijsegem F (1985) *Alcaligenes eutrophus* CH34 is a facultative chemolithotroph with plasmid-bound resistance to heavy metals. J. Bacteriol. 162: 328–334

Misra TK, Brown NL, Haberstroh L, Schmidt A, Goddette D & Silver S (1985) Mercuric reductase structural genes from plasmid R100 and transposon Tn501: functional domains of the enzyme. Gene 34: 253–262

Nakahara H, Ishikawa T, Sarai Y, Kondo I, Kozukue H & Silver S (1977) Linkage of mercury, cadmium, and arsenate and drug resistance in clinical isolates of *Pseudomonas aeruginosa*. Appl. Environ. Microbiol. 33: 975–976

Nies D, Mergeay M, Friedrich B & Schlegel HG (1987) Cloning of plasmid genes encoding resistances to cadmium, zinc, cobalt in *Alcaligenes eutrophus* CH34. J. Bacteriol. 169: 4865–4868

Nucifora G, Chu Lien, Misra TK & Silver S (1989) Cadmium resistance from *Staphylococcus aureus* plasmid pI258 cadA gene results from a cadmium-efflux ATPase. Proc. Natl. Acad. Sci. U.S.A. 86: 3544–3548

Perry RD & Silver S (1982) Cadmium and manganese transport in *Staphylococcus aureus* membrane vesicles. J. Bacteriol. 150: 973–976

Pickett AW & Dean ACR (1976) Cadmium and zinc sensitivity and tolerance in *Klebsiella (Aerobacter) aerogenes*. Microbios. 15: 89–91

Olafson RW, McCubbin WD & Kay CM (1988) Primary- and secondary-structural analysis of a unique prokaryotic metallothionein from a *Synechococcus* sp. cyanobacterium. Biochem. J. 251: 691–699

Salah Azab M & Peterson PJ (1989) The removal of cadmium from water by the use of biological sorbents. In: Lijklema et al. (Eds) Water Pollution Research and Control, Brighton, Part 5, Vol 21 (pp 1705–1706). Water Sci. Technol.

Schottel J, Mandal A, Clark D & Silver S (1974) Volatization of mercury and organomercurials determined by inducible R-factor systems in enteric bacteria. Nature 251: 335–337

Scott JA & Palmer SJ (1988) Cadmium bio-sorption by bacterial exopolysaccharide. Biotechnol. Lett. 10: 21–24

Scott JA & Palmer SJ (1990) Sites of cadmium uptake in bacteria used for biosorption. Appl. Microbiol. Biotechnol. 33: 221–225

Scott JA, Sage GK, Palmer SJ & Powell DS (1986) Cadmium adsorption by bacterial capsular polysaccharide coatings. Biotechnol. Lett. 8: 711–714

Sequin C & Hamer DH (1987) Regulation in vitro of metallothionein gene binding factors. Science 235: 1383–1387

Silver S (1991) Bacterial heavy metal resistance systems and possibility of bioremediation. In: Kamely D et al. (Eds) Biotechnology: Bridging Research and Applications (pp 264–287). Kluwer Academic Publishers, Boston

Silver S & Walderhaug M (1992) Gene regulation of plasmid- and chromosome-determined inorganic ion transport in bacteria. Microbiol. Rev. 56: 195–228

Sprangler WJ, Spigarelli JL, Rose JM, Flippin RS & Miller HH (1973) Degradation of methylmercury by bacteria isolated from environmental samples. Appl. Microbiol. 25: 488–493

Steffens JC, Hunt DF & Williams BG (1986) Accumulation of non-protein metal-binding polypeptides (gamma-glutamyl-cysteinyl)$_n$-glycine in selected cadmium-resistant tomato cells. J. Biol. Chem. 261: 13879–13882

Surowitz KG, Titus JA & Pfister RM (1984) Effects of cadmium accumulation on growth and respiration of a cadmium-sensitive strain of *Bacillus subtilis* and a selected cadmium resistant mutant. Arch. Microbiol. 140: 107–112

Tonomura K, Maeda K & Futai F (1968) Studies on the action of mercury resistant microorganisms on mercurials. II. The vaporization of mercurials stimulated by mercury-resistant bacterium. J. Ferment. Technol. 46: 685–692.

Trevos JT, Oddie KM & Belliveau BH (1985) Metal resistance in bacteria. FEMS Microbiol. Rev. 32: 39–54

Tynecka Z, Gos Z & Zajac J (1981) Reduced cadmium transport determined by a resistance plasmid in *Staphylococcus aureus*. J. Bacteriol. 147: 305–312

Van Bogelen RA, Kelley PM & Neidhardt FC (1987) Differential induction of heat shock, SOS, and oxidation stress regulons and accumulation of nucleotides in *Escherichia coli*. J. Bacteriol. 169: 26–32

Witte W, Green L, Misra TK & Silver S (1986) Resistance to mercury and to cadmium in chromosomally resistant *Staphylococcus aureus*. Antimicrob. Agents Chemother. 29: 663–669

Wong PTS, Chao YK, Luxon PL & Silverberg B (1975) Methylation of lead and selenium in the environment. International Conference on heavy metals in the environment. Toronto, Ontario, Canada, C220–C221

Biodegradation **3**: 171–188, 1992.

The cellulose paradox: pollutant *par excellence* and/or a reclaimable natural resource?

Edward A. Bayer[1] & Raphael Lamed[2]
[1] *Department of Biophysics, The Weizmann Institute of Science, Rehovot, Israel;* [2] *Department of Molecular Microbiology and Biotechnology, George S. Wise Faculty of Life Sciences, Tel Aviv University, Ramat Aviv, Israel*

Key words: cellulose (solubilization of), pollution, cellulosome, *Clostridium thermocellum*, multienzyme systems

Abstract

The various aspects of cellulose as a pollutant are considered in view of its lack of toxicity on the one hand and its recalcitrant durable nature on the other. The microbial degradation of cellulosics is discussed, and the contrast between its success in handling natural cellulosic wastes versus its failure to cope with man-made refuse is described. Research carried out in the past decade has demonstrated that cellulolytic organisms are provided with cell surface multifunctional multienzyme conglomerates, called cellulosomes, which are capable of solubilizing solid cellulosic substrates. The intriguing properties of such complexes include their cohesive nature, their many enzymatic components, and a characteristic glycosylated cellulose-binding, 'scaffolding' component. The latter appears to serve as a substrate-targeting carrier, which delivers the other (hydrolytic) components to the cellulose. Progress in establishing efficient model systems for *in vitro* solubilization of purified cellulose or natural cellulosic substrates has been achieved using purified cellulosome preparations, fortified with β-glucosidase and pectinase. The latter enzymes were required in order to alleviate the phenomenon of product inhibition which reduces the efficiency of the free cellulosome. Such combined enzyme systems are proposed as examples of future tailor-made cellulolytic systems for the degradation of natural cellulosics.

Introduction

Is cellulose, in fact, really a pollutant?

After all, it is common knowledge that paper is easy to degrade, or is it? There appears to be a common myth or delusion as to the biodegradability of cellulose. In any case, cellulose is not particularly toxic to the environment, but is it a serious problem? Figure 1 gives a schematic description of the fate of cellulose in the environment and the consequences of human activities.

Cellulose is the most abundant organic raw material in the world, comprising the major portion of plant matter. In its 'native' form, e.g., as the major (40–50%) component of wood in trees, cellulose is an ecological treasure. In other plant tissues, the cellulose content ranges from 20% in some grasses to over 90% in cotton fiber. But trees and other plants die, bringing the cellulose and other plant components (notably lignin and hemicellulose) to the soil, where, in time, natural forces (particularly microbes) are capable of accomplishing their degradation.

Fortunately, the process is reversible in that trees and other plants grow, thus replenishing the supply of cellulose and other organic polymers at

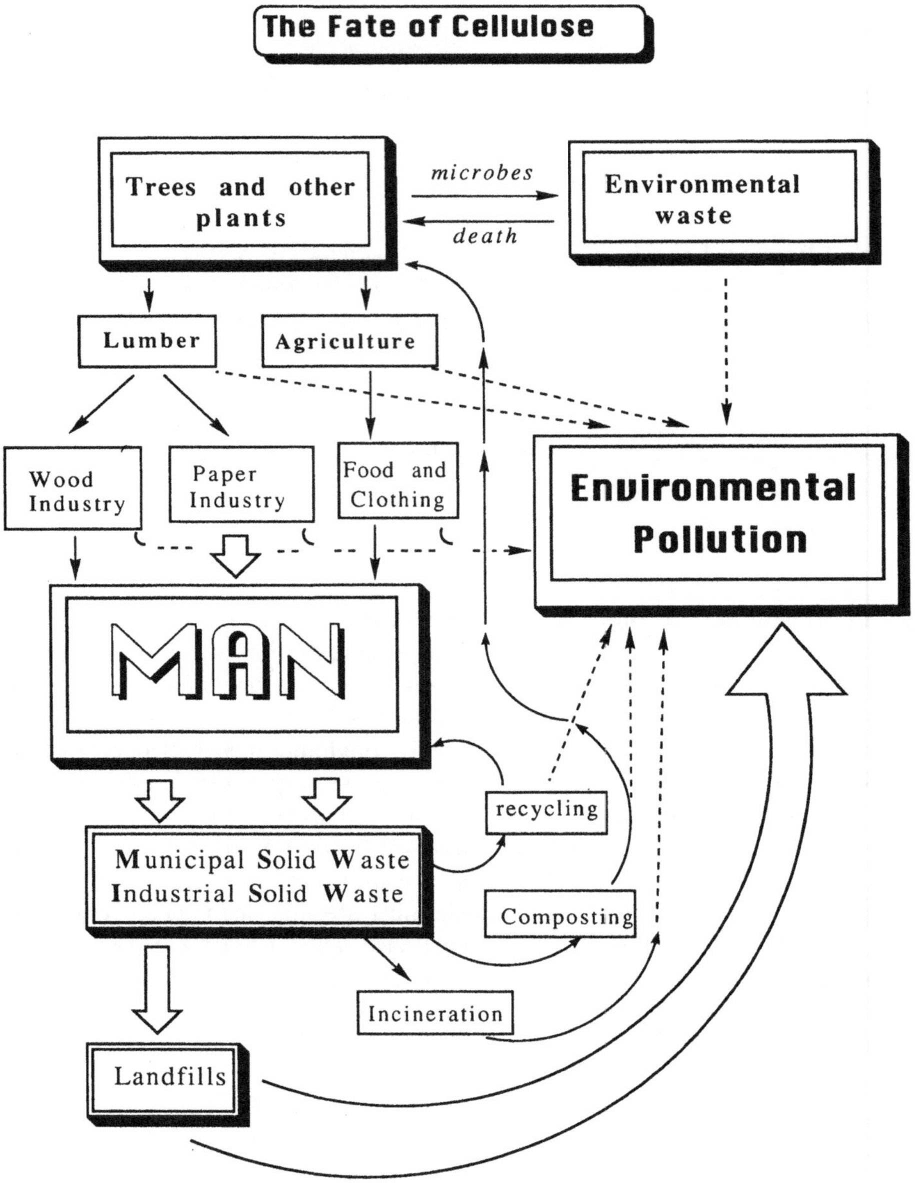

Fig. 1. Schematic view of the fate of cellulose in the environment resulting from both natural (microbial) degradation and human activities.

the expense of atmospheric carbon dioxide. In nature, the evolution of this cycle has been perfected over a period of hundreds of millions of years, and a variety of microbes exist in virtually every niche and clime, which are specifically suited for eliminating cellulosic waste. The facts speak for themselves: in the natural state there is no uncontrollable accumulation of cellulose-containing refuse. In other words, nature takes care of natural waste very nicely.

Human-generated trash is quite another story. Industrial wastes, agricultural wastes, commercial and residential wastes eventually become environmental waste. The industrial utilization of wood

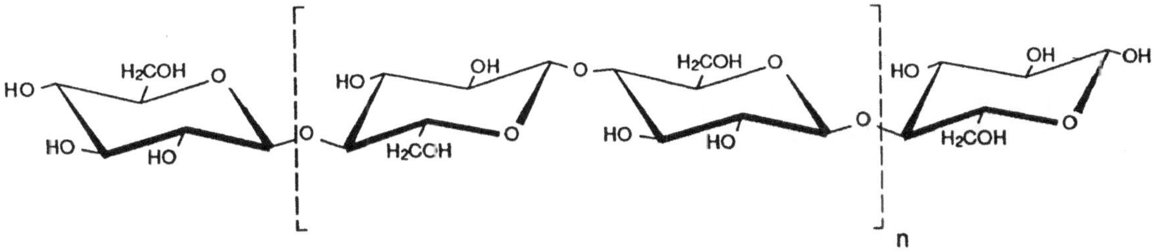

Fig. 2. Cellulose is composed of repeating units of cellobiose, 4-0-(β-D-glucopyranosyl)-D-glucopyranose where n is the degree of polymerization and is commonly about 10,000 units.

and other agricultural products is a serious source for depletion of viable plant matter which ultimately becomes cellulosic waste. Timber is prepared by logging and lumbering, and is used in construction, furniture and so on, before being returned in some way to the soil. In the sphere of agriculture, after crops are harvested, the residue is generally rich in cellulosics. Cotton waste is two-fold; the debris left in the fields are eventually joined by the derelict fabric. But, by far, the major source of cellulose waste in our environment is derived from paper.

On the one hand, the continued industrial processing of paper from wood is one of the important contributions to modern society. The dissemination of information in newspapers and books is dependent on the paper industry. The advent of computer technology has not alleviated this dependence; quite the contrary. Thus, wood is converted to paper, which is eventually converted to a lot of wastepaper.

The reason that paper has become such an important mainstay in our society is actually a function of the physical characteristics of cellulose. Cellulose is a remarkably stable polymer owing to its tertiary structure; the polymer consists of a simple undeviating syndiotactic linear arrangement of glucose residues, coupled exclusively by β [1, 4] glycosidic linkages. The simple disaccharide formed, called cellobiose [4-0-(β-D-glucopyranosyl)-D-glucopyranose], is the repeating unit of the cellulose polymer (Fig. 2). The resultant chains are organized in parallel fashion into fibrils, which are stabilized by interchain hydrogen bonding. Individually, the hydrogen bonding is quite weak, but collectively they result in a very strong associative force. The fibrils are considered to attain a 'paracrystalline' state, which accounts for about 75% of the cellulose fiber, the residual chains comprise 'amorphous' areas. The structural rigidity and stability of paper is thus a direct result of the molecular properties of cellulose.

For the same reasons, microbial attack of the cellulose is very slow. Although many microbes synthesize relevant enzymes, the cellulases, the types and number of individual microbes which can degrade efficiently, or solubilize, cellulose are indeed scarce. Under the best conditions in nature, cellulose degradation occurs at a sluggish pace. For certain uses, this is of course desirable; books are stable on the shelf for generations and sometimes centuries. In contrast, for other (even related uses) the stability of paper is perhaps exaggerated; the information recorded in newspapers is of only transient interest, and the newspapers are discarded on a daily basis. It should therefore come as no surprise that newspapers are the largest single item in landfills! And collectively, paper occupies almost 50% of the space in landfills. Thus, in terms of volume and mass, the accumulated cellulose waste can certainly be considered a problem.

Whether or not cellulose is indeed toxic, it has great potential as a source of biomass for conversion to soluble sugars that can serve as substrates for fermentation to fuels and chemicals. Low value cellulosic materials, such as wheat straw, sawdust, cornstover, sugar cane bagasse, and other agricultural residues are produced at a rate of about one billion tons (dry weight) annually in the United states (Tsao et al. 1987). It has been estimated that about 20% of this biomass can be economically collected and fed into the production of chemicals. Many chemicals can be produced from cellulosic residue, including those useful in the production of energy [such as ethanol (Wright 1988; Fan et al. 1982), ethylene, propylene and related compounds (Hanselmann 1982)]. Other uses include production of high energy animal feeds (Gould 1985), polymer feedstocks and organic solvents (Hanselmann 1982).

The microbial degradation of cellulosics for combined waste management, by-product and energy resource has been a stated goal for over half a century. Cell-free enzymatic hydrolysis is also considered to be a promising approach for obtaining glucose from cellulosics, yet cellulase systems are neither sufficiently inexpensive nor efficient to achieve this objective. To date, the failure to design commercially viable technologies to achieve this goal can largely be attributed to the reliance on engineering finesse, often at the expense of understanding the basics of the cellulolytic phenomenon *per se.* This policy has resulted in decades of applied microbial and enzyme research, but, in retrospect, has led to poor potential with regard to the degradation of cellulosic materials.

In this chapter, we will concentrate on some of the intricacies of the microbial degradation of cellulose. In doing so, we will present some of the new principles discovered in our laboratory which deal with the bacterial degradation of cellulose. The primary role of cell adhesion to the substrate will be discussed. In this context, cellulolytic bacteria appear to produce a cell-associated, multifunctional, multienzyme complex, coined the cellulosome, which mediates the attachment of the bacteria to the cellulose surface, thus targeting the relevant cellulases to their sites of action. Knowledge gained from such studies has contributed to substantial progress in the efficient hydrolysis of pure and complex cellulosic substrates, using the purified cellulosome combined with other enzyme preparations.

The problem with cellulose pollution

As mentioned above, the real problem with cellulose waste is not necessarily one of direct toxicity or of wanton buildup in nature of immense quantities of cellulose which constitutes an environmental threat. Nevertheless, a variety of cellulosic wastes are often associated with the increasing problem of disposal and inherent negative environmental impact. These include residues from crop production, animal husbandry, forestry, paper pulp production, sawn wood and other forest product industries as well as food processing – wastes which range in moisture content from dry combustible straws to slurries and manures of very high water composition. These substances also offer the best potential short-term option as raw materials for energy generation as well as for composting and pulp production.

Significant amounts of non-hazardous industrial solid waste as well as the organic fraction of municipal solid waste are already disposed through energy- and resource-recovery routes, which include electricity generation, gas extraction from landfills, composting and combustion with district heating. Separation and utilization technologies, such as anaerobic digestion, pyrolysis and gasification offer opportunities for a massive increase in the utilization of such waste. Although some crops are specifically intended for the production of feedstock and energy, no emphasis will be placed in this chapter on trees and other crops grown for this purpose.

Ironically, most of the real pollution generated by the paper industry and the pulping process is not cellulosic in essence, but either relates to the chlorination of lignin aromatics or emanates directly from the bleached pulp which is then converted to paper. The chlorinated derivatives and sulfur emissions resulting from the processing of paper are not only toxic themselves, but they represent an added factor in secondarily inhibiting the biodegradation of the cellulose, thus exacerbating the overall problem. Agricultural residues, such as fiber from sweet sorghum, have also been found suitable for paper pulp production, but such straw-based factories have frequently been closed, due, emphatically, to the problems of pollution associated with the large amounts of water and chemicals used in the process (as well as the high costs involved), thus rendering the product noncompetitive by conventional processes.

Processed paper waste not only constitutes the vast majority of cellulosic waste, but comprises a very large part of environmental trash as a whole. These include newspapers and magazines, telephone books, and packaging. By all accounts, present estimates of the paper waste (in terms of percentage bulk content of the total solid waste) currently hover around 40 to 50%. This value has more than doubled in the past 40 years or so and is expected to continue to rise in the future. A pointed example of this trend concerns the analysis of refuse from the city of Birmingham, England, where the content of paper and cardboard in trash rose from slightly less than 8% in 1934, to over 20% in 1960 and over 50% in 1970 (Bridgwater & Mumford 1979).

One of the great popular myths is that paper decomposes rapidly, but this perception is greatly erred; *legible* newspapers, buried up to 40 years in landfills, have been unearthed in archeologically oriented 'garbage digs' (Rathje 1991). Such studies have been designed to provide us with new insights into our society. They are extremely enlightening, and the accumulated data are currently serving both as a warning to the public concerning the nature of our recent and current trash and as guidelines towards future planning for its effective disposal. In fact, newspapers are used to 'date' other items, buried proximately. A combination of reasons may account for this astonishing stability, ranging from the anaerobic conditions of landfills, their deficiency in moisture content, the lack of necessary nutrients and proper agitation, and the presence of toxic wastes (e.g., solvents, preservatives, and the like).

Most municipal solid waste is disposed in landfills (McCarthy 1990). More than 70% of the trash in the United States is being buried in 5,500 active landfills. However, many are reaching capacity, and few new ones are being approved, mainly due to the increasing awareness of leachate-derived pollution of water, air pollution or other adverse effects on the environment. The knowledge gained through research into the mistakes of the past has contributed to more intelligent site selection, design, operation and reclamation of modern landfill sites (Crawford & Smith 1985). Modern alternatives to landfills include recycling and composting. But these too have their problems: recycling is hampered by market instability for certain materials and technical barriers for the recycling of mixed materials as well as pollution problems of its own. Composting can also cause environmental pollution, often compounded by the production of pathogenic organisms (Biddlestone & Gray 1987). Nevertheless, future use of these alternatives to landfill technology should continue to be explored.

In agriculture, perhaps one of the best examples of undue pollution is caused by the commonly employed practice of incineration for eliminating straw, leaves, and other low-moisture high-level cellulosic materials in the fields. Burning reduces waste volume by up to 90%, and the technology is currently available to establish incinerators which would convert the heat from burning garbage to steam or electricity. However, both air pollution as well as toxic ash would remain a problem. Notably, the pollution produced by smoke has instigated a recent law in the United Kingdom which bans this method, creating in turn great pressure for the development of alternative methodologies to digest surplus agricultural waste. Many other such wastes are high in cellulose content and often burned in the fields, some examples being banana husks, coffee peels, cotton straw, lemon grass and sugar cane begasse.

The potential for degradation of cellulose wastes

Three different approaches for the degradation of cellulosics are generally considered: the microbial approach, the use of purified enzymes, and combined chemical and/or physical approaches. In its natural lignocellulose state, cellulose is fairly well protected from microbial degradation, mainly due to the lignin and hemicellulose polymer components. These must be removed from the surface of the highly crystalline polymer. The degradability of lignocellulose under anaerobic conditions is indeed rather low, due to the near-absolute requirement of high levels of oxygen for microbial attack on lignin. The susceptibility of lignocellulose to enzymatic hydrolysis depends on various factors, including particle size or surface area (Grethlein 1985), degree of crystallinity (Walseth 1952), and extent of lignification (Schwald et al. 1988). Different chemical and physical pretreatments, including steam explosion, ball-milling, pyrolysis or swelling in alkali, can markedly increase, in some cases, the extent of degradation (Marsden & Gray 1986). However, such procedures may considerably add to the overall operational costs. Moreover, degradation products of both chemical pretreatment processes and lignin-hemicellulose biodegradation are often toxic to strict cellulolytic microorganisms.

Microbial approach

Many microorganisms are known which can degrade cellulose in nature, including both fungi and bacteria, some of which exist under extreme environmental conditions, e.g., temperature, salts and pH. Microbial growth and decomposition of matter are generally divided into environmental categories (Haigler & Weimer 1991), usually based on temperature (thermophilic, mesophilic, cryophilic) and oxygen levels (aerobic and anaerobic). Other critical environmental factors are humidity, pH, and particularly in man-made environs, the presence of toxicants, such as heavy metal ions, solvents and other toxic chemicals, all of which may severely interfere with normal microbial growth. In any case, it appears that there are cellulolytic microorganisms which can exist in nearly every habitat (Table 1).

It should also be noted that, in nature, cellulolytic microbes do not occupy a given niche in pure culture. Rather, they exist in mixed culture with, perhaps, other cellulolytic organisms as well as those which degrade associated polymers (i.e., lignolytic and xylanolytic microbes). The polymer-degrading members are joined by others which utilize proficiently the polymer breakdown products (e.g., oligosaccharides and simple sugars), end-products (e.g., organic acids and molecular hydrogen), and toxic side-products (e.g., by-products of lignin and xylan hydrolysis).

Despite the anticipated complexity of natural systems, only one or a few strains may dominate the niche for a given time period, whereas other 'satellite' strains may provide assistance in facilitating the activity of the major strain(s). As the degradative process proceeds, the conditions of the microenvironment change, and the prosperity of a given strain would rise and fall according to demand.

The types of organisms necessary to achieve maximum results in a given niche would thus be dictated by the conditions of the given environment

Table 1. Some representative cellulolytic microorganisms.

BACTERIA
Aerobic Mesophiles
Cellulomonas fimi
Cellulomonas uda
Cellvibrio fulvus
Cellvibrio gilvus
Streptomyces falvogriseus
Aerobic Thermophiles
Thermomonospora curvata
Thermomonospora fusca
Anaerobic Mesophiles
Acetivibrio cellulolyticus
Bacteroides cellulosolvens
Clostridium cellobioparum
Clostridium cellulovorans
Fibrobacter succinogenes
Ruminococcus albus
Ruminococcus flavefaciens
Anaerobic Thermophiles
Caldocellum saccharolyticum
Clostridium stercorarium
Clostridium thermocellum
FUNGI
Aerobic Mesophiles
Aspergillus niger
Fusarium solani
Neurospora crassa
Penicillium pinophilum
Sporotrichum pulverulentum
Talaromyces emersonii
Trichoderma reesei
Aerobic Thermophiles
Chaetomium thermophile
Sporotrichum thermophile
Thermoascus aurantiacus
Anaerobic Mesophile
Neocallimastix frontalis

and the nature and complexity of the substrate. In this context, wood is much more difficult a substrate than paper, and, as mentioned above, pure cellulose itself is not so easy a substrate for microbial degradation. More than 100 species of bacteria and fungi have been considered for cellulose studies in the last century (Ljungdahl & Eriksson 1985), but to date none has been successfully applied (either in pure or in combined culture) in a commercially viable process for converting even pure cellulose to a desired product. The reasons for this failure range in character from low hydrolytic yields of the microbial cellulases, to the sensitivity of the microbial cells to the accumulation of end-products (such as ethanol). One recent study has registered some success in growing *C. thermocellum* in continuous culture on either microcrystalline cellulose or pretreated wood (Lynd et al. 1986). The pretreatment step was found to be essential, and substrate utilization was in the range of about 80%. The authors concluded that further efforts in identifying and overcoming the limiting factors would be necessary for greater conversion at higher rates. In any case, there is still a critical need to better understand the scientific aspects of cellulolytic microorganisms and their cellulases *before* allowing technology to proceed.

In view of the above considerations, the design of a successful applied system for the microbial elimination of cellulosics must take into account the composition of the substrate, the characteristics of the desired microbial strain (or strains) and component enzyme systems, the expected product pattern, and, of course, the costs involved. In the final analysis, it would be advantageous to couple the use of microbes to combat cellulose pollution with an economically feasible process. In this regard, certain products (such as soluble sugars, ethanol, acetone, and butanol) are preferred whereas others (acetic acid and lactic acid) are considered undesirable due to the difficulty in their separation and processing.

Application of cell-free cellulases

Individual microbial cellulases have been historically classified in several defined categories. The endoglucanases (endo-1, 4-β-D-glucan 4-glucanohydrolases, EC 3.2.1.4) sever the linear cellulose polymer presumably by random attack at sites within the chain. The exoglucanases (exo-1,4-β-D-glucan cellobiohydrolases, EC 3.2.1.91) remove cellobiose units from the nonreducing end of the chain. The β-glucosidases (β-D-glucoside glucohydrolases, EC 3.2.1.21) hydrolyze the free cellobiose molecules to two molecules of free glucose. Some cellulolytic systems, although rare, purportedly include glucohydrolases (1,4-β-D-glucan 4-glucohydrolases, EC 3.2.1.74), which cleave the terminal glucose unit from the nonreducing end of the cellulose chain. The neatness of this classification scheme is not generally matched by the actual characteristics of the purified, individual cellulases (especially those of bacterial origin), and, in due time, the establishment of a new scheme may prove beneficial.

Individual cellulases are incapable of hydrolyzing the paracrystalline cellulose polymer in an efficient manner. That is, a single cellulase enzyme cannot cause the total solubilization of the native cellulose polymer. One of the important doctrines, which has been almost sacred to the study of cellulolysis, is that the different enzymes are considered to act synergistically with each other. To explain this phenomenon, various hypotheses have been proposed in the past, including (1) the requirement for a noncatalytic hydrogen bond dissociating element (defibrillating enzyme), (2) the simple combination of endo- and exo-acting activities, (3) the requirement of two different cellobiohydrolases to account for the two possible stereochemical forms of the terminal chain glucose residue, and (4) the requirement for simultaneous multiple attack of the chains, which can be accomplished by higher order arrangement of the different enzymes. One thing is clear, there is a consensus that the end-product cellobiose is inhibitory to efficient cellulolysis, and the concerted action of β-glucosidase is essential for maximal synergistic effect. However, the mechanism(s) responsible for synergism is still under dispute, and the reader is referred to other sources for more detailed discourse of this issue.

Before the advent of molecular biology and the

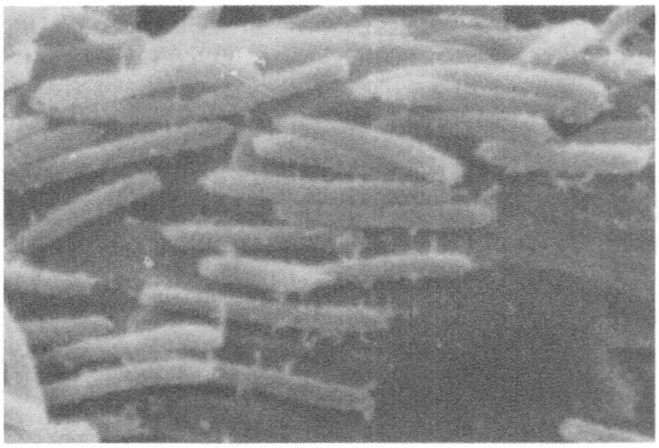

Fig. 3. Scanning electron micrograph of cationized ferritin-stained cells of *Clostridium thermocellum* growing on cellulose. Note the protracted protuberances which connect the cells to cellulose. Magnification: × 15,000.

recent strides in understanding the unique aspects of bacterial cellulase systems, fungal enzymes were the systems of choice. Unlike the bacterial enzymes, those from fungal sources are generally extracellular in nature, and their isolation and characterization have been relatively simple. Pure bacterial cellulases were sought for many years, but only in the last decade was it demonstrated that they are commonly cell-associated in the form of very large multienzyme complexes, called cellulosomes (Lamed et al. 1983a,b; Lamed & Bayer 1988a).

Despite the progress in this area, again, like in the studies using intact microbial systems, the use of cell-free cellulases for ambitious industrial application has been a frustrating endeavor. Why, then, has there been such marked success with amylases on the degradation of starches? The major reasons for this are manifold. First of all, the β-glucoside bond in cellulose is structurally very rigid and the cellulases are subject to strong product inhibition; in contrast, amylase systems generally lack of feedback inhibition, and glucose in the α configuration (which characterizes the starch polymer) is readily available to the enzymes. Incidentally, these properties reflect the natural roles of the respective materials in nature: i.e., cellulose relies on its natural stability for its role as a structural polymer, whereas starch is a storage material, which is subject to enzymatic breakdown when necessary. Thus, these vast differences in structural stability of these polymers and their consequences regarding their function in nature and propensity towards microbial and enzymatic degradation are all due to the simple difference in the configurational state of the anomeric carbon of the repeating glucose unit.

Chemical and physical approaches

Dilute acid hydrolysis combined with elevated temperatures is the only reasonable alternative to date as non-enzymatic methods of cellulose hydrolysis. Although this approach has been seriously considered in the past, especially for unprocessed wastes (e.g., municipal solid wastes), a maximum of only about 60% glucose can be expected after optimal hydrolysis conditions over a 24-h period (Green et al. 1988). The residue includes unwanted decomposition products of glucose and other sugars to furfural and derivatives, levulinic and formic acids and humic substances, in addition to the original acid and its by-products. Mechanical pretreatment, such as hammer-milling, trommeling and grinding are necessary to facilitate hydrolysis. These methods (inclusive) may also be appropriate for pretreatment of cellulosic wastes prior to microbial or cell-free enzymatic strategies.

The cellulosome of *Clostridium thermocellum*

Clostridium thermocellum is one of the few, and certainly the most prevalent, anaerobic thermophilic cellulolytic bacterium in nature (Lamed & Bayer 1991). Such bacteria are generally considered attractive for applicative research, mainly due to the fact that costly refrigeration and aeration procedures are unnecessary for maintaining their active culture. The processes are usually free from contaminating pathogens, and the thermophilic enzymes produced are generally of high activity and structural stability. *C. thermocellum* is considered to be the epitome of cellulolytic microorganisms, and its solubilizing potential of native crystalline cellulosics has been reported to be about 50 times higher than that of microbial systems (Johnson et al. 1982). Indeed, the cellulolytic system of this bacterium has been a standard for decades of previous industrially oriented research and development.

In nature, the utilization of a polymeric substrate, like cellulose, by a microbial system would appear to be most efficient, were the hydrolytic enzymes not randomly secreted leading to free competition for the soluble products; rather, if these enzymes are positioned in the interface between the cell and the insoluble substrate, they would gain a diffusional advantage as well as possible synergistic action amongst the densely packed enzymes. Lastly, regulation by product inhibition would be very important to avoid waste of unused substrate, in cases where the cell is unable to absorb the soluble products generated by the hydrolytic action of the enzyme. It is therefore not surprising that one of our first observations, regarding the interaction of *C. thermocellum* with cellulose, was the very strong adhesion of the bacterium to its substrate prior to its degradation (Fig. 3).

To further study this phenomenon, we employed a combined genetic-immunochemical program which consisted of isolation of an adhesion-defective mutant combined with the preparation of a monospecific antiserum to analyze the cellulose-binding factor on the wild-type cells (Bayer et al. 1983). The factor was isolated and identified as a large 2-MDa multicomponent conglomerate, which not only exhibited the cellulose-binding function but also expressed very high levels of cellulolytic activity (Lamed et al. 1983a), estimated to be the major portion of the cellulolytic action in this organism. Electron microscopic and SDS-PAGE analysis supported the notion that the factor was organized into a cohesive multi-subunit structure. The subunits were examples of very large bacterial proteins, ranging in size from 210 kDa to 48 kDa (Fig. 4). Due to its proposed central and multifunctional role in the degradation of cellulose by *C. thermocellum,* the cellulose-binding factor was termed the cellulosome (Lamed et al. 1983b). The isolated cellulosome was shown to exhibit the same properties as the crude cellulase system from this bacterium; most notably, the purified cellulosome was alone capable of completely

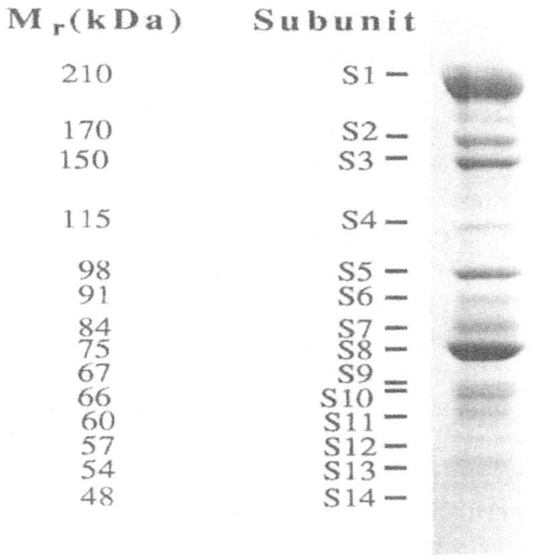

Fig. 4. Subunit composition of the cellulosome of *Clostridium thermocellum.*

solubilizing the paracrystalline structure of the insoluble substrate (Lamed et al. 1985). The presence of a discrete cellulosome complex was proposed to be a definitive property of the cellulase system in this organism, which promotes the synergistic action among its enzyme components.

Since these early reports, further study has been centered on characterizing the molecular and supramolecular nature of the cellulosome, with emphasis on functional and structural descriptions of the individual cellulosomal components. The major findings include the following: (1) descriptions of the disposition of the cellulosome on the cell surface, (2) the elucidation of a major nonenzymatic glycopeptide component (the S1 subunit) which is believed to play a decisive dual function, including both the crucial organizational role and the major cellulose binding role, (3) the analysis of the enzymatic components of the cellulosome, and (4) the isolation and characterization of a major exoglucanase, a cellobiohydrolase, which was found to exhibit many of the distinctive properties of the cellulosome. It was also discovered that many other cellulolytic bacteria, irrespective of their evolutionary or physiological profiles, appear to contain cellulosome-like entities which are similar in content and action to that of *C. thermocellum.*

Throughout the years, one of the major problems in studying the action of the components of the cellulosome is the exceptionally cohesive nature of the protein complex. Only treatment in detergent at high temperatures succeeded in dissociating the cellulosome into its component parts, and isolation of reasonable quantities of the individual subunits for conventional biochemical studies proved to be extremely difficult. Nevertheless, *in situ* zymogram assays of SDS-PAGE-separated components revealed that some subunits expressed very high activity levels on various substrates (Morag et al. 1990). In one case (Morag et al. 1991), limited proteolysis of the intact cellulosome led to the purification of a truncated form of a key subunit (S8), which was classified as a fairly typical cellobiohydrolase – the first and, as yet, the only confirmed example of this important enzyme type for this organism.

A second complementary direction in the study of the cellulase system of *C. thermocellum* has involved the cloning and expression of the relevant genes (Hazlewood et al. 1988; Beguin 1990). More than a dozen genes from this organism have been cloned and many of them have been expressed. By comparing the cellulase genes from *C. thermocellum*, both among themselves and with those from other species, it can be concluded that the cellulases are constructed in a modular fashion composed of multiple domains of defined function (Beguin et al. 1992). Interestingly, many cellulases contain a cellulose-binding domain separate from the catalytic domain. Many contain segments rich in proline, threonine and/or serine of as yet undetermined function.

Table 2. Some properties of the components of the cellulosome from *C. thermocellum* strain YS.

Subunit	M_r(kDa)	$\%_{app.}$	Sugars	EGase	MUCase	CBHase	Xyl'ase
S1	210	24	+++	–	–		–
S2	170	5	–	++++	–		++++
S3	150	9	–	+	++		(++++)
S4	115	<2	–	±	±		–
S5	98	8	–	±	+++		±
S6	91	<2	–	–	?		–
S7	84	7	–	–	+		++
S8	75	29	–	–	–	+++++	(++++)
S9,10	66,67	7	–	–	+++		+++
S11	60	~2	–	(++++)	–	–	+
S12	57	<2	–	–	–		–
S13	54	4	–	+++++	–		+++++
S14	48	<2	–	?	–		?

$\%_{app.}$, the apparent percent of cellulosomal content, contributed by individual subunits, was determined by densitometric tracings of Coomassie Blue-stained SDS-PAGE gels.
Enzymatic activities are denoted as follows: EGase, endoglucanase; MUCase, methylumbelliferyl cellobiohydrolase; CBHase, cellobiohydrolase; Xyl'ase, xylanase.
Genuine CBHase activity was verified only for the purified (truncated) S8 subunit; the cloned endoglucanase D (termed EGD, and identified subsequently as cellulosomal subunit S11) definitely lacked this activity. Other activities were detected on zymograms after SDS-PAGE.
+++++, ++++, etc., the relative activity in each assay is denoted by the number of '+'s.
–, indicates that no activity could be detected in the given assay.
(++++), indicates that the designated activity was undetected in the normal zymogram, but could be verified in other assays or under altered conditions. Thus, EGase activity of the S11 subunit could not be detected in the zymogram, but was very high for the cloned and expressed enzyme (EGD). Likewise, xylanase activity was absent in S8 after SDS-PAGE, but was very high for the isolated subunit. The xylanase activity of S3 was heat sensitive (when samples were incubated above 70° C).
?, due to the low percentage of certain cellulosomal subunits (e.g., S6 and S14), their activities could have been masked by those of their neighboring subunits.

As time progresses, it becomes clear that certain cellulase gene products are equivalent to cellulosomal subunits whereas others are not. Thus, endoglucanase D (EGD), appears to be synonymous with the S11 subunit (Joliff et al. 1986; Lamed & Bayer 1988b), the xynZ gene product has been equated with the S5 subunit (Grepinet et al. 1988) and EGE corresponds to cellulosomal subunit S7 (Durrant et al. 1991). One of the intriguing findings that ensued from the cloning studies was the presence of a 24-amino acid reiterated domain in cellulase genes which seems to be associated with cellulosome-based enzymes (Navarro et al. 1991; Tokatlidis et al. 1991). The reiterated domain appears not to be involved in either binding to cellulose or catalysis thereof; one of the most plausible roles of this domain will be discussed below.

Some of the important enzymatic properties of the cellulosomal subunits are summarized in Table 2. With the exception of cellobiohydrolase (CBHase) activity, the other activities listed in the table were assayed on zymograms following SDS-PAGE of heated cellulosomal samples. Consequently, unless otherwise indicated for isolated components, the activities listed are those which were still recovered after treatment at either 70 °C or 100 °C in the presence of SDS (i.e., conditions which are required for separation of the cellulosome into its components). Endoglucanase activity (EGase) was measured using carboxymethyl cellulose as a substrate, 'mock' exoglucanase activity (MUCase) employed methyl umbelliferyl cellobiose as a substrate, and xylanase activity (Xyl'ase) was assessed using a soluble xylan fraction. The lack of detection of a given activity, under the described assay conditions, does not necessarily indicate the absolute lack of the said activity, since the relatively harsh conditions prior to enzymatic analysis would impede all but the stablest of enzymes. Indeed, in cases where enzymatic components were successfully isolated (i.e., S8 and S11), high levels of enzymatic activities were detected despite their absence in zymograms. Nevertheless, the information given in Table 2 provides some indication as to the character of the cellulosomal enzymes.

It is clear from the table that the different enzymatic subunits of the cellulosome express different patterns of cellulolytic activities. It is interesting to note that the subunits which express very high levels of endoglucanase (e.g., S2, S11 and S13) are all relatively minor components (≤ 5%) of the cellulosomal protein in apparent quantitative terms. In contrast, the confirmed cellobiohydrolase (the S8 subunit) appears to account for almost a third of the cellulosomal protein. As stated above, this subunit is believed to play a central role in cellulose hydrolysis.

Another cellulosomal subunit which accounts for a major proportion of its protein is one of the most intriguing – i.e., the nonenzymatic, glycosylated, 210-kDa S1 subunit. The importance of this subunit to the cellulosome was recognized in the first publications, and, in view of its perplexing lack of enzymatic activity, various other roles have been proposed, including the crucial cellulose-binding activity and the central component carrier function. The reiterated domain, which has been associated with the other, mainly enzymatic, components of the cellulosome, may be responsible for the linkage of the respective component to the S1 subunit and, hence, to the cohesive cellulosomal complex. This would suggest a reciprocal subunit binding domain on the S1 subunit. Recent evidence, based on its anomalous dissociative behavior in SDS-PAGE gels (Morag et al. 1991b), has suggested that upon binding to cellulose, the S1 subunit undergoes dramatic conformational alterations in its structure (Morag et al. in press).

Thus, on the basis of this and other recent data, the general structure of the cellulosome and its initial interaction with cellulose can be envisioned as shown in Fig. 5. The cellulosome consists of a central S1 subunit which contains one or more cellulose binding domains (CBD) and a series of subunit binding domains (SBD). Bound to the latter in cohesive fashion are the other catalytic subunits via their reiterated sequences. The cellulosome first binds to the cellulosic substrate via the high-affinity CBD(s) on the S1 subunit, which undergoes conformational alterations on the surface of the cellulose. Upon binding, low-affinity CBDs on the other (catalytic) cellulosomal subunits may then bind the substrate, perhaps in a defined orienta-

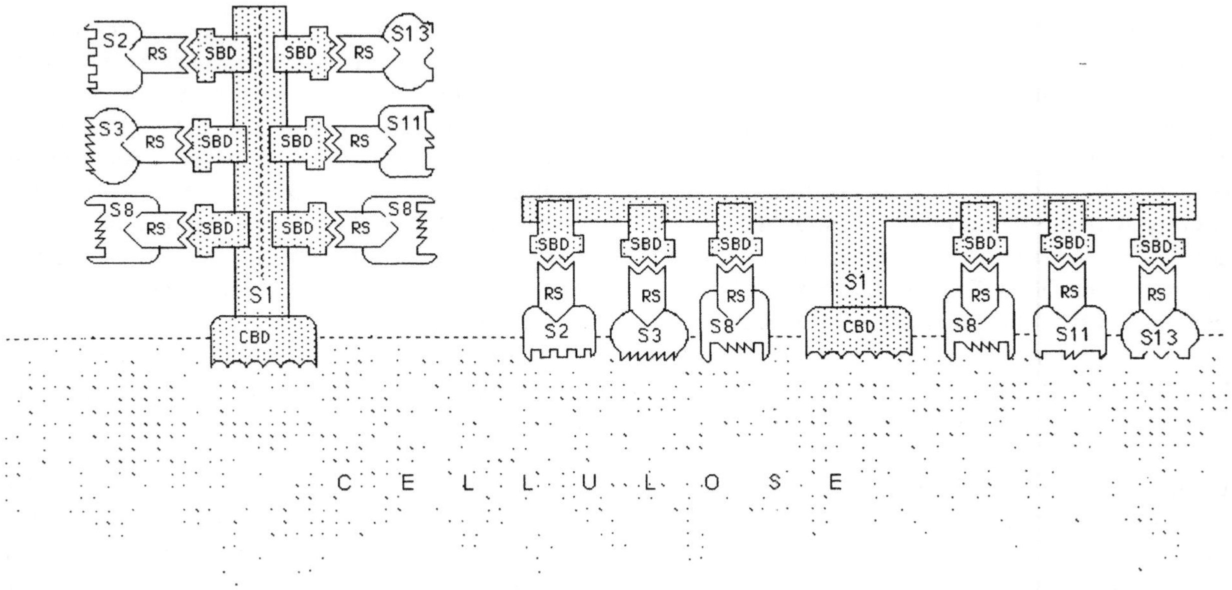

Fig. 5. Proposed model of cellulosome structure. The cellulosome comprises a central assembly or 'scaffolding' component, the S1 subunit, which exhibits a modular structure with multiple domains. The S1 subunit contains multiple subunit binding domains (SBD) which serve as attachment sites for the individual catalytic subunits (the cellulases). The S1 subunit also bears a very strong cellulose binding domain or domains (CBD) which targets the complex to the substrate. The cellulase subunits contain a 24-amino acid reiterated sequence (RS) which appears to be the site of attachment to the S1 (i.e., via the proposed SBDs, each of which may be selective for a specific RS). Upon binding to cellulose, the S1 subunit undergoes a conformational change which may be assisted by low-affinity CBDs known to occur on bacterial cellulases.

tion, and the concerted catalysis of the substrate ensues.

The binding of the cell-free cellulosome to cellulose actually reflects a more important function which the cellulosome performs, i.e., the binding of the cell itself to cellulose by virtue of the cell-

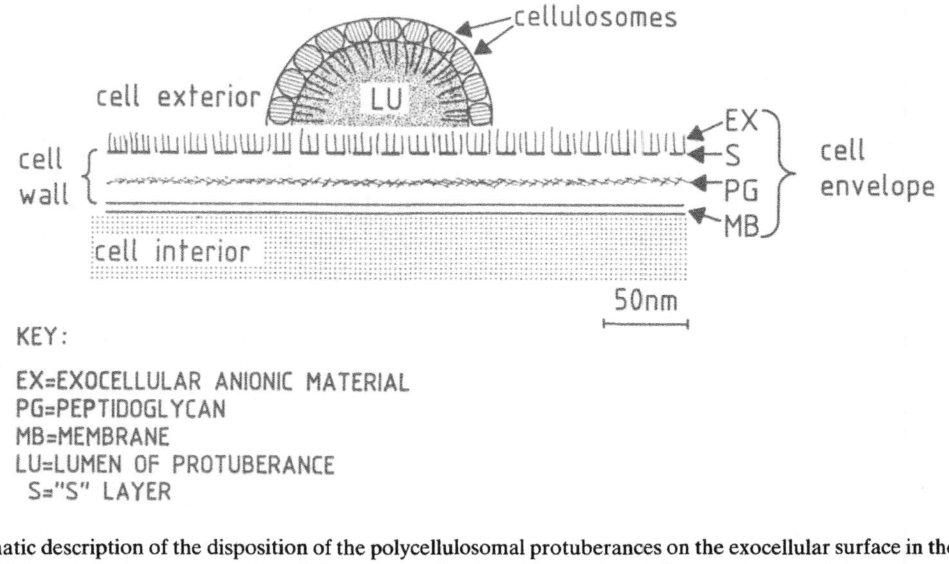

Fig. 6. Schematic description of the disposition of the polycellulosomal protuberances on the exocellular surface in the resting state. In the absence of the cellulosic substrate, the cellulosomes are fixed on the exterior of the protuberance in a compact state.

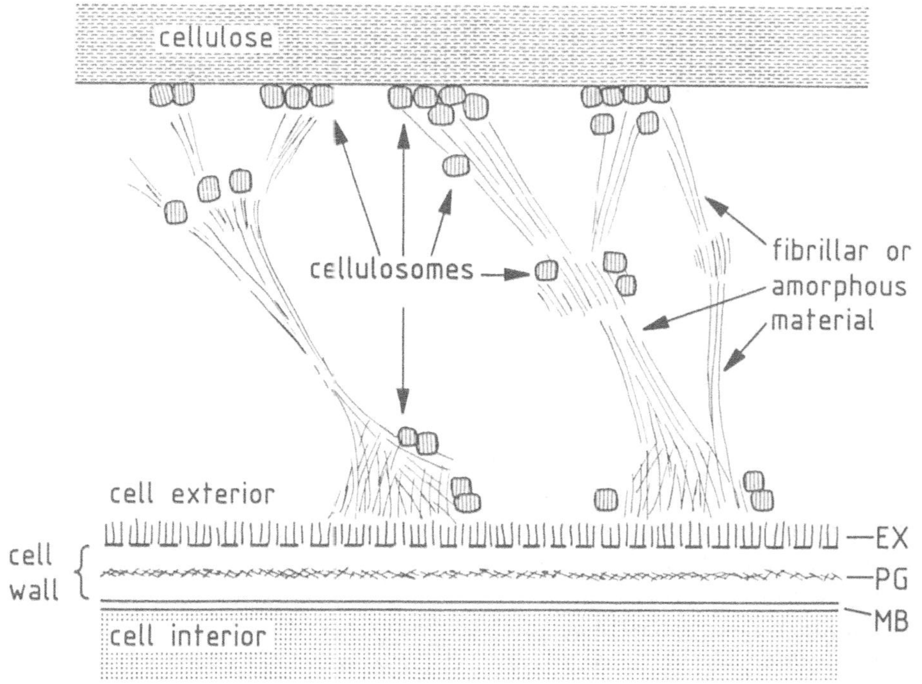

Fig. 7. Schematic description of the interaction of the polycellulosomal protuberance with cellulose. Upon contact with cellulose, the protuberance undergoes a dynamic suprastructural transformation in which the cellulosomes attach to the substrate but remain connected to the cell surface via protracted fibrous material. Thus, depolymerization of the substrate occurs via the synergistic action of a concentrated array of cellulosomal enzymes in the exocellular interface. The resultant soluble sugars, mainly cellobiose, are taken up by the cell by the appropriate transport apparatus.

associated cellulosomes (Bayer et al. 1983). To accomplish this task, the bacterial cells have been provided with a specific organelle, the polycellulosomal protuberance, which appears to contain multiple copies of the cellulosome (Bayer et al. 1985, 1986). In the absence of substrate, cells are inundated with large numbers of protuberances in contracted form (Fig. 6). Upon interaction with the substrate, however, the protuberances protract to form 'contact corridors' which appear to comprise fibrillar or amorphous material (Fig. 7). The cellulosomes which line the surface of the cellulose, begin to degrade the substrate extracellularly, producing large quantities of cellobiose, which is taken up by the cell via the appropriate transport system. Cell-associated β-glucosidase converts the cellobiose to glucose, and glucose is metabolized intracellularly via the usual pathways (Lamed & Bayer 1991). The major end-products in *C. thermocellum* are ethanol, lactic and acetic acids, molecular hydrogen and carbon dioxide.

The existence of the cellulosome in *C. thermocellum* appears not to be an isolated incident. Other unrelated cellulolytic bacteria, such as *Clostridium cellobioparum, Acetivibrio cellulolyticus, Bacteroides cellulosolvens,* and *Ruminococcus albus,* seem to harbor such multienzyme complexes on their cell surfaces, complete with protuberance-like organelles (Lamed et al. 1987). Moreover, S1-related antigenic determinants and similar sugar moieties seem to be shared among the various cellulolytic species. In this regard, the defining oligosaccharide residue of the S1 subunit has been elucidated in *C. thermocellum* (Gerwig et al. 1989, 1991). The sequence is shown in Fig. 8. The structure is characterized by a majority of galactosyl groups, one of which is in the very unusual furanose ring form. The terminal GalNAc residue is 3-0-

D-Gal*p*-α(1-3)
\
3-*O*Me-D-Glc*p*NAc-α(1-2)-D-Gal*f*-α(1-2)-D-GalOH [1]

D-Gal*f* -β(1-3)
\
D-Gal*f*-α(1-3)-D-Glc*p*NAc-α(1-2)-D-Gal*f*-α(1-2)-D-GalOH [2]

Fig. 8. The structure of the major galactose-rich cellulosomal oligosaccharide moieties from two different cellulolytic species of bacteria. [1] is from *Clostridium thermocellum* and [2] is from *Bacteroides cellulosolvens*. The oligosaccharide from *C. thermocellum* has been shown to be attached to the scaffolding S1 subunit. Note the similarities in the respective structures. The three initial sugars (from the nonreducing end) appear to be identical; the oligosaccharides differ in their substitutions, indicating species-specific variation.

methylated to form a unique naturally occurring sugar. In a very different cellulolytic bacterial species, a very similar structure was also found for its defining oligosaccharide, isolated from the cellulosome-like entity of *B. cellulosolvens* (Gerwig et al. 1992; Lamed et al. 1991a). The sequence contained several of the structural elements which characterized the S1-derived oligosaccharide from *C. thermocellum*. Although the precise role(s) of these oligosaccharide residues is currently unknown, the evidence implies that the S1 subunit and its sugars are critical to the cellulolytic action in such bacteria.

Solubilization of purified cellulose by a combined cellulosome-β-glucosidase system

Perhaps the major reason that cell-free enzyme systems have failed so dismally in attempts to efficiently degrade cellulosics is due to the strong inhibition by the major end product, cellobiose. Consequently, in designing a process for the enzymatic solubilization of cellullose, the major consideration is to rid the system of the resultant cellobiose so as to prevent feedback inhibition of the cellulase apparatus.

In nature, the cell precludes such inhibition by a variety of mechanisms (Lamed & Bayer 1991b). First, the substrate is degraded extracellularly by means of the polycellulosomal protuberances, after which part of the cellobiose formed is converted to glucose by the cell-associated β-glucosidase. The glucose and the residual cellobiose are transported into the cell by the corresponding transport systems. In addition, the large amounts of cellobiose formed in natural ecosystems would be further utilized by saccharolytic (but non-depolymerizing) satellite bacteria which occupy the same niche. In this way, cellobiose is effectively removed from the environment, and the cellulase system (i.e., the cellulosome) continues to degrade the substrate efficiently. In contrast, in the cell-free state, the cellulosome produces saturating levels of cellobiose, and cellulolysis is thus inhibited.

In a recent study (Lamed et al. 1991), we demonstrated that a purified β-glucosidase (from *Aspergillus niger*) can be combined with the purified cellulosome from *C. thermocellum,* in order to hydrolyze the accumulated cellobiose. Using this strategy, a dramatic enhancement of cellulose digestion could be achieved. The observed 10-fold enhancement was expressed both in the rate and degree of solubilization of the paracrystalline substrate, compared with that observed for the cellulosome alone (Table 3). Very dense suspensions of cellulose (at least 20%) could be converted to a glucose syrup with almost complete hydrolysis.

The combination of a β-glucosidase with a cellulose system has been reported earlier with varying success (for example, see Chakrabarti & Storey 1989; Kadam & Demain 1989). In general, these processes were economically prohibitive, since catalytic efficiencies were extremely low, and very high enzyme-substrate levels were necessary. The major contribution, in using the cellulosome from *C. thermocellum* with the readily available β-glucosidase from *A. niger,* is that very high rates and levels of solubilization can be achieved with relatively low amounts of enzyme components. Preliminary results involving the hydrolysis of dried alfalfa using the combined enzyme system have shown that the approach is also suitable for natural cellulosic substrates.

Solubilization of citrus albedo using a combined cellulosome-β-glucosidase-pectinase system

As described above, cellulase systems in general are prone to feedback inhibition, and, in most cases, cellobiose is the predominant cause of such inhibition. In nature, cellular uptake of cellobiose and/or its hydrolysis to glucose serve to relieve the inhibitory effect, thus providing a regulatory mechanism to improve the economy of the process.

Pectin is another naturally occurring polymer, which is present in large quantities in nature, particularly in fruit-derived agricultural wastes. Pectin is a polymer consisting of methyl 1,4-α-D-galacturonic acid, wherein the extent of methylation varies. Pectin can be especially problematic in applied processes, due to its gelling properties which are responsible for the high volumetric nature of pectin-containing materials. Pectinase activity was not detected in our cellulosome preparations, and the cellulosome, either alone or in combination with β-glucosidase, was incapable of degrading such substrates. The cellulosome requires the presence of pectinase in order to reach its substrate.

Thus, a combined enzymatic system, which consisted of the cellulosome from *C. thermocellum,* the purified β-glucosidase preparation and a partially purified pectinase (both from *Aspergillus niger,* obtained from Novo Industries a/s, Bagsvared, Denmark), was designed to hydrolyze a model pectin-containing cellulosic substrate (citrus albedo, from pomelo, which contained 27% cellulose and 70% pectin).

We discovered that, in order to hydrolyze albedo, the addition of the cellulosome-β-glucosidase system failed to improve control systems containing the pectinase alone, unless the reaction was performed under conditions, whereby low-molecular-weight breakdown products (which served as inhibitors of the cellulosome) were removed (R. Kenig et al., in preparation). This could be achieved either by dialysis or by introduction of an

Table 3. Enhancement of cellulose solubilization by the cellulosome from *C. thermocellum* supplemented by β-glucosidase from *Aspergillus niger.*

Avicel concentration (mg/ml)	Solubilization (%)	
	Cellulosome alone	Cellulosome + β-glucosidase
20	53	100
50	38	100
200	12	96*

Data after Lamed et al. (1991b)
The assay was performed using 8 μg cellulosome per mg of substrate. β-Glucosidase (0.04 cellobiase units per mg substrate) was added when required. The degree of solubilization was determined following a 9-day reaction period.
* At the highest substrate concentration tested, a second addition of the enzyme mixture (administered 72-h after the onset of the reaction) was required to effect near-total hydrolysis.

Table 4. Inhibition of cellulosome action on microcrystalline cellulose by pectin breakdown products and related acidic sugars.

Additive	Solubilization (%)	
	60° C	50° C
None	95	95
Pectin hydrolysate	9	60
Galacturonic acid	22	88
Glucuronic acid	93	95

Data after Kenig et al. (unpublished results).
Solubilization (%) was determined in a turbidometric assay system, following a 72-h reaction period at the designated temperature, using 0.6 mg/ml of a suspension of Avicel as the substrate. The concentration of pectin hydrolysate used in the competition experiments was 5 mg/ml, and that for both uronic acid derivatives was 6 mg/ml. The concentration of the cellulosome used in the reaction was 8 μg/mg cellulose.

anion exchanger. Galacturonic acid also inhibited the degradation of pectin by the combined system but to a somewhat lesser degree (Table 4), and controls using a similar compound (glucuronic acid) were ineffective inhibitors.

Further experimentation indicated that the inhibitory effect of the pectin hydrolysates was determined to reflect a reduction in the stability of the cellulosome and not a strict example of competitive inhibition. Indeed, the S8 (the crucial cellobiohydrolase subunit of the cellulosome) was eventually found to undergo strong destabilization by the pectin degradation products, thus reducing the efficiency of the entire combined enzyme system.

In order to achieve optimum solubilization of the albedo, the destabilizing effect had to be eliminated. High concentrations of albedo (200 mg dry-weight/ml!) were therefore subjected to the enzymatic action of the combined system at lower temperatures and in the presence of an anion exchanger to remove negatively charged pectin hydrolysis products as soon as they are formed. The results indicated that these conditions can effectively solubilize pectin-cellulosics such as citrus albedo, and that the coupling of extraneous enzymes to the cellulosome can be a valid approach for the degradation of naturally occurring mixed polymer systems. When a greater understanding is achieved concerning the composition of such native substrates and the effect of their breakdown products on the relevant enzyme systems, the design of cost-effective processes for waste disposal and utilization may eventually be realized.

Supercellulosomes – crosslinked enzyme systems

In the above-described systems, the enzymes were applied together in free soluble form. We also considered the possibility of coupling combined enzyme systems, either by chemical or biochemical means, in order to produce supercellulosomes which contain the relevant enzymes associated into complexes for more efficient hydrolysis and economy considerations. The idea behind this approach is that the cellulosome would serve to target the other enzymes to the relevant sites of hydrolysis, and the enzymes would achieve maximum effect where required.

As a model system, the purified cellulosome and β-glucosidase system were selected. Initial experiments in this direction were performed by mixing the desired enzymes in the presence of crosslinking reagents. For this purpose, several different reagents were examined: (1) glutaraldehyde was used to crosslink the enzymes via their lysine residues, (2) sodium periodate was used to oxidize the cellulosomal or β-glucosidase oligosaccharides to aldehydes which would interact secondarily with the lysines of either component, and (3) water-soluble carbodiimide was used to couple aspartate and glutamate residues of one enzyme component to lysines of another. Unfortunately, none of these covalent procedures worked, since in every case, the enzyme components were irreversibly inactivated.

In view of these results, we used a biochemical approach based on the avidin-biotin complex (Wilchek & Bayer 1990). Biotin was coupled to both the cellulosome and β-glucosidase; the addition of avidin (which has 4 biotin-binding sites per molecule) served to produce complexes which contained both enzyme species. In this case, both components could be biotinylated with over 70% of the enzymatic activity remaining. In the standard experiment, the biotinylated cellulosome was first applied alone to purified cellulose samples. Next, avidin was introduced to the resultant suspension, and the residue was washed. Finally, the biotinylated β-glucosidase was applied to the adsorbed cellulose sample, and again washed. In this manner, the combined enzyme system would comprise an active cellulolytic layer which produces glucose directly from microcrystalline cellulose on the surface of the insoluble substrate. The rate and extent of cellulolysis obtained using this targeted system was about three times more effective in its solubilization of cellulose than the cellulosome-absorbed system alone. These results were similar to those obtained using an excess of β-glucosidase (without washing). We are currently examining an *in situ* approach wherein the components are not applied in sequence, rather the crosslinked supercellulosome constructs are preformed prior to application to the insoluble substrate.

Other strategies for combining the appropriate enzyme species can also be explored in the future. One particularly appealing approach would be to use genetic engineering techniques, e.g., gene fusion or insertion of desired domains into a single entity. Thus, cellulose-binding domains can be combined at the gene level with catalytic domains of the desired enzyme species to enhance synergism and overall substrate hydrolysis.

As our knowledge of the structure and composition of natural cellulosic systems progresses, and our understanding of the various cellulolytic and related enzyme systems is broadened, it is hoped that we will be able to design and develop 'smart' cellulase systems for use in future economically viable industrial processes.

References

Bayer EA & Lamed R (1986) Ultrastructure of the cell surface cellulosome in *Clostridium thermocellum*. J. Bacteriol. 167: 828–836

Bayer EA, Kenig R & Lamed R (1983) Studies on the adherence of *Clostridium thermocellum* to cellulose. J. Bacteriol. 156: 818–827

Bayer EA, Setter E & Lamed R (1985) Organization and distribution of the cellulosome in *Clostridium thermocellum*. J. Bacteriol. 163: 552–559

Beguin P (1960) Molecular biology of cellulose degradation. Annu. Rev. Microbiol. 44: 219–248

Beguin P, Millet J, Chauvaux S, Salamitou S, Tokatlidis K, Navas J, Fujino T, Lemaire M, Raynaud O, Daniel M-K & Aubert J-P (1992) Bacterial cellulases. Biochem. Soc. Transactions 20: 42–46

Biddlestone AJ & Gray KR (1987) Production of organic fertilizers by composting: In: Moriarty DJW & Pullin RSV (Eds) Detritus and Microbial Ecology in Aquaculture (pp 151–180). International Center for Living Aquatic Resources Management (Conference Proceedings 14), Manila, Philippines

Bridgwater AV & Mumford CJ (1979) Waste Recycling and Pollution Control Handbook. George Godwin, Ltd, London

Chakrabarti AC & Storey KB (1989) Enhanced glucose production from cellulose using coimmobilized cellulase and β-glucosidase. Appl. Biochem. Biotechnol. 22: 263–278

Cowling EB (1975) Physical and chemical constraints in the hydrolysis of cellulose and lignocellulosic materials. Biotech. Bioeng. Symp. 5: 163

Crawford JF & Smith PG (1986) Landfill Technology. Butterworths, London

Durrant AJ, Hall J, Hazlewood GP & Gilbert HS (1991) The noncatalytic C-terminal region of endoglucanase E from *Clostridium thermocellum* contains a cellulose binding domain. Biochem. J. 273: 289–293

Fan LT, Lee YH & Gharpuray MM (1982) The nature of lignocellulosics and their pretreatments for enzymatic hydrolysis. Adv. Biochem. Eng. 23: 157–183

Gerwig G, de Waard P, Kamerling JP, Vliegenthart JFG, Morgenstern E, Lamed R & Bayer EA (1989) Novel *O*-linked carbohydrate chains in the cellulase complex (cellulosome) of *Clostridium thermocellum*. J. Biol. Chem. 264: 1027–1035

Gerwig G, Kamerling JP, Vliegenthart JFG, Morgenstern E, Lamed R & Bayer EA (1991) Primary structure of *O*-linked carbohydrate chains in the cellulosome of *Clostridium thermocellum* strains. Eur. J. Biochem. 196: 115–122

—(1992) Novel oligosaccharide constituents of the cellulase complex of *Bacteroides cellulosolvens*. Eur. J. Biochem. 205: 799–808

Gould JM (1985) Enhanced polysaccharide recovery from agricultural residues and perennial grasses treated with alkaline hydrogen peroxide. Biotech. Bioeng. 27: 893–896

Green M, Kimchie S, Malester AI, Rugg B & Shelef G (1988) Utilization of municipal solid wastes (MSW) for alcohol production. Biol. Wastes 26: 285–295

Grepinet O, Chebrou M-C & Beguin P (1988) Purification of *Clostridium thermocellum* xylanase Z expressed in *Escherichia coli* and identification of the corresponding product in the culture medium of *C. thermocellum*. J. Bacteriol. 170: 4582–4588

Grethlein HE (1985) The effect of pore size distribution on the rate of enzymatic hydrolysis of cellulosic substrates. Bio/Technology 3: 155–160

Haigler CH & Weimer PJ (Eds) (1991) Biosynthesis and Biodegradation of Cellulose. Marcel Dekker, Inc., New York

Hanselmann KW (1982) Lignochemicals. Experientia 38: 176–185

Hazlewood GP, Romaniec MPM, Davidson K, Grepinet O, Beguin P, Millet J, Raynaud O & Aubert J-P (1988) A catalogue of *Clostridium thermocellum* endoglucanase, β-glucosidase and xylanase genes cloned in *Escherichia coli*. FEMS Microbiol. Lett. 51: 231–236

Johnson EA, Sakojoh M, Halliwell G, Madia A & Demain AL (1982) Saccharification of complex cellulosic substrates by the cellulase of *Clostridium thermocellum*. Appl. Environ. Microbiol. 43: 1125–1132

Joliff G, Beguin P, Juy M, Millet J, Ryter A & Aubert J-P (1986) Isolation, crystallization and properties of a new cellulase of *Clostridium thermocellum* overproduced in *Escherichia coli*. Bio/Technology 4: 896–900

Kadam SK & Demain AL (1989) Addition of cloned β-glucosidase enhances the degradation of crystalline cellulose by the *Clostridium thermocellum* cellulase complex. Biochem. Biophys. Res. Commun. 161: 706–711

Lamed R & Bayer EA (1988a) The cellulosome of *Clostridium thermocellum*. In: Laskin A (Ed) Advances in Applied Microbiology, Vol 33 (pp 1–46). Academic Press, San Diego

—(1988b) The cellulosome concept: exocellular/extracellular enzyme reactor centers for efficient binding and cellulolysis.

In: Aubert J-P, Beguin P & Millet J (Eds) Biochemistry and Genetics of Cellulose Degradation (pp 101–116). Academic Press, London
—(1991) Cellulose degradation by thermophilic anaerobic bacteria. In: Haigler T & Weimer P (Eds) Biosynthesis and Biodegradation of Cellulose and Cellulose Materials (pp 377–410). Marcel Dekker, Inc., New York
Lamed R, Setter E & Bayer EA (1983a) Characterization of a cellulose-binding, cellulase-containing complex in *Clostridium thermocellum*. J. Bacteriol. 156: 828–836
Lamed R, Setter E, Kenig R & Bayer EA (1983b). The cellulosome – a discrete cell surface organelle of *Clostridium thermocellum* which exhibits separate antigenic, cellulose-binding and various cellulolytic activities. Biotechnol. Bioeng. Symp. 13: 163–181
—(1985) The major characteristics of the cellulolytic system of *Clostridium thermocellum* coincide with those of the purified cellulosome. Enzyme Microb. Technol. 7: 37–41
Lamed R, Naimark J, Morgenstern E & Bayer EA (1987) Specialized cell surface structures in cellulolytic bacteria. J. Bacteriol. 169: 3792–3800
Lamed R, Morag (Morgenstern) E & Bayer EA (1991a) Cellulosome-like entities in *Bacteroides cellulosolvens*. Current Microbiol. 22: 27–33
Lamed R, Kenig R, Morag E, Calzada JF, de Micheo F & Bayer EA (1991b) Efficient cellulose solubilization by a combined cellulosome-β-glucosidase system. Appl. Biochem. Biotechnol. 27: 173–183
Ljungdahl LG & Eriksson K-E (1985) Ecology of microbial cellulose degradation. Adv. Microbiol. Ecology. 8: 237–299
Lynd LR, Wolkin RH & Grethlein HE (1986) Continuous fermentation of Avicel and pretreated mixed hardwood by *Clostridium thermocellum*. Biotech. Bioeng. Symp. 17: 265–274
Marsden WL & Gray PP (1986) Enzymatic hydrolysis of cellulose in lignocellulosic materials. Crit. Rev. Biotechnol. 3: 235–272
McCarthy JE (1990) Solid waste management. Congressional Research Service Issue Brief, Order Code IB87176 (pp 1–13). The Library of Congress, Washington
Morag E, Bayer EA & Lamed R (1990) Relationship of cellulosomal and noncellulosomal xylanases of *Clostridium thermocellum* to cellulose-degrading enzymes. J. Bacteriol. 172: 6098–6105
Morag E, Halevy I, Bayer EA & Lamed R (1991a) Isolation and properties of a major cellobiohydrolase from the cellulosome of *Clostridium thermocellum*. J. Bacteriol. 173: 4155–4162
Morag (Morgenstern) E, Bayer EA & Lamed R (1991b) Anomalous dissociative behavior of the major glycosylated component of the cellulosome of *Clostridium thermocellum*. Appl. Biochem. Biophys. 30: 129–136
Morag E, Bayer EA & Lamed R (1992) Unorthodox intra-subunit interactions in the cellulosome of *Clostridium thermocellum:* Identification of structural transitions induced in the S1 subunit. Appl. Biochem. Biophys. (in press)
Navarro A, Chebrou M-C, Beguin P & Aubert J-P (1991) Nucleotide sequence of the cellulase gene *celF* of *Clostridium thermocellum*. Res. Microbiol. 142: 927–936
Rathje WL (1991) Once and future landfills. National Geographic 179: 116–134
Schwald W, Chan M, Brownell HH & Saddler JN (1988) Influence of hemicellulose and lignin on the enzymatic hydrolysis of wood. In: Aubert J-P, Beguin P & Millet J (Eds) Biochemistry and Genetics of Cellulose Degradation (pp 252–302). Academic Press, London
Tokatlidis K, Salamitou S, Beguin P, Dhurjati P & Aubert J-P (1991) Interaction of the duplicated segment carried by *Clostridium thermocellum* cellulases with cellulosome components. FEBS Lett. 291: 185–188
Tsao GT, Ladisch MR & Bungay HR (1987) Biomass refining. In: Bungay HR & Belfort G (Eds) Advanced Biochemical Engineering (pp 79–101). Wiley Interscience, New York
Walseth CS (1952) Influence of the fine structure of cellulose on the action of cellulases. Tappi 35: 228–232
Wilcheck M & Bayer EA (Eds) (1990) Avidin-Biotin Technology, Methods in Enzymology, Vol 184. Academic Press, San Diego
Wright JD (1988) Ethanol from biomass by enzymatic hydrolysis. Chem. Eng. Prog. 84: 62–73

Biodegradation 3: 189–205, 1992.

Utilization of lignocellulosic waste by the edible mushroom, *Pleurotus*

Yitzhak Hadar, Zohar Kerem, Barbara Gorodecki & Orly Ardon
The Otto Warburg Center for Biotechnology in Agriculture, Faculty of Agriculture, P.O. Box 12, Rehovot, 76100 Israel

Key words: lignin degradation, *Pleurotus*, lignocellulose utilization, solid state fermentation

Abstract

Lignocellulosic waste represents huge amounts of unutilized renewable resource. The use of the polysaccharides in the lignocellulosic complex is limited due to their high lignin content. White rot fungi are capable of selectively degrading lignin, thereby upgrading it. The focus of this article is on the potential utilization of edible mushrooms of the genus *Pleurotus*, via solid state fermentation, using cotton plant stalks as a substrate. This material poses agrotechnical problems since the stalks have a fibrous structure similar to that of hardwood. Potential uses for this material are as a fuel in rural areas, a substrate for mushrooms, an animal feed and substrate for paper making. In this study, degradation of cotton stalks by *Pleurotus* is described using chemical analyses and scanning electron microscopy. During four weeks of solid state fermentation, lignin content significantly decreased and *in vitro* digestibility was increased. The fermentation product was consumed by ruminants at a level of up to 40% of their diet.

Introduction

The lignocellulosic complex constitutes a major portion of the total carbon fixed by photosynthesis. However, only a small part of the cellulose, hemicellulose and lignin produced as agricultural or forestry by-products is utilized, while most of it is considered waste material. In Europe, for example, the amount of unexploited lignocellulosic by-products is immense. Cereal straw, the major by-product, shows a yearly surplus of $24 \cdot 10^6$ tons (Zadrazil & Reinger 1988). Most of the straw is either burned in the field creating environmental and ecological problems, or wasted by being ploughed back into the soil.

The direct use of lignocellulosic residue as animal feed, or as a component of such feeds, represents one of its oldest and most widespread applications and as such it plays an important role in the ruminant diet. However, lignocellulosic residues are not high value feeds: they are classified instead as low quality roughage, i.e. high in fiber, low in protein, vitamins and minerals.

The lignin component creates the barrier to efficient utilization, conversion or degradation of the polysaccharides in lignocellulose to useful products or high value animal feed (Buswell & Odier 1987). Lignin is defined as a polymeric product arising from an initial enzymatic step followed by chemically driven dehydrogenative polymerization of primary precursors possessing a p-hydroxycinnamyl alcohol structure (Monties 1989). Due to its complex and heterogeneous structure, lignin degradation is slow and limited to a relatively small number of microorganisms (Buswell & Odier 1987).

The white rot fungi can degrade all the major components of wood and are generally considered to be the main agents of lignin degradation in nature (Buswell & Odier 1987). The best studied organism of this group is *Phanerochaete chrysosporium* (Kirk & Farrell 1987; Schoemaker & Leisola

1990). Most of the knowledge regarding physiological, biochemical and genetic factors associated with lignin biodegradation as well as their practical uses, mainly in the pulp and paper industry, has been obtained from this model organism.

Another group of interesting white rot fungi that can utilize lignocellulose is comprised of the edible mushrooms. These saprophytic basidiomycetes have been successfully cultivated at a commercial level worldwide (Wood & Smith 1987). This paper focuses on the edible mushrooms of the genus *Pleurotus* and their potential in the utilization and upgrading of lignocellulosic wastes during solid state fermentation (SSF).

Cultivation of *Pleurotus* spp. on lignocellulosic waste

Pleurotus spp. are wood-degrading saprophytic fungi which are widely distributed and cultivated in many countries throughout the world. *Pleurotus* cultivation is gaining popularity in Europe, America and the Far East, and from the standpoint of annual production, it has become the third most important mushroom, after *Agaricus bisporus* and *Lentinus edodes* (Wood & Smith 1987). Commercial production techniques for this basidiomycete are well developed (Tautorus 1985; Wood & Smith 1987; Zadrazil 1978) and relatively simple as compared to those used with the most commonly cultivated mushroom, *A. bisporus*. *Pleurotus ostreatus*, popularly known as the "oyster mushroom" is the most common artificially cultivated species. The substrate is usually partially shredded, mixed with water (up to 70%) and placed in containers such as bags, trays or frames. However, unlike *A. bisporus*, no composting or casing layer is required. Because *P. ostreatus* is a wood-degrading fungus, it was first cultivated on logs (Falck 1917). Today it has become common practice to prepare *Pleurotus* substrate from shredded wheat straw, which can be supplemented with protein-rich materials such as alfalfa meal or soybean flour. Since *Pleurotus* spp. can decompose lignocellulose efficiently without chemical or biological pre-treatment, a large variety of lignocellulosic wastes can be utilized and recycled. Some examples of the agricultural wastes studied as substrates for *Pleurotus* spp. are presented here.

In coffee-growing regions, coffee pulp is considered to be one of the most abundant, as well as one of the hardest to handle agricultural wastes. To date, the majority of the by-product remains unused, causing contamination problems (Guzman & Martinez 1986). A most economical way of dealing with the excessive amount of pulp and of solving the environmental problem would be to use it for animal feed. However, chemical components such as lignin, phenolics and caffeine hamper its utilization by animals. Caffeine and phenolics are known to exert detrimental effects to both the rumen's microflora and the host animal. Biological pre-treatment could, however, improve the value of this waste product. The use of coffee by-products as a substrate for *Pleurotus* has been studied by Leon et al. (1983). Guzman & Martinez (1986) successfully cultivated *P. ostreatus* on coffee pulp on a semi-industrial scale, at a coffee farm in Mexico.

Sharma (1987) studied the utilization of flax shive, a woody by-product of little value. For every ton of flax fiber produced, 2.5 tons of shive are left over. Shive has a high lignin content (24.9%), preventing its direct utilization. *Pleurotus* grown on this substrate was able to reduce the lignin content of the shive to 12.4%.

Cassia spp. are tropical shrubs with medicinal, e.g. laxative, properties. The residue of extracted cassia plants was tested for cultivation of *Pleurotus* by Muller (1987). The utilization of *Pleurotus* in the recycling of medicinal plants could be of additional importance in cases where toxic components are present in the material. The reduction of such components by the fungus would simplify the problem of waste disposal (Muller 1987).

Oriaran et al. (1989) compared lignin degradation of hard and soft wood chips by *P. ostreatus*, L. *edodes* and *Phanerochaete chrysosporium*. All three white rot fungi selectively degraded the lignin on glucose-supplemented chips, an observation which could be useful in biopulping by the paper industry.

A large variety of other wastes, such as corn cobs

and leaves, sugar cane bagasse and leaves, citronella bagasse, rice hulls, water hyacinth leaves, cocoa shells and cotton gin waste, could easily be used as substrates for *Pleurotus* spp. The substrates used in each region would depend on the available agricultural wastes.

In our study, the use of cotton stalks as a substrate for SSF by *P. ostreatus* was analyzed. Platt et al. (1981) and Balasubramanya (1981) have successfully grown *P. ostreatus* and *P. sajor-caju* on this substrate.

Utilization of cotton stalks as a substrate for *Pleurotus* fermentation, as animal feed and in paper production

Cotton is Israel's main field crop and generates the largest proportion of local agricultural waste. Its worldwide importance is illustrated by the 10 million tons of cotton stalks reported for India yearly (Balasubramanya et al. 1989). Those remaining in the field, 5 tons/ha, are ploughed under the soil surface. In addition to wasting a potential agricultural resource, this treatment could lead to an increase in cotton diseases and pests, as well as to difficulties in cultivation due to slow decomposition in the soil (mainly in dry regions). The major obstacle to using cotton stalks as a mushroom substrate lies in its difficult preservation. This is due mainly to the high water content and to the high level of soluble carbohydrates (2–4%, as compared to 0.4–1.4% in wheat straw) (Silanikove & Levanon 1986). The substrate is rapidly overgrown by molds, resulting in spoilage and aerobic degradation. The anaerobic preservation of cotton stalks and the production of silage has been studied by Silanikove & Levanon (1986), Levanon et al. (1988) and Danai et al. (1989). Cotton stalks were harvested and chopped into 2–3 cm particles with a forage harvester originally designed to cut corn. The material was taken to a concrete silo with a storage capacity of 450 tons, pressed with a heavy tractor, and then covered with black plastic sheets. After one month of storage, the pH of the preserved cotton stalks stabilized at 5.5 and the material was successfully utilized for commercial *Pleurotus* cultivation, up to nine months after harvest (Danai et al. 1989; Levanon et al. 1988).

Chemical treatment for the upgrading of cotton stalks and their subsequent utilization as animal feed has also been investigated (Ben-Ghedalia et al. 1983; Ben-Ghedalia et al. 1980). Apparent digestibility of organic matter in ozone-treated cotton stalks used in a sheep diet was 75%, as compared to 30 and 39% in control and NaOH-treated substrates. Cotton stalks could also potentially be used as raw material for the manufacture of some grades of paper (Balasubramanya et al. 1989; Pandy & Shaikh 1987).

The use of *Pleurotus* to improve digestibility of cotton stalks in ruminants

The lignocellulose complex in straw and other plant residues is degraded very slowly by ruminants because of the physical barrier imposed by lignin polymers, which prevent free access of hydrolytic enzymes such as cellulases and hemicellulases to their substrates. Normally, the rates of decay of plant debris are proportional to their lignin content. Biological delignification of straw seems to be the most promising way of improving its digestibility (Kamra & Zadrazil 1986; Streeter et al. 1982; Zadrazil & Reinger 1988). Several authors have examined this possibility, using mainly wheat straw and *Pleurotus* spp. under different conditions and substrate pre-treatments (Kamra & Zadrazil 1986; Lindenfelser et al. 1979; Streeter et al. 1982; Zadrazil 1980). A significant increase in *in vitro* dry matter digestibility of wheat straw by *Pleurotus* was reported by Lindenfelser et al. (1979), who used commercial cellulase to compare the release of glucose from untreated straw to that of straw following *Pleurotus* fermentation. They found that lignin content decreased by 51% during the incubation period (90 days). Similar results were reported by Zadrazil (1985) showing an increase of 23–32% in the *in vitro* digestibility following fermentation by *P. eryngii*, *P. sajor-caju*, *P. ostreatus* and *P. serotinus*.

These results encouraged us to examine the effect of *Pleurotus* on cotton stalks during SSF. SSF

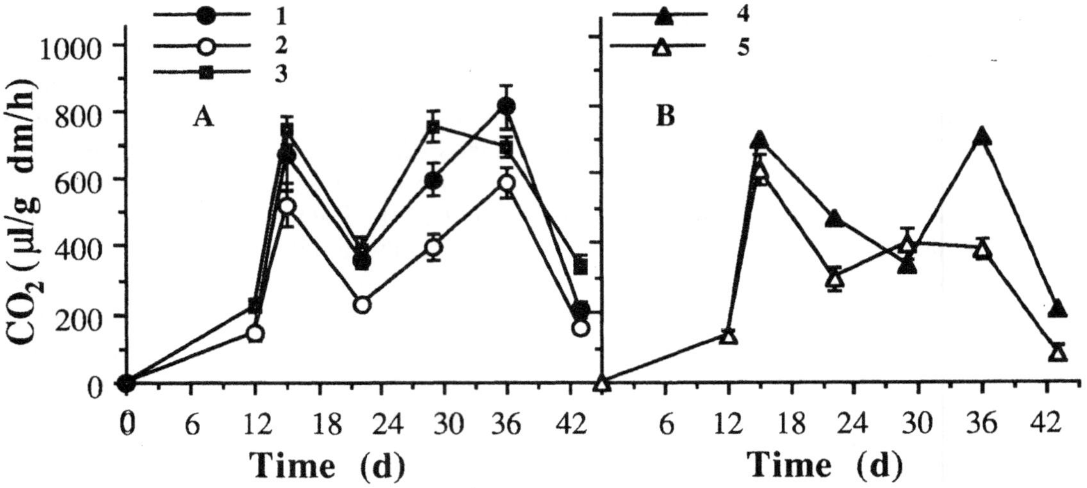

Fig. 1. Pleurotus respiration during SSF on ground (A) and chopped (B) cotton stalks. 1–5 are the treatments as described in the text. Bars represent standard errors.

seems to be the most appropriate method, both for conducting basic research and for applied, large-scale processes involving selective lignin biodegradation. SSF was carried out in the laboratory to define optimal conditions for increasing the *in vitro* digestibility of cotton stalks and improving the SSF process. Five pre-treatments were compared:

1. ground cotton stalks (particle size 2–5 mm), 70% water content;
2. ground cotton stalks, 70% water content plus the following nutrients: (g/Kg dry cotton stalk) 6.0 K_2HPO_4; 2.0 $MgSO_4 \cdot 7H2O$; 4.0 $CaCO_3$; 0.008 thiamine·HCl; 4.0 Bacto-peptone, pH 5.6;
3. ground cotton stalks, 75% water content;
4. chopped cotton stalks (particle size 2–7 cm), 70% water content;
5. chopped cotton stalks, 70% water content plus the nutrients as in treatment 2.

The chopped cotton stalks were harvested in the

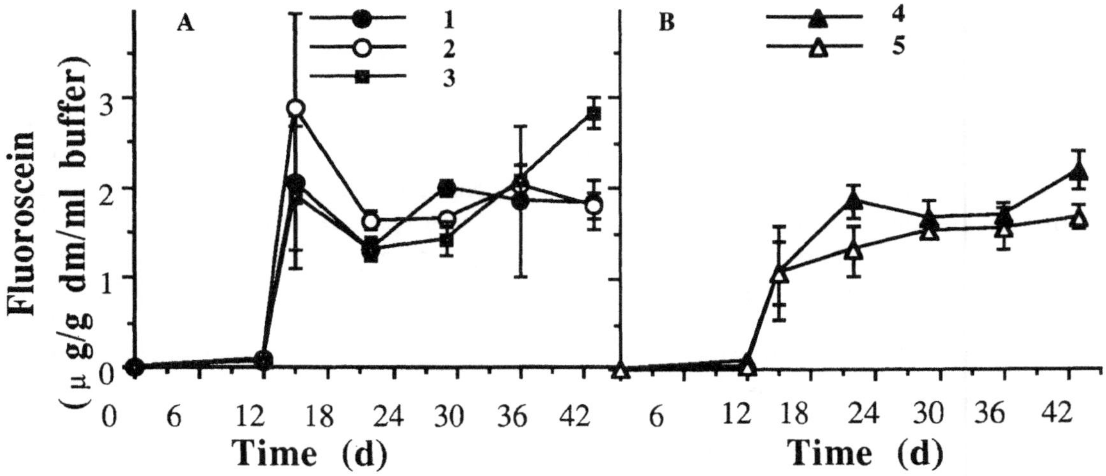

Fig. 2. Fluoroscein diacetate hydrolysis during SSF of *Pleurotus* on ground (A) and chopped (B) cotton stalks. 1–5 are the treatments as described in the text. Bars represent standard errors.

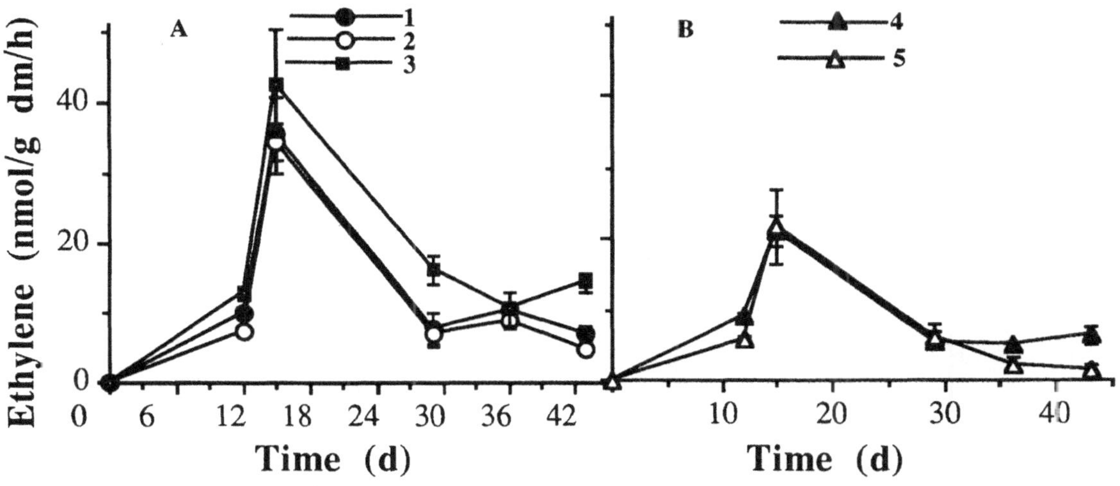

Fig. 3. Ethylene generation from KTBA during SSF of *Pleurotus* on ground (A) and chopped (B) cotton stalks. 1–5 are the treatments as described in the text. Bars represent standard errors.

field using agricultural equipment, with no further treatment.

Polystyrene bags containing 95 g (d.w.) of cotton stalks were either wetted with water or supplemented with nutrient solution and sterilized once for one hour, then again 24 h later. At various stages after inoculation with *P. ostreatus*, bags from each treatment were sampled for enzymatic and chemical analyses. Methods used in this study have been described previously (Kerem et al. 1992). Fungal activities and chemical composition of the fermentation product were followed.

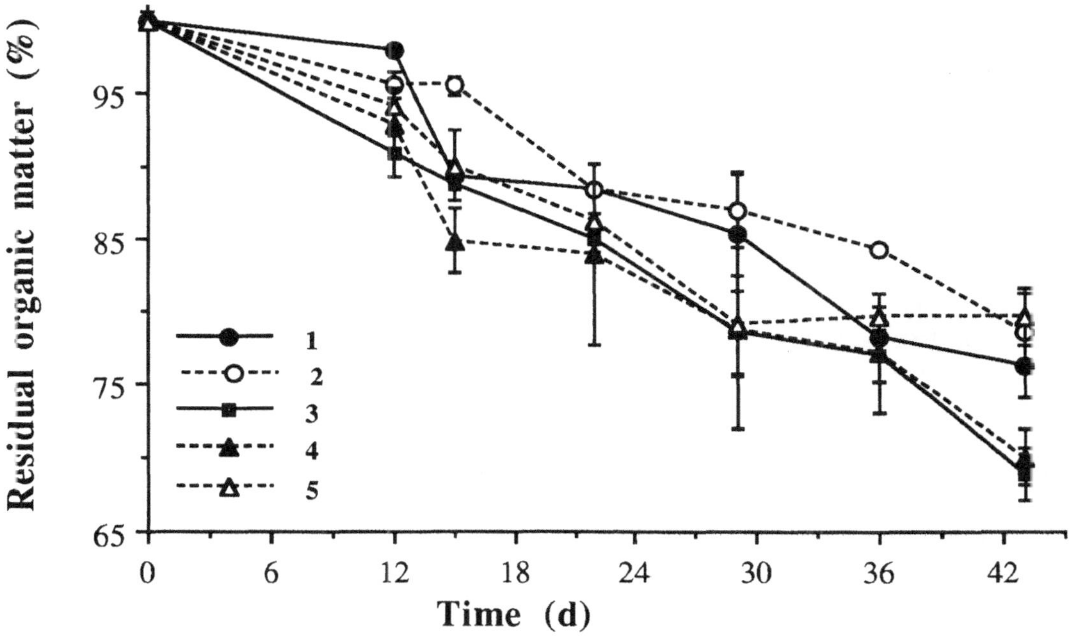

Fig. 4. Mineralization of organic matter during SSF of *Pleurotus* on cotton stalks. 1–5 are the treatments as described in the text. Bars represent standard errors.

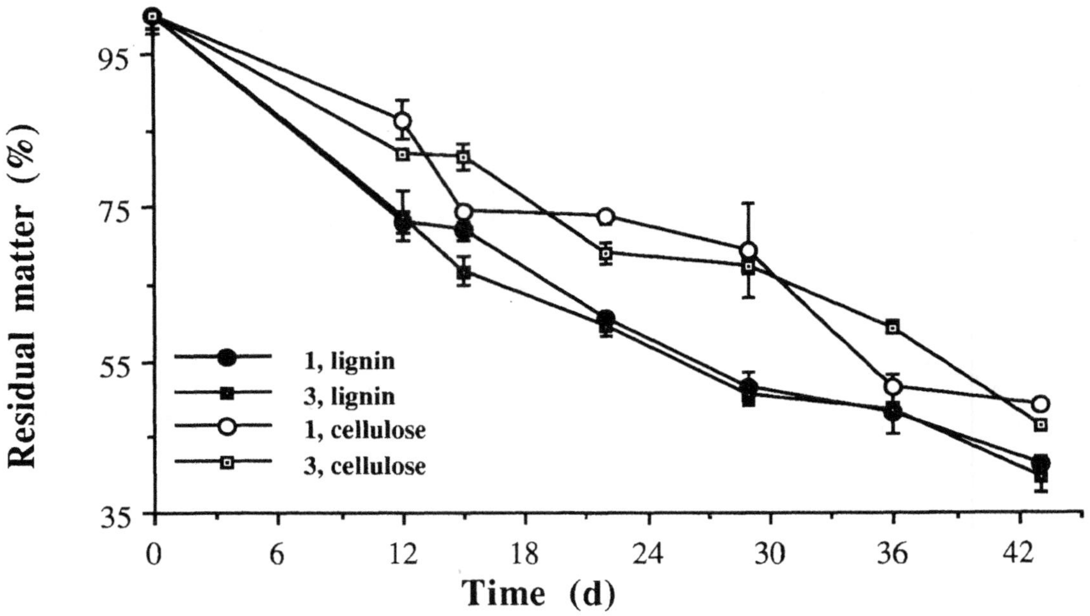

Fig. 5. Lignin and cellulose degradation by *Pleurotus* during SSF of ground cotton stalks with 70% (1) and 75% (3) water content. Bars represent standard errors.

Fungal activities related to colonization: Fungal respiration was recorded as CO_2 evolution per hour, as determined by gas chromatograph (Fig. 1). Respiration rates rose for the first 15 days, then stabilized. On day 18, the contents of the bags were mixed to disperse the colonized bulks and on day 22, respiration rates had decreased. This could have been due to dispersion of the already colonized bulk. However, mixing showed no effects on the biodegradation process. Treatment 5 showed no resumption of fungal activity after mixing. Fungal activity defined by extracellular hydrolysis of fluorescein diacetate (Schnurer & Rosswall 1982) showed a similar trend for all treatments (Fig. 2). There was a slow rise in the extracellular activity towards day 12, a definite increase towards day 15 and stabilization towards the end of the experiment (day 42). Treatment 3 showed a second peak towards the end of the experiment.

Ethylene generation from KTBA (α-keto-γ-thio-ethyl butyric acid): KTBA is a water-soluble molecule which, in the presence of free (·OH centered) radicals, is cleaved to release gaseous ethylene (Kelley 1988). Cleavage activity was observed from day 12, reaching a maximum on day 15 and continuing at a low level from day 29 to the end of the experiment (day 42) (Fig. 3). Treatment 3 showed the highest activity level. Treatments 1 and 2 produced slightly lower activities than treatment 3 and treatments 4 and 5 were found to have the lowest levels of activity.

Organic matter degradation: Mineralization of organic matter was calculated from changes in the ash content of the substrate throughout the experiment. In treatments 2, 3 and 5, a steady mineralization rate was observed, with treatment 3 producing the highest rate (Fig. 4). In treatments 1 and 4 an acceleration was observed between days 12 and 15.

By the end of the experiment (day 42), 35% of the organic matter had been mineralized in treatments 3 and 4, as compared to 30% in treatment 1 and 20% in treatments 2 and 5. A correlation was evident between mineralization patterns and respiration rates. Differentiation of fiber composition and *in vitro* digestion: Cellulose and lignin, the major fiber components of the substrate, were analyzed. Since chemical analysis of these components

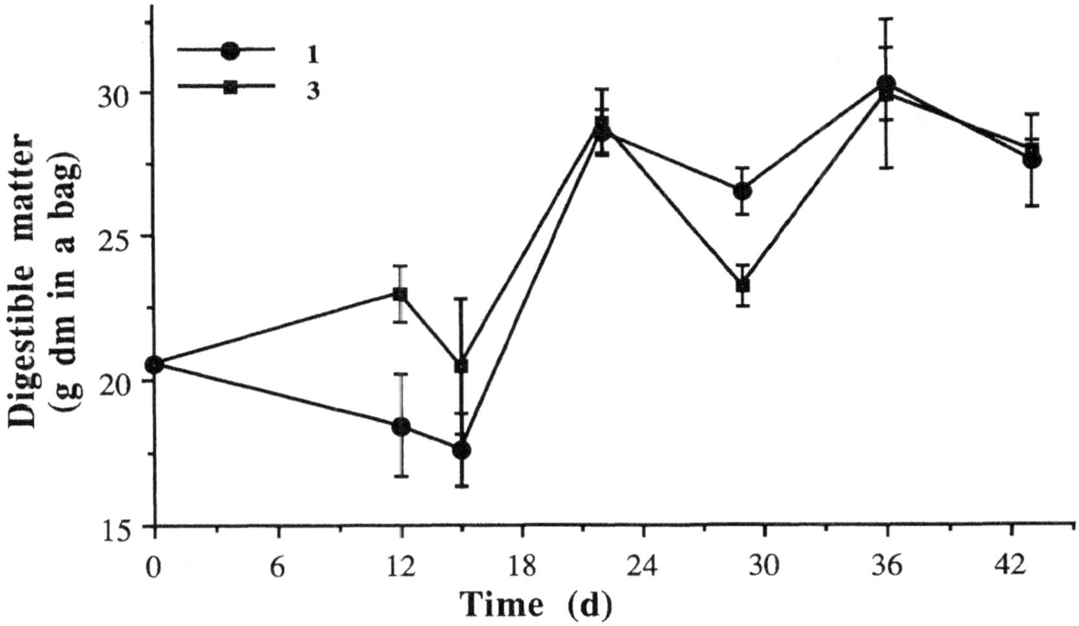

Fig. 6. In vitro digestibility of ground cotton stalks with 70% (1) and 75% (3) water content during SSF of *Pleurotus*. Bars represent standard errors.

is time-consuming, only treatments 1 and 3 were studied in detail. Results are shown as residual lignin and cellulose (Fig. 5). Lignin degradation patterns were similar in both treatments, as were those of cellulose. During the first 15 days, lignin degradation rates were highest, whereas cellulose degradation accelerated from day 22. Thus, after 21 days of fermentation, the amount of degraded lignin was higher than that of cellulose. *In vitro* digestibility of the substrate was measured for treatments 1 and 3 (Fig. 6). Results are presented as dry weight of digested matter per bag. Digested matter in treatment 1 (70% water) decreased for the first 15 days, increased thereafter until day 22 and then stabilized until the end of the experiment. The level of digested matter in both treatments peaked on days 22 and 36. On those days, the amounts of digested matter per bag were found to be similar in both treatments. It was concluded that fermentation for 22 to 36 days produces the highest levels of digestibility, rendering the fermentation product suitable for use as animal feed. Day 22 represents the lowest percentage of digested matter, while day 36 represents the longer fermentation period and a higher ash content due to overall organic matter degradation by the fungus.

Chemical analyses of the substrates from the five treatments were therefore conducted on days 22 and 36. Acid insoluble fibers, cellulose, lignin and ash contents, are presented in Table 1, along with *in vitro* digestibility and the amount of residual dry substrate per bag. Since treatments 2 and 5 were supplemented with nutrients, their initial composition differed from that of the other three treatments, leading to higher ash and nitrogen contents in the fermentation products.

During the first 22 days, a 5% decrease of the fibers was observed in treatments 1–4 and a 10% decrease in treatment 5. An additional 4% decrease in treatment 1 and 1% in treatment 2 were observed during a further 15 days of fermentation. Cellulose content decreased slightly in the five treatments throughout the fermentation period, at a rate similar to that of overall organic matter degradation. A decrease of 5–6% in substrate lignin content was observed in treatments 1, 3, 4 and 5, 3% in treatment 2, during the first 22 days. On day 36, lignin content reached levels of between 15.7

and 17.5%, with the lowest level being recorded in treatment 4. *In vitro* digestibility was increased by 12–15% in all treatments during the first period, and by an additional 5% during the following two weeks.

Protein content and composition were analyzed on days 22 and 36 in the five treatments (Table 2). Crude protein was determined by the Kjeldahl method. An increase in its nitrogen content shows that no crude protein was lost from the system. Furthermore, the content of the true protein fraction, i.e. the amino acid fraction (Table 2) increased beyond the increase in crude protein content, indicating the fungus' ability to decompose nonprotein nitrogen, probably from lignoprotein compounds, and to incorporate the released nitrogen into its own true protein. A source of nonprotein nitrogen in the final product could be chitin, a major component of the fungal cell wall.

Digestibility of the five treated substrates was similar. Treatment 4, requiring the least pre-treatment, was therefore chosen for experiments on a larger scale. One ton of chopped cotton stalks was fermented in an industrial mushroom production plant. The fermentation product was fed to ruminants - goats and heifers, at levels of up to 40% of their diets. No side effects were observed in animals consuming this diet. The *in vivo* nutritional value of this feed remains to be evaluated.

Cotton stalks, harvested using commercial equipment, preserved using silage technology and fermented with *P. ostreatus*, exhibited increased digestibility and could therefore be utilized by ruminants as a dietary supplement.

Table 1. Ash, ADF, cellulose and lignin contents, *in vitro* digestibility and residual dry weight in treated cotton stalks after 0, 22, and 36 days of fermentation by *P. ostreatus*.

Treatment	Ash (%)	ADF (%)	Cellulose (%)	Lignin (%)	*In-vitro* digestibility (%)	Residual dry weight (g/bag)
A – 0 days	6.4	68.2	46.5	22.1	21.8	100
B – 22 days						
1. Ground, 70% water content	8.4	63.4	46.4	17.2	34.4	83
2. Ground, 70% water content + nutrients	9.8	59.3	40.3	19.1	35.1	83
3. Ground, 75% water content	7.7	63.2	45.3	17.6	36.2	80
4. Chopped, 70% water content	8.5	59.4	42.4	16.9	35.4	78
5. Chopped, 70% water content + nutrients	10.4	58.8	42.2	16.3	37.0	81
C – 36 days						
1. Ground, 70% water content	8.9	59.4	41.5	17.5	41.1	73
2. Ground, 70% water content + nutrients	10.3	57.7	40.5	17.4	38.8	80
3. Ground, 75% water content	9.4	62.4	45.2	16.6	41.4	72
4. Chopped, 70% water content	8.3	59.8	44.2	15.7	38.1	73
5. Chopped, 70% water content + nutrients	10.8	57.5	40.6	16.8	40.7	75

Ultrastructural changes in cotton stalks during *Pleurotus* growth

Pleurotus spp. can degrade lignin in a variety of lignocellulosic waste materials (Kamra & Zadrazil 1986; Kerem et al. 1992; Platt et al. 1984). In addition, *Pleurotus* has been found to degrade ^{14}C-labeled DHP to $^{14}CO_2$ (Platt et al. 1983b; Trojanowski & Huttermann 1987) and to decolorize the polymeric dye Poly-B411 (Platt et al. 1985). Even though lignin degradation by white rot fungi has received a great deal of attention over the last two decades, a direct method for following this degradation is not available. One of the techniques which is increasingly being used in studies of the micromorphological changes taking place during the delignification process is scanning electron microscopy (SEM) (Blanchette 1991). Otjen and Blanchette (1986) reviewed variations in wood decay by observing its microscopic appearance. They demonstrated that white rot fungi cause several recognizable patterns of cell wall decomposition, influenced by the host cell type and the fungal species. With the aid of SEM they distinguished between three decay patterns: a white pocket rot, a mottled decay pattern and a uniform white rot. Blanchette (1984) identified selective lignin degradation in wood in his study of 26 white rot fungi. SEM was also used by Agosin et al. (1990) to characterize ultrastructural changes in the Palo Podrido. Ultrastructural changes during lignin degradation by various white rot fungi in different lignocellulosic materials such as sugar cane bagasse and wheat straw, have also been reported (Agosin et al. 1990; Agosin et al. 1987; Johnsrud et al. 1987).

We followed the pattern of lignin degradation in cotton stalks during SSF. Sterile cotton stalks were inoculated with *P. ostreatus* and incubated for six months. Each week a sample was prepared as follows: radial and tangential sections of the cotton stalks were mounted on brass stubs and fixed by vapors of osmium oxide (OsO_5) and glutaraldehyde.

A radial section of sterile cotton stalk which had not been incubated with *P. ostreatus* (control) is shown in Fig. 7. There are no signs of lignin degradation in the domains where lignin is abundant. In cotton stems lignin is distributed throughout the cell wall layers. Greater quantities of lignin are found in the tracheal walls, the middle lamella and between the pith and the xylem. The xylem layers (Fig. 7-X) are intact to the pith (Fig. 7-P) and there is no degradation of the middle lamella. A tangential section after one week of incubation with *P. ostreatus* clearly shows fungal hyphae inhabiting the meristematic cells (Fig. 8). A whole radial section of a cotton stalk, incubated with *P. ostreatus* for two weeks, can be seen in Fig. 9. The pith is detached from the xylem and many holes appear due to lignin degradation, since this is a tissue

Table 2. Amino acid and crude protein content of cotton stalks inoculated with *P. ostreatus* after 0, 22 and 36 days of fermentation.

Treatment	Total amino acids (mg/g substrate)	Essential amino acids (mg/g substrate)	Crude protein (mg/g substrate)
A – 0 days	22.7	10.6	64.5
B – 22 days			
1. Ground, 70% water content	26.9	13.6	79.9
2. Ground, 70% water content + nutrients	33.0	16.4	87.1
3. Ground, 75% water content	28.4	13.5	77.7
4. Chopped, 70% water content	31.7	15.7	66.8
5. Chopped, 70% water content + nutrients	34.6	17.2	83.2
C – 36 days			
1. Ground, 70% water content	34.2	17.3	85.6
2. Ground, 70% water content + nutrients	31.9	15.5	88.7
3. Ground, 75% water content	27.4	13.7	95.1
4. Chopped, 70% water content	25.4	12.4	70.8
5. Chopped, 70% water content + nutrient	35.0	17.9	85.0

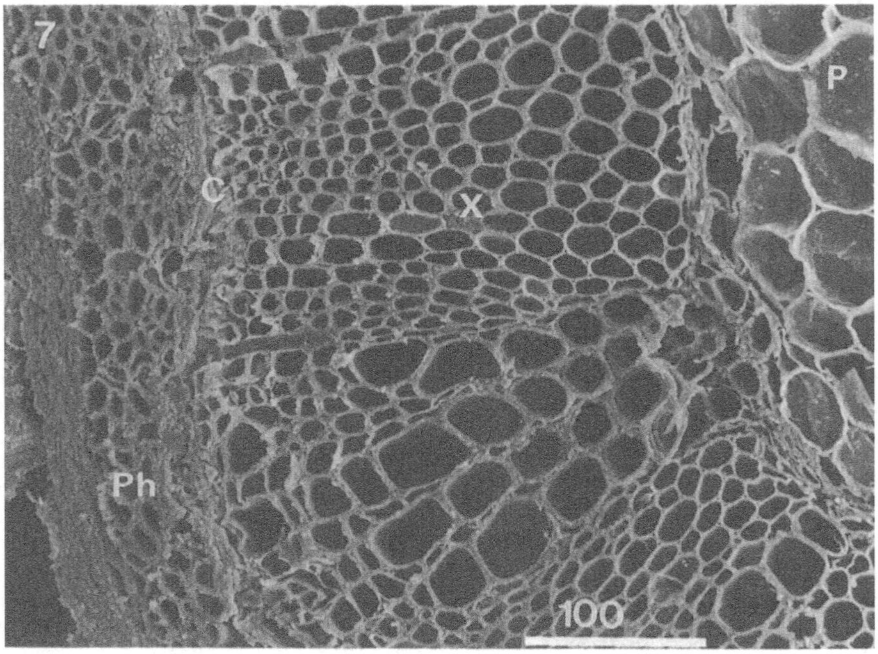

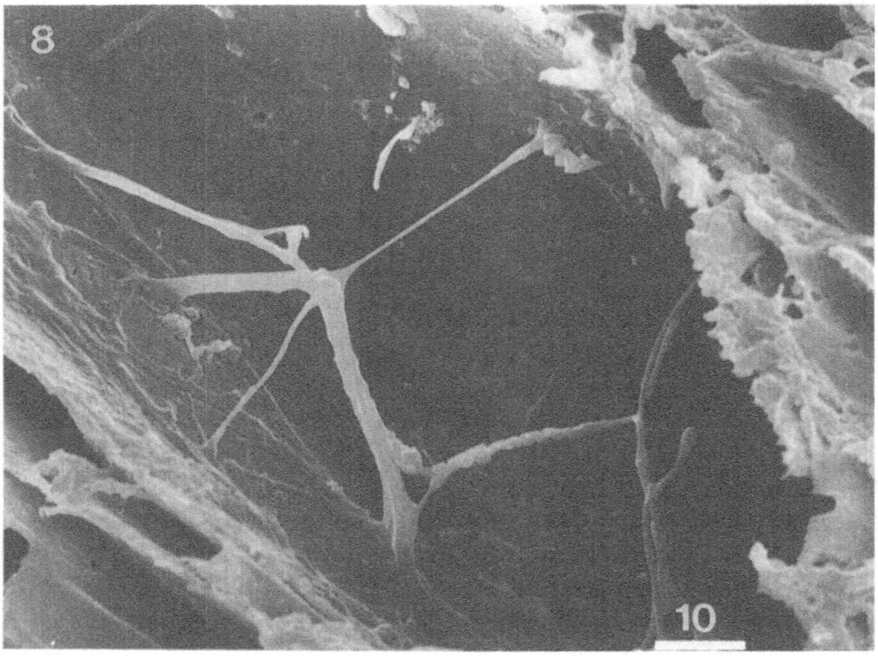

Fig. 7. Radial section of a control cotton stalk demonstrating the different areas, which have not been degraded. P – pith ; X – xylem ; C – cambium ; Ph – phloem.
Fig. 8. Tangential section of a cotton stalk , after one week of incubation with *P. ostreatus*. The fungal hyphae populate the cells and grow through them.

which is rich in lignin. Figure 10 shows radial sections in which areas of delignification in the middle lamella can be observed. Cells are separated from each other and degradation of the cell walls is apparent. Another domain of lignin deposition is in the tracheal cell walls. A tangential section of the trachea cells, after a three-week incubation with *P. ostreatus* (Fig. 11), shows areas of complete decay. Fungal hyphae populate the cells and are destroying the tracheal cell wall (Fig. 11-DT). Figure 12 is a radial section of a cotton stalk after six months of incubation with *P. ostreatus*. Most of the lignin has been degraded and the cells have lost their shape and collapsed.

These observations provide us with a record of the sequence of events occurring during *P. ostreatus* colonization on cotton stalks. A section of a uninoculated cotton stem shows a complete structure with no signs of delignification. After one week of incubation, fungal hyphae were colonizing the cells and lignin degradation was taking place. After the third week delignification had proceeded to the lignin-rich tissues. The pith was detached from the xylem and the trachea cells were degraded. The same phenomenon was also observed in the middle lamella. These observations are in agreement with results of chemical analyses of fermented cotton stalks and with the rate of [^{14}C]lignin mineralization (Kerem et al. 1992).

Enzymes involved in lignin degradation by *Pleurotus* spp.

Extracellular enzyme activity related to lignin degradation has been reported by Platt et al. (1984) and Kerem et al. (1992) during SSF of cotton stalks by *Pleurotus*. The major activity was that of laccase, peaking between days 6 and 8. Similar trends were observed by Sharma (1987), who followed *Pleurotus* activities on flax shive and by Kannan et al. (1990), growing *P. sajor-caju* on paper-mill sludge. Sannia et al. (1986) purified and partially characterized laccase from *P. ostreatus*. Lignin peroxidase, the most widely studied enzyme in relation to lignin degradation, was not detected in *Pleurotus* cultures. This observation is in agreement with others who were unable to detect lignin peroxidase in *Pleurotus* at the protein level using antibodies (Waldner et al. 1988) or by using DNA probes (Kimura et al. 1990). The importance of laccase in cleaving phenolic β-0-4 lignin substructure model compounds has been demonstrated by Higuchi (1990) who suggested that the same chemical principle, i.e. a phenoxy radical as intermediate, is involved in the degradation of phenolic substructure model compounds by both lignin peroxidase and laccase. Platt et al. (1984) suggested that some correlation exists between the level of laccase activity and lignin degradation. *P. ostreatus* 'florida' activities were compared to those of *P. ostreatus* 'P3' (Table 3). The results demonstrated a possible relationship between lack of laccase activity and the rate and mode of lignin degradation, DHP mineralization or Poly-B411 decolorization. Nevertheless, laccase is not the only enzyme responsible for lignin degradation in the genus *Pleurotus* since both strains were capable of degrading lignin.

Acidic peroxidase, neutral peroxidase laccase and veratryl alcohol oxidase activities were investigated in cultures of *P. sajor-caju* (Fukuzumi 1987). The crude extracellular enzymes reduced the brown color of chlorinated oxi-lignin.

Bourbonnais and Paice (1988) found two veratryl alcohol oxidizing enzymes in the culture medium of *P. sajor-caju*. 4-Methoxybenzyl alcohol was oxidized the most rapidly, followed by veratryl alcohol. Not all aromatic alcohols were oxidized. Oxidases may play a role in biodegradation by producing H_2O_2 during the oxidation of lignin fragments (Bourbonnais & Paice 1988). These findings are supported by those of Guillen et al. (1990) who found H_2O_2 production by aryl alcohol oxidase from *P. eryngii*. The most rapidly oxidized substrates were anisyl alcohol and veratryl alcohol. It seems that the enzymes described in these two reports have similar substrate preferences.

A similar enzyme was discovered in cultures of *P. ostreatus* (Sannia et al. 1991) and purified. The enzyme is a glycoprotein containing FAD as a prosthetic group. Cinnamyl alcohol was the most rapidly oxidized substrate. Sannia et al. (1991) also suggested that veratryl alcohol oxidase plays a central role in the biodegradation of lignin by *Pleuro-*

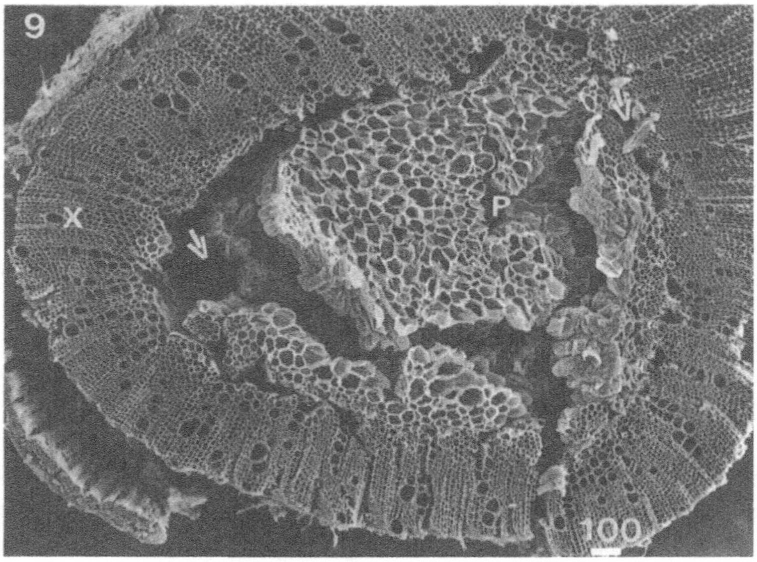

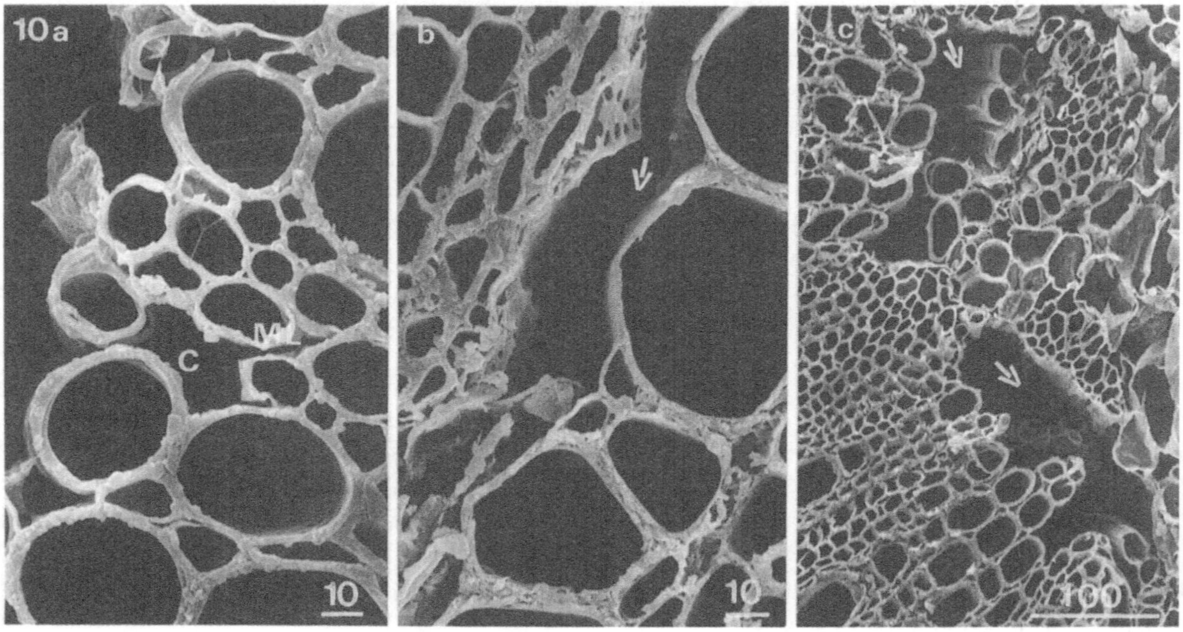

Fig. 9. Radial section of a cotton stalk, two weeks after inoculation, shows clear signs of degradation. The pith is detached from the xylem layers. P – pith ; X – xylem ; → – areas of degraded lignin.

Fig. 10. Radial sections of a cotton stalk, three weeks after inoculation. In all three sections there are clear domains of degraded lignin in the middle lamella. ML – degraded areas of the middle lamella ; C – complete degradation of the cell wall ; → – degraded areas.

tus. Kim et al. (1986) demonstrated the activity of polyphenol oxidase in *P. ostreatus*, which is able to oxidize dihydroxyphenylalanine (DOPA). Enzyme activity was evident only after induction with phenolic compounds, such as ferulic and gallic acids.

The mechanism of lignin degradation by *Pleurotus* has been studied significantly less than those of *Phanerochaete chrysosporium* and some other

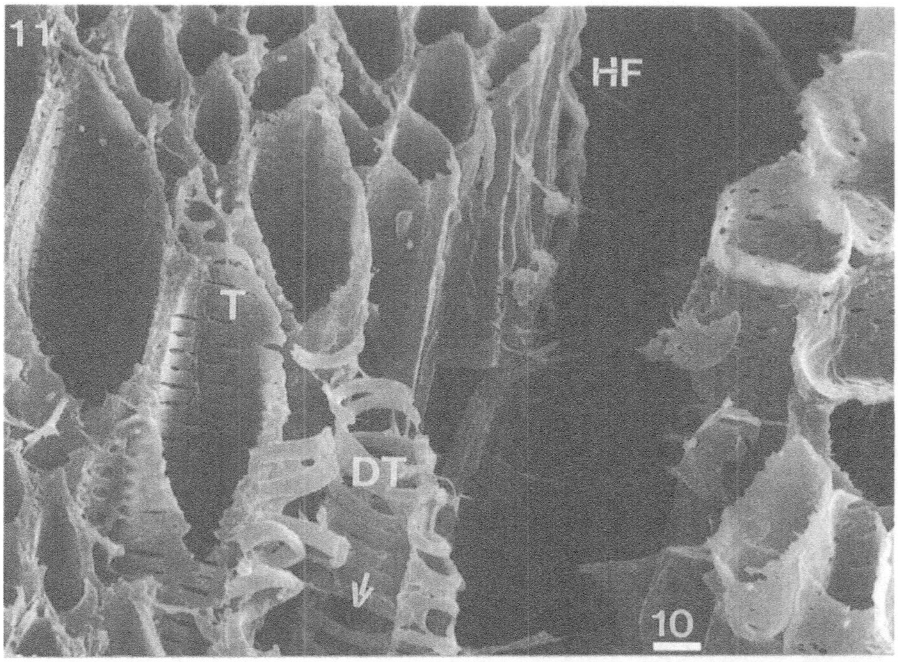

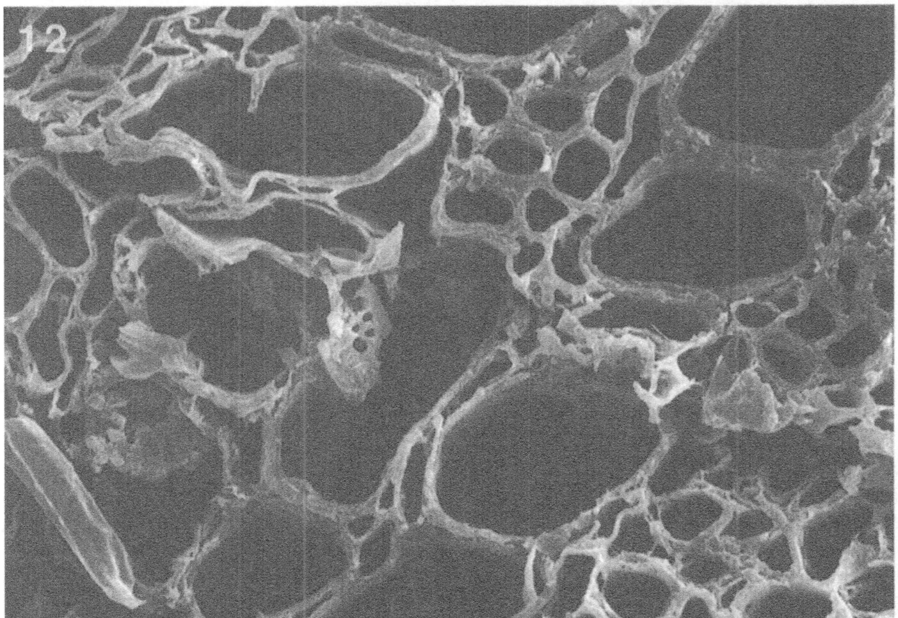

Fig. 11. Tangential section of trachea cells in a cotton stalk, three weeks after inoculation. HF – fungal hyphae which are seen in a degraded area ; → – degraded areas ; T – undegraded trachea cell wall ; DT – degraded trachea cell wall.
Fig. 12. Radial section of a cotton stalk after six months of incubation with *P. ostreatus*, revealing complete destruction of the cells' shape. The cell walls have collapsed, having no lignin to support them.

white rot fungi. From studies conducted to date however, a different enzymatic system seems to be responsible for lignin degradation in *Pleurotus*.

The effect of cotton stalk extract on *Pleurotus* growth and activity

Growth and some activities of some *Pleurotus* strains have been found to be enhanced by the phenolic constituents found in crude water extracts of some plants. Zadrazil (1975) reported the metabolism of flavonoid-type phenolic compounds existing in straw substrates by *Pleurotus* spp. in the early stages of degradation. Sharma (1987) reported that flax shive extract enhanced degradation of flax shive by four strains of *Pleurotus*. The extract also induced a significant increase in the number of primordia produced by these fungi. The infrared spectrum of a chromatogram of flax shive extract showed the presence of a flavonoid-type compound in the aqueous extract of the flax shive.

Platt et al. (1981, 1983a) showed that *P. ostreatus* 'florida' grows faster on cotton straw than on other substrates. After 21 days, 17.1% of the initial dry weight was reduced by the fungi, as compared to a 10.8% reduction in the dry weight of wheat straw. The addition of crude aqueous cotton stalk extract to wheat straw increased its degradation by 33%. When the fungus was grown on water-extracted wheat straw, degradation of both lignin and straw was slower than in the native straw. The crude extract also induced high laccase activity, similar to activity found after induction with known phenolic compounds (Platt et al. 1984). The active fraction in the extract was characterized as a flavonone or a dihydroxyflavonol-type flavonoid (Platt et al. 1983a).

When cotton stalk extract was incorporated into a solid synthetic growth medium, the linear growth rate of the fungal colonies increased, by 14% for *P. ostreatus* 'florida' F6, 20% for *P. ostreatus* IMI 341688, 14.3% for *P. salmoneo stramineos* and 17.6% for *P. pulmonarius* P3014. Other white rot fungi were also studied: Growth rate of *Ganoderma applanatum* increased by 17%, and *Phlebia tremellosus* by 12%. Other fungi, such as *Trametes versicolor* and *Gleophyllum striatum*, were not affected, whereas in the case of *L. edodes*, a 9% inhibition of growth was observed. A similar trend was noted when fungal biomass grown in stationary liquid cultures was monitored. Laccase activity was studied in liquid medium amended with cotton stalk extract. In preliminary studies, all strains tested showed different levels of laccase induction, ranging from a three to fivefold increase in activity

Table 3. Fungal activities and lignin degradation by *P. ostreatus*.

Activity	*P. ostreatus* 'florida' F6	*P. ostreatus* P3	Reference
Cotton degradation (%)	17	9	Platt et al. (1984)
Lignin degradation (%)	56	10	Platt et al. (1984)
Laccase activity (OD/ml/min)	0.86	0	Platt et al. (1984)
Poly-B411 decolorization: absorbance (486nm:553nm) ratio change in 7 days (initial ratio = 3)	0.95	2.9	Platt et al. (1985)
Release of $^{14}CO_2$ from ^{14}C-ring DHP (% of initial)	7	0.7	Platt et al. (1983b)
	7.5	4	Trojanowski & Hüttermann (1987)
Release of $^{14}CO_2$ from DHP-O-$^{14}CH_3$ (% of initial)	21	9	Platt et al.(1983b)
	18.1	7.5	Trojanowski & Hüttermann (1987)
$^{14}CO_2$ release from ^{14}C organosolve lignin (% of initial)	9.4	5	Trojanowski & Hüttermann (1987)
^{14}C-water solubles in culture (% of initial)	35.3	11.5	Trojanowski & Hüttermann (1987)

level within 6 h. *T. versicolor* and *G. applanatum* showed a similar increase, while other laccase-producing fungi, such as *Rhizoctonia solani*, were not affected by the added extract (Ardon & Hadar, unpublished).

Mineralization of [^{14}C]lignin by *P. ostreatus* in the presence of cotton stalk extract was studied. A twofold increase was observed after two weeks of growth, changing only slightly thereafter until the end of the experimental period (75 days). Forty-one and 47% of the total radioactivity was released as $^{14}CO_2$, from the control and the treated substrate, respectively. These results suggest a change in fungal lignin metabolism, affected by the cotton stalk extract. The role of plant extracts in lignin degradation by *Pleurotus* remains to be elucidated.

Conclusions

The utilization of lignocellulosic agricultural wastes as well as forest products via SSF by white rot fungi could be applied in several industries (Buswell & Odier 1987). Its major potential advantages are the upgrading of under utilized resources (agricultural waste) and the introduction of environmentally sound biotechnologies to the pulp and paper industry.

Such technologies could be introduced in several areas:

1. delignification of wood, straw and bagasse to increase digestibility by ruminants, and for enzymatic hydrolysis of cellulose;
2. delignification of wood to reduce energy consumption and the utilization of harmful chemicals in pulping (biopulping);
3. delignification, bleaching and modification of pulp fibers (biobleaching);
4. modification of lignin to produce useful chemicals;
5. treatment of waste bleach water to reduce color and toxicity;
6. treatment of soil or waste water to remove toxic pollutants such as PCB and DDT.

Pleurotus degrades lignin efficiently and selectively, as evidenced by electron microscopy and chemical analyses. Our study on its SSF of cotton stalks suggests the usefulness of the fermentation product as a supplement to ruminant diets. Since *Pleurotus* has been cultivated commercially for human consumption for decades, it can be considered safe for animals as well. This conclusion is in line with findings on SSF of *Pleurotus* on wheat straw (Zadrazil & Reinger 1988) and other wastes.

The mechanism of lignin degradation by *Pleurotus* should be targeted for future research. The enzymes involved seem to be laccase and aryl alcohol oxidases. However, related activities need to be investigated before any definitive conclusions can be drawn.

References

Agosin E, Rouau X & Brillouet JM (1987) Fermentation of wheat straw xylan by the white rot fungus *Dichomitus squalens*. Can. J. Microbiol. 33: 1050–1054

Agosin E, Blanchette RA, Silva H, Lapierre C, Cease KR, Ibach RE, Abad AR & Muga P (1990) Characterization of Palo Podrido, a natural process of delignification in wood. Appl. Environ. Microbiol. 56: 65–74

Balasubramanya RH (1981) An edible mushroom crop on cotton stalks. Indian Soc. Cotton Imp. J. 6: 104–105

Balasubramanya RH, Pai DY, Saikh AJ & Khandeparker VJ (1989) Biological softening of spent cotton-plant stalks for the preparation of pulp. Biol. Wastes 30: 317–320

Ben-Ghedalia D, Shefet G & Miron J (1980) Effect of ozone and ammonium hydroxide treatments on the composition and in vitro digestibility of cotton straw. J. Sci. Food Agric. 31: 1337–1342

Ben-Ghedalia D, Shefet G & Dror Y (1983) Chemical treatments for increasing the digestibility of cotton straw. J. Agric. Sci. 100: 393–400

Blanchette RA (1984) Screening wood decayed by white rot fungi for preferential lignin degradation. Appl. Environ. Microbiol. 48: 647–653

Blanchette RA (1991) Delignification by wood decay fungi. Annu. Rev. Phytopathol. 29: 381–398

Bourbonnais R & Paice MG (1988) Veratryl alcohol oxidases from the lignin-degrading basidiomycete *Pleurotus sajor-caju*. Biochem. J. 255: 445–450

Buswell JA & Odier O (1987) Lignin biodegradation. Crit. Rev. Biotechnol. 6: 1–60

Danai O, Levanon D & Silanikov N (1989) Cotton straw silage as a substrate for *Pleurotus* sp. cultivation. Mush. Sci. 12: 81–90

Falck R (1917) Uber die Waldkultur des Austernpilzes (*Agar-*

icus ostreatus) auf laubholzstubben. Zeitschrift fur Forst-und Jagdwesen 49: 159–165

Fukuzumi T (1987) Ligninolytic enzymes of *Pleurotus sajor-caju*. Coll. INRA 40: 137–142

Guillen F, Martinez AT & Martinez AJ (1990) Production of hydrogen peroxide by aryl-alcohol oxidase from the ligninolytic fungus *Pleurotus eryngii*. Appl. Microbiol. Biotechnol. 32: 465–469

Guzman G & Martinez D (1986) *Pleurotus* growing on coffee pulp in semi-industrial plant - a new promising mushroom cultivation technology in the subtropics of Mexico. Mushroom Newsletter for the Tropics 6: 7–10

Higuchi T (1990) Lignin biochemistry: biosynthesis and biodegradation. Wood Sci. Technol. 24: 23–63

Johnsrud SC, Fernandez P, Lopez I, Gutierrez A, Saez A & Eriksson KE (1987) Properties of fungal pretreated high yield bagasse pulps. Nord. Pulp Paper Res. J. 2: 47–52

Kamra DN & Zadrazil F (1986) Influence of gaseous phase, light and substrate pre-treatment on fruit-body formation, lignin degradation and in vitro digestibility of wheat straw fermented with *Pleurotus* spp. Agric. Wastes 18: 1–17

Kannan K, Oblisami G & Loganathan BG (1990) Enzymology of ligno-cellulose degradation by *Pleurotus sajor-caju* during growth on paper-mill sludge. Biol. Wastes 28: 1–8

Kelley RL (1988) Ligninolytic activity of *Phanerochaete chrysosporium* measured as ethylene production from α-keto-γ-methylthiolbutyric acid. In: Wood AW & Kellog ST (Eds) Biomass: Lignin, Pectin and Chitin, Vol 161b (pp 79–82). Academic Press, San Diego, CA

Kerem Z, Friesem D & Hadar Y (1992) Lignocellulose degradation during solid-state fermentation: *Pleurotus ostreatus* versus *Phanerochaete chrysosporium*. Appl. Environ. Microbiol. 58: 1121–1127

Kim KJ, Shin KS & Hong SW (1986) Induction of extracellular polyphenol oxidase from two white rot fungi. Korean J. Mycol. 14: 43–47

Kimura Y, Asada Y & Kuwahara M (1990) Screening of basidiomycetes for lignin peroxidase genes using a DNA probe. Appl. Microbiol. Biotechnol. 32: 436–442

Kirk TK & Farrell RL (1987) Enzymatic 'combustion': the microbial degradation of lignin. Ann. Rev. Microbiol. 41: 465–505

Leon RD, Morales E, Agreda LD & Rolz C (1983) Coffee byproducts and citronella bagasse as substrates for *Pleurotus* production. Mushroom Newsletter for the Tropics 4: 13–16

Levanon D, Danai O & Masaphy S (1988) Chemical and physical parameters in recycling organic wastes for mushroom production. Biol. Wastes 26: 341–348

Lindenfelser LA, Detroy RW, Ramstack JM & Worden KA (1979) Biological modification of the lignin and cellulose components of wheat straw by *Pleurotus ostreatus*. Dev. Indust. Microbiol. 20: 541–551

Monties B (1989) Lignins. In: Harborne JB (Ed) Plant Phenolics, Vol 1 (pp 113–157). Academic Press Ltd., London

Muller J (1987) Cultivation of the oyster mushroom, *Pleurotus ostreatus* (Jack. ex Fr.) Kummer, on cassia-substrate. Mush. J. Tropics 7: 89–95

Oriaran TP, Laboski P Jr. & Royse DJ (1989) Lignin degradation capabilities of *Pleurotus ostreatus*, *Lentinula edodes* and *Phanerochaete chrysosporium*. Wood Fiber Sci. 21: 183–192

Otjen L & Blanchette RA (1986) A discussion of microstructural changes in wood during decomposition by white rot basidiomycetes. Can. J. Bot. 64: 905–911

Pandy SN & Shaikh AJ (1987) Utilization of cotton plant stalk for production of pulp and paper. Biol. Waste 21: 63–70

Platt MW, Chet I & Henis Y (1981) Lignocellulose degradation during growth of the mushroom *Pleurotus* sp. 'florida' on cotton straw. Eur. J. Appl. Microbiol. Biotechnol. 13: 194–195

Platt MW, Hadar Y & Chet I (1984) Fungal activities involved in lignocellulose degradation by *Pleurotus*. Appl. Microbiol. Biotechnol. 20: 150–154

Platt MW, Hadar Y & Chet I (1985) The decolorization of the polymeric dye Poly-Blue (polyvinylamine sulfonate-anthroquinone) by lignin degrading fungi. Appl. Microbiol. Biotechnol. 21: 394–396

Platt MW, Hadar Y, Henis Y & Chet I (1983a) Increased degradation of straw by *Pleurotus ostreatus* sp. 'florida'. Eur. J. Appl. Microbiol. Biotechnol. 17: 140–142

Platt MW, Trojanovski J, Chet I & Huttermann A (1983b) Differences in the degradation of specifically ^{14}C-labelled lignin model compounds within the *Pleurotus* genus group. Microbiol. Lett. 23: 19–21

Sannia G, Giardina P, Luna M, Rossi M & Buonocore V (1986) Laccase from *Pleurotus ostreatus*. Biotechnol. Lett. 8: 797–800

Sannia G, Limongi P, Cocca E, Buonocore F, Nitti G & Giardina P (1991) Purification and characterization of a veratryl alcohol oxidase enzyme from the lignin degrading basidiomycete *Pleurotus ostreatus*. Biochim. Biophys. Acta 1073: 114–119

Schnurer J & Rosswall T (1982) Fluoroscein diacetate hydrolysis as a measure of total microbial activity in soil and in litter. Appl. Environ. Microbiol. 43: 1256–1261

Schoemaker HE & Leisola MSA (1990) Degradation of lignin by *Phanerochaete chrysosporium*. J. Biotechnol. 13: 101–109

Sharma HSS (1987) Comparative study of the degradation of flax shive by strains of *Pleurotus*. Appl. Microbiol. Biotechnol. 25: 542–546

Silanikove N & Levanon D (1986) Cotton straw: composition, variability and effect of anaerobic preservation. Biomass 9: 101–112

Streeter CL, Conway KE, Horn GW & Mader TL (1982) Nutritional evaluation of wheat straw incubated with the edible mushroom *Pleurotus ostreatus*. J. Anim. Sci. 54: 183–188

Tautorus TE (1985) Mushroom fermentation. Adv. Biotechnol. Proc. 5: 227–273

Trojanowski J & Huttermann A (1987) Screening of wood inhabiting fungi for their capacity to degrade and to solubilize ^{14}C-labeled lignin. Microbios. 50: 91–97

Waldner R, Leisola MSA & Fiechter A (1988) Comparison of

ligninolytic activities of selected white-rot fungi. Appl. Environ. Microbiol. 29: 400–407

Wood DA & Smith JF (1987) The cultivation of mushrooms. In: Norris JR & Pettipher GL (Eds) Essays in Agricultural and Food Microbiology (pp 309–343). John Wiley & Sons Ltd, New York

Zadrazil F (1975) Die Zersetzung des Stroh-zellulose-Lignin-Komplexes mit *Pleurotus* florida und dessen Nutzung. Z. Pflantzenernahr Bodenk 138: 263–278

Zadrazil F (1978) The cultivation of *Pleurotus*. In: Chang S & Hayes W (Eds) The Biology and Cultivation of Edible Mushrooms (pp 521–524). Academic Press, New York

Zadrazil F (1980) The conversion of different plant wastes into feed by basidiomycetes. Eur. J. Appl. Microbiol. Biotechnol. 9: 243–248

Zadrazil F (1985) Screening of fungi for lignin decomposition and conversion of straw into feed. Angew. Botanik. 59: 433–452

Zadrazil F & Reinger P (1988) Treatments of Lignocellulosics with White Rot Fungi. Elsevier Applied Science Publishers Ltd, London

Biodegradation 3: 207–218, 1992.

Delignification of wood pulp by a thermostable xylanase from *Bacillus stearothermophilus* strain T-6

Yuval Shoham,[1] Zeev Schwartz,[1] Alexander Khasin,[1] Orit Gat,[1] Zinaida Zosim[2] & Eugene Rosenberg[2]
[1] *Department of Food Engineering and Biotechnology, The Technion, Technion City, Haifa 32000, Israel;*
[2] *Department of Molecular Microbiology and Biotechnology, Tel Aviv University, Ramat Aviv, Israel*

Key words: xylanase, biobleaching, thermostable, *Bacillus stearothermophilus*, constitutive mutants

Abstract

During the bleaching of wood pulp for the paper industry, large amounts of chlorinated aromatic compounds are produced and released into the environment. These compounds are extremely toxic and are a major source of pollution. The paper and pulp industry is seeking for alternative methods for bleaching pulp. One such method involves the use of hemicellulases to release the colored lignohemicellulose. We have isolated and characterized several thermophilic bacteria which produce xylanases. One such strain, T-6, produced high levels of extracellular xylanase, free of cellulase and proteinase activities. Strain T-6 was classified as a strain of *Bacillus stearothermophilus* and was able to grow on defined medium containing xylose, methionine and asparagine at 65 °C. Xylanase activity was induced by either xylose or xylan; no activity was detected with other carbon sources, such as glycerol, acetate, lactose, glucose, maltose, fructose, mannose, galactose or sucrose. Xylanase constitutive mutants were obtained following mutagenesis and detection on p-nitrophenol β-D-xylopyranoside containing agar plates. Xylanase T-6 was produced on large scale, and was purified and concentrated by a single adsorption-desorption step from a cation exchanger. The overall purification yield of a 1000 liter fermentation was 45%, resulting in a 98% pure enzyme. Xylanase T-6 was shown to partially remove lignin from unbleached pulp at 65 °C and pH 9.0, without loss in pulp viscosity. The enzyme-treated pulp was used to make handsheets that had higher brightness than untreated pulp.

Introduction

One of the major sources of environmental pollution today is the pulp and paper industry. During the process of making white paper pulp, large amounts of toxic effluents are released into the environment. Although the major raw material of this industry, wood, is a biodegradable organic material, it is only in the last decade or so that the power of biotechnology has been applied to solve problems in the pulp and paper industry (Eriksson & Kirk 1985; Hakulinen 1988; Jurasek & Paice 1986, 1988; Kirkpatrick 1991). To understand the possible applications of biotechnology in the pulp industry concerning pollution, we would like first to briefly summarize some of the principle technologies involved in making pulp and paper.

Wood is composed of three major components: cellulose, hemicellulose and lignin. These compounds, respectively, comprise about 35–50%, 20–39% and 20–30% of the dry weight of woody plants (Browning 1963). Cellulose is a linear macromolecule consisting of anhydro-β-glucopyranose (glucose) units joined together through β-1,4 glycosidic linkages. Cellulose molecules have a fully extended flat conformation and they are usually present in tightly packed microfibrils that form insoluble crystalline material. Hemicellulose is composed of linear and branched polysaccharides, that can be either hetero- or homopolymers made of various five or six carbon sugars. These sugars include, for example, xylose, arabinose, mannose and galactose; the homopolymer saccharides for these sugars would be xylan, arabinan, mannan and galactan, respectively. Xylan is more flexible than cellulose, and the xylan backbone can be substituted with arabinose, glucuronic acid and methylglucuronic acid. Hemicelluloses are bound through hydrogen bonds to cellulose and by covalent bonds to lignin. Lignin, the third component of wood, is an amorphous, three-dimensional, aromatic polymer composed of oxyphenylpropane units. The three main precursors of lignin are coumatyl alcohol, coniferyl alcohol and sinaphyl alcohol which are polymerized randomly forming a variety of carbon-carbon and ether linkages. The biological role of lignin in living plants is to form, together with celluloses and hemicelluloses, a tissue with enhanced strength and durability against physical and biological degradation. It is the complex, random structure of lignin that allows it to hold and protect the cellulose and hemicellulose fibers.

Wood is converted to pulp for the manufacture of paper and paperboard. Paper, in essence, is a two-dimensional cross-linked cellulolytic fibers that form random hydrogen bonds. In wood, the cellulose fibers are attached to lignin and hemicellulose. Therefore, the cellulose must first be released to give separate fibers that can form intermolecular hydrogen bonds. The process of fiber separation, called pulping, results in paper pulp which is a slurry of wood fibers. Pulping procedures can be divided into four main classes: mechanical pulping, chemical pulping, semichemical pulping and chemimechanical pulping, where the last two procedures are, in essence, combinations of the first two (Sanyer & Chidester 1963). In mechanical pulping, fiberilization is accomplished by grinding bolts of wood with a grindstone, or, by disk milling of chipped wood. The separation of the fibers takes place entirely by mechanical means producing high yield pulp. In this procedure, lignin modification and removal is negligible and the yield of the pulp is about 95%. In chemical pulping the wood is treated with harsh chemicals at high temperature and pH values which degrade and solubilize most of the lignin (delignification). The most common processes are sulfate (Kraft), soda and sulfite pulping. In the Kraft process, wood chips are cooked in a solution of $Na_2S/NaOH$ at about 170 °C for two hours resulting in degradation and solubilization of lignin (Sanyer & Chidester 1963). The lignin in the chemical pulping process is extensively modified, and 60% to 90% of it is removed from the wood. The pulp yield varies from 45 to 60% of the wood weight.

In all of the pulping processes, the resulting pulp has a characteristic brown color which is primarily due to the residual lignin and lignin derivatives. The intensity of the pulp color is a function of the amount and chemical state of the remaining lignin. To obtain white and bright pulp suitable for the manufacture of printing and tissue grade papers, it is necessary to bleach the pulp and to remove most of the constituents of the unbleached pulp, such as lignin and its degradation products, resins and metal ions (Singh 1979). The methods of bleaching pulp with high lignin content (mechanical pulp), differ fundamentally from those of bleaching chemical pulps. Bleaching mechanical pulps is usually conducted in such a way as to avoid loss of lignin. Bleaching agents such as hydrogen peroxide and sodium hydrosulfite are used. The bleaching of mechanical pulp, which improves the eye-appeal of the pulp, does not yield a high brightness paper with stability against brightness reversion on aging. To obtain pulp of a very high brightness and brightness stability, all of the lignin must be removed from the pulp. For this reason, chemical pulping (where most of the lignin is removed during the cooking process), is the method of choice for producing high quality white paper. The residual lignin in chemical pulp is very dark in color because it has been extensively oxidized and modified in the cooking process. This residual lignin is difficult to remove due to the fact that it is covalently bound to the hemicellulose and perhaps to cellulose fibers. Covalent lignin-carbohydrate linkages in wood consist mainly of ester and ether linkages through sugar hydroxyl groups to the a-carbonyl of phenylpropane subunits in lignin (Eriksson et al. 1980; Jeffries 1990). In addition, unnatural covalent bonds between lignin and carbohydrates are thought to form during the chemical pulping process (Iversen & Waennstroem 1986). The process of bleaching chemical pulp is the main source of pollution in the paper industry (Trubacek & Wiley 1979). The bleaching can be regarded as a purification process involving the destruction, alteration or solubilization of the lignin, colored organic matters, and other undesirable residues on the fibers (Singh 1979).

The traditional bleaching of chemical pulp is performed in a number of stages, known as a bleaching sequence. It has become common practice to use a single letter as a symbol for a particular stage. Some of the symbols and the stages which they represent are, for example: C – chlorination; E – alkaline extraction; H – hypochlorite bleaching; D – chlorine dioxide bleaching; P – peroxide bleaching; O – oxygen bleaching; A – acid treatment; Z – ozone bleaching. The exact choice of the bleaching sequence depends upon the type of pulp to be bleached and the level of brightness to be achieved (Singh 1979; Sanyer & Chidester 1963). The most relevant step regarding pollution is chlorination. Chlorination of pulp, as it is usually carried out, does not show any decoloring effect, and in fact, the color of the pulp may increase with chlorination (Loras 1980). The dominant role of chlorine in bleaching is to convert the residual lignin in the pulp to water or alkali soluble products (bleaching of chemical pulps to a higher brightness without complete removal of lignin has not been successful so far). At low pH the main reaction of chlorine is chlorination rather than oxidation. That is, chlorine selectively chlorinates and degrades lignin compounds rather than the carbohydrates moieties in the unbleached pulp. It is for this reason that chlorine is so effective in the bleaching process. Following the chlorination step, the chlorinated lignins can be efficiently removed from the pulp by alkaline extraction. Under basic conditions, the phenolic compounds are more soluble and, in addition, at high pH, further degradation of chlorinated lignins takes place. Other treatments that usually follow the chlorination and alkaline extraction stages include bleaching with hypochlorite, chlorine dioxide, peroxide and oxygen. These agents are oxidative in nature and perform true bleaching of the pulp; however, they are not as selective as chlorine and can react with the polysaccharides. Today, modern Kraft pulp bleaching is carried out in five or six stages. Common bleaching sequences are, for example, CEDED, CEHDED and CEDEDP.

The effluents that are produced during the bleaching process, especially those following the chlorination and the first extraction stages, are the major contributors to waste water pollution from the pulp industry (Trubacek & Wiley 1979; Eriksson & Kirk 1985). These effluents, which contain large amounts of chlorinated phenols and Kraft lignin, are usually treated biologically by processes such as activated sludge or aerated lagoons. However, some of these harmful chemicals pass the treatment plant and find their way into the recipient water way. A study presented by Larsson et al. (1988) showed that bleached Kraft mill effluents, disposed into the sea, exert biological effects on fish, caught even 10 km from the pulp plant. Typical symptoms were: reduced gonad growth, enlarged liver, strong induction of enzymes in the hepatic mixed function oxidase system, disturbed ion balance and marked effects on the red and white blood cell pattern.

Another aspect of bleaching pulp with chlorine involves the paper itself. Following the bleaching process, some of the chlorinated compounds are likely to be associated with the paper. The most worrying chemical is the animal carcinogen dioxin, 2,3,7,8-tetrachlorodibenzo-p-dioxin, which often finds its way into finished paper products. There is some concern, therefore, as to how safe it is to use chlorine-bleached chemical pulp paper for purposes such as baby diapers and food packaging.

The pulp and paper manufactures are testing different approaches to reduce and, if possible, to eliminate the use of chlorine in the bleaching process. While paper mills in the United States have preferred to focus on waste water treatment, European companies, particularly in Sweden, Finland and Germany, have addressed process changes instead (Samdani 1990).

For over a decade, it is recognized that white-rot fungi can remove color out of effluents coming from Kraft pulp mills (Fukuzumi et al. 1977) and degrade ^{14}C-labelled chlorolignins (Lundquist et al. 1977). Based on these findings, several methods have been developed for the treatment of pulp mill effluents. Patented procedures include the MYCOR and the MYCOPOR processes which utilize the white-rot fungus *Phanerochaete chrysosporium*. (Chang et al. 1987; Messner et al. 1989; Eriksson & Kirk 1985; Pellinen et al. 1988a,1988b). Other white-rot fungi that were used for bleaching and purifying bleach plant effluents include: *Coriolus versicolor* (Livernoche 1981); *Phlebia radiata* and *Merlius tremellosus* (Lankinen et al. 1991). For more details on biotechnology in waste water treatments see: Jurasek & Paice 1986; Lankinen et al. 1991; Heinzle et al. 1992; Eriksson & Kirk 1985; Hakulinen 1988.

A different biotechnological approach for reducing or even eliminating the use of chlorine in the bleaching process involves the use of enzymes or fungi for treating the pulp. Two general enzyme categories were considered and tested; ligninases and hemicellulases. Ligninases are produced mainly by white-rot fungi, which are the natural lignin degraders. The complicated biology and chemistry of ligninases had been elucidated in the last years (Kirk & Farrell 1987). Ligninases are peroxidases that catalyze the formation of oxygen radicals from H_2O_2. These oxygen radicals attack and oxidize the aromatic nuclei of lignin and form unstable cation radicals, which decompose through several different reactions (Kersten et al. 1985). Because the mechanism of lignin degradation involves the formation of oxygen radicals, ligninases are nonspecific in the substrate they attack. For example, Kraft lignin is readily metabolized by white-rot fungi (Lundquist et al. 1977; Eriksson & Kirk 1985).

Kraft pulp can be bleached directly with white rot fungi by incubating the fungi with the pulp for several days (Kirk & Yang 1979; Tran & Chambers 1987; Paice et al. 1989; Fujita et al. 1991). Kirk & Yang (1979) achieved with *Phanerochaete chrysosporium* a reduction of 50–75% in Kappa number (a measure of lignin content) during 6–8 days of incubation. Paice et al. (1989) reported that hardwood Kraft pulp can be bleached by the fungus *Coriolus versicolor*. Following a five-day treatment, the pulp brightness increased by 15 points (ISO) and the Kappa number decreased from 11.6 to 7.9. The bleaching effect of *C. versicolor* appeared to be limited to hardwood since experiments with spruce (softwood) pulp failed to show a bleaching effect. However, Reid et al. (1990) recently reported the use of this fungus for bleaching softwood Kraft pulp. Biobleaching directly with fungi is probably not a viable approach in industry for several reasons. Among them are the time required for treatment and the need for nutrients for the growth of the fungi.

The use of purified or crude preparation of ligninases was also considered and, in fact, a US patent has been published, claiming that this enzyme was able to bleach Kraft pulp when used with a subsequent alkaline extraction procedure (Farrell 1987). Trials in other laboratories failed to reproduce biobleaching using enzyme(s) from *P. chrysosporium* (Kirkpatrick 1991; Viikari et al. 1987), and the effectiveness of the ligninase treatment is not clear. One possible explanation could be, that following the ligninase treatment the lignin components polymerize back onto the pulp. A recent patent application by International Paper Company claims, that lignolytic enzymes can be used to bleach pulp in a process which maintains low steady-state concentration of hydrogen peroxide (0.001 to 0.1 mM). The hydrogen peroxide concentration may be maintained by *in situ* enzymatic generation, for example, through the action of glucose oxidase on glucose (Enzymatic delignification of lignocellulosic material, European patent application EP 0 406 617 A2; 1991).

Hemicellulases are another type of enzymes that were tested for biobleaching. Hemicellulases comprise a group of enzymes that can hydrolyze the hemicellulose polymer chains at various positions or act on side groups attached to the backbone polymer. These enzymes are produced by various microorganisms, among them are fungal yeast and bacterial species and include, for example: endoxylanases, β-xylosidase, arabinfuranosidase and acetyl-xylanesterase. (Dekker & Richards 1976; Reily

1979; Dekker 1985). Viikari et al. (1986, 1987) were first to demonstrate that hemicellulases can be used to enhance delignification and bleaching. Fungi hemicellulases were used to treat unbleached pine sulfate pulp. Following the enzymatic treatment the pulps were bleached with hydrogen peroxide. In these experiments, although the enzymatic treatment resulted in a reduction in chlorine consumption, there was also an unacceptable loss of yield and viscosity. The yield loss was mainly a result of hemicellulose solubilization, where the viscosity drop was probably a result of contaminating cellulases. Further experiments with purified and cloned hemicellulases indicated that enhanced bleaching can be achieved without yield and viscosity loss (Paice et al. 1988; Farrell 1991; Senior et al. 1988; Clark et al. 1990). Comparison between different hemicellulases indicated that endo-β-xylanases have the major impact on delignification, even in softwood pulp, where mannan is the major component. It was suggested by Kantelinen et al. (1991) that two types of phenomena are involved in the enzymatic pretreatment. The major effect is due to hydrolysis of reprecipitated and readsorbed xylan or xylan-lignin complexes, that are separated during the cooking process. As a result of the enzymatic treatment, the pulp becomes more accessible to the bleaching chemicals. A minor effect is due to the hydrolysis of the residual, non-dissolved hemicellulose. In this latter phenomenon other enzymes beside endoxylanases can be used.

From the number of patents and patent applications filed in the last years, it is evident that hemicellulases, and xylanases in particular, are considered a viable option to facilitate pulp bleaching. Indeed, full scale mill trials of enzyme pre-bleaching are already under way (Grant 1991; Koponen 1991; Viikari et al. 1991a, 1991b). The results from these trials indicate that the enzymatic treatment can reduce chlorine requirement by 30% and allow a brightness of 89% ISO to be reached without chlorine.

Most of the hemicellulases studied so far are active at neutral or acidic pH and their optimum temperature is below 45 °C. Hemicellulases that are active at higher temperatures and basic pH are of great potential since they can be introduced more freely in the different stages of the bleaching line without the need for cooling and pH changes. We have recently isolated several thermophiles that produce thermostable xylanases (Shoham et al. 1993). In this report we describe the characterization of these isolates and a large scale production and bleaching ability of one of these enzymes, xylanase T-6.

Materials and methods

Organisms

Bacillus stearothermophilus T-6 was isolated following an enrichment procedure for bacteria capable of producing extracellular thermostable xylanases (Shoham et al. 1993). Strain T-6 was identified as *B. stearothermophilus* by the NCIMB (National Collection of Industrial and Marine Bacteria, England), and was designated as NCIMB 40221. Fatty acid composition of strain T-6 was determined by Microbial ID Inc., Newark, DE, USA. Strain M-7 is a xylanase constitutive mutant of T-6 isolated in this work.

Growth conditions

Growth media were composed of a basic salt medium (BSM) supplemented with various carbon sources. BSM contained (g/l): K_2PO_4 $7H_2O$, 0.75; KH_2PO_4, 0.15; $MgSO_4$ $7H_2O$, 0.1; $(NH_4)_2SO_4$, 2; MOPS buffer (3-N-morpholino propane sulfonic acid), 10.4; and 1 ml of trace elements (trace elements solution contained (g/l): $CaCl_2$ $2H_2O$, 0.37; $CuSO_4$ $5H_2O$, 0.62; $FeSO_4$ $7H_2O$, 0.60; $MnSO_4$ $4H_2O$, 0.59; $ZnSO_4$ $7H_2O$, 0.42; $CoCl_2$ $6H_2O$, 0.79; Na_2 MoO_4, 0.696; the pH of the solution was adjusted to 2.0 with sulfuric acid). The pH of the medium was adjusted to 7.0 and the carbon sources were added after sterilization. ND medium was BSM with 0.5% xylose and 0.3% vitamin free casamino acids. Growth was carried out in 125 ml flasks containing 25 ml medium and reciprocally shaking (180 strokes/min) in a water bath (Tuttnauer, Jerusalem Israel) at 60 °C.

Large scale purification

A large scale fermentation (2 × 500 l) was carried out with strain M-7 in the fermentation facilities of the Hebrew University, Hadasa, Jerusalem. The growth medium was BSM containing 0.1% urea instead of $(NH_4)_2SO_4$, 0.5% glucose and without MOPS buffer. The pH was maintained between 6.5–7.0 with NaOH and H_2SO_4 and the growth temperature was 60 °C. At the end of the fermentation the cell free supernatant fluid contained about 4.5 xylanase units/ml. Fifteen kilograms of sulphoxyethyl cellulose (SE-52, Whatman, Maidstone, England) were added to the cell free broth (500 l) with gentle mixing for one hour. After allowing the adsorbent to settle, it was collected (about 40 l), washed with 20 mM phosphate buffer, pH 7.0, and resuspended in 60 l of 1.0 M KCl. The adsorbent was removed by filtering through a Whatman No. 3 filter paper, and the enzyme was concentrated by ultrafiltration using a Pellicon Cassette Filter Acrylic Holder (Millipore, Bedford, MA) containing an 8K 5 sq. ft. cassette (Filtron, Clinton MA) (flow rate of 300 ml/min with retentate:filtrate ratio of 1:4).

Enzymatic assays

Xylanase was assayed by mixing an aliquot of appropriately diluted enzyme with 0.25 ml of 2% oat spelt xylan (Sigma, St Louis, Mo), 0.5 ml of 0.1 M phosphate buffer, pH 7, and water to a final volume of 1 ml (xylan was prepared by sonicating a 2% xylan solution for 3 min at output 7 on an Ultrasonic W375 sonicator, Heat System Ultrasonics, Plainview, NY). Aliquots of 0.1 ml were taken out of this mixture and placed in four 16 ml glass tubes. Two tubes served as time zero control and were kept at room temperature or ice, the two other tubes were placed in a water bath at 65 °C for 10 or 15 min. The reaction was terminated by placing the tubes in a water bath at room temperature. The reducing sugars content in the tubes was determined by the DNSA method (Miller 1959), with D-xylose as a standard (0.004% xylose was added to the DNSA reagent just before the color reaction). One unit of xylanase activity was defined as that amount of enzyme which produces 1 micro mol of xylose equivalent per min. Cellulytic activity was measured essentially as for xylanase with carboxymethyl cellulose (CMC) as a substrate and glucose as standard. Proteolytic activity was assayed by incubating 0.1 ml of an enzyme sample together with 4.5 ml 0.2% casein in 100 mM phosphate buffer, pH 7.4, at 65 °C. The reaction was terminated by addition of 0.22 ml cold 50% TCA. After centrifugation, the absorbance at 280 nm was de-

Table 1. Characterization of extracellular xylanase activity from several thermophilic isolates.

Strain	Xylanase				CMCase	Protease
	U/ml	Half life[a] (h)	pH 9 (%)[b]	pH 10 (%)	(U/ml)	(U/ml)
T-1	0.63	0.90	67	58	0.026	nd[c]
T-2	0.89	0.71	59	53	0.014	15.7
T-3	0.51	2.04	58	51	0.031	nd
T-4	0.90	0.80	59	45	0.015	7.3
T-5	1.50	0.60	54	27	0.017	nd
T-6	2.33	4.62	66	54	0.021	nd

[a] Half life of xylanase activity at 70° C.
[b] Percentage activity at pH 9 and 10 compared to pH 7.
[c] nd – not detected.

termined. One unit of activity is defined as 0.001 A_{280}/min.

Protein

Protein content was determined by the method of Bradford (1976) using the Bio-Rad protein assay kit (Bio-Rad, Richmond, CA) with bovine albumin fraction V (Sigma) as standard.

Thermostability

Eppendorf tubes (1.5 ml) containing 0.2 ml of purified enzyme solution (40 units/ml in 10 mM phosphate buffer, pH 7) were incubated at 70 °C. At various times the tubes were removed and placed at −20 °C for 10 to 24 h. The residual enzymatic activity in each tube was determined with the standard assay.

Isolation of constitutive mutants

Mutagenesis with N-methyl-N′-nitro-N-nitrosoguanidine (NTG) was performed at 60 °C according to Adelberg et al. (1965). Mutagenized cells were plated on LB agar plates (Difco or Oxid Bacto Agar) incubated for 12 h at 60 °C and then overlaid with 5 ml of soft agar (0.7% agar) containing 2 mg of p-nitrophenyl β-D-xylopyranoside (PNPX). Colonies of constitutive mutants produced yellow color after 10 min at room temperature.

Table 2. Fatty acids profile of strain T-6.

Fatty acid	%
9:0	0.64
12:0	0.46
14:0	1.92
15:0 iso	46.20
15:0 anteiso	2.35
16:0 iso	4.15
16:1	5.53
16:0	6.14
17:1 iso H	2.60
17:0 iso	22.58
17:0 anteiso	5.82
18:0	1.09

Enzymatic delignification

Enzymatic delignification of softwood pulp was done as described by Shoham et al. (1993).

Results and discussion

Initial characterization of extracellular xylanase activity from several thermophilic isolates

Following enrichment procedures, several different thermophilic strains capable of producing extracellular xylanase were isolated (Shoham et al. 1993). To identify the best strain for the production of xylanase, several criteria were tested including: level of extracellular xylanase activity at high pH, thermostability and lack of proteolytic and cellulolytic activities in the supernatant fluid. Table 1 summarizes the results obtained with 6 different isolates. Strain T-6 gave the highest xylanase activity (2.33 units/ml); its enzyme was the most thermostable and had the highest activities at pH 9 and 10. In addition, the extracellular fluid of strain T-6 had no detectable level of proteolytic activity. All of the strains showed very low cellulolytic activity. Based on these results, strain T-6 and its xylanase were further characterized.

Table 3. The effect of phosphate on the growth of strain T-6.

Phosphate (mM)	Growth rate[a] (1/h)	Extent of growth (Turbidity units KU)
2.7	0.94	256
4.1	0.96	260
5.5	0.84	160
8.1	0.26	87
10.8	No growth	

[a] Growth was carried out on ND medium in 125 ml shake flasks at 60° C.

Classification of strain T-6

Strain T-6 was a Gram positive, aerobic, spore-forming rod, capable of growing at 65 °C. The spore shape was either elliptical or cylindrical and spore position terminal or subterminal. The strain failed to grow at 25 °C or 37 °C or in the presence of 5–10% NaCl, was catalase and oxidase positive and was capable of decomposing both casein and gelatin. Based on these results and its fatty acid profile (Table 2), strain T-6 was assigned to the heterogeneous species *Bacillus stearothermophilus*. Strain T-6 was also identified as a *Bacillus stearothermophilus* strain by the National Collection of Industrial & Marine Bacteria (NCIMB), England, and was designated as NCIMB 40221.

Growth characteristics

Strain T-6 can readily grow on rich or semi defined media. On the semi defined media, ND, the optimum temperature and pH for growth are 65 °C and 6.5, respectively, with a doubling time of about 40 min. Strain T-6 is inhibited by high phosphate concentrations (Table 3). At concentrations above 4 mM, there was some inhibition of growth; at concentrations above 10 mM, there was no growth at all. A defined medium was also constructed for strain T-6. Although the strain can grow to some extent on basic salt medium, BSM, in the presence of either methionine or asparagine, addition of the two amino acids together strongly stimulated growth (Table 4). Addition of trace elements to either the defined or semi defined media improved growth, and the most important metals for growth were Zn, Fe and Mn. The growth yield of strain T-6 on glucose and xylose were 0.27 and 0.17 g dry cell weight per g sugar, respectively.

Xylanase production

Strain T-6 produces extracellular xylanase activity when grown in the presence of either xylan or xylose. When grown on ND medium, strain T-6 reached a turbidity of 600 KU and xylanase activity was about 2 units/ml. No activity was detected with other carbon sources, such as glycerol, acetate, lactose, glucose, maltose, fructose, mannose, galactose or sucrose. The rate of xylanase production was usually highest towards the stationary phase; enzyme synthesis continued during the stationary phase if enough xylose was present. Combinations of xylose, together with other readily metabolized carbon sources such as glucose or glycerol, resulted in lower levels of xylanase production. These results suggest that xylanase production is inducible, can be repressed by other carbon sources and probably can be controlled by stationary phase genes such as degQ and degU (Fisher & Sonenshein 1991; Sonenshein 1989). In *B. subtilis*, enzymes responsible for the degradation of xylose and xylan are induced when grown in xylose-containing medium and repressed when rapidly metabolized carbon sources are available (Gartner et al. 1988).

Isolation of xylanase constitutive mutants

We were interested in obtaining xylanase constitutive mutants for the following reasons: a) xylose is a relatively expensive carbon source; economical large scale production of the enzyme would require a less expensive source; b) in many cases, constitutive mutants are over-produces of the desired protein; and c) constitutive mutants can facilitate some understanding on the regulation of the xylanolytic system in strain T-6. Screening for constitutive mutants on xylan containing agar plates (looking

Table 4. Growth of strain T-6 on basic salt medium (BSM) with amino acids.

BSM + amino acid[a]	Growth rate[b] (1/h)	Extent of growth (Turbidity units KU)
Met	0.32	61
Asn	0.35	55
Asn, Met	0.66	253
Asn, Met, Asp, Lys, Thr, Ile	0.71	260

[a] Amino acid concentration was 50 mg/l.
[b] Growth was carried out in 125 ml shake flasks at 60 °C.

for clear zones) was not possible since xylan is an inducer. We, therefore, took advantage of the fact that in many cases, xylanase and xylosidase are under the same regulatory control (repressor). Xylosidase producing strains can be easily detected on agar plates containing the chromagenic substrate p-nitrophenyl β-D-xylopyranoside (PNPX). Only constitutive mutants will produce yellow color on agar plates in the absence of xylose or xylan. To obtain xylosidase constitutive mutants, cells were mutagenized with NTG and then plated on LB agar plates. Plates containing 12 to 24 h old colonies were then overlaid with 5 ml of soft agar containing 2 mg PNPX. Colonies of constitutive mutants were identified after 10 min at room temperature by the yellow color they produced. Out of about 30,000 colonies screened, 30 colonies that produced yellow color were isolated. All of the isolates produced both xylosidase and xylanase activities in LB liquid media. From the relatively high frequency of mutants obtained and the mode of action of NTG (NTG induces multiple mutations in localized regions), it is likely that the mutations are in the repressor (inactivation of a gene) rather than mutations in the promoter/operator region. One of the constitutive mutants, strain M-7, was used for further studies. Strains M-7 and T-6 showed identical growth patterns on LB or ND media. However, on DM containing glucose instead of xylose, strain M-7 produced about 2 units/ml of xylanase, whereas strain T-6 gave no detectable levels of xylanase activity.

Large scale purification of xylanase T-6

Preliminary results during the purification of xylanase T-6, indicated that the enzyme can be highly purified and concentrated after a single adsorption step to a cation exchanger (CM-11 or CM-52 Whatman). Batch adsorption has the advantage of speed and ease of scaling the process. However, unless the partition coefficient P_c is well above 0.98, some losses are inevitable (Scopes 1987) (the coefficient P_c is defined as $P_c = Q/(C + Q)$ where Q is the concentration of the adsorbed protein and C is the concentration of the free protein). To assess the possibility of using batch adsorption for purifying xylanase T-6, it was necessary to determine the partition coefficient (P_c) and the maximum adsorbent capacity (Qmax) of CM-52. According to the mathematical model for an ion exchange process formulated by Cowan et al. (1986), the solid phase concentration of the adsorbed material will vary with the concentration of the adsorbate in a manner described by the equation $Q^* = C^* \times Qmax/(Kd + C^*)$, where Qmax is the maximum adsorbent capacity, C^* and Q^* are the equilibrium protein concentrations in liquid and solid phase, respectively, and Kd is the dissociation constant of the equilibrium reaction. The maximum adsorbent capacity and dissociation constant of an adsorbent can be estimated using the linear forms of this equation.To determine Qmax and Kd for xylanase T-6 and CM-52, different amounts of purified enzyme were mixed together with a constant amount of CM-52 adsorbent. Flasks containing 0.02 to 0.1 mg/ml of xylanase T-6 in 10 ml of phosphate buffer (20 mM, pH 6.5) and 0.1 g CM-52 were shaken for 12 h to reach the equilibrium. The adsorbent was filtered and the protein content in the solutions was determined using the Bio-Rad assay. The amount of the enzyme bound to the adsorbent (Q^*) was calculated from the difference between the amount

Table 5. Purification of xylanase T-6 from a 1000 liter fermentation of strain M-7.

Steps	Volume (l)	Protein (mg/ml)	Act. (U/ml)	Sp. act. (U/mg)	Yield (%)
Supernatant	1150	0.23	0.23	17.4	100
Residual broth[a]	1000	0.22	0.22	4.03	20
KCl elution	63	–	37.8	–	45.5
Ultra-filtration	7	0.78	3410	437	46.2

[a] Residual broth after adsorption.

of the enzyme present in the beginning of the experiment and the amount still left in the soluble phase. From the linear forms of adsorption isotherms, Qmax and Kd were calculated to be 25 mg/g and 0.017 mg/ml, respectively. The partition coefficient was calculated as $P_c = Q^*/(C^* + Q^*)$ (1.05 g CM-52 ~ 1 ml of packed bed) for the four experimental conditions, and the average value was 0.999. This high partition coefficient implies that batch adsorption of xylanase T-6 by CM-52 Cellulose is a viable approach. Another commercially available cation exchanger is SE-52 (the negative groups are sulfoxyethyl attached to cellulose). SE-52 is known to have a larger adsorption capacity then CM-52 and since the purification process is also a concentrating step, it is advantageous to use as little as possible of the adsorbent. We found that 2–3% of SE-52 will result in a yield of 48% of the purified enzyme. Based on these results, a simple scaleable process was developed for purifying xylanase T-6 directly from the fermentation broth. The process was based on the use of a single adsorption step for purifying and concentrating the enzyme. The results from a 1000 l purification process are summarized in Table 5. The overall yield of the process was 45%, and the enzyme obtained was over 98% pure as estimated by SDS-PAGE and FPLC gel filtration. The purified xylanase had a molecular weight of 43,000 and a pI of 9.0. At pH 7 and 65 °C, the enzyme was stable for over 10 h; at pH 9 and 65 °C, the half life of the enzyme was about 6 h.

Delignification

The delignification activity of pure xylanase T-6 was tested in laboratory using partially oxygen bleached soft wood pulp obtained from Korsnas paper and pulp company. The net release of lignin was 18% and 10% for 5 units/ml and 2 units/ml, respectively, for 4 h at pH 9.0 and 65 °C (Table 6). Larger quantities (10 g) of the pulp were then treated with 5 units/ml xylanase T-6 at pH 9.0 and 65 °C for 2 h and sent to Korsnas for analysis (Table 7). Compared to the control (no enzyme), the enzyme reduced the lignin (Kappa no.) and pentosan content without loss in the viscosity of the cellulose. The slight but significant increase in viscosity was probably due to the loss of lignohemicellulose, which does not contribute to the viscosity. Handsheets prepared from the pulps demonstrated that the enzyme-treated pulp yielded higher brightness with no significant loss of fiber strength. More significantly, the ClO_2, Cl_2 required to completely bleach the enzyme-treated pulp was reduced by 35%.

The xylanase T-6 treatment is now being optimized and scaled up. We anticipate that xylanase, as well as possibly other hemicellulases, will soon be used in the pulp industry to reduce or eliminate the use of chlorine-containing compounds for preparation of bleached pulp.

Table 6. Xylanase T-6 treatment of softwood Kraft pulp[a].

Treatment[b]	% Lignin released	
	Total	Net
Buffer, 2 h	4	–
Buffer, 4 h	5	–
Xylanase, 2 U/ml, 2 h	10	6
Xylanase, 5 U/ml, 2 h	17	13
Xylanase, 20 U/ml, 4 h	15	10
Xylanase, 50 U/ml, 4 h	23	18

[a] Partially oxygen bleached softwood Kraft pulp (Kappa No. = 17.5) was used at a final concentration of 5% dry weight pulp.
[b] pH 9.0, 65° C, 20 mM $(NH_4)_2SO_4$.

Table 7. Analysis of xylanase T-6 treated Kraft pulp.

Parameter	Control	Enzyme treated[a]
Kappa no.	17.1	15.4
Viscosity (dm^3/kg)	1061	1072
Pentosan (%)	7.3	7.3
Handsheets		
Brightness (% ISO)	32.6	34.4
Tensile index (Nm/g)	21.2	20.7
Zero-span (Nm/g)	100	103

[a] pH 9.0 and 65° C for 2 h with 50 U/ml xylanase T-6.

Acknowledgements

This work was supported by a grant from Biovik AB and Korsnas AB. Technical support was provided by the Technion-Otto Meyerhof Biotechnology Laboratories.

References

Adelberg EA, Mandel M & Chen GCC (1965) Optimal conditions for mutagenesis by N-methyl-N′-nitro-N-nitrosoguandine in *Escherichia coli* K12. Biochem. Biophys. Res. Commun. 18: 788–795

Bradford MM (1976) A rapid and sensitive method for the quantitation of microgram quantities of protein utilizing the principle of protein-dye binding. Anal. Biochem. 72: 248–254

Browning BL (1963) The composition and chemical reactions of wood. In: Browning BL (Ed) The Chemistry of Wood (pp 58–101). John Wiley & Sons, New York

Chang HM, Joyce TW & Kirk TK (1987) Process of treating effluent from a pulp or paper making operation. US Pat. 4, 655, 926

Clark TA, McDonald AG, Senior DJ & Mayer PR (1990) Mannanase and xylanase treatment of softwood chemical pulps: effects on pulp properties and bleachability. In: Kirk TK & Chang HM (Eds) Biotechnology in Pulp and Paper Manufacture (pp 153–167). Butterworth-Heinemann, Boston

Cowan GH, Gosling IS, Laws FJ & Sweetenhm WP (1986) Physical and mathematical modeling to aid scale-up of liquid chromatography. J. Chromatogr. 363: 37–56

Dekker RFH (1985) Biodegradation of the hemicelluloses. In: Higuchi T (Ed) Biosynthesis and Biodegradation of Wood Components (pp 505–533). Academic Press, Orlando

Dekker RFH & Richards GN (1976) Hemicellulases: their occurrence purification properties, and mode of action. In: Tipson RS & Horton D (Eds) Advances in Carbohydrate Chemistry and Biochemistry, Vol 32 (pp 277–352). Academic Press, New York

Eriksson KE & Kirk TK (1985) Biopulping, biobleaching and treatment of kraft bleaching effluents with white-rot fungi. In: Moo-Young M (Ed) Comprehensive Biotechnology, Vol 4 (pp 271–294). Pergamon Press, New York

Eriksson O, Goring DAI & Lindgren BO (1980) Structural studies on the chemical bond between lignins and carbohydrates in spruce wood. Wood Sci. Technol. 14: 267–279

Farrell RL (1987) Use of RLDM 1–6 and other ligninolytic enzymes in the bleaching of kraft pulp. US Patent No 4, 690, 895

Farrell RL (1991) Chlorine free bleaching with carazyme HS treatment. Xylans and Xylanases, Int. Symp. Wageningen, The Netherlands, December 8–11 (p L29)

Fisher HF & Sonenshein AL (1991) Control of carbon and nitrogen metabolism in *Bacillus subtilis*. Annu. Rev. Microbiol. 45: 107–135

Fujita K, Kondo R, Sakai K, Kashino Y, Nishida T & Takahara Y (1991) Biobleaching of kraft pulp using white rot fungus IZU-154. Tappi J. 74(11): 123–127

Fukuzumi T, Nishida A, Aoshima K & Minami K (1977) Decolourization of kraft waste liquor with white-rot fungi. I. Screening of fungi and culturing condition for decolourization of kraft waste liquor. Mokuzai Gakkaishi 23: 290–298

Gartner D, Geissendorfer, M & Hillen W (1988) Expression of the *Bacillus subtilis* xyl operon is repressed at the level of transcription and is induced by xylose. J. Bacteriol. 170: 3102–3109

Grant R (1991) First mill-scale trials get underway. Pulp & Paper Int. 33(6): 61–63

Hakulinen R (1988) The use of enzymes for wastewater treatment in the pulp and paper industry – a new possibility. Wat. Sci. Tech. 20: 251–262

Heinzle E, Geiger F, Fahmy M & Kut OM (1992) Integrated ozonation-biotreatment of pulp bleaching effluents containing chlorinated phenolic compounds. Biotechnol. Prog. 8: 67–77

Iversen T & Waennstroem S (1986) Lignin carbohdrate bonds in a residual lignin isolated from pinekraft pulp. Holzforschung 40: 19–22

Jeffries TW (1990) Biodegradation of lignin-carbohdrate complexes. Biodegradation 1: 163–176

Jurasek L & Paice M (1986) Pulp, paper and biotechnology. CHEMTECH 16(6): 360–365

Jurasek L & Paice M (1988) Biological treatments of pulps. Biomass 15: 103–108

Kantekinen A, Sundquist J, Linko M & Viikari L (1991) The role of reprecipatated xylan in the enzymatic bleaching of kraft pulp. The 6th International Symposium on Wood and Pulping Chemistry, Melbourne, April 29–May 3 (pp 493–500)

Kersten PJ, Tien B, Kalyanaraman B & Kirk TK (1985) The ligninase of *Phanerochaete chrysosporium* generates cation radicals from methoxybenzenes. J. Biol. Chem. 260: 2609–2612

Kirk TK & Farrell RL (1987) Enzymatic 'combustion': the microbial degradation of lignin. Ann. Rev. Microbiol. 41: 465–505

Kirk TK & Yang HH (1979) Partial delignification of unbleached kraft pulp with ligninolytic fungi. Biotechnol. Lett. 1: 347–352

Kirkpatrick N (1991) Biological bleaching of wood pulps – a viable chlorine-free bleaching technology? Wat. Sci. Tech. 24: 75–79

Koponen R (1991) Enzyme systems prove their potential. Pulp & Paper Int. 33(9): 81–83

Lankinen VP, Inkeroinen MM, Pellinen J & Hatakka AI (1991) The onset of lignin-modifying enzymes, decrease of AOX and color removal by white-rot fungi grown on bleach plant effluents. Wat. Sci. Tech. 24: 189–198

Larsson A, Anderson T, Forlin L & Hardig J (1988) Physiological disturbances in fish exposed to bleached kraft mill effluents. Wat. Sci. Tech. 20(2): 67–76

Livernoche D, Jurasek L, Desrochers M & Veliky IA (1981) Decolorization of a kraft mill effluent with fungal mycelium immobilized in calcium alginate gel. Biotechnol. Lett. 3: 701–706

Loras V (1980) Bleaching of chemical pulps. In: Casey JP (Ed) Pulp and Paper Chemistry and Chemical Technology, 3rd ed, Vol 1 (pp 663–702). John Wiley & Sons, New York

Lundquist K, Kirk TK & Connors WJ (1977) Fungal degradation of kraft lignin and lignin sulfonates prepared from synthetic ^{14}C-lignins. Arch. Microbiol. 112: 291–296

Messner K, Ertler G & Jaklin-Farcher S (1989) Treatment of bleach effluents by the MYCOPOR system. In: Fourth Int. Conf. of Biotechnology in the Pulp and Paper Industry, Raleigh, NC, USA May 16–19 (pp 67–68)

Miller GL (1959) Use of dinitrosalicilic acid reagent for determination of reducing sugars. Anal. Chem. 31(3): 426–428

Paice M, Bernier M & Jurasek L (1988) Viscosity enhancing bleaching of hardwood kraft pulp with xylanase from cloned gene. Biotechnol. Bioeng. 32: 235–239

Paice MG, Jurasek L, Bourbonnais CHoR & Archibald F (1989) Direct biological bleaching of hardwood kraft pulp with the fungus *Coriolus versicolor*. Tappi J. 72(5): 217–221

Pellinen J, Joyce TW & Chang HM (1988a) Dechlorination of high-molecular weight chlorolignin by the white rot fungus *P. chrysosporium*. Tappi J. 71(9): 191–194

Pellinen J, Yin CF, Joyce TW & Chang HM (1988b) Treatment of chlorine bleaching effluent using a white-rot fungus. J. Biotechnol. 8: 67–76

Reid ID, Paice MG, Ho C & Jurasek L (1990) Biological bleaching of softwood kraft pulp with the fungus *Trametes* (*Coriolus*) *versicolor*. Tappi J. 73(8): 149–153

Reily PJ (1981) Xylanases; structure and functions. In: Hollaender AE & Robson R (Eds) Trends in the Biology of Fermentations for Fuels and Chemicals (pp 111–129). Plenum, New York

Samadni G (1991) Pulp bleaching – The race for safer methods. Chem. Eng.(Int. Ed.) 98(1): 37–43

Sanyer N & Chidester GH (1963) Manufacture of wood pulp. In: Browning BL (Ed) The Chemistry of Wood (pp 58–101). John Wiley & Sons, New York

Scopes RK (1987) Protein Purification, Principles and Practice 2nd ed. Springer-Verlag, New York

Senior DJ, Mayers PR, Miller D, Sutcliffe R, Tan L & Saddler JN (1988) Selective solubilization of xylan in pulp using a purified xylanase from *Trichoderma harzianum*. Biotechnol. Lett. 10: 907–912

Singh PS (1979) Principles of pulp bleaching. In: Singh PS (Ed) The Bleaching of Pulp, 3rd ed (pp 15–28). Tappi Press, Atlanta

Shoham Y, Zosim Z & Rosenberg E (1993) Partial decolorization of Kraft pulp at high temperature and at high pH values with an extracellular xylanase from *Bacillus stearothermophilus*. J. Biotech. (in press)

Sonenshein AL (1989) Metabolic regulation of sporulation and other stationary phase phenomena. In: Smith I, Slepecky A & Setlow P (Eds) Regulation of Procaryotic Development (pp 109–130). American Society for Microbiology, Washington, DC

Tran AV & Chambers RP (1987) Delignification of an unbleached hardwood kraft pulp by *Phanerochate chrysosporium*. Appl. Microbiol. Biotechnol. 25: 484–490

Trubacek I & Wiley A (1979) Bleaching and pollution. In: Singh PS (Ed) The Bleaching of Pulp, 3rd ed (pp 423–461). Tappi Press, Atlanta

Viikari L, Ranuae M, Kantelien A, Linko M & Sundquist J (1986) Application of enzymes in bleaching. Proc. 3rd Int. Conf. Biotechnology in the Pulp and Paper Industry, Stockholm (pp 67–69)

Viikari L, Ranuae M, Kantelien A, Linko M & Sundquist J (1987) Application of enzymes in bleaching. 4th International Symposium Wood and Pulping Chemistry, EUCEPA, April 27–30, 1987, Paris, Vol 1, Oral Presentations (pp 151–154)

Viikari L, Kantelinen A, Ratto M & Sundquist J (1991a) Enzymes in Pulp and Paper Processing. ACS Symp. Ser. 460 (Enzymes Biomass Conversion) (pp 12–21)

Viikari L, Sundquist J & Kettunen J (1991b) Xylanase enzymes promote pulp bleaching. Paperi ja Puu – Paper and Timber 73(5): 384–389

Biodegradation **3**: 219–238, 1992.

Biotechnology in the degradation and utilization of lignocellulose

Paul Broda
Department of Biochemistry & Applied Molecular Biology, University of Manchester Institute of Science and Technology, Manchester M60 1QD, UK

Key words: cellulase, gene expression, lignin, *Phanerochaete chrysosporium*, *Streptomyces cyaneus*, xylanase

Abstract

Lignocellulose is the predominant renewable resource. Its uses include fuel, as the feedstock for the pulp and paper industry, and for animal nutrition. It also constitutes a large proportion of agricultural and urban waste. Biotechnology has roles in its efficient production and utilisation. The types of lignin substrates available for study of lignin biodegradation are described. The white rot fungus *Phanerochaete chrysosporium* is the archetypal system for the study of lignocellulose degradation, since it mineralises lignin and degrades both cellulose and hemicellulose. The salient features of the *P. chrysosporium* system are described. The lignin peroxidases are a family of proteins, and it is shown that expression of their genes is differential. *P. chrysosporium* is heterokaryotic with two gene equivalents that have abundant RFLPs. A set of basidiospore-derived strains with genetic compositions defined by such RFLPs provided the potential basis for a strain improvement programme for lignin degradation. However, analysis of this system using radiolabelled synthetic lignin (DHP) as the substrate confirmed previous evidence that both the substrate and the fungal cultures displayed much variation, so that it was difficult to quantify performance for this property. The cellobiohydrolase I enzymes are also coded for by a family of genes, and evidence is also presented for allelic variants, for differential expression and for differential splicing. In contrast, the cellobiohydrolase II function is encoded at a unique genetic locus. Approaches to an homologous integrative transformation system are discussed. Some actinomycete bacteria represent an alternative system for lignin solubilisation in which strains differ in their spectra of activities on lignocellulose substrates. The xylanase system of *Streptomyces cyaneus* is shown to include three enzymes, two of which are inducible by xylan. A novel assay method was developed and used to demonstrate that the third is constitutive and also non-repressible by glucose. It is proposed that this acts as a sensor for xylans in the environment that can yield breakdown products that are taken up and can then act as inducers of the other two enzymes. The studies on microbial lignocellulose degradation from different laboratories have allowed the formulation of specific biotechnological goals, and some of the problems and opportunities in this area are identified.

Introduction

Lignocellulose is the principal source of biomass and therefore of renewable resources; it is a complex of three classes of polymer. Cellulose, which consists of repeating glucose units largely in crystalline fibres, is chiefly what is sought in chemical pulping processes. The hemicelluloses are a more mixed group of polysaccharides, in which the monomers are mainly pentoses. In the case of lignin the monomers are three cinnamyl alcohols. Both cellulose and lignin are difficult to disrupt, but for fundamentally different reasons. Because cellulose is largely crystalline, it is substantially more difficult to attack than, for instance, starch or the hemicelluloses; in contrast, lignin is assembled by an oxidative polymerisation of free radical species of its monomers to yield a highly irregular three-dimensional polymer with a number of intermonomeric bond types. The predominant class is the β aryl ether bond, but others include direct linkages and ethers between aromatic rings.

The objective of this article is to survey existing forms of lignocellulose utilisation, which have little biotechnological input, and to identify aspects where biotechnology might in the future make contributions to improved utilisation. It will be shown that such innovations will depend upon improved understanding of the biodegradation of these polymeric components, and work from the author's laboratory will be described in this context.

Utilization of whole lignocelluloses

The range of biomass materials includes trees and other crops, and also wastes from forests, agriculture, stock raising and towns (Coombs 1986). Much lignocellulose is and will be used in an unfractionated form. Thus, mechanical pulping involves the milling of the total lignocellulose to give high yields that can be used for newsprint and for brown papers and board, e.g. for packaging materials; this process requires a very high energy input. Production of lignocellulose as an energy crop is an alternative to the utilisation of fossil fuels that has the merit of being CO_2-neutral. Crude burning is likely to be supplanted by chemical processes that yield various mixtures of gas, liquid and solid waste; a potential return for investment comparable to that for oil is suggested. There will be changes in the appropriateness of different technologies depending on economic, resource-availability and environmental considerations. Nevertheless, there are processes that are being developed and it will be for biotechnology to find a competing or complementary role in such processes (Hartley et al. 1987; Anon. 1992).

Biological utilization of whole lignocellulose

It should be recognised that at least three biologically based processes already exist that utilise whole lignocellulose. First, it is a major component of animal feeds such as grasses, hay and straw. Its precise composition varies between species and varieties of plant, and also in its proportion of the total material during the life cycle of the plant. Whereas cellulose can be degraded by microorganisms in the ruminant gut to glucose, the hemicelluloses have an as yet poorly defined nutritional role (Williams 1989). The lignin component is especially recalcitrant to microbial degradation and constitutes a physical barrier to the extraction and utilization of the other components of plant material. Biotechnology might thus improve the nutritional value of lignocellulose for instance through upgrading of silage and ultimately by depolymerisation of the polysaccharides so that such materials provide a feed for non-ruminants such as pigs and poultry (Coughlan 1989).

The second biologically based process is the production of biogas, which has been practised for centuries in China. There are substantial efforts to develop such technology in India, and it has a future in the utilization of wastes in landfills and on farms in developed countries (Ferrero et al. 1984). A particular problem with industrial farming is the disposal of animal waste, especially from pig production.

The third process is production of mushrooms in solid-substrate fermentation systems (Flegg et al. 1985). The major problems for the process are in the physiology of the organisms and in protection against disease. For some species, e.g. *Agaricus bisporus*, the preparation and management of the starting compost and the casing layer substrates are critical (Derikx et al. 1990), but a very wide range of lignocellulosic starting materials might be used for other species, e.g. *Pleurotus ostreatus* (Hadar et al., this issue). There is much work to be done at the physiological and process engineering levels on the analysis of the conditions that could be employed for different fungal strains and substrates. This is therefore a model that is already operating on a large scale (Reid 1989; Karem et al. 1992).

Fractionation of lignocellulose

By far the largest process currently in operation is chemical pulping for both the paper industry and the production of cellulose fibre for textiles. There are a number of pulping methods, but in each what is obtained is cellulose, with a proportion of the hemicelluloses. In the Kraft process (the dominant process worldwide) the bulk of the hemicelluloses are retained in a modified form whereas lignin is converted into soluble products (Kraft lignins) that as yet have little value other than for burning; however, this means that the process requires no external energy source and there is extra material that could be made available for other uses. Thus in the overall balance sheet principally the cellulose and the hemicellulose emerge from the process. Furthermore, the bleaching and subsequent paper-making processes create major environmental problems to do with effluents and requirement for water (Shoham et al., this issue).

Industrial lignins surplus to the needs of chemical pulping processes have been considered as a source of aromatic chemicals such as phenols and catechols, and have been used for fibreboard, as an emulsion stabilizer, as an additive in oil drilling, and for cement processing, but none of these is a high value product. Novel uses must be sought for derivatives of lignin that would match the enormous scale on which they are produced in nature and then industrially.

Possible processes for utilization of the monomers of the different components of lignocellulose, whether chemical or biological, are visualised in three stages: pretreatment to disrupt the gross structure, depolymerization, and utilization of the monomers. In the pretreatment, mechanical chipping is followed in a number of schemes that are in the pilot stage by either steam explosion or solvent extraction. Such methods render the mixtures of lignin, cellulose and hemicellulose amenable to further treatment. The methods used to fractionate the carbohydrates include treatment with dilute acid, which removes the hemicelluloses from the celluloses, or with concentrated acids, which have high material costs. Both steam explosion and solvent extraction yield hemicellulose that could serve as a substrate for enzymatic hydrolysis (Overend & Chornet 1987; Ragg & Fields 1987). Relatively mild treatments with ammonia are used to upgrade the digestibility of straw by ruminants.

It seems that pretreatment is likely to be common to all processes, but that biological methods can then be considered in competition with chemical and physical methods (or as complements to them). In general, biological methods will operate at lower temperatures and pressures, with cheaper starting materials and the possibility of operating on a smaller scale; on the other hand, the throughput time may be much slower.

Biologically based processes

There are at least six areas in which scientific progress might lead to biologically based processes. Three of these are concerned with the polymers, namely disruption of the crystallinity of the cellulose allowing more rapid action by cellulases, the depolymerisation of hemicellulose, and the solubilization of lignin. The other three are concerned with the utilization of monomers, i.e. the fermentation of pentose sugars, the production of ethanol in high yield together with its separation from the aqueous medium, and the conversion of solubilized lignin-derived monomers into high added-value aromatics. The solubilization of the polymers is the key to biological processes and therefore these will be discussed. Considerable effort has been expended on cellulases, which now represent a commercial product (Coughlan 1985), and as will be described in this issue and elsewhere (Shoham et al. 1992; Visser et al. 1992), hemicellulases are assuming importance in the pulp and paper industry. In contrast, no industrial processes yet exist involving biological delignification (but see Kirk & Chang 1990).

Cellulose biodegradation

Cellulose in lignocellulose is a much larger resource than starch for breakdown to sugars. But it is also represents a much more complex problem because of its crystallinity and its admixture with hemicellulose and lignin. Most work on its biodegradation has been done with the filamentous fungus *Trichoderma reesei*, which, given optimum culture conditions, can produce yields of 20 g extracellular cellulase protein from 60 g cellulose (Coughlan 1985). It is now evident that in this organism cellulase is a complex of at least eight proteins that act synergistically and are subject to complex controls (Eveleigh 1987). *T. reesei* produces five or more endo β-glucanases which convert the crystalline cellulose to oligocellodextrins and amount to 20–35% of the protein by weight. In addition, there are at least two cellobiohydrolase components (65–80% of total protein), which convert these to cellodextrins and cellobiose. Both classes of enzymes are subject to end product inhibition by cellobiose. Finally there is a cellobiase that converts the cellodextrins and cellobiose to glucose; this enzyme constitutes a mere 1% of the protein, and its action is inhibited by glucose. The whole complex is also subject to catabolite repression by glucose.

The pilot studies show that with preparations of cellulase that contain these various activities it is possible in stirred tanks, under realistic conditions, to hydrolyze various types of cellulose to soluble sugars with substrate concentrations of 10–30% and yields of 30–60%. One could convert 1000 kg of lignocellulose of substrate (wood chips, straw, corn stover, or fibre from urban waste) to 350 kg of glucose in 24 h. But the economic barrier is the very large amounts of enzyme complex needed to hydrolyze the cellulose (perhaps a hundred times more than the amylase needed to break down starch).

The reasons have to do with the recalcitrant and impure nature of cellulose, with the requirement for a synergistic complex, with product inhibition, and with the difficulty of desorbing the enzyme complex from the substrate. However, it seems premature to dismiss the biological attack of cellulose in bulk materials on grounds of present economics. The appropriate view to take is that the methods of molecular genetics, allowing cloning of individual enzymes, will result in a far better understanding of how, in one organism, the system works. There will then be opportunities for site-directed mutagenesis to remove product inhibition, to improve the rate at which the enzymes are produced (very little is secreted while the fungus has sufficient cellulose for growth; it occurs as cellulose and nitrogen are approaching depletion), perhaps to improve the activities of the enzymes themselves, and for the different synergistic enzymes to be produced in the optimum ratio.

Other fungi have also been studied, but none is qualitatively more attractive than *T. reesei*, and this organism is an excellent model for this difficult system. However, other organisms should also be studied further. Among these is the bacterial species *Clostridium thermocellum*, which is the topic of a separate article (Bayer & Lamed, this issue).

Lignin biodegradation

Discussions on lignin breakdown and utilization are still primarily based on chemical engineering approaches (Ragg & Fields 1987) with the lignin being recovered in polymeric form. The achievement of this step by biological methods would require the efficient release of the cellulose and hemicelluloses and then the lignin might in principle be recovered either in polymeric form or as monomers that could be converted into useful chemicals. In nature the function of lignin solubilization may be primarily to allow the degrading organism to obtain access to the carbohydrates since the energetic investment in lignin solubilization is rather substantial; thus the basidiomycete white-rot fungus *Phanerochaete chrysosporium* will only degrade lignin if it is depleted for carbon or nitrogen sources (Keyser et al. 1978).

We must recognise that the natural substrate is lignocellulose and hence it is more complex even than lignin. Nevertheless, lignocellulose degradation is a unitary problem because the materials involved (e.g. wood, straw and sugar cane bagasse) are as a first approximation similar to each other. It is the interactions between degrader organisms and the substrate that must be studied, analogous to the cases of microbe-plant and microbe-animal host interactions. These interactions must be understood at different levels that include biochemistry, gene expression and physiology. This involves studying the synergy between different enzymes attacking different components, and indeed between those attacking individual components. As stated above, for cellulose it is already evident that all microbial systems being studied involve a number of synergistic enzymes (Coughlan 1985), and this is likely to be the case for the other components as well. With respect to the substrates, better analytical techniques will allow greatly improved analysis of both the substrates and the products (Kirk & Farrell 1987; Chesson & Orskov 1989).

There is a need to continue to concentrate on a few organisms, so as to obtain a complete description of the enzymes involved and how they act together, their control and topographical relationships. The white rot fungi, e.g. *P. chrysosporium*, remain the only organisms for which the mineralisation of lignin to CO_2 is known to be the primary process; nevertheless other organisms, notably the actinomycetes, also have important roles in lignocellulose degradation (Broda et al. 1989; Zimmerman 1990; Magnuson & Crawford 1992) and should therefore also be studied. The recent results from our own laboratory that will be described are concerned with both *P. chrysosporium* and actinomycete bacteria.

Different groups use different assays for lignin biodegradation. The problems are two-fold: first, in nature lignin is intimately associated with polysaccharides, probably being covalently bound to hemicellulose. Therefore any extensively purified lignin is different from lignin in native lignocellulose. Industrial lignin isolated from the pulping process has substantial chemical modifications. Second, purified lignin is insufficient as a growth substrate: some other carbon co-substrate, e.g. glucose, must be provided, so that growth cannot be used as an assay for lignin degradation. For these reasons a critical discussion on the substrates and assays that are used is an essential component of any review of lignin degradation. Wood & Kellogg (1988a,b) is an invaluable source of information on this area.

Lignocellulose substrates

Studies designed to screen for organisms that are capable of degrading and/or utilizing lignin have employed either synthetic or natural substrates. Synthetic model compounds of lignin include ethoxy and/or methoxy substituted phenols and aromatic alcohols joined together to form dimers either linked via an oxygen (ether bond) or directly through the beta group (Kirk & Farrell 1987). Such substrates are appropriate for the study of the mechanism of purified enzymes, such as the lignin peroxidases of white rot fungi that have been implicated in lignin degradation. However degradation of such model compounds is not by itself proof of degradation of lignin in lignocellulose. The oxidation of veratryl alcohol (3,4 dimethoxybenzyl alcohol) has been used as the standard assay for lignin peroxidases.

In an attempt to reflect the complexity of the lignin molecule, the degradation of polymers obtained from the *in vitro* polymerization of radiolabelled aromatic alcohols (e.g. coniferyl alcohol; Haider & Trojanowski 1975; Tanahashi & Higuchi 1981; Kirk & Brunow 1988) have also been investigated. Such 'dehydrogenative polymerizate' (DHP) has a variety of bonds and is regarded as a good model for lignin; the molecular weights of these compounds are in the range 2 to 10 $\times$ 10^3. These studies have been designed to measure their mineralization to $^{14}CO_2$ during incubation with the organisms.

Radiolabelled natural substrates have also been used; they have been prepared either from cereal grasses (e.g. wheat and barley) grown in the presence of ^{14}C phenylalanine (McCarthy et al. 1984; Mason et al. 1990), or ^{14}C cinnamic acid (Benner et al. 1984), or by dipping cut stems of woody plants in a buffer containing ^{14}C ferulic acid (Crawford & Crawford 1978). In each case the radiolabelled acids provide a precursor for lignin biosynthesis. The removal of non-lignocellulose materials using inorganic and then organic solvents results in a lignocellulose substrate which contains 85% of the radioactivity in the lignin component of the complex, and is therefore sometimes referred to as (^{14}C lignin) lignocellulose; such determinations are based on the Klason definition of lignin, namely the material that is insoluble in 72% H_2SO_4 (Efland 1977). The proportion of ^{14}C lignin may be improved to 90% by the removal of some residual proteinaceous material using proteases. The resulting substrates have a specific activity (approximately 10^3 dpm/mg) sufficient to allow the screening of organisms for the ability to degrade lignin (Phelan et al. 1979; Haider & Trojanowski 1980; McCarthy & Broda 1984; Agosin et al. 1985). The use of such substrates has revealed two quite different types of activity: one is the solubilization of the ^{14}C lignin, giving rise to soluble radiolabelled products in culture supernatants of actinomycetes (McCarthy & Broda 1984); Mason et al. 1988), and the other is the complete degradation of the ^{14}C lignin to $^{14}CO_2$, as exhibited by cultures of *P. chrysosporium* (McCarthy et al. 1984; Agosin et al. 1985).

Given the complexity of lignocellulose and indeed of lignin itself, it is likely that both solubilization and degradation require that a number of enzymes act synergistically. As a result of this synergism, factors affecting the composition of the lignocellulose substrate, such as plant maturity, will determine the contribution that each enzyme makes in the fate of lignin from lignocellulose. This was shown in a recent study (Mason et al. 1990) in which (U-^{14}C) phenylalanine was injected into spring barley plants at frequent intervals during 140 d growth. Autoadiography of the plants revealed that, apart from high concentrations around the injection sites, radioactivity was evenly distributed throughout the plants. (^{14}C lignin) lignocellulose was prepared from plants cropped at 30, 58, 82 and 140 d after sowing, using organic and aqueous solvents followed by enzymic hydrolysis with commercial protease; portions of each were then also treated with polysaccharidase preparations. Weight loss due to solvent and enzymic treatments was greatest in the youngest plants, resulting in a preparation of ^{14}C lignocellulose of specific activity 2-8-fold higher than that of preparations from more mature plants.

When subjected to crude preparations of extracellular protein from either *P. chrysosporium* or the actinomycete *Streptomyces cyaneus*, the substrates derived from the youngest plants (30 d) were particularly susceptible to solubilization of lignocellulose by polysaccharidases (determined by comparison of their action on substrates with and without previous polysaccharidase treatment). Substrates prepared from plants after 58 d growth (i.e. fully grown) showed significant levels of lignocellulose solubilization that were not due to the action of polysaccharidases. When used in similar assays the oldest crops (82 and 140 d) were solubilized at very low levels. Thus three different types of result were obtained, depending on the age of the cropped material.

The *Phanerochaete chrysosporium* lignin degradation system

A dozen years ago the situation was that it had been shown that the white rot fungi could mineralise lignin to CO_2 and that *P. chrysosporium* had been identified as a good experimental organism. Two major facts were known about such degradation (Kirk & Farrell 1987).

1. Overall, it is an oxidative rather than a hydrolytic process. Since no enzyme mechanism had yet been identified one hypothesis was that lignin degradation depended upon the production of free radicals (e.g. singlet oxygen or hydroxyl radicals) that would effect a chemical degradation (Hall 1980).
2. As mentioned earlier, lignin degradation is not inducible by lignin itself but is triggered by severe carbon or fixed nitrogen limitation (Keyser et al. 1978). The biological rationale is that lignin is degraded as a stress response, so that the organism can obtain access to further sources of nutrients and energy previously made inaccessible by the presence of the lignin. Note that lignin does not contain nitrogen.

A number of groups were already involved in studies on *P. chrysosporium* and other became involved at this time. In the intervening years the following major developments have occurred:

1. The groups of Kirk and Gold have described

lignin peroxidases; these extracellular haem-containing enzymes can cleave model dimers representing some of the different bond types present in lignin. *P. chrysosporium* strain BKM-F-1767 produces a family of such enzymes; they have molecular weights of around 40,000 and are glycosylated (Kirk & Farrell 1987).

2. The discovery of these enzymes immediately raised the issue of how the H_2O_2 that they require is generated. This could be derived from the breakdown of carbohydrates and/or of the products of lignin de-polymerisation; therefore the process of lignin degradation is likely to be coupled to the degradation of cellulose and/or hemicellulose.
3. As a development of the discovery of lignin peroxidases, a mechanism involving the formation of substrate radical intermediates was proposed (Kirk & Farrell 1987); such radicals might invade the lignin molecule and be the immediate effectors of its degradation. A mechanism of this type would account for the range of bonds cleaved, and makes these enzymes of particular interest.
4. A second class of enzymes, the manganese peroxidases, have been defined (Kuwahara et al. 1984). These oxidise Mn(II) to Mn(III) and it is proposed that such Mn(III), chelated to organic acids, functions as an active radical that can mediate oxidative depolymerisation of lignin in wood (Glenn et al. 1986).
5. Both cDNA and genomic sequences of lignin peroxidase genes from *P. chrysosporium* have been published by many groups. It has been shown that there are families of lignin peroxidase and Mn peroxidase-related genes (Brown et al. 1988; Godfrey et al. 1990).
6. It has been suggested that peroxidases can depolymerize lignin more effectively in non-aqueous solvents (Dordick et al. 1986). If this is indeed the case (Lewis et al. 1987), it has relevance to the technological exploitation of lignin peroxidases and lignins.
7. Since peroxidases are implicated in polymerisation as well as depolymerisation of lignin, it has been particularly important to establish whether lignin peroxidases and Mn peroxidases do indeed have *in vitro* depolymerising activity. This has now been established using DHP and each of these enzymes (Hammel & Moen 1991; Wariishi et al. 1991).
8. It has been possible to elucidate the sexual cycle, and establish a mating system, using auxotrophic markers (Alic & Gold 1991). A genetic map has been constructed for *P. chrysosporium* using RFLP analysis (Raeder & Broda 1986; Raeder et al. 1989a), and it has been shown that the lignin peroxidase-related sequences are clustered (Raeder et al. 1989a). This has been largely confirmed using pulse field gel electrophoretic methods (Gaskell et al. 1991).
9. *P. chrysosporium* and *T. reesei* clearly have different strategies of action on lignocellulose, since *T. reesei* does not attack lignin. However, is *P. chrysosporium* typical among the white rot fungi? Interest exists in the range of extracellular enzymes produced by other fungal species such as *Phlebia radiata*, (Niku-Paavola et al. 1990), *Coriolis versicolor* (Eriksson et al. 1990), *Rigidosporus lignosus* (Galliano et al. 1991) and *Dichomitus squalens* (Pham et al. 1990).

The present model for lignin degradation in *P. chrysosporium* is that the system consists of manganese and lignin peroxidase enzymes together with the H_2O_2 generation system that they require, and that this is coupled to carbohydrate breakdown. This model is useful in that it is open to a number of *in vitro* and *in vivo* tests.

Lignin peroxidase genes: multiple sequences and differential expression

As stated above, a number of groups have reported lignin peroxidase and Mn-peroxidase gene sequences. In our case, the cloned sequences carrying putative lignin peroxidase genes were found as a set among clones isolated from *P. chrysosporium* strain ME446 because they carried genes expressed specifically in secondary metabolism (Raeder & Broda 1986). In order to establish which of this set were expressed, oligonucleotide probes were made that would distinguish between them by hybridising differentially to their mRNA transcripts. The surprising result obtained was that none of the four sequences that we had characterised was expressed (James et al. 1992).

An alternative approach was then used to resolve this paradox; this was to construct a cDNA library (using mRNA prepared from mycelium grown on ball milled straw), and partially sequence lignin peroxidase related cloned sequences identified within it. It was found that only one class of cDNA was present, with a sequence that was different from any of those previously identified. The conclusion that only one of the sequences is expressed under our conditions is different from the situation found by others, with another strain of *P. chrysosporium* (Leisola et al. 1987; Stewart et al. 1992). However, it is consistent with our own finding that only a single lignin peroxidase protein peak is present on chromatography of supernatants of strain ME446 (Birch 1988), another observation that has been different from that published by others (Kirk & Farrell 1987). In a situation where only a single gene is expressed, in principle it should be possible to test the hypothesis that lignin peroxidase is necessary for lignin degradation. This requires the disruption of the expressed sequence, which in turn depends upon the development of homologous integrative transformation (see later).

Genetic factors affecting extracellular lignin peroxidase activity in P. chrysosporium

Earlier work referred to above had shown that *P. chrysosporium* ME446 has two genome equivalents with abundant RFLPS; 53 strains derived from basidiospores obtained from the parental strain each had a single genome equivalent, with different combinations of alleles for the RFLP markers. Analysis of the RFLP composition of this set of strains yielded a genetic map, the first RFLP-based map for a fungus (Raeder et al. 1989a).

Is the extensive polymorphism at the DNA level (evidenced by RFLPs) mirrored by polymorphism at the level of performance? The basidiospore-derived strains that had been used for map construction now provided a set (with known compositions of alleles) with which this question could be asked. It was shown that there was indeed substantial variation in levels of extracellular lignin peroxidase levels between these strains (Raeder et al. 1989b). The amount of such enzyme activity was independent of the allele distributions of the lignin peroxidase gene clusters, but correlated with the allele distribution of another locus. This locus appears to control the spread of the lignin peroxidase-active state within the mycelial mat, and may be the mating type locus.

Other parameters examined with these strains were weight loss during the onset of the ligninolytic phase and the appearance of foci of brown coloration that spread, and that correlate with the amount of extracellular lignin peroxidase activity. A model was put forward for the onset of ligninolytic activity: this proposes two switches, the first resulting in either more than 42% or less than 28% weight loss in the mycelium, and the second (dependent on there having been extensive weight loss) resulting in more or less rapid spread of the brown coloration.

Improvement of performance

In principle, at least three routes to improvement in performance can be used separately or together. First, growth conditions can be optimised, whether it is for individual activities, e.g. extracellular lignin peroxidase production, or for the overall process of mineralisation of lignin to CO_2. For instance, we have identified conditions under which lignin, cellulose and hemicellulose are degraded simultaneously; these use ball-milled straw as the substrate, and sodium acetate buffer, pH6 (Birch 1988). Second, one can introduce mutations. This has been done for cellulase-deficiency (Eriksson et al. 1983; Eriksson et al. 1990); cellulase-deficient mutants that are still able to degrade lignin (perhaps obtaining the required H_2O_2 from hemicellulose degradation) might give a nutritionally upgraded feed for cattle since the cellulose is spared. Another class of mutant (that would require the development of transformation methods) would be to couple lignin peroxidase expression to an inducible promoter such as that for a cellobiohydrolase I (CBH I) gene. The third approach is to exploit the genetic diversity of *P. chrysosporium* strains and of their basidiospore-derived progeny in a programme of crosses. This system could provide a model for fungal strain improvement, in which the question that is asked is how far can one go exploiting such diversity, compared with the classical strain improvement programmes such as those for penicillin production and that of cellulase production by *T. reesei*, which have consisted of mutagenesis and screening in single lineages.

Such a strain improvement programme would involve crosses between basidiospore derived strains of known composition and of opposite mating types to establish strains with two genome equivalents, from which further basidiospore-derived strains would be obtained by fruiting. Correlations would then be made between levels of individual activities, overall performance and genetic composition, and strains for further crosses would then be chosen. Such a programme depends upon there being rapid, reproducible and convenient assays for the various parameters of performance, e.g. supernatant lignin peroxidase activity and lignin mineralisation.

We have found that optimal lignin mineralisation and optimal levels of extracellular lignin peroxidase are produced under different growth conditions. Indeed, under the conditions used for optimal lignin degradation, measuring $^{14}CO_2$ release from ^{14}C-labelled DHP or (^{14}C-lignin) lignocellulose (dimethyl succinate as buffer, pH4.5, and 2% glucose), no extracellular lignin peroxidase activity is found (A. Wyatt, P. Broda, unpublished results). However, this result may arise from entrapment of lignin peroxidase in the polysaccharide slime which is abundantly produced in the high glucose media found to be optimal for lignin mineralisation. Thus the measurement of extracellular enzyme levels may not be a good measure of the amount of enzyme available for lignin degradation and it may actually be misleading. Oligonucleotide probes for mRNA provide an alternative means of assaying for lignin peroxidase synthesis (James et al. 1992). An approach already being used by others is enzyme localisation studies using specific antibodies with either transmission electron microscopy or light microscopy (Blanchette et al. 1989; Daniel et al. 1989; Lackner et al. 1991).

We have asked whether there are differences with respect to ^{14}C DHP mineralisation between strains (A. Wyatt, P. Broda, unpublished results). A set of homokaryotic basidiospore-derived strains were compared with the parental strain ME446 and with a number of putative heterokaryotic strains obtained by crosses of the homokaryotic strains. There was considerable variation in the values obtained with independent cultures of each of the individual strains. A possible reason for such variation was the particulate and insoluble nature of the DHP substrate. However, such independent cultures of the same strain were more variable in performance than could be accounted for by this limitation on the experiment. Even so, there were still reproducible differences between the strains, so that in principle a strain improvement programme for DHP mineralisation is possible, but it would involve much effort because of the experimental variability in the system.

Expression of cellulase genes

One cellulase sequence from *P. chrysosporium*, from a CBH I gene, has been published (Sims et al. 1988). The argument for the nature of the CBH I sequence was based on its close homology with that of the archetypal CBH I sequence of *T. reesei*.

The genomic library of strain ME446 was also probed with endoglucanase I (EG I) DNA from *T. reesei*. The clones that were isolated and characterised contained CBH I-like sequences rather than EG I-like sequences; this result can be explained on the basis that in *T. reesei* there is a sequence that is common to the genes for these two types of enzyme (Knowles et al. 1987). These sequences would code for a second class of CBH I enzyme, different both from that previously reported by us and those described by another group (Covert et al. 1992). Probing with the two classes of CBH I DNA sequences revealed yet more related sequences in *P. chrysosporium* DNA digests. A similar result was obtained with DNA digests from a monokaryotic derivative of strain ME446, excluding an explanation for such complexity based on any RFLPs between allelic sequences in the heterokaryotic parent (P.F.G. Sims, unpublished).

Sequences of both CBH I classes were expressed when ball-milled straw was used as substrate, but not when glucose was used. A cDNA clone of the original (CBH I.1) type had 12 nucleotide differences (eleven of them silent) from the archetypal genomic sequence, indicating that it was a closely related sequence, perhaps an allelic variant, that was expressed. Two internal introns and two upstream introns were demonstrated. These latter introns are therefore located within the probable regulatory region of this gene and their precise role and influence on its expression may be of considerable importance. Similarly, cDNA of a possibly allelic variant of the CBH I.2 genomic sequence was found to have multiple base differences from the original sequence (mostly silent), and an upstream intron and an internal intron had both been spliced out. In addition, in this case differential splicing was also indicated. Thus *P. chrysosporium* has a CBH I system that has multiple non-allelic sequences, possible allelic differences, differential splicing, and differential regulation. The continued failure to isolate EG I like-sequences raises the question of whether such genes are present in *P. chrysosporium*.

A CBH II probe from *T. reesei* was used to isolate homologous sequences in the *P. chrysosporium* genomic library. Probing against a digest of total DNA of *P. chrysosporium* showed that in this case there was a unique locus, unlike with CBH I.1 and CBH I.2 (see above). Nevertheless, sequencing of both genomic and cDNA clones showed that there were two variants of this sequence, which can therefore be regarded as allelic (C. Tempelaars, P.F.G. Sims, P. Broda, unpublished results). This gene contains 6 introns, all of which are in the coding region. There are 5 single base differences within the intron sequences.

Towards a transformation system

A transformation system that allows insertional mutagenesis of *P. chrysosporium* is critical if we are to fully exploit the possibilities offered by DNA manipulation techniques. A site-directed mutagenesis experiment has allowed the isolation of the CBH I.1 promoter in a form that can be precisely coupled to other coding sequences. We have taken advantage of this in the development of a transformation vector for use in *P. chrysosporium*. This contains the coding sequence of a selectable bacterial marker, phleomycin resistance, coupled to the CBH I.1 promoter (P. Birch, P.F.G. Sims, P. Broda, unpublished results). Since phleomycin is toxic to *P. chrysosporium* as well as to bacteria, we hoped that this would facilitate controllable expression of drug resistance and thus selection of transformants. Experiments using the monokaryotic *P. chrysosporium* strain referred to above showed that stable apparently transformed isolates can indeed be recovered. However, analysis of these putative transformants suggests that transforming DNA is not integrated into the genome as is typically seen in filamentous fungi, but may be maintained extrachromosomally. Other groups have reported both integrative but non-homologous transformation (Alic & Gold 1991) and autonomous transformants (Randall et al. 1991; Randall & Reddy 1992).

We have also begun to evaluate other systems for integrative transformation. We have focused on those that offer a clear route to an homologous selectable marker and for which suitable recipient strains can themselves be easily isolated by positive selection strategies. Three such systems have been or are being assessed:

1. In a number of fungi strains deficient in nitrate reductase have been positively selected by their resistance to chlorate and also the nitrate reductase genes have been identified using heterologous probes. Although mutants of *P. chrysosporium* were isolated, we were unable to demonstrate homology between the *P. chrysosporium* genome and the *A. nidulans* nitrate reductase gene; therefore work on this system was not taken further.

2. The TrpC gene of *P. chrysosporium* was sequenced (Schrank et al. 1991). Sequence analysis of this gene has defined restriction fragments that contain the whole coding sequence and the putative promoter; such fragments could therefore be incorporated into vectors for use as a selectable marker in transformation experiments. The genetic organisation of the multifunctional TrpC gene has novel features compared with ones known from other fungi. We have used Polymerase Chain Reaction technology to demonstrate that it is unique in that it contains an intron. The same technique showed that another *prima facie* candidate for an intron was in fact not one, but instead codes for an additional amino acid sequence not found in other fungi.
3. We have also isolated and characterised the acetyl-CoA synthetase (FacA) gene of *P. chrysosporium* (Birch et al. 1992) as well as mutant strains lacking this activity, which is essential for growth on acetate. This will allow the evaluation of a second homologous marker system.

Actinomycete bacteria

There have also been studies on the role of actinomycete bacteria in lignocellulose biodegradation (McCarthy 1987). Actinomycetes, which have the same invasive habit as fungi, are common in decaying lignocellulose systems such as compost. Early studies (Phelan et al. 1979) indicated that a *Streptomyces* strain, in addition to being able to degrade the cellulose from lignocellulose and thereby contributing to a substantial weight loss of the substrate, is also able to degrade lignin, although to a much lesser extent. We began a research programme by screening actinomycetes for ligninolytic activity (McCarthy & Broda 1984). Organisms of several genera, *Streptomyces*, *Micromonospora*, *Thermomonospora* and *Actinomadura*, were identified as being able to grow on grass lignocellulose (McCarthy & Broda 1984). Whereas in liquid cultures of *P. chrysosporium* 30–40% of the (^{14}C lignin) lignocellulose used was converted to $^{14}CO_2$, the levels achieved by the actinomycetes were much lower (typically 10%, McCarthy & Broda 1984). However, in addition, some strains solubilize up to 30% of the total lignin content; that part which is not mineralized to $^{14}CO_2$ accumulates in the supernatant of cultures, and may represent an end product of lignin degradation.

This soluble product has been termed acid precipitable polymeric lignin or APPL (Crawford et al. 1983; Pometto & Crawford 1986) because it is insoluble when the pH of the medium is lowered. It has been partially characterized using high performance liquid chromatography (HPLC) (Borgmeyer & Crawford 1985; Mason 1988) and nuclear magnetic resonance (NMR) spectroscopy (McCarthy et al. 1986). APPL is a population of components in the apparent molecular weight range 1 to 3 $\times 10^5$ and the size distribution of the APPL products is specific for each organism cultured on ball milled straw (Mason 1988).

Differential degradation of lignocellulose components by different actinomycete species

Do different actinomycete strains have similar profiles of activity against the different components of lignocellulose? This was studied for ground and ball milled barley straw samples in 3-week incubations in liquid cultures of *Streptomyces cyaneus*, *Thermomonospora mesophila* and *Actinomadura* sp. (Zimmermann & Broda 1989). High weight losses from both substrates were found after 3 weeks of incubation. Water-soluble components released from the substrates did not accumulate in the supernatants but were partially utilised.

Because of the partial solubility of the lignin-carbohydrate component of straw in water at higher temperatures and pressures, part of the substrate was solubilised during the initial autoclaving; such water-soluble material contains not only low molecular weight compounds but also lignin-carbohydrate complexes of higher molecular weight. Although all three strains caused further weight losses from the substrates, they differed with respect to the removal of lignin. In addition to that removed by autoclaving, only *S. cyaneus* and *T. mesophila* removed a further part of the lignin from the substrate residues. This work demonstrates the importance of taking into account the effect of autoclaving and that, as with the fungi, different strains have different strategies for attacking lignocellulose.

Elemental analysis confirmed that pronounced chemical and structural changes in the lignin-carbohydrate polymer had occurred during incubation; these included oxidative reactions and the removal of methoxyl groups from the lignin. Gel chromatography of the lignin-carbohydrate components showed solubilisation of lower molecular size components, particularly by *S. cyaneus* and *T. mesophila*.

Degradation of a non-phenolic lignin model compound by S. cyaneus

Although we have not made extensive use of low molecular weight lignin model compounds, we have made one contribution in this area (Zimmermann et al. 1988a). It is known that many actinomycetes can utilise aromatic substrates including lignin-related phenols; some can perform transformations that include hydroxylations, demethylations and aromatic ring cleavages. Although an arylglycerol-β-phenyl ether was readily degraded by various actinomycetes, the presence of a substituent on the phenyl ether moiety inhibited or completely prevented its breakdown. We were able to show that cultures of *S. cyaneus* can degrade a non-phenolic dimeric β-O–4 model compound; by identifying the main products, we were able to show that the mechanism involved was cleavage of the C-α-C-β bond.

Cell-free solubilisation activities

An important step towards elucidating the system in actinomycetes was to develop a cell-free system (McCarthy et al. 1986). We showed that supernatants from a *T. mesophila* strain yielded a soluble product that was a complex of lignin, carbohydrates (especially pentoses derived from the hemicellulose) and protein. Chemical analysis and solid-state ^{13}C NMR spectroscopy both indicated similarities with humic acid, suggesting that actinomycetes might have an important role in the process of humification. The NMR spectra revealed an increase in the number of α-aryl ether linkages, one of the most common linkages in lignin, but there was no evidence for extensive demethoxylation.

It was found that *T. mesophila*'s solubilizing activity may be induced by either xylan or ball-milled straw, but not to a significant extent by indulin AT (an industrial lignin). An obvious possibility was that the solubilizing activity is primarily xylanolytic in nature. However, supernatants from strains of two other species (*Thermomonospora fusca* and *Saccharomonospora viridis*), which gave much higher levels of xylanolytic activity with xylan showed no significant levels of ^{14}C solubilization, suggesting that this is not the case, at least with these strains.

Later work involved the resolution of different activities from *S. cyaneus* MT813 that had been grown on ball-milled straw (Mason et al. 1988). This was achieved using ammonium sulphate precipitation and gel filtration of extracellular proteins. Material with an apparent molecular weight of about 20,000 accounted for almost all of the solubilization not attributed to cellulase activity. Since this had neither cellulase or xylanase activity, it was termed a lignin solubilizing activity.

Extracellular xylanases from actinomycetes

As well as examining activity against lignin in lignocellulose, we tested a number of thermophilic isolates for extracellular xylanase production. In each, xylanase activity was optimal within the temperature range 60–75 °C and was maintained between pH 5 and pH 8. In all cases, the xylan was initially hydrolysed to a mixture of oligomeric products, indicating that the main activity was of the endoxylanase type (McCarthy et al. 1985).

One xylanolytic strain of *S. viridis* was unusual in that it had no detectable cellulase activity. This was judged to be of some industrial interest, e.g. as a way of removing xylans from pulp to produce pure cellulose. It was also of value as a way of assessing the contribution of xylans to the papermaking properties of pulp (Roberts et al. 1990). Application of crude enzyme preparations to birch wood pulp resulted in the selective removal of approximately 20% of the total xylan. Paper handsheets from pulp treated in this way demonstrated modifications in a number of physical properties, of which a decrease in burst and long-span tensile strength were the most significant. The results suggested that specific removal of xylan reduces the extent of the inter-fibre bonding but does not significantly weaken the fibres themselves.

Xylan is mainly composed of 1,4 linked β-xylopyranosyl units. The presence of various substituents including arabinofuranosyl, glucuronyl and acetyl groups has a pronounced influence on the chemical and structural properties, and also on the enzymatic degradability of xylan in lignocellulose. We screened for xylanase, acetyl esterase and α-arabinofuranosidase activities in a number of actinomycetes, using oat spelt xylan (a de-acetylated preparation) and ball milled straw (Zimmermann et al. 1988b). The latter substrate was used as the source of a more complex and native form of hemicellulose. Each of the enzyme activities could be detected in at least some of the strains after different times of incubation; their production was also dependent on the growth medium.

Xylanolytic activities of S. cyaneus

In preliminary experiments, we sought to optimise the conditions for expression of cellulase and lignin solubilisation-activities as well as xylanase activity (Wang et al. 1992a). Eighteen possible carbon sources (mono-, di- and polysaccharides, aromatic compounds and ball milled straw) were tested for their ability to elicit these activities. Of these, only ball milled straw and xylan had this effect, and xylan was chosen as the carbon source in most subsequent experiments. pH 7.0 was also found to be the optimum for the expression of all activities in cultures.

The fact that both xylan and ball milled straw can induce all three lignocellulose solubilisation systems raises questions about their regulation. Both substrates are too large and complex to enter cells, so how does induction occur? Why should xylan induce all three activities? It is possible that the xylan used (oat spelt) is not pure, but it also could indicate that xylan degradation is a key step. The prediction was made that *S. cyaneus* produces a 'sensor' xylanase constitutively and that this degrades any xylan present in the environment to smaller molecules that can act as intracellular inducers. The testing of this prediction is described below.

Demonstration of three xylanases from S. cyaneus

Supernatants from xylan-grown cultures were precipitated with ammonium sulphate, and gel filtration and SDS-PAGE allowed the resolution of three xylanases and their existence was confirmed using a zymogram method. The apparent sizes were 37 (Xylanase I), 34 (II) and 45 (III) kD. After further purification, these were characterised for their temperature optima (72 °C, 65 °C and 53 °C, respectively) and pH optima. Xylanase I and III had activity between pH 4 and 10.5 whereas xylanase II only had activity between pH 5.5–7.5.

Novel turbidimetric assay for xylanase activity

Xylanase activities are usually determined by measuring the release of reducing sugars from the substrate. However, this method cannot offer accurate results where the presence of both endo- and exo-xylanases results in synergism. Moreover, it is not sensitive enough to monitor very low levels of xylanase activity or activity in the presence in the mixture of reducing sugars. The oxidation or reduction of products can also affect the values obtained. An alternative method was required for the detection of sensor xylanase activity.

The alternative method developed (Wang et al. 1992a), involves the use of a stable xylan preparation that consists of xylan chains with sizes in the range of 70–100 residues. Such molecules are in suspension rather than solution (like for instance *E. coli* cells), and their concentration can therefore be measured by the deflection of light (e.g. at 600 nm); that is, turbidimetrically. Hydrolysis of such particles by endoxylanase results in their disintegration into soluble products, and this can be measured as a decrease in turbidity. We have shown that, over a range, linear kinetics are obtained.

Regulation of synthesis of S. cyaneus *xylanases*

The kinetics of xylanase induction by xylan were studied over a 17-day period. It was observed that there was an increase in reducing sugar concentration before any large amount of xylanase was induced, and that at this point the reducing sugar concentration declines. A possibility is that there is induction of transport proteins at this point. Mixed incubation with xylan and glucose resulted in the delay of xylanase induction until the glucose concentration had declined to a low level, suggesting that glucose repressed xylanase induction. However, in the presence of glucose, as in its absence, there was an increase in reducing sugar concentration and a decrease in xylan concentration, suggesting the existence of a constitutively produced low level of activity. It was shown that xylanase III is detectable before xylanases I and II on induction, and the novel assay was used to show that with glucose-grown cultures this is the only xylanase activity that is present. It is therefore proposed that xylanase III is the predicted constitutively produced enzyme and that xylanases I and II are inducible.

Cloning of S. cyaneus *xylanase genes*

Investigation of the lignocellulose degrading enzymes from *S. cyaneus* has involved the preparation of genomic libraries in *E. coli* using both lambda and plasmid vectors. However, the introduction of DNA constructs containing *S. cyaneus* DNA into *E. coli* resulted in very low transformation frequencies.

Strains of *E. coli* derived from K12 contain a methylcytosine-specific restriction system encoded by *mcrA* and *mcrB* and we hypothesised that the incoming *S. cyaneus* DNA was being restricted. Strains mutated in the *mcrAB* region could be transformed by constructs containing *S. cyaneus* DNA at a much higher frequency than before, suggesting that in previous *E. coli* hosts the incoming *S. cyaneus* DNA had indeed been cleaved at frequent methylcytosine residues. The improved transformation frequencies thus obtained made possible the construction of gene libraries with adequate-sized inserts (Wang et al. 1992b).

As stated above, three xylanases are produced by *S. cyaneus*. One is expressed constitutively whereas the other two are only expressed when xylosaccharides are present in the medium. An *S. cyaneus* DNA library was constructed in Lambda gt11 by inserting *Eco*RI-digested DNA into the *Eco*RI site of the *lacZ* gene in the vector. The expression library was screened by xylanase activity using RBB-xylan as substrate. Two classes of xylanase-producing clones were identified; they contained different inserted fragments that had no DNA homology with each other. These inserts were then subcloned into pUC18, mapped with restriction enzymes and examined for xylanase production. A subclone derived from one of the classes of insert expressed xylanase at a very low level whereas a subclone derived from the other class of lambda insert expressed a very high level of xylanase activity, which was secreted from the *E. coli* host cells (Wang 1991).

The first report of cloning and sequencing of *Streptomyces* xylanase genes was that of Shareck et al. (1991). The protein sequences inferred from the DNA sequence showed similarities with those of one or other of two sets of glycanases that had already been described. DNA sequencing of the two clones that we have isolated is in progress and shows significant amino acid similarities between the derived protein sequences and conserved regions of both xylanases and cellulases from other prokaryotes.

Conclusions

Biotechnology has been primarily concerned with the health care and food industries, and agricultural biotechnology is now assuming greater importance. However, the use of biotechnology to solve environmental problems and for the rational utilization of renewable resources are still in their infancy. The reasons include problems of what the precise objectives are, the complexity of the ecosystems, concern about release of genetically manipulated organisms, the generally low value and bulk of the materials involved, the lack of public and private funding for the basic research needed, and perhaps inadequate recognition of the true environmental costs of existing processes, in terms of pollution and inefficient utilisation of resources.

There are strong arguments for improved and enhanced production and utilisation of biomass, which include minimising the release of CO_2 by switching from once-off utilisation of fossil fuels to use of renewable resources. Properly managed, biomass production can also minimise soil degradation and enhance employment. However, each of the global regions has different problems and opportunities, and in each the overall process from production to utilization and also disposal of wastes must be appraised in terms of Clean Technology. That is, solutions to pollution problems should not be 'end of pipe', but should incorporate the avoidance or minimisation of creation of pollutants within the process in the first place.

Some crops, e.g. wood and straw, will be used as fuels, especially in countries such as Denmark that have a heavy tax on the domestic use of fossil fuels, and they can be used on the farm, e.g. for grain drying and greenhouse heating. Such processes require very large storage facilities. Science and technology has already made a major input to improving productivity of biomass, e.g. in tree propagation, silviculture and cropping. However, the overall process of paper production involves many subsequent steps, including utilisation and disposal of bark, branches and roots, transport, milling, thermomechanical or chemical pulping, bleaching, and disposal of wastes, and also re-cycling of waste paper and board into fresh pulp. The reappraisal of the whole of this complex process must be in terms that allow real environmental costs to be included in the balance sheet; the same holds for the conversion and utilisation of lignocellulose for animal nutrition. Such analyses would identify the pressing aspects from the point of view of devising novel technologies, which might, but need not, have biological components.

Biotechnology has two roles: the obvious one is to design processes with a biological component, whether that involves whole organisms or enzymes; the other is to provide the understanding that is still lacking on the interaction of degrading and modifying organisms with their substrates. Specific scientific questions include how in nature is encrustation by lignin disrupted, and what are the immediate fates of lignin and the other components. Because of the progress made in a number of laboratories, including our own, each adopting its distinctive approach, this is a good time to identify some of the major problems and opportunities for the immediate future (Broda et al. 1989).

1. Because of the chemical and structural complexity of the lignocellulose substrate, which includes water-soluble and insoluble components, a description of the overall process of lignocellulose degradation does not yet exist for any substrate or organism. Thus the degree of solubilization and modification of the different components, the further fate of the material that is solubilized and to what extent particular fungi or actinomycetes have similar effects on the lignocellulose substrate are not known at the chemical level. With respect to the lignin component specifically, in this article the use of model compounds, DHP and radiolabelled lignin have been emphasized. However, it is important to complement these with other approaches such as chemical and gravimetric analysis of the fate of the substrate, improved liquid chromatography methods, and use of NMR spectroscopy and other non-destructive techniques.
2. It is well known that in both prokaryotic and eukaryotic systems degradation of cellulose involves a number of enzymes; in some organisms a large number are involved (Bayer & Lamed, this issue). Hemicellulose degradation involves enzymes for both main-chain and side-chain degradation, though little is yet known of the details, largely because hemicelluloses

themselves are so poorly understood, and model compounds are generally not yet available. The degradation of each of these classes of substrate will be better understood with the availability of individual enzymes, in a pure form, made by genetic engineering methods. It will then be possible to do experiments to determine the synergistic actions of different enzymes, confident that each is free of contaminating activities. The availability of better-defined carbohydrate substrates and assays are also essential for such studies.

3. *P. chrysosporium* is the archetypal organism for lignin degradation studies. Lignins from a number of sources (wood, straw, bagasse) are all degraded efficiently and furthermore molecular and classical genetics as well as biochemical methods can be used, as has been discussed. Moreover, it has a sharp switch from growth to the ligninolytic idiophase. In this, it may be atypical, and it will be of interest to what extent the white rot fungi have common systems with respect to the enzymes and their regulation. However, it is possible that lignin degradation by *P. chrysosporium* simply requires lignin peroxidases, manganese peroxidases and the complementary H_2O_2 generating system. It will be important to determine whether there is any spatial organization of the component enzymes of the system (see e.g. Ruel & Joseleau 1991). Other questions of importance are the nature of the coupling of lignin degradation to carbohydrate metabolism at the level of gene expression, the reason for the presence of multiple genes and the basis for the variability in performance discussed earlier.
4. The actinomycete system is at an earlier stage of study, mainly because with the available techniques lignin solubilization and degradation have been more difficult to demonstrate rigorously than with the fungal system. However, it is a promising system which is significantly different from the *P. chrysosporium* system, and where different strains have differing characteristics. Parallel studies of lignin degradation with the white-rot fungi and actinomycetes might provide interesting comparisons and suggest avenues of exploitation utilizing these differences. If the aim is the complete destruction of the lignin component of lignocellulose then the choice of a fungal system would seem obvious. However, as described above, such degradation of lignin only occurs during the idiophase, following severe nitrogen and carbon depletion. In contrast, with actinomycetes lignin solubilization occurs during primary growth and has no requirement for the presence of H_2O_2. Thus although lignin degradation may be more limited, it may have the advantage in making available solubilized products of lignin degradation (APPL) rather than its total destruction to CO_2.
5. The progress made in this area means that we can now begin to assemble a specification for a useful organism to exploit particular lignocellulose resources. One objective of particular relevance to the U.K. is the improved utilization of lignocellulose for animal nutrition. In this case, one can envisage an organism that degrades lignin and hemicellulose efficiently but has no activity against cellulose, with the ability to grow in the low pH and high temperature commonly encountered in silage. Alternatively, one might use enzyme preparations from such an organism grown under industrial fermentation conditions. However, it is still too early to predict which group of micro-organisms, e.g. white-rot fungi or actinomycetes, will produce the best candidate for exploitation in this manner.
6. Lignin biodegradation provides a good model for the biotechnology of open systems of an intermediate scale. It is thus different both from the fermenter-scale systems in use for e.g. production of interferons and also from typical large chemical engineering processes. A possible mode for lignocellulose utilization is solid substrate fermentation (comparable in complexity to mushroom composting or silage production) which would be effective on the farm and suitable for use in some developing countries. The emphasis would not be on major capital costs, but on devising robust organisms appropriate to the local context.

7. As stated at the beginning of this article, rational discussion of whether biotechnology can contribute to the utilization of lignocellulose requires a better understanding of lignocellulose biodegradation. Until this is achieved, it will not be possible to assess the economics of such processes in particular contexts. *P. chrysosporium* has served as an excellent model system; it degrades a wide range of lignocellulose substrates (hardwood, softwood, straw and bagasse) and a substantial body of biochemical and molecular genetic knowledge exists for it. Nevertheless, it is important that other fungi should be assessed for their performance on lignocellulose. Heterologous probing using genes from *P. chrysosporium* and *T. reesei* will allow the rapid development of molecular genetics for such organisms. One can envisage the production of enzymes either by the original organism or by cloning the genes into alternative hosts such as yeast or *Trichoderma*, which is able to produce exceptionally large amounts of extracellular cellulases. In some cases the objective would be to tailor the organism to produce a set of enzymes in relative amounts that have been optimized for the intended process.
8. Preliminary experiments have already been reported on the use of cellulase-defective mutants of *P. chrysosporium* as part of a pulping process for papermaking of bagasse. In this case presumably residual sugars and/or the hemicelluloses are utilized by the organisms for the production of H_2O_2. The different ways of strain improvement that are possible include genetic crosses and cloning techniques. An obvious example is to attempt to obtain mutants of *P. chrysosporium* that are able to degrade lignin in normal growth rather than only in starvation. Traditional methods of strain improvement involving mutagenesis may not yield organisms that are sufficiently robust for solid substrate fermentation conditions. The prospects with strains obtained using genetic crosses and/or site directed mutagenesis may be better.
9. The development of particular organisms or of ways of producing enzymes in high yield are only the start of the development of industrial processes. There are many aspects of the problem which will require substantial inputs from microbial physiologists and chemical engineers, in particular, before one can arrive at viable processes.
10. The use of whole organisms would have the advantage of low price compared with the use of enzymes, but the problems of survival and release of whole organisms must be addressed, especially if *in vitro* genetic manipulation techniques are used to construct such organisms.

Acknowledgements

Research in the author's laboratory was supported jointly by British Petroleum and the Agricultural and Food Research Council.

References

Agosin E, Daudin JJ & Odier E (1985) Screening of white-rot fungi on (^{14}C) lignin-labelled and (^{14}C) whole-labelled wheat straw. Appl. Microbiol. Biotechnol. 21: 6811–6817

Alic M & Gold MH (1991) Genetics and molecular biology of the lignin-degrading basidiomycete *Phanerochaete chrysosporium*. In: JW Bennett & LL Lasure (Eds) More Gene Manipulations in Fungi (pp 320–341). Academic Press, San Diego

Anon (1992) Proceedings of 13th Symposium on Biotechnology for Fuels and Chemicals, 1991. In: Appl. Biochem. & Biotechnol. Vol 35–5

Bayer E & Lamed R (1992) The cellulose paradox: pollutant *par excellence* and/or a reclaimable natural resource? Biodegradation 3: 171–188 (this issue)

Benner R, Maccubbin AE & Hodson RE (1984) Preparation, characterization and microbial degradation of specifically radiolabelled ^{14}C-lignocelluloses from marine and freshwater macrophytes. Appl. Environ. Microbiol. 47: 988–1004

Birch OM (1988) Extracellular enzymes from the lignin-degrading fungus *Phanerochaete chrysosporium*. PhD thesis, University of Manchester

Birch PRJ, Sims PFG & Broda P (1992) Nucleotide sequence of a gene from *Phanerochaete chrysosporium* that shows homology to the *facA* gene of *Aspergillus nidulans*. DNA sequence 2: 319–323

Blanchette RA, Abad AR, Farrell RL & Leathers TD (1989) Detection of lignin peroxidase and xylanase by immunocytochemical labelling in wood decayed by basidiomycetes. Appl. Environ. Microbiol. 55: 1457–1465

Borgmeyer JR & Crawford DL (1985) Production and characterization of polymeric lignin degradation intermediates from two different *Streptomyces* spp. Appl. Environ. Microbiol.49: 273–278

Broda P, Sims PFG & Mason JC (1989). Lignin Biodegradation: a molecular biological approach. In: Marshall RD & Tipton KF (Eds) Essays in Biochemistry 24 (pp 82–114). Academic Press, London

Brown A, Sims PFG, Raeder U & Broda P (1988) Multiple ligninase genes from *Phanerochaete chrysosporium*. Gene 73: 77–85

Chesson A & Orskov ER (Eds) (1989) Physico-chemical Characterisation of Plant Residues for Industrial and Feed Use. Elsevier Applied Science, London & New York

Coombs J (1987) EEC resources and strategies. In: Hartley BS et al. (pp 1–16)

Coughlan MP (1985) Cellulases with comment on their production and application. In: Russell GE (Ed) Biotechnology and Genetic Engineering Reviews, Vol 3 (pp 39–109). Intercept, Ponteland, Newcastle upon Tyne

Coughlan MP (Ed) (1989) Enzyme Systems for Lignocellulose Degradation. Elsevier Applied Science, London & New York

Covert SF, Vanden Wymelenberg A & Cullen D (1992) Structure, organization, and transcription of a cellobiohydrolase gene cluster from *Phanerochaete chrysosporium*. Appl. Environ. Microbiol. 58: 2168–2175.

Crawford RL & Crawford DL (1978) Radioisotopic methods for the study of lignin biodegradation. Dev. Ind. Microbiol. 19: 35–49

Crawford DL, Pometto AL & Crawford RL (1983) Lignin degradation by *Streptomyces viridosporus*: isolation and characterization of a new polymeric lignin degradation intermediate. Appl. Environ. Microbiol. 45: 898–904

Daniel G, Nilsson T & Pettersson B (1989) Intra- and extracellular localization of lignin peroxidase during the degradation of solid wood and wood fragments by *Phanerochaete chrysosporium* by using transmission electron microscopy and immunogold labelling. Appl. Environ. Microbiol. 55: 871–881

Derikx PJL, Op den Camp HJM, van der Drift C, van Griensven LJLD & Vogels GD (1990) Biomass and biological activity during the production of compost used as a substrate in mushroom cultivation. Appl. Environ. Microbiol. 56: 3029–3034

Dordick J, Marletta MA & Klibanov AM (1986) Peroxidases depolymerize lignin in organic media but not in water. Proc. Nat. Acad. Sci. U.S.A. 83: 6255–5257

Effland MJ (1977) Modified procedure to determine acid-insoluble lignin in wood and pulp. TAPPI 60: 143–144

Eriksson K-E, Johnsrud SC & Vallander L (1983) Degradation of lignin and lignin model compounds by various mutants of the white rot fungus *Phanerochaete chrysosporium*. Arch. Microbiol. 135: 161–168

Eriksson K-EL, Blanchette RA & Ander P (1990) Microbial and Enzymatic Degradation of Wood and Wood Components. Springer, Berlin, Heidelberg, New York

Eveleigh DL (1987) Cellulase: a perspective. In: Hartley et al. (pp 31–43)

Ferrero GL, Ferranti MP & Naveau H (1984) Anaerobic Digestion and Carbohydrate Hydrolysis of Waste. Elsevier Applied Science, London & New York

Flegg PB, Spencer DM & Wood DA (Eds) (1985) The Biology and Technology of the Cultivated Mushroom. Wiley, Chichester

Galliano H, Gas G, Seris JL & Boudet AM (1991) Lignin degradation by *Rigidoporus lignosus* involves synergistic action of two oxidising enzymes: Mn peroxidase and laccase. Enzyme Microb. Technol. 13: 478–482

Gaskell J, Dieperink E & Cullen D (1991) Genomic organization of lignin peroxidase genes of *Phanerochaete chrysosporium*. Nucleic Acids Res. 19: 599–603

Glenn JK, Akileswaran L & Gold MH (1986) Mn(II) oxidation is the principal function of the extracellular Mn-peroxidase from *Phanerochaete chrysosporium*. Arch. Biochem. Biophys. 251: 688–696

Godfrey BJ, Mayfield MB, Brown JA & Gold MH (1990) Characterization of a gene encoding a manganese peroxidase from *Phanerochaete chrysosporium*. Gene 93: 119–124

Hadar Y, Karem Z, Gorodecki B & Ardon O (1992) Utilization of lignocellulosic waste by the edible mushroom, *Pleurotus*. Biodegradation 3: 189–205 (this issue)

Haider K & Trojanowski J (1975) Decomposition of specifically ^{14}C-labelled phenols and dehydropolymers of coniferyl alcohol as models for lignin degradation by soft and white rot fungi. Arch. Microbiol. 105: 33–41

Haider K & Trojanowski J (1980) A comparison of the degradation of ^{14}C-labelled DHP and cornstalk lignins by micro and macrofungi and by bacteria. In: Kirk TC, Higuchi T & Chang HM (Eds) Lignin biodegradation: Microbiology, Chemistry and Potential Applications (pp 111–134). CRC Press, Boca Raton

Hall PL (1980) Enzymatic transformations of lignin: 2. Enzyme Microb. Technol. 2: 170–176

Hammel KE & Moen MA (1991) Depolymerization of a synthetic lignin *in vitro* by lignin peroxidase. Enzyme Microb. Technol. 13: 15–18

Hartley BS, Broda PMA & Senior PJ (Eds) (1987) Technology in the 1990s: utilization of lignocellulosic wastes. The Royal Society, London

James CM, Felipe MSS, Sims PFG & Broda P (1992) Expression of a single lignin peroxidase-encoding gene in *Phanerochaete chrysosporium* strain ME446. Gene 114: 217–222

Kerem Z, Friesem D & Hadar Y (1992) Lignocellulose degradation during solid-state fermentation: *Pleurotus ostreatus* versus *Phanerochaete chrysosporium*. Appl. Environ. Microbiol. 58: 1121–1127

Keyser P, Kirk TK & Zeikus JG (1978) Ligninolytic enzyme system of *Phanerochaete chrysosporium*: synthesized in the absence of lignin in response to nitrogen starvation. J. Bacteriol. 135: 790–797

Kirk TK & Brunow G (1988) Synthetic ^{14}C-labelled lignins. In: Wood & Kellogg (1988b) (pp 65–73)

Kirk TK & Chang H-M (Eds) (1990) Biotechnology in Pulp and Paper Manufacture. Butterworth-Heinemann, Boston

Kirk TK & Farrell RL (1987) Enzymatic 'Combustion': the microbial degradation of lignin. Ann. Rev. Microbiol. 41: 465–505

Knowles J, Lehtovaara P, Teeri T, Pentilla M, Salovuori I & Andre L (1987) The application of recombinant-DNA technology to cellulases and lignocellulosic wastes. In: Hartley et al. (pp 45–50)

Kuwahara M, Glenn JK, Morgan MA & Gold MH (1984) Separation and characterization of two extracellular H_2O_2-dependent oxidases from ligninolytic cultures of *Phanerochaete chrysosporium*. FEBS Lett. 169: 247–250

Lackner R, Srebotnik E & Messner K (1991) Immunogold-silver staining of extracellular ligninases secreted by *Phanerochaete chrysosporium*. Canad. J. Microbiol. 37: 665–668

Leisola MSA, Kozulic B, Meussdoerffer F & Fiechter A (1987) Homology among multiple extracellular peroxidases from *Phanerochaete chrysosporium*. J. Biotechnol. 2: 379–382

Lewis NG, Razal RA & Yamamoto E (1987) Lignin degradation by peroxidase in organic media: a reassessment. Proc. Nat. Acad. Sci. U.S.A. 84: 7925–7927

McCarthy AJ (1987) Lignocellulose-degrading actinomycetes. FEMS Microbiol. Rev. 46: 145–163

McCarthy AJ & Broda P (1984) Screening for lignin-degrading actinomycetes and characterization of their activity against ^{14}C-lignin-labelled wheat lignocellulose. J. Gen. Microbiol. 130: 2905–2913

McCarthy AJ, MacDonald MJ, Paterson A & Broda P (1984) Degradation of ^{14}C lignin-labelled wheat lignocellulose by white-rot fungi. J. Gen. Microbiol. 1340: 1023–1030

McCarthy AJ, Peace E & Broda P (1985) Studies on the extracellular xylanase activity of some thermophilic actinomycetes. Appl. Microbiol. Biotechnol. 21: 238–244

McCarthy AJ, Paterson A & Broda P (1986) Lignin solubilisation by *Thermomonospora mesophila*. Appl. Microbiol. Biotechnol. 24: 347–352

Magnuson TS & Crawford DL (1992) Comparison of extracellular peroxidase and esterase-deficient mutants of *Streptomyces viridosporus* T7A. Appl. Environ. Microbiol. 58: 1070–1072

Mason JC (1988) HPLC analysis of solubilized products from lignocellulose degradation by actinomycetes. Biotechnol. Techn. 2: 95–100

Mason JC, Richards M, Zimmermann W & Broda P (1988) Identification of extracellular proteins from actinomycetes responsible for the solubilization of lignocellulose. Appl. Microbiol. Biotechnol. 28: 276–280

Mason JC, Birch OM & Broda P (1990) Preparation of ^{14}C radiolabelled lignocellulose from spring barley of different maturities and their solubilisation by *Phanerochaete chrysosporium* and *Streptomyces cyaneus*. J. Gen. Microbiol. 136: 227–232

Niku-Paavola M-L, Karhunen E, Kantilinen A, Viikari L, Lundell T & Hatakka A (1990) The effect of culture conditions on the production of lignin modifying enzymes of the white-rot fungus *Phlebia radiata*. J. Biotechnol. 13: 211–221

Overend RP & Chornet E (1987) Fractionation of lignocellulosics by steam-aqueous pretreatments. In: Hartley et al. (pp 119–132)

Pham TTT, Maaroufi A & Odier E (1990) Inheritance of cellulose- and lignin-degrading ability as well as endoglucanase isozyme pattern in *Dichomitus squalens*. Appl. Microbiol. Biotechnol. 33: 99–104

Phelan MB, Crawford DL & Pometto AL (1979). Isolation of lignocellulose-decomposing actinomycetes and degradation of specifically ^{14}C-labelled lignocellulose by six selected *Streptomyces* strains. Canad. J. Microbiol. 25 1270–1276

Pometto AL & Crawford DL (1986) Catabolic fate of *Streptomyces viridosporus* T7A-produced, acid-precipitable polymeric lignin upon incubation with ligninolytic *Streptomyces* species and *Phanerochaete chrysosporium*. Appl. Environ. Microbiol. 51: 171–179

Raeder U & Broda P (1986) Meiotic segregation analysis of restriction site polymorphisms allows rapid genetic mapping. EMBO J. 5: 1125–1127

Raeder U, Thompson W & Broda P (1989a) RFLP-based genetic map of *Phanerochaete chrysosporium* ME446: lignin peroxidase genes occur in clusters. Molec. Microbiol. 3: 911–918

Raeder U, Thompson W & Broda P (1989b) Genetic factors influencing lignin peroxidase activity in *Phanerochaete chrysosporium*. Molec. Microbiol 3: 919–924

Ragg PL & Fields PR (1987) The development of a process for the hydrolysis of lignocellulosic waste. In: Hartley et al. (pp 133–143)

Randall TA & Reddy CA (1992) The nature of extra-chromosomal maintenance of transforming plasmids in the filamentous basidiomycete *Phanerochaete chrysosporium*. Curr. Genet. 21: 255–260

Randall TA, Reddy CA & Boominathan K (1991) A novel extrachromosomally maintained transformation vector for the lignin degrading basidiomycete *Phanerochaete chrysosporium*. J. Bacteriol. 173: 776–782

Reid ID (1989) Solid-state fermentations for biological delignification. Enzyme Microb.Technol. 11: 786–803

Roberts JC, McCarthy AJ, Flynn NJ & Broda P (1990) Modification of paper properties by pretreatment of pulp with *Saccharomonospora viridis* xylanase. Enzyme Microb. Technol. 12: 210–213

Ruel K & Joseleau J-P (1991) Involvement of an extracellular glucan sheath during degradation of *Populus* wood by *Phanerochaete chrysosporium*. Appl. Environ. Microbiol. 57: 374–384

Schrank A, Tempelaars C, Sims PFG, Oliver SG & Broda P (1991) The *trpC* gene of *Phanerochaete chrysosporium* is unique in containing an intron but nevertheless maintains the order of functional domains seen in other fungi Molec. Microbiol. 5: 467–476

Shareck F, Roy C, Yaguchi M, Morosoli R & Kluepfel D (1991) Sequences of three genes specifying xylanases in *Streptomyces lividans*. Gene 107: 75–83

Shoham et al. (1992) Delignification of wood pulp by a thermostable xylanase from *Bacillus stearothermophilus* strain T-6. Biodegradation 3: 207–218 (this issue)

Sims PFG, James C, & Broda P (1988) The identification, molecular cloning and characterisation of a gene from *Phanerochaete chrysosporium* that shows strong homology to the exo-cellobiohydrolase I gene from *Trichoderma reesei*. Gene 74: 411–422

Stewart P, Kersten P, vanden Wymelenberg A, Gaskell J & Cullen D (1992) Lignin peroxidase gene family of *Phanerochaete chrysosporium*: complex regulation by carbon and nitrogen limitation and identification of a second dimorphic chromosome. J. Bacteriol. 174: 5036–5042

Tanahashi M & Higuchi T (1981) Dehydrogenative polymerization of monolignols by peroxidase and H_2O_2 in a dialysis tube. I. Preparation of highly polymerized DHPs. Wood Research 67: 29–42

Visser J, Kusters MA, Beldman G & Voragen AG (Eds) (1992) Xylans and Xylanases. Elsevier Science Publishers, Amsterdam

Wang P (1991) Enzymology and molecular biology of lignocellulose degradation systems of *Streptomyces cyaneus*. Ph.D. thesis, University of Manchester

Wang P, Ali S, Mason JC, Sims PFG & Broda P (1992a) Xylanases from *Streptomyces cyaneus*. In: Visser et al. (pp 225–234)

Wang P, Harvey SS, Sims PFG & Broda P (1992b) The construction of *Streptomyces cyaneus* genomic libraries in *Escherichia coli* is dependent upon the use of Mcr-deficient strains. Gene 119: 127–129

Wariishi H, Valli K & Gold MH ((1991) *In vitro* depolymerization of lignin by manganese peroxidase of *Phanerochaete chrysosporium*. Biochem. Biophys. Rec. Comm. 176: 269–275

Williams AG (1989) Hemicellulose utilization by microorganisms in the alimentary tract of ruminant and non-ruminant animals. In: Coughlan MP (1989) (pp 183–219)

Wood WA & Kellogg ST (1988a) Methods in Enzymology, Vol 160: Biomass Part A, Cellulose and Hemicellulose. Academic Press, San Diego

Wood WA & Kellogg ST (1988b) Methods in Enzymology, Vol 161: Biomass Part B, Lignin, Pectin & Chitin. Academic Press, San Diego

Zimmermann W (1990) Degradation of lignin by bacteria. J. Biotechnol. 13: 119–130

Zimmermann W & Broda P (1989) Utilization of lignocellulose from barley straw by actinomycetes. Appl. Microbiol. Biotechnol. 30: 103–109

Zimmermann W, Broda P, Umezawa T & Higuchi T (1988a) Degradation of a non-phenolic arylglycerol β-aryl ether by *Streptomyces cyaneus*. FEBS Letts. 239: 5–7

Zimmermann W, Winter B & Broda P (1988b) Xylanolytic enzyme activities produced by mesophilic and thermophilic actinomycetes grown on graminaceous xylan and lignocellulose. FEMS Microbiol. Letts. 55: 181–186

Biodegradation **3**: 239–254, 1992.

The use of natural bacterial populations for the treatment of sulphur-containing wastewater

J.G. Kuenen & L.A. Robertson
Kluyver Laboratory for Biotechnology, Delft University of Technology, Julianalaan 67, 2628 BC Delft, The Netherlands

Key words: sulphur, sulphide, wastewater treatment, pollution

Abstract

Pollution by inorganic and organic sulphur compounds is increasing and, because of the many environmental hazards associated with these compounds (e.g. toxicity, acidification of rain and freshwater, increase of COD, the greenhouse effect), must be taken seriously. There is a wide variety of sulphur oxidizing bacteria available in nature, and these can be used for the effective control of such pollution. The best way to break the sulphur cycle is to stop it at sulphur which, being insoluble, can be easily recovered (e.g. $SO_4^{2-} \rightarrow S^{2-} \rightarrow S^0$). (Eco)physiological knowledge about the sulphur oxidizing bacteria has proved very useful in the prediction of the performance of sulphur oxidizing communities in actual wastewater treatment systems. Appropriate reactor design, based on this type of study, is essential if such bacterial communities are to function efficiently, especially when toxic sulphides must be treated. This paper reviews the natural and anthropogenic sources of sulphur pollution, its consequences and possible solutions.

Introduction

Human activities are disturbing the sulphur cycle on a local, or even global, scale. There are two major causes: the formation of SO_2 and other sulphur compounds by the burning of fossil fuels, and the boosting of the sulphur cycle by various forms of eutrophication. This boosting can result in local pollution as a result of extensive sulphate reduction.

The biological sulphur cycle (Fig. 1) is normally kept in balance by a number of processes that make the cycle, as a whole, very resilient. Thus if high concentrations of hydrogen sulphide accumulate, it may be oxidized (biologically or chemically) to elemental sulphur, and thus removed from the general cycle. That this is sometimes the case on a geological scale is demonstrated by the huge sulphur deposits in some geological formations. Similarly, under suitable conditions, over-production of hydrogen sulphide can result in the formation of metal sulphide masses, generally to be found as the various pyritic ores. The oxidized form of sulphur, sulphate, also occurs on the geological scale, as calcium sulphate deposits.

In spite of this impressive resilience and storage capacity, the cycles can be observed to be out of balance, both globally and in localized areas. Of cause, this may be due to natural causes, such as a volcanic explosion or 'natural' eutrophication. However, the greatest threat to the balance of the sulphur cycle is human activity, particularly SO_2 generation which is causing acid rain (see below). When it is asked what biology, and especially microbiology, can do to abate these problems, it must always be emphasized that prevention is better

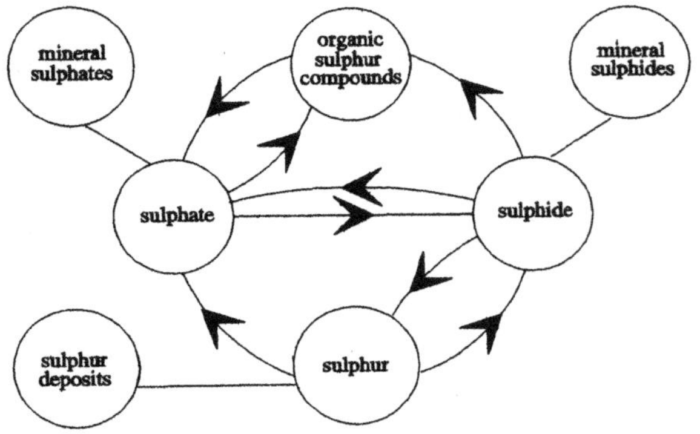

Fig. 1. The biological sulphur cycle.

than a cure. Microbiological contributions to the solution of the sulphur pollution problem can be made along three lines:

1. Expand our understanding of the sulphur cycle, and the ways in which human activity is unbalancing it.
2. On the basis of this information, develop prevention methods.
3. Should prevention not be possible, or fail, provide methods of remediation, especially for localized problems.

This review will catalogue the main natural and anthropogenic causes of sulphur pollution, and discuss existing and conventional techniques for treating reduced and partially oxidized sulphur-containing wastes. Although most of these processes are 'end of the pipe' solutions, some emphasis will be placed on recycling processes which can be integrated into a total process design.

General pollution problems

Sources

Natural sources of reduced and (partially) oxidized sulphur compounds

The most spectacular sources of sulphur and sulphides are undoubtedly geothermal, whether by free dispersal into the atmosphere as a volcano erupts, by diffusion from geothermal springs (e.g. Yellowstone Park), or within the hydrothermal areas of the sea. It is, of course, difficult to quantify these contributions to the global sulphur cycle, estimates of SO_2 emission from volcanoes, for example, vary by an order of magnitude (from 1.5 to 34 Tg S as SO_2 year^{-1}; Lein 1983). The amount emitted will depend on the type of volcano involved. Andesitic volcanoes emit high concentrations of water vapour and lower amounts of SO_2. The more common basaltic volcanoes release greater concentrations, with the basaltic volcanoes of the Hawaiian Islands producing the most (Lein 1983). Fumarole activity, while less spectacular than full scale eruptions, also contributes a certain amount. Interested readers should consult Lein (1983) for an extensive review of the data. Sulphur compounds are also physicochemically released from sedimentary rocks and mineral deposits (e.g. pyrites) by weathering and from the sea by aerosol formation.

Biological activity also contributes significantly to the global sulphur cycle. Thus, the activity of sulphate reducing bacteria releases sulphate from aquatic sediments and biofilms, stratified seas and lakes and, of course, bogs and marshes as hydrogen sulphide. If, however, the area is undisturbed, this H_2S rarely reaches the atmosphere because of the formation of mats of sulphur oxidizing organisms at the sulphide:oxygen interface. The behaviour of the community of phototrophs and chemotrophs (see below) in such a system was described by Jørgensen (1982). As can be seen from Fig. 2, the position of the various species within the mat was determined by the amount of light available for photosynthesis, as well as the position of the sulphide:oxygen boundary. The importance of the organic sulphides being produced within algal blooms in the oceans is only now beginning to be appreciated, and measured (see, for example, de Zwart & Kuenen 1992).

Anthropogenic sources of reduced and (partially) oxidized sulphur compounds

Probably one of the greatest anthropogenic contributors to sulphur pollution is the burning of fossil fuels. Sulphur is present in coal in the form of pyrites crystals, sulphate and, distributed within the matrix of the coal, as organic compounds. Sulphides are also present in natural gas and oil deposits. The actual amounts in fossil fuels vary with the

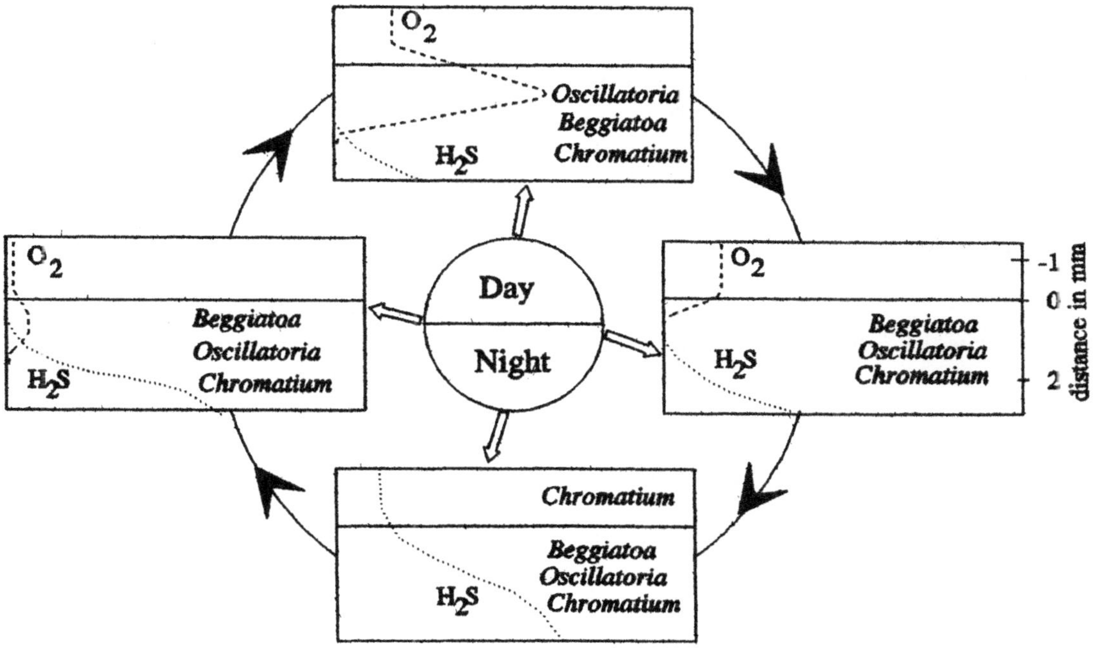

Fig. 2. Diurnal cycle of oxygen and sulphide gradients and microbial distribution in a marine sulfuretum (adapted from Jørgensen 1982).

source. Thus most natural gas sources are fairly low in sulphur content (around 0.05% S by weight), but sources containing 80% H_2S are known (Ryaboshapko 1983). The sulphur content of oil is known to vary from 0.3–0.5% in African oils to 5% in Venezuelan oil. The sulphur content of coal can vary from less that 1% to as much as 8%. Another major source of SO_2 is the mining and smelting of sulphidic metal ores in which sulphides can be as much as 45% of the dry material. Table 1 summarizes the estimates of SO_2 emissions, and Table 2 the sulphide emissions contributed by different anthropogenic activities and collated by Ryaboshapko (1983).

Anthropogenic activity has also boosted biological production of reduced sulphur compounds. Everyone is familiar with the smells associated with the local sewage works, and this is extended by sulphides produced during methanogenisis from organic wastes, as well as sulphate reduction within rubbish dumps. Locally, H_2S and the methylated sulphides (CH_3SH, $(CH_3)_2S$ and $(CH_3)_2S_2$; MAC values 10, 0.5, 20 and < 20 ppm, respectively) may

Table 1. Estimates of SO_2 emission (TgS year^{-1}) from different processes in various countries (before the political boundaries were changed). Adapted from Ryaboshapko (1983). * = included in Industrial total.

Process type	USA	Great Britain	West Germany	USSR
Energy	10.65	1.53	1.53	7.32
Petroleum processing	1.65	0.2	0.15	0.3
Iron production	4.05	0.25	0.25	1.3
Industrial	0.90	0.61	0.55	3.4
Others	2.1	0.3	*	1.4

Table 2. Estimates of sulphide emissions due to industrial activity (adapted from Ryaboshapko (1983).

Process	Sulphide	Emission (as TgS year^{-1})
Paper making	DMS, H_2S, mercaptans	0.1–0.65
coke production	H_2S	0.07–0.7
smelting of iron	H_2S	0.1–0.5
cement production	HH_2S	1.2

be produced in industrial effluent, notably from paper mills and the nylon industry.

The activities of metal pyrite-oxidizing bacteria were previously limited by the accessibility of the minerals. Mining has exposed enormous areas of minerals to microbial activity, with the resultant acidification and sulphate pollution of ground water and soils in the surrounding area.

Problems associated with sulphur pollution

Health

The only two sulphur compounds mentioned in the World Health organization (WHO) guidelines for drinking water quality (WHO 1984a) are sulphate and hydrogen sulphide. Their recommended levels are 400 mg SO_4^{2-} per litre and 'not detectable by consumer', respectively. In the volume on health criteria (WHO 1984b), dimethyl sulphide is also discussed, but no recommended level is given. At very low concentrations, it is, of course, an important constituent of some beer flavours.

The accepted sulphate level is primarily determined from its taste threshold which is in the range between 200 and 500 mg SO_4^{2-} per litre, depending on the associated cation. However, sulphate can also act as a laxative, especially when combined with Mg^{2+}. Normal adults are affected at levels around 1000 mg per litre, but children and new users can be affected by less. Hydrogen sulphide is very toxic. It inhibits various enzyme systems (WHO 1984b), and irritates the epithelium of the mucous membranes. If inhaled, it paralyses the respiratory centre. Instant death occurs at a concentration of 2800 mg per m^3, levels of 1400–2100 mg per m^3 will result in death after 30 minutes. Indeed, sudden deaths have occurred when workers cleaning sewers have inhaled lethally high concentrations of H_2S from the sewage. At lower concentrations, H_2S causes depression (0.12 mg per m^3), conjunctivitis and visual problems ((1.5–43 mg per m^3), and psychic changes, dizziness and vomiting, among other problems (70–700 mg per m^3). Perhaps one of its greatest dangers is that hydrogen sulphide also affects the sense of smell, and concentrations above 225 mg per m^3 can thus not be detected in the usual manner. Swallowing alkali sulphides can cause nausea, vomiting and pain. 10–15 g NaS is fatal. Hydrogen sulphide has been detected in drinking water from stagnant sources. However, its taste and odour thresholds are much lower (0.05–0.2 mg per l) than its toxic levels, making poisoning from ingestion unlikely.

Organic sulphides can also be extremely hazardous. At very low concentrations, DMS is an important component of foods and beverages, contributing to the flavour of, for example, beer and milk. It also contributes to the 'smell of the sea'. At higher concentrations, but still in the ppb range, DMS has a very offensive odour. At even higher concentrations, the toxic effects are similar to those described above for H_2S.

The atmosphere

Atmospheric sulphur compounds are dominated by the organic sulphides (especially DMS), SO_2, and (localized) H_2S. An indicator of anthropogenic sulphur pollution in the atmosphere can be found by comparing the levels of sulphate in ice cores taken from the Arctic and Antarctic. Because the southern hemisphere is thinly populated and industrialized (compared to the north), it can be assumed that most, if not all, sulphate in the Antarctic ice is due to natural causes, and indeed a correlation between volcanic eruptions and sulphate levels has been observed (Legrand & Delmas 1987; Stauffer & Neftel 1987). The steady increase in sulphate (from below 40 ng/g ice in 1900 to above 100 ng/g ice in 1980) in the Arctic does not correlate with volcanic activity, and were thought to be almost entirely due to human activities. The sulphate (as H_2SO_4) comes from the oxidation of SO_2 with H_2O_2, OH and O_3 in clouds, and is one of the primary causes of acid rain (Isakson 1987). SO_2 is one of the few gaseous pollutants not associated with the Greenhouse Effect (it absorbs at roughly the same wavelength as ozone), but another sulphur compound, COS, does contribute to the 'dirty window effect' as it absorbs at in the same region (although to a much lesser extent) as the CFCs (Ramanathan 1987). The organic sulphides, especially dimethyl sulphide (DMS), are also radiatively significant (de Zwart & Kuenen 1992) through the formation of the more oxidized dimethyl sulphonate. The occurrence of dimethyl sulphide, itself, corresponds with various types of algal blooms. Indeed, recent investigations indicate that some of the sulphate in the Antarctic ice may originate from dimethyl sulphonate precipitation.

Aquatic

One of the most obvious effects of reduced sulphur compounds in bodies of water is anoxia. In most situations, layers of sulphur oxidizing bacteria practically separate the sulphide and oxygen gradients. However, in extreme situations (e.g. with very high sulphide concentrations or after turnover of a normally stratified lake), chemical oxidation of sulphide can proceed rapidly enough to consume all of the available oxygen. Even an increase in the sulphate content of freshwater may be sufficient to boost the sulphur cycle, promoting sulphate reduction and, again, anoxia Of course, it is essential that sufficient (organic) electron donors are available in addition to sulphate. In such cases, sulphate reducing bacteria may out-compete the slower-growing bacteria that ferment lower fatty acids and the H_2 metabolizing methanogens, because of the higher affinity for reduced substrates of the sulphate reducing bacteria (Nanninga 1987).

Many sulphidic water bodies are highly acidic, largely because of the sulphuric acid produced by the oxidation of the sulphur compounds. Even in the hot, acid springs of volcanic areas, the sulphur cycle continues, with sulphur oxidizers such as *Sulfolobus* and sulphur reducers such as *Acidianus* active at 80° C and above.

Terrestrial

Most of the terrestrial problems associated with sulphur-linked pollution are, again, linked to acidity. Damage to forests caused by acid rain has been much in the news recently, as has the erosion of famous buildings such as the Parthenon in Athens. In the vicinity of mines or exposed mineral deposits, this takes on another dimension since the acidity combined with bacterial activity can result in heavy metals forming soluble salts, and leaching into the ground water (this facility is actually employed in leaching process for metal recovery). Bacteria with resistance to relatively high concentrations of mercury, for example, have been isolated from such streams, and the isolation from an acid mine stream of a strain able to reduce chromate as an electron acceptor has been described. Sulphate reducing and sulphur oxidizing bacteria (the latter producing H_2SO_4) have also been associated with the corrosion of pipes, tanks and other metallic structures. The sulphate reducers are thought to cause depolarization of the metal, resulting in electrochemical corrosion of structures such as oil platforms (Kuenen & Bos 1989). The sulphur oxidizing bacteria produce sulphuric acid which may also stimulate electrochemical corrosion. This is admirably illustrated by the example of a fairly new ship loaded with wet, pyrite-containing coal from a waste pond, and then travelled from the USA to South America. The journey was rough, so that the contents of the holds were well mixed. Not long after this, corrosion was noted in the holds. Not only was this greatest in the holds nearest the engines, where it was warmer, but there was also a greenish liquid appearing. The captain tasted this, and found it to be acid. It was eventually concluded that the production of sulphuric acid by the pyrite oxidizing bacteria had created the electrochemical conditions for the corrosion of the stainless steel hull of the ship, also promoting chemical corrosion (P. Bos, pers. comm.). This type of damage is not limited to metallic structures. For example, considerable damage has been observed in concrete sewers This is due to the fact that the sewer pipes contain anaerobic sewage with active sulphate reduction, but an aerobic head space. H_2S leaves the sewage and may be precipitated as elemental sulphur on the aerobic, wet surfaces above the water level. There it is further oxidized to sulphuric acid, which dissolves the calcium carbonate in the concrete. Indeed, a strain of *Thiobacillus thiooxidans* was formerly known as *T. concretivorans* (Kuenen et al. 1992).

The current state of the art

The colourless sulphur oxidizing bacteria

It is clear that many of the problems associated with sulphur pollution cannot be solved by microbiological means. At the same time, it is also obvious that microbial activity will exacerbate, rather than diminish sulphur-related pollution problems. Fortunately, there are many ways in which microbial activity can be beneficially exploited. Microbiological methods of dealing with problems associated with reduced sulphur compounds depend on bacteria able to oxidize these compounds. The group of bacteria able to oxidize reduced sulphur compounds is extremely heterogeneous, and the most obvious sub-division is that between the phototrophs and the non-phototrophic (generally called colourless) groups. An overview of the phototrophs is given in Table 3. There are a number of reasons why the phototrophic sulphur oxidizing bacteria would not be the first choice for biocatalysts in biotechnological processes. The most obvious problems are associated with reactor design. For example, the phototrophs require light, and many of them retain molecular sulphur within the cell, giving separation problems. An African lake, with a very large surface to volume ratio, was observed to have *Chlorobium* species oxidizing the available sulphide to sulphur at a very slow rate (Butlin & Postgate 1954). Moreover, the sulphide oxidation is strictly coupled to growth in these organisms, with the generated electrons being required for CO_2 reduction (i.e. fixation). Thus during S^{2-} oxidation, *Chlorobium* cultures (which excrete their sulphur) and *Chromatium* cultures (which accumulate it intracellularly) can be expected to generate approximately 1–2 grams S^0 per gram of cells produced. In *Thiobacillus* species, the oxidation of sulphur compounds is not obligately coupled to growth as the main electron acceptor is oxygen (or nitrate) rather than CO_2. In fact, these chemolithotrophs can oxidize sulphur compounds (e.g. extracellular elemental sulphur) without any appreciable growth when nutrients other than their energy source are limiting (Kuenen et al. 1985). They can therefore produce at least 20 grams S^0 per gram of cells. Since biotechnological processes tend to deal with much more concentrated sulphide solutions, and require higher specific rates of oxidation and low sludge production, the colourless sulphur bacteria from the genus *Thiobacillus* are clearly more appropriate for most applications.

The ‘colourless sulphur bacteria’ is a general term of physiological, but little taxonomical significance (Kuenen 1989; Kelly & Harrison 1989). All of the physiological types outlined in Table 4 can be found within this group. With such a great diversity, there has been some debate as to the taxonomic

Table 3. Physiological types found among the phototrophic sulphur oxidizing bacteria (adapted from Kuenen et al. 1985). obl. = obligate; fac. = facultative.

Family	Metabolic definition	Electron donor
Chlorobiaceae	obl. phototroph, fac. (photo)autotroph, external S^0 produced, obl. anaerobe	S^{2-}, $S_2O_3^{2-}$, S^0, H_2, organic acids
Chloroflexaceae	fac. phototroph, fac. photoautotroph, external S^0 produced, thermophilic	S^{2-}, organic acids
Chromatiaceae	phototroph, fac. photoautotroph, internal S^0 produced, fac. aerobe	S^{2-}, $S_2O_3^{2-}$, H_2, organic acids
Rhodospirillaceae	fac. phototroph, fac. photoautotroph, external S^0 produced	S^{2-}, $S_2O_3^{2-}$, organic acids

Table 4. Definition of the physiological types of bacteria able to oxidize reduced sulphur compounds. Synonyms: [A] obligate autotrophs. [B] facultative autotrophs, mixotrophs.

Physiological type	Carbon source		Energy source	
	Inorganic	Organic	Inorganic	Organic
Obligate chemolithoautotroph[A]	+	–	+	–
Facultative chemolithoautotroph[B]	+	+	+	+
Chemolithoheterotroph	–	+	+	+
Heterotroph	–	+	–	+

validity of the group. Indeed, evidence from 5S and 16S RNA analysis (Lane et al. 1990) has indicated that some species (e.g. the members of the genus *Thiomicrospira*) are more closely related to *Escherichia coli* than to other colourless sulphur bacteria (e.g. *T. ferrooxidans*). The precise relationships require further elucidation of the significance of RNA analysis in taxonomy, after all, as pointed out recently (Woese 1991), there is greater difference in terms of RNA analysis between species in the same bacterial genus than there is between higher eukaryotes such as frogs and humans. Taxonomic problems aside, the ability to oxidize reduced sulphur compounds and gain energy from the reaction requires special enzymes, and the grouping thus has physiological significance. Table 5 provides examples of the many types found among the colourless sulphur bacteria, together with some of their environmental requirements. In addition to these, it appears that many symbiotic bacteria, including those of the spectacular hydrothermal vent worms, *Riftia*, are sulphide oxidizing bacteria related to *Thiomicrospira* (Lane et al. 1990).

The obligate and facultative autotrophs are of most obvious interest for wastewater treatment but, as will be seen below, chemolithoheterotrophs (e.g. *Thiobacillus* Q (Gommers et al. 1988), *Hyphomicrobium* sp. (Suylen & Kuenen 1986)) can be important, particularly where organic sulphides are involved. The type of organism in the system will be primarily defined by the system itself – a wastewater treatment reactor running at 25° C, neutral pH and with a feed containing sulphide and nitrate will obviously enrich for bacteria resembling *Thiobacillus denitrificans*. Similarly, a reactor oxidizing pyrites in hot, acid solutions will probably give rise to a population similar to *Sulfolobus*. The selective pressures determining whether obligate or facultative autotrophs will dominate a community were investigated some time ago. Using chemostat cultures of *T. neapolitanus* (obligate autotroph), *Thiobacillus* A2 (now *T. versutus*, a facultative autotroph) and a spirillum, G7 (obligate heterotroph), Gottschal et al. (1979, 1981) showed that the final population (or community in some cases) was determined by the ratio between the turnover of inorganic and organic substrates in the feed to the chemostat, as outlined in Fig. 3. That this model was a simplification was emphasized by the fact that *T. neapolitanus* was found to excrete glycolate in sufficient quantities to support the growth of *T. versutus* in 3-membered cultures when it was expected to wash out. However, the model

Table 5. Survey of the metabolic types found among the colourless sulphur bacteria. Obl. = obligate; Fac. = facultative; *T.* = *Thiobacillus*; *Tms.* = *Thiomicrospira; Tsa.* = *Thiosphaera; Tx.* = *Thermothrix*; [a] = microaerophilic; [b] = also gains energy from iron; [c] = denitrifys heterotrophically; * = known to produce extracellular sulphur.

Representative species	Autotrophy		Denitrification		pH range	Temp. range
	Obl.	Fac.	to NO_2^-	to N_2		
*T. neapolitanus**	+	–	–	–	6–8	<37° C
*T. thioparus**	+	–	+	–	6–8	<37° C
T. denitrificans	+	–	+	+	6–8	<42° C
*Tms. denitrificans**	+	–	+	+[a]	6–8	<37° C
*T. intermedius**	–	+	–	–	5–7	<37° C
Tsa. pantotropha	–	+	+	+	6–10	<42° C
Beggiatoa sp.	–	+	–	–	7	<37° C
Beggiatoa sp.	–	+	+	+	7	<37° C
T. ferrooxidans	+[b]	–	–	–	2–4	<37° C
T. acidophilus	–	+	–	–	2–4	<37° C
T. tepidarius	+	–	+	–	6–8	20–52° C
Tx. thioparus	–	+	+	+[c]	6–8	40–80° C
*S. acidocaldarius**	–	+	–	–	1–6	55–85° C

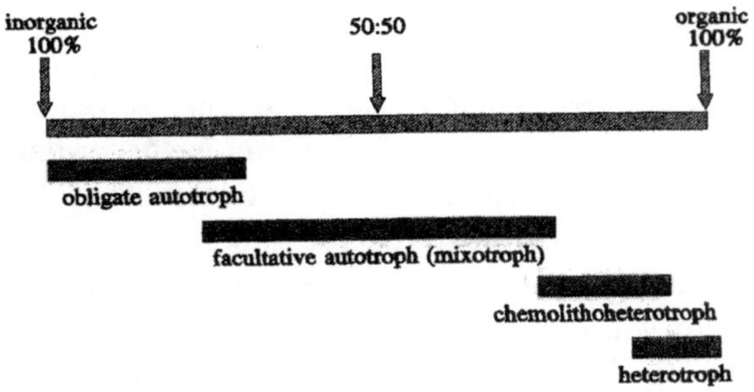

Fig. 3. The selection of different physiological types according to the organic:inorganic ratio in the medium supply (Gottschal et al. 1979).

also predicted that facultatively autotrophic bacteria such as *T. versutus* should dominate continuous cultures limited by equimolar amounts of organic and reduced sulphur compounds. This has now been observed for a number of chemostat enrichment cultures (see, for example, Beudeker & Kuenen 1982; Kuenen et al. 1985).

Possible solutions

There are some problems to which purely biological solutions will not be practical or even possible. Into this category might fall the super-heated, sulphide-containing gas exhausts from smelting. In other situations, a combined physico-chemical and biological treatment might be more feasible. Such systems might include the cleaning of refinery gases where a chemical method is simple and relatively cheap for the bulk of the sulphur, but biological treatment is necessary to attain required standards. Similarly, the cleaning of coal might efficiently combine separation techniques to remove the greatest part of the pyrites, with biological treatment removing the rest (Bos 1991). Such a system will be discussed briefly below. If the effluent contains toxic compounds, it may be necessary to remove them before biotreatment can take place. For example, high concentrations of sulphide which might be toxic for nitrification could be stripped off before the water is nitrified. In principle, the H_2S might then be reused for sulphuric acid production or, elsewhere, be converted biologically or chemically into sulphur or sulphate. Alternatively, compounds can be removed for treatment elsewhere. For example, it might be simpler to treat the ammonia in photographic waste if the ammonia is first stripped from the effluent as the waste also contains thiosulphate, thiocyanate, cyanide and various complex organic compounds (Robertson & Kuenen 1984). Another major problem with photographic waste is the presence of quinones which not only resist treatment, but inhibit the conversion of other compounds. The simplest way to make their effluent easier to handle might be for the photographic industry to replace these compounds with something less recalcitrant and less toxic.

Biofilters (Diks & Ottengraf 1991) provide a simple solution for effluent gas streams that contain volatile reduced sulphur compounds. These may take the form of a peat filter, or a system where the gas flow is sparged through water, which is then treated elsewhere. For the treatment of aqueous, sulphur-containing waste streams, many options are available, depending on the type of waste to be treated and the requirements of the biomass. Systems using fluidized beds, in which the bacteria form biolayers on sand which is kept suspended in the reactor by the upward flow of the effluent, and biodiscs, in which the biomass is attached to surfaces which rotate in and out of the effluent stream, will be discussed below.

Research at Delft

Coal desulphurization

As mentioned above, the sulphur content of some coals can add enormously to atmospheric sulphur pollution. Methods of removing the sulphur from the coal before combustion have therefore been attracting a great deal of attention. There are various physical methods which can be used for the removal of large pyrites crystals, but biological oxidation using the acidophilic *Thiobacillus* or *Sulfolobus* species (ironically, the same species that can cause environmental problems such as acid mine water and pipe corrosion) is particularly suitable

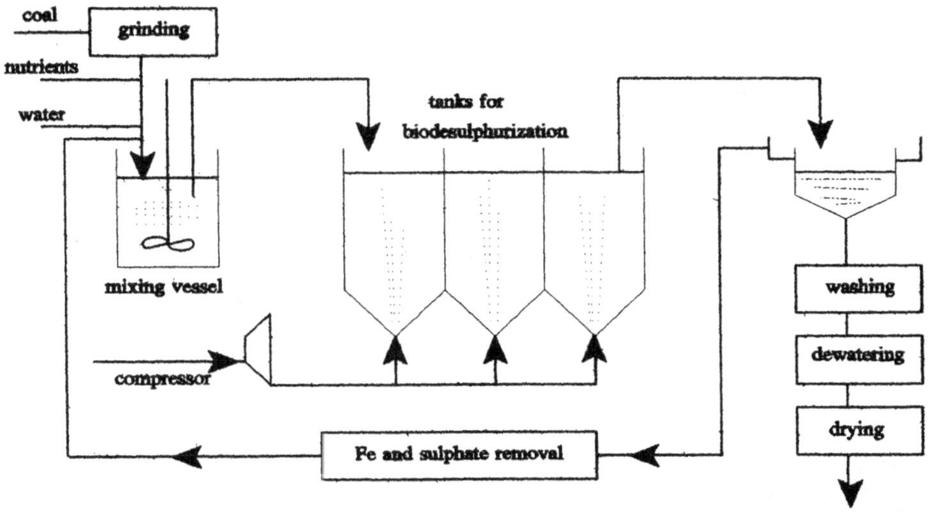

Fig. 4. Simplified system for the microbial desulphurization of coal in Pachuca tanks (biodesulphurization reactors) The coal is treated as a slurry of particles $< 100\ \mu m$ in diameter in acid, nutrient solution (from Bos & Kuenen 1990).

for the treatment of coal containing finely distributed pyrite. It has been shown that such a process could be effective and feasible, especially for coals with a relatively low sulphur content (Bos et al. 1988; Bos & Kuenen 1989). Laboratory studies showed that the process was most effective if carried out in two steps. In the first stage, the pulverized coal suspension is mixed with a suspension of suitable bacteria growing on pyrite at a (very low) pH (around 1.8). This mixture is then pumped through a series of reactors to create a plug flow process where most of the pyrites would be oxidized. At the end of the process, the coal and process water are separated. The water can be recirculated a number of times, until the iron and/or salt concentration reach a critical level. A suitable reactor configuration for this type of process was described by Bos et al. (1988), and is shown in Fig. 4. The desulphurization reactors are of a type know as Pachuca tanks (in their simplest form, an inverted cone with aeration at the bottom of the tank). These are simple, relatively cheap and, because the up-flowing air keeps the suspension well mixed while maintaining aerobic conditions, do not require complex stirring equipment. It is essential that the first reactor has a residence time of at least 1 day to allow the inoculation of freshly introduced coal with bacteria attached to partially treated coal particles. This residence time is based on the observed growth rate of *T. ferrooxidans* ($0.06\ h^{-1}$, a doubling time of approximately 10 hours) in the system. Once a few bacteria are attached to the exposed surfaces of the pyrite crystals, growth and degradation can take place in the succeeding reactors. The obvious advantage of using a plug flow system is that treated and untreated coals are not mixed. Of course, the nature of the organisms involved demands a certain amount of care in choice of materials – *T. ferrooxidans* has been known to corrode stainless steel, and sulphur-containing concrete should clearly also be avoided. The efficiency of the process depends on the grinding of the coal to fine particles (maximum diameter 100 µm) in order to expose the sulphide crystals to bacterial activity. Grinding the coal to this particle size is completely compatible with its later use as an oil-like slurry in the combustion process. This system is only effective against the inorganic sulphides, because organic sulphur compounds occur within the matrix, rather than as localized crystals. In order to prevent SO_2 being produced from organic sulphur in the coal, lime is added to the slurry It is also necessary to treat the wastewater from the process to remove iron and other heavy metals, as well as to neutralize the acid.

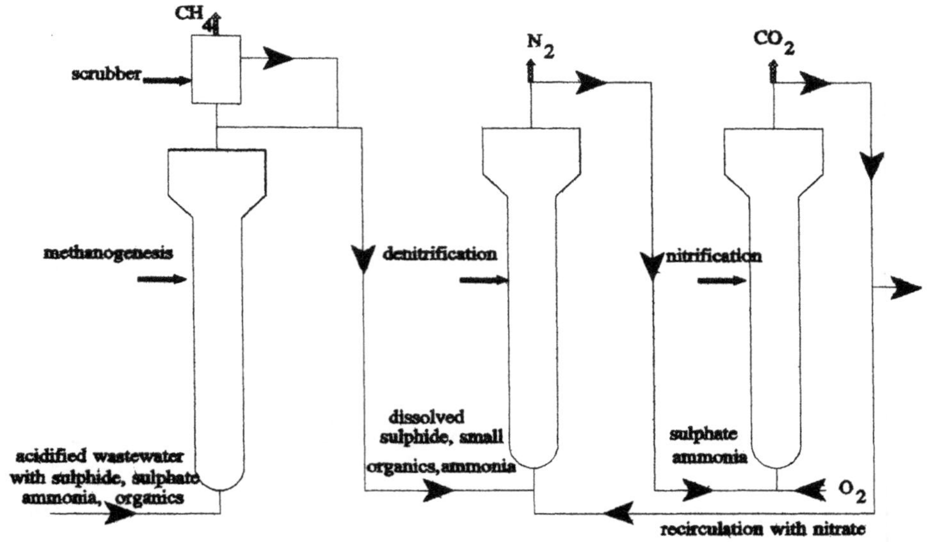

Fig. 5. A combination of reactors used to generate methane from organic industrial wastes, and then remove the nitrogen and sulphur content by means of nitrification and sulphide oxidation with denitrification.

Wastewater treatment

Sulphate production with denitrification
Reduced sulphur compounds are an almost inevitable component in wastewater streams, especially if the system is completely or partially anaerobic and the wastewater contains significant amounts of sulphate. The most frequent (and conspicuous) sulphidic component in wastewater is sulphide, although thiosulphate and sulphur may also occur. As can be seen from the following reactions, oxidation of sulphide or thiosulphate with oxygen results in the equimolar production of sulphuric acid:

$$H_2S + 2O_2 \rightarrow H_2SO_4$$
$$NaS_2O_3 + 2O_2 + H_2O \rightarrow Na_2SO_4 + H_2SO_4$$

This problem can be overcome by using denitrifying bacteria to oxidize the sulphur compounds:

$$5H_2S + 8KNO_3 \rightarrow 4K_2SO_4 + H_2SO_4 + 4N_2 + 4H_2O$$

Using denitrifiers has the added advantage that the nitrogen content of the wastewater can be reduced at the same time, if it is present in an oxidized form. A denitrifying, sulphide oxidizing, fluidized bed reactor was part of a system patented by a Dutch company in the early '80s, and illustrated in Fig. 5. The wastewater came primarily from bakers yeast production, and therefore had a high COD. This was reduced by methanogenesis after acetogenesis, both anaerobic processes which permitted the reduction of the sulphate in the wastewater to sulphide. To deal with the sulphide, and also remove any organics not consumed in the methanogenic reactor (e.g. small organic acids, protein from dying bacteria), a denitrifying reactor was next in the sequence. The necessary nitrate was provided by recirculation from the last reactor, in which the ammonia in the wastewater was oxidized by aerobic nitrifying bacteria. The overall performance of, and the microbial community in the denitrifying reactor received detailed attention in this laboratory over a period of about five years. Studies on the actual reactor appeared to support the model for microbial selection outlined in Fig. 3. The influent contained similar amounts of sulphide and small organic molecules (mostly acetate). A large proportion of the population was heterotrophic, but all of the sulphur bacteria isolated were facultative autotrophs (Robertson & Kuenen 1984). However, later experiments with a laboratory-scale model of the column revealed that this was an over-simplification as obligate autotrophs were never isolated, even when the feed to the system was wholly inorganic (Robertson & Kuenen 1991) Sulphur balances made over long periods showed that a steady state was never reached by the system, there were periods of sulphur accumulation within the bed of the reactor, and periods of sulphate production. Although the reasons for this were never clear, it can be speculated that the fluctuations might be explained by the 'chaotic' behaviour of the hydraulic regime. All of the sulphide and small organics were consumed in the lowest part of the reactor, and the other, necessary, reactions took place higher up. For example, an accidental overload of the system resulted in the expansion of the sulphide and sulphur containing area within the bed, sometimes resulting in the production of N_2O rather than N_2. This is most probably due to the inhibition of the last enzyme in the denitrifying pathway, N_2O reductase, which is known to be sulphide-sensitive.

The maximum sulphide and sulphur oxidizing capacities of the biomass after growth with different sulphide (or thiosulphate) to acetate ratios was measured, and the results are shown in Table 6. At all times, there was an over-capacity for sulphide oxidation to sulphur, but the sulphur-oxidizing capacity was only sufficient to cope with the load. A step-wise increase of the sulphide supplied to the column receiving the highest amount of acetate revealed that the biomass increased its sulphur and sulphate forming activity to match the sulphide supply (Gommers et al. 1988). Further activity measurements revealed that the sulphide/sulphur oxidizing activity of the bed was stratified. Biomass taken from different heights in the fluidized bed was supplied with sulphide, acetate and nitrate, and the progress of the conversions followed. The sulphide was depleted within the first 20 minutes in all three samples, but the sulphate production rate was much slower. Clearly, during the rapid initial sulphide oxidation, sulphur was produced, and subsequently oxidized to sulphate. Only in the sample from the top of the bed was all of the expected sulphate produced. In the samples from the bottom and middle, only about 50% of the expected sulphate was found. These experiments explain the observation that, when the load on the reactor was increased above the sulphur oxidizing capacity of the biomass, elemental sulphur accumulated (temporarily) in the biofilm. Indeed, the ability to switch from complete sulphide oxidation to sulphate (producing 8 electrons) to sulphur accumulation (2 electrons) provided the system with a very high resilience for sulphide overloads.

Table 6. The maximum capacities of biomass grown at different S:acetate ratios in a denitrifying, fluidized bed reactor, together with the S-loads necessary to obtain the correct ratios. * = thiosulphate used in place of sulphide for experimental convenience. The thiosulphate oxidizing capacity was equivalent to that of sulphide.

Compound used	S:acetate ratio	S-load	Capacity $S^{2-} \rightarrow S^0$	Capacity $S^0 \rightarrow SO_4^{2-}$
S^{2-}	0.1	1.1	20	0.8
S^{2-}	1.1	2.6	25	2.6
S^{2-}	0.3	3.3	18	2.9
$S_2O_3^{2-}$*	9.0	10.3	54	28
$S_2O_3^{2-}$*	15	22	187	33

To sum up, this research concluded that the biomass in the reactor could adapt to changes in its feed, and adjust its sulphur-oxidizing capacity to suit its requirements. In addition, it could adjust to deal with sulphide overloading. However, overloading with sulphide could result in N_2O production until adaption had taken place. Moreover, the nitrate supply to the system had to be carefully monitored because nitrite, if allowed to accumulate, was strongly inhibitory.

Sulphur production

As mentioned above, high sulphate levels are undesirable as they boost the natural sulphur cycle and can cause taste and alimentary problems in drinking water. It has been suggested that sulphur in wastewater could be recovered as S^0 (Butlin & Postgate 1954; Postgate 1982). If this was the case, it should be possible to prevent increased sulphate concentrations and reduce the sulphuric acid production by aerobic reactors if the sulphide oxidation could be stopped at S^0 rather than proceeding to sulphate:

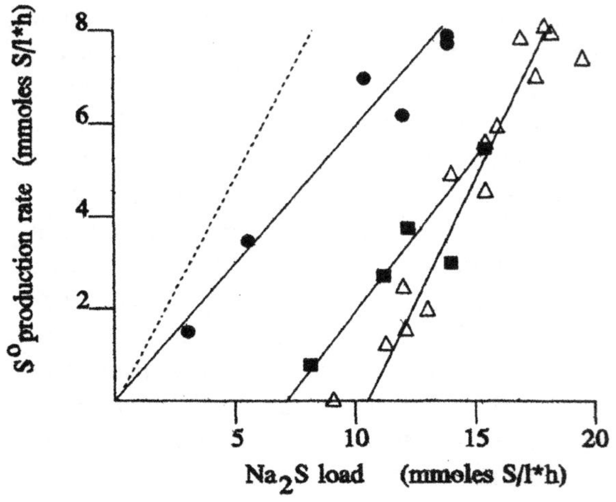

Fig. 6. The effect of the dissolved oxygen concentration on sulphur production by sulphide-oxidizing chemostat cultures of *Thiobacillus neapolitanus*. The dissolved oxygen concentrations quoted as percentage air saturation. Closed circles = oxygen limitation; closed squares = 20 % air saturation; open triangles = 50–60% air saturation; broken line = 'theoretical maximum' indicating conversion of all sulphide to sulphur, with no other products (G.C. Stefess, L.A. Robertson & J.G. Kuenen, unpublished results).

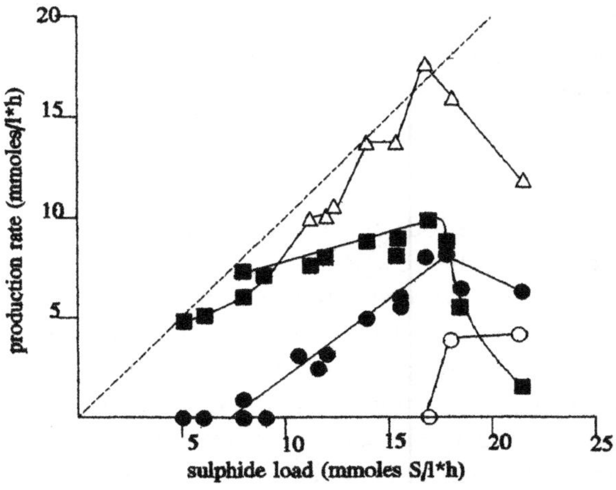

Fig. 7. The effect of increasing loads on sulphur production by sulphide-oxidizing chemostat cultures of *Thiobacillus neapolitanus*. Open triangles = total S calculated; closed squares = sulphate; closed circles = S^0; open circles = H_2S stripping; broken line = 100% S recovery (G.C. Stefess, L.A. Robertson & J.G. Kuenen, unpublished results).

$$2H_2S + O_2 \rightarrow S^0 + 2H_2O$$

Research at Delft (as part of a combined project with the Agricultural University of Wageningen) has revealed that high sulphide loads and electron acceptor limitation both favour sulphur (rather than sulphate) production, even when strains of *T. neapolitanus* from a culture collection were used (Stefess & Kuenen 1989). Figure 6 shows the sulphur production obtained with steady state *T. neapolitanus* cultures at different dissolved oxygen concentrations. It can be seen that relatively high sulphide loads were necessary before the cultures grown at significant (20–60% air saturation) dissolved oxygen concentrations began to accumulate sulphur. Under oxygen limitation, sulphur was produced at all loads. The effect of varying sulphide loads on *T. neapolitanus* grown at 50–60% air saturation can be seen in Fig. 7. It can be seen that, as might be expected, sulphate production levels stabilized after sulphur accumulation started. At a critical load of about 17 mmoles $S \cdot l^{-1} \cdot h^{-1}$, even the sulphur producing capacity of the culture was inadequate and sulphide began appearing in the off-gas (G.C. Stefess, L.A. Robertson & J.G. Kuenen, unpublished results). A pilot plant based on this reaction is being used to treat the effluent from a paper mill (Buisman 1989). The behaviour of a new autotrophic strain isolated from this reactor, currently designated *Thiobacillus* W1, was similar to that of *T. neapolitanus*, but differed in that above a certain sulphide load, sulphur production increased at the expense of sulphate production, rather than a constant level of sulphate being generated. Thus, sulphur production by *Thiobacillus* W1 under oxygen limitation could account for up to 70% of the sulphide added. With *T. neapolitanus* this never exceeded 50%. produced a greater proportion of sulphur.

The recovery of elemental sulphur is now being considered for integration in another process in which sulphate reducing bacteria are employed to generate sulphide from sulphate-rich wastewater. This sulphide is then used to precipitate dissolved metals. However, the sulphide generated from reduction of the sulphate is generally more than is required, and this must be dealt with. Sulphur production holds obvious attractions for this process, as well as for others where the dissolved sulphate level must be reduced.

Treatment of organic sulphides

As mentioned above, organic sulphides contribute significantly to environmental problems. The bacteria able to metabolize these compounds and gain energy from the sulphide oxidation appear to fall into two groups, chemolithoautotrophs that fix their carbon via the Calvin cycle (e.g. *Thiobacillus thioparus*; Kanagawa & Kelly 1986), and chemolithoheterotrophs that use the serine pathway (e.g. *Hyphomicrobium* S; de Bont et al. 1981). Another *Hyphomicrobium* species, *Hyphomicrobium* EG was isolated from a sample from a paper mill waste treatment system, and studied in detail (Suylen & Kuenen 1986; Suylen et al. 1986). Despite the fact that the cultures were grown on methyl mercaptan, dimethyl sulphide or hydrogen sulphide, all three substrates remained toxic to the culture, apparently inhibiting the enzymes of organic sulphide breakdown (Suylen 1988). Sulphide oxidation was least affected (Table 7). pH was also critical, with the toxicity of DMS and MM increasing as the pH fell. Any wastewater treatment system employing such organisms should therefore be designed in such a way as to prevent sudden environmental changes or pulses of sulphide or overloads. For the same reasons, the reactor should be well mixed, a plug flow type would not be suitable. The treatment of gases containing these compounds may not be such a problem, provided that high concentrations of biomass can be maintained in the reactor, because the gases will gradually dissolve as they pass through the water phase rather than entering the reactor as a concentrated solution. Indeed, such a system has been used on the laboratory scale to deodourize gases containing H_2S and organic sulphides (Kanagawa & Mikami 1989; Tanji et al. 1989).

Table 7. The effect of dimethyl sulphide (DMS), methyl mercaptan (MM) and hydrogen sulphide on the oxidation of dimethyl sulphoxide (DMSO), DMS, MM and H_2S by *Hyphomicrobium* EG (data from Suylen 1988).

Inhibitor	Maximum inhibition (as % of original)				Approx. concentration giving 50% of maximum inhibition (mM)			
	DMSO	DMS	MM	H_2S	DMSO	DMS	MM	Na_2S
DMS	60	40	40	35	0.5	1.25	1	0.5
MM	100	100	70	25	0.1	0.1	<0.1	<0.1
Na_2S	90	70	20	20	0.1	0.18	0.18	0.2

Future prospects

The human race is steadily increasing, and with it our demands on the resources of the planet. If we are not to emulate a batch bacterial culture and perish in a mess of waste products, it is essential that we increase the efficiency of our activities, while ensuing that they have as little further impact on the environment as possible. The recent reports of holes in the ozone layer over the Northern as well as the Southern hemispheres and the apparent correlation between algal blooms and sea mammal mortality serve to emphasize that we can no longer see the oceans, the atmosphere, or the planet as a whole as bottomless pits into which our wastes can be dumped with impunity. It will be essential that renewable resources are used, and that 'clean technology' processes are developed and constructed as (almost) closed cycles. Of course, no process will ever be completely without waste products, and tailor-made, dedicated wastewater treatment systems will have to become a natural part of industrial operations, either within the cycle or as an 'end of the pipe' solution. To this effort, those working on the treatment of sulphidic wastes can greatly contribute, both in preventative measures and, when necessary, for cures.

Coal and other minerals

As discussed, bacterial processes are feasible for the removal of inorganic sulphides from coal (for a review see Bos et al. 1993). The process described here is about to be tested on the pilot plant scale. The economic feasibility of the process, however, must be regarded as borderline. One of the strongest economic factors is the cost of energy – as oil prices go up, coal cleaning becomes more attractive. It is especially appropriate for small-scale industrial use when coal-water mixtures are needed. The coal can be centrally treated and then transported. This has the added attraction of reducing the amount of corrosion likely to occur during transportation because of the activities of the sulphur oxidizing bacteria. Thermophilic bacteria have not proved appropriate because their environmental requirements also promote the formation of jarosite-like precipitates, defeating the object of the exercise. The solution to the problem posed by organic sulphides is likely to be more problematical. Although some promising results have been obtained with model substrates (for a review, see Bos et al. 1993), the dispersal of the sulphur within the coal macromolecules (making it inaccessible) remains a major difficulty.

Bacterial leaching of low quality ores appears to have a brighter future. Acidophilic sulphur pyrite-oxidizing bacteria have long been used for the extraction of copper. 15% of the copper produced in the USA comes from large scale leaching operations. (Brierly 1982) Sulphuric acid produced from pyrite by acidophiles is also used to leach uranium from low grade ores. Finally, a recent development is that acidophilic, pyrite oxidizing bacteria are being used to remove pyrites from gold-bearing ores, making the gold more accessible to chemical extraction.

Wastewater treatment and sulphur production

Sulphur-based denitrification is still occasionally mentioned in the literature, but the need to reduce sulphate output is making this less attractive. Similarly, sulphide removal by means of complete oxidation to sulphate is equally unattractive. Research into the possibilities of removing the sulphide as sulphur is continuing, together with investigations into the feasibility of removing sulphate from waste streams by conversion to sulphur via a sulphate reduction step. The problems associated with specialized waste streams (e.g. photographic waste) still remain, and will probably require a less than simple solution (i.e. alternative formulations).

Research into the treatment of the organic sulphides is now attracting a great deal of attention, and a coordinated, inter-University research programme has begun in The Netherlands. This programme will look at both natural and anthropogenic organic sulphide generation, environmental impact and treatment.

Concluding remarks

Much remains to be done, and it is really up to governments at the national and international level to set realistic targets for controlling sulphur emissions. If this is not done, industry will not implement treatment procedures which will add to their costs, while their competitors do not. It is to the credit of some industrial concerns that they are already carrying out research into the further control of sulphur emissions, but it is now time to bring the others into line.

References

Beudeker RF, Gottschal JC & Kuenen JG (1982) Reactivity versus flexibility in thiobacilli. Antonie van Leeuwenhoek 48: 39–51

Bos P (1991) Microbial desulfurization of coal: a feasible process. In: Hecq W & Crine M (Eds) Microbial Processes for Heavy Metal Recovery (pp 63–78). University de Liège

Bos P & Kuenen JG (1989) Microbial treatment of coal. In: Brierley C & Ehrlich H (Eds) Microbial Metal Recovery (pp 343–377). McGraw Hill Book Company, New York

Bos P & Kuenen JG (1990) Microbial treatment of coal. In: Erlich H & Brierley C (Eds) Microbial Mineral Recovery (pp 344–377). McGraw-Hill Book Company, New York

Bos P, Huber TF, Luyben KChAM & Kuenen JG (1988) Feasibility of a Dutch process for microbial desulphurization of coal resources. Conservation and Recycling 1: 279–291

Bos P, Boogerd FC & Kuenen KG (1993) Microbial desulfurization of coal. In: Mitchell R (Ed) New Concepts in Environmental Microbiology. John Wiley & Sons, Chichester (in press)

Brierly CL (1982) Microbiological mining. Scientific American 247 (August): 42–51

Buisman CJN (1989) Biotechnological sulphide removal with oxygen. PhD Thesis, Agricultural University of Wageningen

Butlin KR & Postgate JR (1954) The microbiological formation of sulphur in Cyrenaican lakes. In: Cloudsley-Thompson J (Ed) Biology of Deserts (pp 112–122). Institute of Biology, London

de Zwart JMM & Kuenen JG (1992) C_1 cycle of sulfur compounds. Biodegradation 3: 37–59

de Bont JAM, van Dijken JP & Harder W (1981) Dimethyl sulphoxide and dimethyl sulphide as a carbon, sulphur and energy source for growth of *Hyphomicrobium* S. J. Gen. Microbiol. 127: 315–323

Diks RM & Ottengraf SP (1991) Process engineering aspects of biological waste gas purification. In: Verachtert H & Verstraete W (Eds) Environmental Biotechnology 1991 (pp 353–367). European Federation of Biotechnology

Gommers PJF & Kuenen JG (1988) *Thiobacillus* strain Q, a chemolithoheterotrophic sulphur bacterium. Arch. Microbiol. 150: 117–125

Gommers PJF, Bijleveld W, Zuijderwijk FJM & Kuenen JG (1988) Simultaneous sulfide and acetate oxidation in a denitrifying fluidized bed reactor-II. Measurements of activities and conversion. Water Res. 22: 1085–1092

Gottschal JC, de Vries S & Kuenen JG (1979) Competition between the facultatively chemolithotrophic *Thiobacillus* A2, and obligately chemolithoptrophic *Thiobacillus* and a heterotrophic spirillum for inorganic and organic substrates. Arch. Microbiol. 121: 241–249

Gottschal JC, Nanninga HJ & Kuenen JG (1981) Growth of *Thiobacillus* A2 under alternating growth conditions in the chemostat. J. Gen. Microbiol. 126: 85–96

Isakson ISA (1987) Is the oxidizing capacity of the atmosphere changing? In: Rowland FS & Isaksen ISA (Eds) The Changing Atmosphere (pp 141–158). John Wiley & Sons, Chichester

Jorgensen BB (1982) Ecology of the bacteria of the sulphur cycle with special reference to anoxic-oxic interface environments. Philosophical Transactions of the Royal Society of London B298: 543–561

Kanagawa T & Kelly DP (1986) Breakdown of dimethyl sulphide by mixed cultures and by *Thiobacillus thioparus*. FEMS Microbiol. Lett. 34: 13–19

Kanagawa T & Mikami E (1989) Removal of methanethiol, dimethyl sulfide, dimethyl disulphide and hydrogen sulfide from contaminated air by *Thiobacillus thioparus* TK-m. Appl. Environ. Microbiol. 55: 555–558

Kelly DP & Harrison AP (1989) The Genus *Thiobacillus*. In: Staley JT, Bryant MP, Pfennig N & Holt JG (Eds) Bergey's Manual of Systematic Bacteriology 3 (pp 1842–1858). Williams & Wilkins, Baltimore

Kuenen JG (1989) Colorless sulfur bacteria. In: Staley JT, Bryant MP, Pfennig N & Holt JG (Eds) Bergey's Manual of Systematic Bacteriology 3 (pp 1834–1837). Williams & Wilkins, Baltimore

Kuenen JG & Bos P (1989) Habitats and ecological niches of chemolitho(auto)trophic bacteria In: Schlegel HG & Bowien B (Eds) Autotrophic Bacteria (pp 53–80). Brock/Springer Series in Contemporary Bioscience, Science Tech. Publ., Madison, WI, Springer-Verlag, Berlin

Kuenen JG, Robertson LA & van Gemerden H (1985) Microbial interactions among aerobic and anaerobic sulfur-oxidizing bacteria. Adv. Microb. Ecology 8: 1–60

Kuenen JG, Robertson LA & Tuovinen OH (1992) The colourless sulphur bacteria In: Balows A, Truper HG, Dworkin M, Harder W & Schlefer KH (Eds) The Prokaryotes, Vol III (pp 2636–2657) Springer-Verlag, New York

Lane DJ, Harrison AP, Stahl DA, Pace B, Giovannoni SJ, Olsen GJ, Heller DJ & Pace NR (1992) Evolutionary relationships among sulfur and iron oxidizing eubacteria. J. Bacteriol. 174: 269–278

Legrand MR & Delmas RJ (1987) A 220-year continuous record of volcanic H_2SO_4 in the Antarctic ice sheet. Nature 327:671–676

Lein AYu (1983) The sulphur cycle in the lithosphere, part II, cycling. In: Ivanov MV & Freney JR (Eds) The Global Biogeochemical Sulphur Cycle (pp 95–128). John Wiley, Chichester

Nanninga HJ (1987) Sulfate reduction and fermentation of amino acids in industrial waste water. PhD Thesis, University of Groningen, The Netherlands

Postgate JR (1982) Economic importance of sulphur bacteria. Philosophical Transactions of the Royal Society of London B298: 583–600

Ramanathan V (1987) The radiative and climatic consequences of the changing atmospheric concentrations of trace gases. In: Rowland FS & Isaksen ISA (Eds) The Changing Atmosphere (pp 159–186). John Wiley & Sons, Chichester

Robertson LA & Kuenen JG (1984) Anaerobic and aerobic denitrification by sulphide oxidizing bacteria from waste water. In: van den Brink WJ (Ed) Anaerobic Waste Water Treatment (pp 3–12). AWWT Symposium Secretariat (pubs)

Robertson LA & Kuenen JG (1992) Aerobic sulphur oxidizing bacteria. In: Balows A, Truper HG, Dworkin M, Harder W & Schlefer KH (Eds) The Prokaryotes, Vol I (pp 385–413). Springer-Verlag, New York

Ryaboshapko AG (1983) The atmospheric sulphur cycle. In: Ivanov MV & Freney JR (Eds) The Global Biogeochemical Sulphur Cycle (pp 203–277). John Wiley, Chichester

Stauffer BR & Neftel A (1987) What have we learned from the ice cores about the atmospheric changes in the concentrations of nitrous oxide, hydrogen peroxide and other trace species? In: Rowland FS & Isaksen ISA (Eds) The Changing Atmosphere (pp 63–78). John Wiley & Sons, Chichester

Stefess GC & Kuenen JG (1989) Factors influencing elemental sulphur production from sulphide or thiosulphate by autotrophic thiobacilli. Forum Microbiologie 12: 92

Suylen GMH (1988) Microbial metabolism of dimethyl sulphide and related compounds. PhD Thesis, Delft University of Technology, The Netherlands

Suylen GMH & Kuenen JG (1986) Chemostat enrichment and isolation of *Hyphomicrobium* EG. A dimethyl-sulphide oxidizing methylotroph and reevaluation of *Thiobacillus* MS1. Antonie van Leeuwenhoek 52: 281–293

Suylen GMH, Stefess GC & Kuenen JG (1986) Chemolithotrophic potential of a *Hyphomicrobium* species, capable of

growth on methylated sulphur compounds. Arch. Microbiol. 146: 192–198
Tanji Y, Kanagawa T & Mikami E (1989) Removal of dimethyl sulfide, methyl mercaptan and hydrogen sulfide by immobilized *Thiobacillus thioparus* TK-m. Journal of Fermentation and Bioengineering 67: 280–285
WHO (1984a) Guidelines for drinking water quality. Vol 1. Recommendations. Pubs WHO, Geneva
WHO (1984b) Guidlines for drinking water quality. Vol 2. Health criteria and other supporting information. Pubs WHO, Geneva
Woese CR (1991) Prokaryote systematics: the evolution of a science. In: Balows A, Truper HG, Dworkin M, Harder W & Schlefer KH (Eds) The Prokaryotes, Vol I (pp 3–18). Springer-Verlag, New York

Biodegradation 3: 255–264, 1992.

Transformations of ammonia and the environmental impact of nitrifying bacteria

Aharon Abeliovich
Environmental Microbiology Unit, The Jacob Blaustein Institute for Desert Research, Ben-Gurion University, Israel

Key words: ammonia, nitrification, nitrogen cycling

Abstract

In the sequence of events leading from ammonia to N_2 during the process of biotransformation of inorganic nitrogen compounds, the weakest link, with respect to our knowledge and understanding of the organisms involved, is nitrification. In particular, this is true for the oxidation of ammonia to nitrite. The enzymes have not been thoroughly studied, and the enzymatic mechanisms have not been identified. Almost any biochemical and physiological aspect studied proved to be controversial, and major ecological questions still remain unanswered. Unless the structure and function of the various components of the process are worked out, progress in developing means for controlling nitrification will depend mainly on laborious trial and error and not on knowledgeable manipulation of this group of bacteria.

Abbreviations: AMO – ammonia monooxygenase; HAO – hydroxylamine oxidoreductase; MPN – most probable number; TCE – trichloroethylene

General aspects

All major transformations of inorganic nitrogen in the environment, such as nitrogen assimilation, nitrification and denitrification are carried out exclusively by microorganisms. While considerable research effort is put into attempts to increase nitrogen assimilation for agricultural purposes, much less is invested in studies on the dissimilatory processes, particularly natural nitrification. This activity is carried out by a small group of bacteria which, in spite of their key role in global nitrogen cycling, is still poorly understood.

Environmental pollution by inorganic nitrogen is the result of a disequilibrium between input of fixed nitrogen (biotic and abiotic), usually in the form of ammonia, and its output, usually in the form of N_2, the only form of nitrogen which can be considered as environmentally safe. All other forms of nitrogen, which can accumulate due to the establishment of unbalanced fluxes, may create problems ranging in their severity from mild nuisances to serious ecological hazards.

The global biogenic fluxes of inorganic nitrogen are governed by 3 major processes: nitrogen fixation, nitrification and denitrification. Annual global input of fixed nitrogen (natural and industrial) is estimated at $1.75 \cdot 10^8$ tons (Bock et al. 1989). Although it may exist in a global equilibrium with natural denitrification, on smaller scales we find greatly unbalanced fluxes of inorganic nitrogen due to ever increasing anthropogenic activities. Only by manipulating and maintaining environmental factors to allow microorganisms to transform fixed nitrogen to dinitrogen gas, can we avoid the consequences of the accumulation of biologically active inorganic nitrogen species. Success depends on an in depth understanding of the processes involved, at all levels, from biochemistry to ecology, as with any other biological activity.

Pollution of the environment with nitrogen differs from pollution by other biogenic pollutants such as carbon and phosphorous, because not only can it trigger eutrophication of water bodies or affect soils and atmosphere, but some forms (ammonia, NO_2^-, NO_3^-) are, to varying degrees, toxic to various life forms, depending on the organisms exposed and upon environmental factors such as pH and temperature.

When compared to other environmentally important biogenic inorganic ions (PO_4^{3-}, Fe^{3+}, S^{2-}, CO_3^{2-}, SO_4^{2-} and others) the major nitrogenous end products of degradation of organic matter also differ in that all forms of inorganic nitrogen except for dinitrogen are very soluble, and can therefore reach high concentrations.

Of the various forms of nitrogen, ammonia and nitrite are the most toxic to aquatic wildlife (fish, crustaceans, mollusks). Concentrations of 0.1–1.0 mg NH_3^- $N \cdot l^{-1}$ are usually considered lethal (Kinne 1976). Because biological membranes are relatively impermeable to the ammonium ion, but readily permeable to unionized ammonia, toxicity of ammonia is a function of the pH. Nitrite concentrations should not exceed 0.1 $mg \cdot l^{-1}$ in closed water systems (King & Spotte 1974). Nitrate is less dangerous because it is toxic at much higher concentrations: 250–350 $mg \cdot l^{-1}$ nitrate were found to be lethal to many fish (DeGraff 1964). The US standard for nitrates in drinking water limits its concentration to 45 $mg \cdot l^{-1}$ (USEPA 1976), while European standard is 50 $mg \cdot l^{-1}$ (EEC 1980).

The maximum contaminant level allowed by recent EPA regulations for nitrite in drinking water is 1.0 $mg \cdot l^{-1}$ (Pontius 1991). It is therefore obvious that for all practical reasons nitrogen assimilation and dissimilatory denitrification must be optimally balanced. If not, either excess assimilation will cause accumulation of toxic inorganic nitrogen, or, on the other hand, denitrification might reduce primary productivity, affecting agriculture and the capacity of any ecosystem to sustain life.

Main inputs of ammonia and nitrite into the environment are through microbial nitrogen fixation and anthropogenic sources such as wastewater discharge and the widespread use of industrial fertilizers. The main sources of NO_x originate from microbial nitrification and denitrification processes, as well as from various industrial and urban activities.

Comprehensive reviews have been recently published on nitrogen fixation (e.g. Gresshoff et al. 1990) and on microbial denitrification (e.g. Revsbech & Sorensen 1990). Therefore, this review focuses primarily on nitrification and nitrifying bacteria, with an emphasis on the group of chemolithotrophic bacteria oxidizing ammonia to nitrite. Although they play a key role in cycling of inorganic nitrogen in the environment, they are the least studied group involved in this process. The reasons are that ammonia oxidizing bacteria are difficult to cultivate, and the enzymes involved in the oxidation of ammonia are difficult to work with. The information presently available is sketchy, incomplete, and subject to controversy.

Nitrification

Although nitrification is a general term used conventionally to describe the oxidation of reduced nitrogenous compounds by chemolithotrophic bacteria, as sole or primary energy source, it is by no means a sole oxidative pathway. Heterotrophic nitrification, not coupled to any energy yielding metabolic pathways, has been demonstrated in many organisms, including bacteria, such as *Arthrobacter* sp. (Witzel & Overback 1979), *Alcaligenes faecalis* (Papen et al. 1989), *Thiosphaera pantotropha* (Robertson & Kuenen 1988), and fungi (for a review see Killham 1986). Even mammals have been shown to be nitrifiers (Green et al. 1985). A possible mechanism for the accumulation of nitrite by heterotrophic organisms is the cooxidation of ammonia to nitrite, coupled to the xanthine oxidase reaction (Nagano & Fridovich 1985).

Chemolithotrophic nitrification is a two stage oxidation of reduced inorganic nitrogen to nitrite and then to nitrate. Different groups of chemolithotrophic bacteria carry out the 2 reactions, and no single known bacterium is capable of complete oxidation of ammonia to nitrate in a single step.

Enzymes of nitrification

The first step in nitrification is the oxidation of ammonia to nitrite, best studied in various species of the genus *Nitrosomonas*. Two enzymes catalyze two key reactions:

$NH_3 + 2H^+ + 2e^- + O_2 \rightarrow NH_2OH + H_2O$
(ammonia monooxygenase)

$NH_2OH + H_2O \rightarrow HNO_2 + 4H^+ + 4e^-$
(hydroxylamine oxidoreductase)

Only the second reaction provides ammonia oxidizing bacteria with their energy requirements.

The second step is the oxidation of nitrite to nitrate:

$HNO_2 + H_2O \rightarrow HNO_3 + 2H^+ + 2e^-$
(nitrite dehydrogenase)

carried out by members of the *Nitrobacter* and *Nitrospira* genera.

The first two key enzymes involved in these reactions proved to be difficult to work with. Ammonia monooxygenase (AMO) is a membrane bound protein. It has not been isolated, and all the information presently available regarding its structure and mode of action is therefore purely circumstantial. Hymen & Wood (1985) showed that acetylene irreversibly inactivates the oxidation of ammonia, and in a crude extract of *N. europaea* they also showed the presence of a single 28 kd protein which reacts with acetylene. This is, at present, the only clue to the identification of the protein responsible for the AMO activity. Hydroxylamine oxidoreductase (HAO) is a soluble 180,000–200,000 kd periplasmic protein. The present state of information with regard to its structure and mode of action leaves much to be desired (Hooper et al. 1978; Yamanaka et al. 1979; Terry & Hooper 1981; Olson & Hooper 1983; Hooper 1984; Hooper et al. 1984; Hooper et al. 1990). The situation with regard to nitrite dehydrogenase is better, as the enzyme was purified and described in detail by Yamanaka & Fukumori (1988).

Oxidation of ammonia to nitrite requires one atom of oxygen derived from O_2 (Hollocher et al. 1981), while the source of the second atom of oxygen is water (Andersson & Hooper 1983). Water is also the source of the oxygen incorporated into nitrate during the oxidation of nitrite by *Nitrobacter* (Aleem et al. 1965).

Habitats of nitrifying bacteria

As true chemolithotrophic microorganisms, nutritional requirements of nitrifiers are minimal, and in practice, in any aerobic ecological niche where traces of ammonia are present, nitrifiers will also be found. Thus they penetrate rocks and sandstone of historical monuments to the depth, causing acidification and deterioration (Bock 1989). Likewise, nitrifiers have been found responsible for destroying asbestos-cement roofs of buildings (Wasserbauer et al. 1988). Nitrifiers have been isolated, and nitrifying activity has been described in Antarctic sea ice (Priscu et al. 1990). They have been found responsible for the depletion of oxygen under Arctic winter ice (Knowles & Lean 1987). Nitrifiers have also been implicated in soil acidification (Van Miegroet & Cole 1984, 1985). The impact of nitrifiers can be significant even when pollution sources are remote and contact indirect. For instance, although there is very little human activity on the shores of Lake Superior, its levels of nitrates increased 4-fold during the past century, primarily due to loading of nitrogenous compounds from the atmosphere (Bennet 1986). Nitrifiers also cause specific problems in disinfecting facilities for drinking water. When nitrite accumulates, due to incomplete nitrification or denitrification, it reacts readily with chlorine, decreasing effective residual chlorine concentrations. It also interferes with chloramination (Wolf et al. 1988). Heavy agricultural fertilization frequently results in massive nitrification and leaching of nitrates to ground water, increasing concentrations of nitrates to levels beyond permitted standards (Soares et al. 1991).

Nitrification as a source for nitric and nitrous oxides

NO_x gases play an important role in the stratospheric ozone cycle and appear to be much more effective green house gases than CO_2. Their reaction with volatile organic carbon also contributes to tropospheric ozone accumulation. However, the significance of nitrifiers as NO_x producers is not clear. Ammonia oxidizing bacteria are known to be capable of reducing nitrite to NO and N_2O (Hooper 1968). Ritchie & Nicholas (1972) attributed part of the NO_x produced aerobically to direct oxidation of NH_2OH. Later studies did not resolve the question whether NO_x are released by nitrifiers as a product of a denitrification reaction under microaerophilic or anaerobic conditions, or also by direct oxidation of ammonia. Although Poth & Focht (1985) stated that *Nitrosomonas* releases N_2O only as a denitrifying reaction, Hooper et al. (1990) also showed that some of the N_2O may be produced through oxidation of ammonia. Also, the data presented by Kim & Craig (1990) suggested that nitrification is the origin of some N_2O in deep ocean water. A possible mechanism of accumulation of NO and N_2O during nitrification by *Nitrosomonas* is that under certain conditions (e.g. the presence of organic reductants) excess hydroxylamine is released by the cells and reacts chemically with nitrite to form NO and N_2O (E. Bock, pers. comm.). There is also a controversy as to the contribution of nitrifiers vs. heterotrophic denitrifiers to total NO_x produced in natural habitats. While the data presented by Klemendtsson et al. (1988) and Downes (1988) suggested that nitrifiers play a major role in N_2O production in soils and freshwater lakes, according to Yoshida (1988) the role of nitrifiers in the production of N_2O in the oceans has been overestimated. Interestingly, Remde & Conrad (1990) showed that NO is produced by ammonia oxidizers and that in contrast to the release of N_2O, this production of NO is insensitive to oxygen concentration.

Biodegradation of hydrocarbons

While early attempts to demonstrate oxidation of methane, methanol and CO by *Nitrosomonas* failed (Drozd 1976), it has since been demonstrated that hydrocarbons are metabolized by a wide range of chemolithotrophic ammonia oxidizers (Hyman & Wood 1983; Jones & Morita 1983; Voysey & Wood 1987; and see Bedard & Knowels 1989 for a review), apparently due to the nonspecificity of ammonia monoxygenase. Although attempts to grow ammonia oxidizers with methane as a sole energy source failed, methane was completely oxidized to CO_2. On the other hand, some methanotrophs are capable of oxidation of ammonia to nitrite (Hutton & ZoBell 1953; Wittenbury et al. 1970). Other hydrocarbons are also co-metabolized by ammonia oxidizers. Acetylene is a suicidal substrate (Hyman & Wood 1985) to *Nitrosomonas*, apparently due to irreversible covalent bonding to the AMO. Ethylene is oxidized by *N. europaea* to ethylene oxide, a reaction sensitive to inhibitors of ammonia oxidation. Ethylene oxide is also further metabolized (Hyman & Wood 1984). Hyman et al. (1988) also demonstrated that n-alkanes (C_1–C_8) were oxidized to their respective alcohols, with increasing rates from C_1 to C_4 and a marked decreased rate above C_4. This oxidation depends on the presence of ammonia as a reductant for the AMO reaction, or, to a limited extent, on endogenous reducing power. All these oxidations were inhibited by AMO inhibitors (acetylene and thiourea) except for some residual activity when 1-pentene was tested. On the other hand, alkynes (C_2–C_{10}) were inhibitory to ammonia but not to hydrazine oxidation. Halogenated hydrocarbons are also degraded by ammonia oxidizers. Vannelli et al. (1990) demonstrated ammonia dependent, AMO mediated, oxidation of a range of halogenated hydrocarbons by *N. europaea,* including dichloromethane, dibromomethane, trichloromethane, cis and trans, dichloro-and dibromo-ethylene, trichloroethylene and trichloropropane (but not tetrachloroethylene) without the inactivation of the enzyme. AMO inhibitors inhibited these reactions by at least 70%. The halogenated hydrocarbons competed with ammonia for the AMO active site. Likewise, Rasche et al. (1990a) demonstrated that all haloethanes are oxidized (F, Cl, Br, I) to acetaldehyde, and that all tested n-chlorinated alkanes (C_1–C_4) were also oxidized to the corresponding aldehydes. Soil nitrifiers (*Nitrosomonas europaea* and *Nitrosolobus multiformis*) degraded nematicidic soil fumigants (methyl bromide, 1,2,dichloropropane, 1,2,dibromo-3-chloropropane) (Rasche et al. 1990b). These degradations were also inhibited by AMO inhibitors such as acetylene and allylthiourea. However, Rasche et al. (1991) demonstrated, using *N. europaea* and trichloroethylene (TCE), that active protein synthesis during the reaction is required to prevent irreversible inactivation of AMO by TCE. Other chlorocarbons varied with respect to their effects of AMO activity.

Aromatic hydrocarbons are also oxidized by the AMO of *N. europaea*. Hymen et al. (1985) showed that benzene is oxidized first to phenol, which was further oxidized to hydroquinone, but not to catochol or resorcinol.

Environmental studies

As mentioned above, oxidation of ammonia to nitrite/nitrate is a key reaction in nitrogen cycling, because there is no biological reciprocal of the nitrogen fixation reaction, back from ammonia to dinitrogen. The majority of nitrogen fertilizer added in agricultural practice is in the form of ammonia, which is adsorbed by the soil particles and then slowly released. However, in the presence of nitrifiers, ammonia is also rapidly oxidized to nitrate which is then lost either through leaching to groundwater (polluting aquifers) or to surface waters (polluting lakes and rivers) or it is oxidised to nitrous and nitric oxides (see above). Under anaerobic conditions, if electron donors are available in sufficient concentrations, nitrates may be lost by their reduction to dinitrogen. Therefore, direct measurements of nitrification potential, number of nitrifiers, and actual nitrification rates, in agricultural lands, as well as in forest ecosystems, submerged soils, and heath soils are needed for accurate estimates of nitrogen fluxes through these habitats.

Many difficulties arise when attempts are made to correlate laboratory studies to actual field conditions, some of them due to two practical factors. Their growth rates are very slow in liquid media, and they cannot be counted by plating on solid media. Therefore the numbers of viable and culturable nitrifying organisms can be estimated only by the most probable number method (MPN), being then followed by measuring concentrations of nitrite and nitrate in the dilution series. Their slow growth rate prolongs incubations. The most common incubation period for MPN counts is 3 weeks, although Matulewich et al. (1975) showed that 8 weeks incubations yield much more accurate results; and Hashimoto & Hattori (1987) suggested even longer incubation periods. Therefore, counts of nitrifiers can serve only as minimal estimates of their true concentrations. Other parameters used, such as measuring nitrification potential, are also questionable, this because of the difficulties in simulating *in situ* conditions in the laboratory. For example, Berg & Rosswall (1985) reported that in their study on arable soils in Sweden, actual nitrification rates were 5–25 times lower than potential rates determined in the laboratory.

Inhibition of nitrification

Phenolic acids excreted by vegetation were cited as inhibitors of nitrification (Rice & Pancholy 1972). On the other hand, McCarty et al. (1991) found no basis for the assumption that phenolic acids inhibit ammonia oxidation in terrestrial ecosystems. Others claim that fulvic and humic acids enhance oxidation of ammonia to nitrite, but inhibit its further oxidation to nitrate (Tan & Lopez-Falcon 1987a, 1987b). Low pH, as well as organic matter, are frequently cited as factors that limit nitrification in forest ecosystems (White & Gosz 1987). However, although nitrification is considered to be pH sensitive, there are several reports describing autotrophic nitrification in acidic environments such as an acid heath soil (pH 3.6–4.0) (DeBoer et al. 1988), as well as ureolytic autotrophic nitrification by *Nitrospira* sp. at pH 4.8 (DeBoer 1989). Hankinson & Schmidt (1988) have also isolated an acidophilic *Nitrobacter* from acid forest land. Although nitrification is also considered to be sensitive to excess ammonia and nitrite, Blouin et al. (1989) reported a complete oxidation of very high concentrations of ammonia nitrogen (1,000 $mg \cdot l^{-1}$) in swine wastes in 5 days to nitrite, and another 10 days were needed for oxidation of all nitrite to nitrate.

In spite of their ubiquitous distribution, nitrifiers do suffer from various unidentified inhibitors, and the addition of powdered activated carbon to activated sludge wastewater treatment process enhances nitrification by adsorbing inhibitory compounds (Ng & Stenstrom 1987). The issue of protection of nitrifiers from various inhibitors by their attachment to surfaces is also a subject of controversy. While there are claims that surface attachment appears to protect nitrifying bacteria from a range of inhibitors (Ratnayake & Audas 1978), others have shown that surface growth alone does not protect soil nitrifiers from the effect of inhibitors (Underhill & Prosser 1987).

Another factor believed to have a detrimental effect upon nitrification is light. Hooper & Terry (1973, 1974) have shown that visible light (420 nm) inhibits oxidation of ammonia but not of hydroxylamine, at a rate constant proportional to its intensity. Inhibition by light also abolishes the ability of the putative AMO peptide to bind acetylene (Hyman & Arp 1992). U.V. light inactivated both oxidation of ammonia and NH_2OH. Protection against photoinactivation could be achieved either by fully inhibiting the nitrification process, as by anaerobiosis, or, conversely, by allowing rapid oxidation of ammonia during illumination. Similarly, Shears & Wood (1985) reported that protection against photoinactivation could be provided by ammonia monooxygenase substrates as well as by anaerobiosis. While Hooper & Terry (1974) reported that recovery of ammonia oxidation by photoinhibited cells took 6 hours, others reported very different recovery rates, up to 120–350 days (Yoshioka & Saijo 1984). From various reports it seems that photoinhibition is not a universal phenomenon. There have been reports on variable or adaptive responses (Horrigan & Springer 1990), while others did not find any inhibitive effect of light (Ward 1984) or even report on light stimulation of ammonia oxidation (Miyazaki et al. 1973). Starvation of *Nitrosomonas cryotolerans* also seemed to sensitize them to photoinhibition (Johnstone & Jones 1988). However, as will be described in the following section, nitrifying bacteria have frequently been found in ecosystems that are permanently exposed to direct sunlight.

Ecophysiology of nitrifying bacteria

In a world where many suffer from food shortages due to poor agricultural practices and insufficient fertilization, where efforts are made to increase microbial nitrogen fixation and to transform staple food plants genetically to become nitrogen fixers as well, the question of inorganic nitrogen pollution seems somewhat out of place. Nevertheless, inorganic nitrogen pollution is a real problem in many parts of the world. It originates either from excess nitrogen fertilization, or as a product of catabolic deamination, releasing ammonia into the environment. This is either through biodegradative processes, or as part of intensive treatment of domestic sewage or agricultural wastes. It is these anthropogenic sources of discharged ammonia which cause a wide range of environmental problems.

The sequence of events leading to the elimination of excess nitrogen from the environment is a complex string of reactions performed by a consortium of microorganisms. For example, during the treatment of domestic wastewater, microbial mineralization of organic matter ends with large quantities of ammonia (80–120 mg/l) in the effluents, and unless a nitrification-denitrification stage is added to the process, these are discharged as such to the environment. When climatic conditions permit and land is available, oxidation ponds are the method of choice for treatment of sewage, due to their low operational costs and high efficiency in eliminating pathogens. Here, ammonia plays a cardinal role in reducing the efficiency of the process. It has been long known that ammonia uncouples electron flow in isolated chloroplasts (Avron & Shavit 1965) and inhibits the whole photosynthetic process in intact algal cells (Abeliovich & Azov

1976). For example, 2–3 mM ammonia at pH 8.1–8.4, will cause 50–90% inhibition in the rate of O_2 generation by any algal population present in the ponds. This phenomenon has far reaching consequences. Efficient and nuisance free operation of oxidation ponds calls for uninterrupted and optimal oxygenation of the water. This is done almost exclusively by the algal population which is expected to dominate the pond. However, un-ionized ammonia penetrates freely through the cell membrane and, unless kept below ca 2 mM, reaches intracellular concentrations which strongly inhibit photosynthesis and oxygenation in a pH dependent manner. Uninterrupted photosynthesis can, in eutrophic water bodies, elevate pH to above 10, but, in the presence of ammonia, it will elevate the pH only to that level at which unionized ammonia will reach intracellular concentrations inhibitory to photosynthesis. Theoretically, respiration can lower the pH to levels at which the concentration of unionized ammonia will not be inhibitory to photosynthesis. But, for respiration to lower pH, oxygen is needed. This vicious circle can be interrupted only by enhancing nitrification (Abeliovich 1983).

Nitrification of the ammonia in domestic sewage is easily achieved in the laboratory in experiments aimed at simulating oxidation pond or wastewater reservoir conditions (Abeliovich 1985). But nitrification in real oxidation ponds is a very slow process (Abeliovich 1987), with measured rates usually within the range of 0.5–1% of the ammonia being oxidized every day.

The reasons for this slow rate of nitrification are not clear. Light, although effective at inhibiting nitrification in mineral medium in the laboratory, had variable effects when tested with wastewater reservoir effluents, and had no effect upon viability of nitrifiers in the field (Abeliovich 1987). In wastewater reservoirs during the early summer months, a phenomenon of unbalanced nitrification is frequently observed, leading to accumulation of nitrite.

Unexpectedly, we found that nitrifiers, both *Nitrosomonas* sp. and *Nitrobacter* sp., are abundant in the anaerobic hypolimnion of these reservoirs, as well as in anaerobic stabilization ponds (Abeliovich 1987). In the laboratory, *Nitrosomonas europaea*, under strict anaerobic conditions, utilized nitrite as an electron acceptor, and pyruvate as energy source. Labeled pyruvate did not penetrate the cell membrane and was not incorporated into cell material, but its presence was essential for incorporation of $^{14}CO_2$ into cell material. Either pyruvate participates in a phosphoroclastic reaction at the outer surface of the cell membrane, or the extracted protons generate a membrane potential used for generating the energy needed by *Nitrosomonas* to survive anaerobic conditions (Abeliovich & Vonshak, 1992). Earlier, Poth (1986) described a *Nitrosomonas* producing N_2 from nitrite under microaerophilic conditions, and Yamanaka & Sakano (1980) demonstrated oxidation of hydroxylamine to nitrite, and its further disappearance by extracts of *N. europaea* under anaerobic conditions. The ecological significance of these observations is not yet clear.

Denitrification

The end product of nitrification is, as mentioned earlier, an environmental contaminant and a health hazard in its own right, and further treatment is required to eliminate nitrates through denitrification. This is a term usually used to describe respiration of microorganisms with nitrate or nitrite replacing oxygen as a terminal electron acceptor, with N_2O or N_2 accumulating as end products. Although it is most common under anaerobic conditions, microbial aerobic denitrification has been shown in recent years to be a wide spread phenomenon (Robertson & Kuenen 1990). With the currently increasing levels of aquifer pollution with nitrates, primarily due to leaching of fertilizers, values often reach levels which are prohibitive to human consumption.

Elimination of nitrates can be achieved by processes based on membrane technology which are, for practical applications, possible only when nitrate is the dominant ion, since there are no nitrate selective membranes. Therefore any process aimed at filtering nitrates will have to filter all soluble ions too, which makes it economically impractical. Microbial denitrification using various carbon sources is, at present, the only selective process, with the added advantage that the end product is gaseous.

Microbial denitrification has been thoroughly reviewed in a book published recently (Revsbech & Sorensen 1990) and will not be discussed here. Attempts in Israel to apply enhanced *in situ* microbial denitrification to treat the contaminated coastal aquifer proved to be impractical, both in the laboratory and in field studies (Soares et al. 1989, 1991). A prerequisite for successful *in situ* denitrification is an environment which permits the free flow of water on one hand, and the free exchange of gases between air and water in a phreatic aquifer on the other. The coastal aquifer in Israel is made of sand, silt and clay which traps the gas liberated in the denitrification process as bubbles which clog and prevent the free flow of water. Therefore, the only practically feasible process in the case of a sandy aquifer should be built as an efficient above ground installation.

Summary

In the sequence of events leading from ammonia to N_2 during the process of biotransformation of inorganic nitrogen compounds, the weakest link, with respect to our knowledge and understanding of the organisms involved, is nitrification. In particular, this is true for the oxidation of ammonia to nitrite. The enzymes have not been thoroughly studied, and the enzymatic mechanisms have not been identified. Almost any biochemical and physiological aspect studied proved to be controversial, and major ecological questions still remain unanswered. Unless the structure and function of the various components of the process are worked out, progress in developing means for controlling nitrification will depend mainly on laborious trial and error and not on knowledgeable manipulation of this group of bacteria.

References

Abeliovich A & Azov Y (1976) Toxicity of ammonia to algae in sewage oxidation ponds. Appl. Environ. Microbiol. 31: 801–806

Abeliovich A (1983) The effects of unbalanced ammonium and BOD concentrations on oxidation ponds. Wat. Res. 17: 299–301

Abeliovich A (1985) Nitrification of ammonia in wastewater. Field observations and laboratory studies. Wat. Res. 19: 1097–1099

Abeliovich A (1987) Nitrifying bacteria in wastewater reservoirs. Appl. Environ. Microbiol. 53: 754–760

Abeliovich A & Vonshak A (1992) Anaerobic metabolism of *Nitrosomonas europaea*. Arch. Microbiol. 158: 267–270

Aleem MIH, Hoch GE & Varner JE (1965) Water as the source of oxidant and reductant in bacterial chemosynthesis. Proc. Natl. Acad. Sci. USA 54: 869–873

Andersson KK & Hooper AB (1983) O_2 and H_20 are each the source of one O in NO_2^- produced from NH_3 by *Nitrosomonas*: ^{15}N N-NMR evidence. FEBS 164: 236–239

Avron M & Shavit N (1965) Inhibitors and uncouplers of photophosphorylation. Biochim. Biophys. Acta 376: 97–104

Bedard C & Knowles R (1989) Physiology, biochemistry and specific inhibitors of CH_4, NH_4^+, and CO oxidation by methanotrophs and nitrifiers. Microbiol. Rev. 53: 68–84

Bennet E B (1986) The nitrifying of Lake Superior. Ambio. 15: 272–275

Berg P & Rosswall T (1985) Ammonium oxidizer numbers, potential and actual oxidation rates in two Swedish arable soils. Biol. Fert. Soils 1: 31–140

Blouin M, Bisaillon JG, Beaudet R & Ishaque M (1989) Nitrification of swine waste. Can. J. Microbiol. 36: 273–278

Bock E, Koops HP & Harms H (1989) Nitrifying bacteria. In: Schlegel HG & Bowien B (Eds) Autotrophic Bacteria (pp 81–96). Science Tech. Publishers, Madison, WI, and Springer Verlag, Berlin.

DeBoer W & Laanbroek HJ (1989) Ureolytic nitrification at low pH by *Nitrospira* sp. Arch. Microbiol. 152: 178–181

DeBoer W, Duyts H & Laanbroek HJ (1988) Autotrophic nitrification in a fertilized acid heath soil. Soil Biol. Biochem. 20: 845–850

DeGraff F (1964) Maintenance problems in large public aquaria. Arch. Neerl. Zool. 16: 142–143

Downs MT (1988) Aquatic nitrogen transformations at low oxygen concentrations. Appl. Environ. Microbiol. 54: 172–175

Drozd JW (1976) Energy coupling and respiration in *Nitrosomonas europaea*. Arch. Microbiol. 110: 257–262

EEC Directive of 15.7.80 relating to the quality of water intended for human consumption, 80-778-EEC. (1980) Office J. Eur. Commun. 23, L229: 11–29

Green LC, Tannenbaum SR & Goldman P (1981) Nitrate synthesis in the germfree and conventional rat. Science 212: 56–58

Gresshoff PM, Roth LE, Stacey G, & Newton WE (1990) Nitrogen Fixation: Achievements and Objectives. Chapman and Hall, New York and London

Hankinson TR & Schmidt TL (1988) An acidophilic and neutrophilic *Nitrobacter* strain isolated from the numerically predominant nitrite oxidizing population of an acid forest soil. Appl. Environ. Microbiol. 54: 1536–1540

Hashimoto T & Hattori T (1987) Length of incubation for the

estimation of the most probable number of nitrifying bacteria in soil. Soil Sci. Plant Nutr. 33: 507–509
Hollocher TC, Tate ME & Nicholas DJD (1981) Oxidation of ammonia by *Nitrosomonas europaea*. Definitive ^{18}O tracer evidence that hydroxylamine formation involves a monooxygenase. J. Biol. Chem. 256: 10834–10836
Hooper AB (1968) A nitrite reducing enzyme from *Nitrosomonas europaea*. Preliminary characterization with hydroxylamine as electron donor. Biochim. Biophys. Acta 162: 49–65
Hooper AB (1984) Ammonia oxidation and energy transduction in the nitrifying bacteria. In: Strohl WR & Tuovinen OH (Eds) Microbial Chemoautotrophy (pp 133–167). Ohio State University Press, Columbus, Ohio
Hooper AB & Terry KR (1973) Specific inhibitors of ammonia oxidation in *Nitrosomonas*. J. Bacteriol. 115: 480–485
Hooper AB & Terry KR (1974) Photoinactivation of ammonia oxidation in *Nitrosomonas*. J. Bacteriol. 119: 899–906
Hooper AB, Maxwell PC & Terry KR (1978) Hydroxylamine oxidoreductase from *Nitrosomonas*: Absorption spectra and content of heme and metal. Biochem. 17: 2984–2989
Hooper AB, DiSpirito AA, Olson KA, Anderson A, Cunningham W & Taaffee LR (1984) Genertation of the proton gradient by a periplasmic dehydrogenase. In: Crawford RL & Hanson RS (Eds) Microbial Growth on C1 Compounds. (pp 53–58). Society for Microbiology, Washington, DC
Hooper AB, Arciero DM, DiSpirito AA, Fuchs J, Johnson M, Mundfrom G & McTavish H. (1990) Production of nitrite and N_20 by the ammonia oxidizing nitrifiers. In: Gresshoff PM, Roth LE, Stacey G & Newton WE (Eds) Nitrogen Fixation: Achievement and Objectives (pp 387–392). Chapman and Hall, New York and London
Horrigan SG & Springer AL (1990) Oceanic and estuarine ammonium oxidation: effects of light. Limnol. Oceanog. 35: 479–482
Hutton WE & ZoBell CE (1953) Production of nitrite from ammonia by methane oxidizing bacteria. J. Bacteriol. 65: 216–219
Hyman MR & Wood PM (1983) Methane oxidation by *Nitrosomonas europaea*. Biochem. J. 212: 31–37
Hyman MR & Wood PM (1984) Ethylene oxidation by *Nitrosomonas europaea*. Arch. Microbiol. 137: 155–158
Hyman MR & Wood PM (1985) Suicidal inactivation and labeling of ammonia monooxygenase by acetylene. Biochem. J. 227: 719–725
Hyman MR & Arp DJ (1992) $^{14}C_2H_2$ and $^{14}CO_2$ labeling studies of the de novo synthesis of polypeptides by *Nitrosomonas europaea* during recovery from acetylene and light inactivation of ammonia monooxygenase. J. Biol. Chem. 267: 1534–1545
Hyman MR, Sansome-Smith AW, Shears JH & Wood PM (1985) A kinetic study of benzene oxidation to phenol by whole cells of *Nitrosomonas europaea* and evidence for the further oxidation of phenol to hydroquinone. Arch. Microbiol. 143: 302–306
Hyman MR, Murton JB & Arp DJ (1988) Interaction of ammonia monooxygenase from *Nitrosomonas europaea* with alkanes, alkenes and alkynes. Appl. Environ. Microbiol. 54: 3187–3190
Johnstone BH & Jones RD (1988) Effects of light and CO on the survival of a marine ammonium oxidizing bacterium during energy source deprivation. Appl. Environ. Microbiol. 54: 2890–2893
Jones RD & Morita RY (1983) Methane oxidation by *Nitrosomonas oceanus* and *Nitrosomonas europaea*. Appl. Environ. Microbiol. 45: 401–410
Killham K (1986) Heterotrophic nitrification. In: Prosser JI (Ed) Nitrification (pp 117–126). IRL Press, Oxford, Washington, DC
Kim K & Craig H (1990) Two isotope characterization of N_2O in the Pacific Ocean and constraints on its origin in deep water. Nature 347: 58–61
King JM & Spotte SH (1974) Marine aquariums in the research laboratory. Aquarium Systems, Inc. East Lake, Ohio
Kinne O (1976) Cultivation of marine organisms: water quality management and technology. In: Kinne O (Ed) Mar. Ecol. Vol 3 (pp 19–268). John Wiley & Sons, London, New York, Sydney, Toronto
Klemendtsson L, Svensson BH & Rosswall T (1988) Relationship between soil moisture content and nitrous oxide production during nitrification and denitrification. Biol. Fertil. Soils 6: 106–111
Knowles R & Lean DRS (1987) Nitrification: a significant case of oxygen depletion under winter ice. Can J. Fish. Aquat. Sci. 44: 743–749
Matulewich VA, Storm PF & Finstein M (1975) Length of incubation for enumerating nitrifying bacteria present in various environments. Appl. Microbiol. 29: 265–268
McCarty GW, Bremmer JM & Schmidt EL (1991) Effects of phenolic acids on ammonia oxidation by terrestrial autotrophic nitrifying microorganisms. FEMS Microbiol. Ecol. 85: 345–349
Miyazaki T, Wada E & Hattori A (1973) Capacities of shallow waters of Sagami Bay for oxidation and reduction of inorganic nitrogen. Deep Sea Res. 20: 571–577
Nagano T & Fridovich I (1985) The co-oxidation of ammonia to nitrite during the aerobic xanthine oxidase reaction. Arch. Biochem. Biophys. 241: 596–601
Ng AS & Stenstrom MK (1987) Nitrification in powdered activated sludge process. J. Environ. Engn. 113: 1285–1301
Olson TC & Hooper AB (1983) Energy coupling in the bacterial oxidation of small molecules: an extra cytoplasmic dehydrogenase in *Nitrosomonas*. FEMS Microbiol. Lett. 19: 47–50
Papen H, von Berg R, Hinkel I, Thoene B & Rennenberg H (1989) Heterotrophic nitrification by *Alcaligenes faecalis*: NO_2^-, NO_3^-, N_2O and NO production in exponentially growing cultures. Appl. Environ. Microbiol. 55: 2068–2072
Pontius FW (1991) Phase III organic and inorganic contaminant regulations. J. AWWA 83: 20–22, 77–79
Poth M (1986) Dinitrogen production from nitrite by *Nitrosomonas* isolate. Appl. Environ. Microbiol. 52: 957–959
Poth M & Focht DD (1985) ^{15}N kinetic analysis of N_2O production by *Nitrosomonas europaea*: an examination of nitrifier denitrification. Appl. Environ. Microbiol. 49: 1134–1141
Priscu JC, Downes MT, Priscu LR, Palmisano C & Sullivan CW

(1990) Dynamics of ammonium oxidizer activity and N_2O within and beneath Antarctic sea ice. Mar. Ecol. Prog. Ser. 62: 37–46

Rasche EM, Hicks RE, Hyman MR & Arp DJ (1990a) Oxidation of monohalogenated ethanes and n-chlorinated alkanes by whole cells of *Nitrosomonas europaea*. J. Bacteriol. 172: 5368–5373

Rasche EM, Hyman MR & Arp DJ (1990b) Biodegradation of halogenated hydrocarbon fumigants by nitrifying bacteria. Appl. Environ. Microbiol. 56: 2568–2571

Rasche EM, Hyman MR & Arp DJ (1991) Factors limiting aliphatic chlorocarbon degradation by *Nitrosomonas europaea*: cometabolic inactivation of ammonia monooxygenase and substrate specificity. Appl. Environ. Microbiol. 57: 2986–2994

Ratnayake M & Audas LJ (1978) Studies on the effects of herbicides on soil nitrification. Pesticide Biochem. Physiol. 8: 170–185

Remde A & Conrad R (1990) Production of nitric oxide in *Nitrosomonas europaea* by reduction of nitrite. Arch. Microbiol. 154: 187–191

Revsbech NP & Sorensen J (Eds) (1990) Denitrification in Soil and Sediment. Plenum Press, New York and London

Rice EL & Pancholy SK (1972) Inhibition of nitrification by climax ecosystems. Am. J. Bot. 59: 1033–1040

Ritchie GAF & Nicholas DJD (1972) Identification and characterization of nitrous oxide produced by oxidative and reductive processes in *Nitrosomonas europaea*. Biochem. J. 126: 1181–1191

Robertson LA & Kuenen JG (1988) Heterotrophic nitrification in *Thiosphaera pantotropha*: oxygen uptake and enzyme studies. J. Gen. Microbiol. 134: 857–863

Robertson LA & Kuenen JG (1990) Physiological and ecological aspects of aerobic denitrification, a link with heterotrophic nitrification? In: Revsbech NP & Sorensen J (Eds) Denitrification in Soil and Sediment (pp 91–104). Plenum Press, New York and London

Shears JH & Wood PM (1985) Spectroscopic evidence for a photosensitive oxygenated state of ammonia mono-oxygenase. Biochem. J. 266: 499–507

Soares MIM, Belkin S & Abeliovich A (1989) Clogging of microbial denitrification in sand columns: gas bubbles or biomass accumulation. Was.-Abwas. Forsch. 22: 20–24

Soares MIM, Braester C, Belkin S & Abeliovich A (1991) Denitrification in laboratory sand columns: carbon regime, gas accumulation and hydraulic properties. Wat. Res. 25: 325–332

Tan KH & Lopez-Falcon RA (1987a) Effect of fulvic and humic acids on nitrification. Part I. *In vitro* production of nitrite and nitrate. Commun. Soil Sci. Plant Anal. 18: 835–853

Tan KH & Lopez-Falcon RA (1987b) Effect of fulvic and humic acids on nitrification. Part II. The nitrifying potential of cecil and davidson soils. Commun. Soil Sci. Plant Anal. 18: 855–873

Terry KR & Hooper AB (1981) Hydroxylamine oxidoreductase: a 20-heme, 200,000 molecular weight cytochrome c with unusual denaturation properties which forms a 63,000 molecular weight monomer after heme removal. Biochem. 20: 7026–7032

Underhill SE & Prosser JI (1987) Surface attachment of nitrifying bacteria and their inhibition by potassium ethyl xanthate. Microb. Ecol. 14: 129–139

US Environmental Protection Agency: Quality criteria for water. 1976. Publication No. 440–976–023. Cincinnati, OH, USA

Van Miegroet H & Cole DW (1984) The impact of nitrification on soil acidification and cation leaching in a red alder ecosystem. J. Environ. Qual. 13: 586–590

Van Miegroet H & Cole DW (1985) Acidification sources in red alder and Douglas fir soils - importance of nitrification. Soil Sci. Soc. Amer. J. 49: 1274–1279

Vannelli T, Logan M, Arciero DM & Hooper AB (1990) Degradation of halogenated aliphatic compounds by the ammonia oxidizing bacterium *Nitrosomonas europaea*. Appl. Environ. Microbiol. 56: 1169–1171

Voysey PA & Wood PM (1987) Methanol and formaldehyde oxidation by an autotrophic nitrifying bacterium. J. Gen. Microbiol. 33: 283–290

Ward BB (1984) Combined autoradiograph and immunofluorescence for estimation of single cell activity by ammonium oxidizing bacteria. Limnol. Oceanogr. 29: 402–410

Wasserbauer R, Zadak Z & Novotny J (1988) Nitrifying bacteria on the asbestos cement roofs of stable buildings. Internat. Biodeterior. 24: 153–165

White CS & Gosz JR (1987) Factors controlling nitrogen mineralization and nitrification in forest ecosystems in New Mexico. Biol. Fert. Soil. 5: 195–202

Wittenbury RK, Phillips C & Wilkinson JF (1970) Enrichment, isolation and some properties of methane utilizing bacteria. J. Gen. Microbiol. 61: 205–218

Witzel KP & Overbeck HJ (1979) Heterotrophic nitrification by *Arthrobacter* sp (strain 9006) as influenced by different cultural conditions, growth state and acetate metabolism. Arch. Microbiol. 122: 13–143

Wolfe RL, Means EG, III, Davis MK & Barrett SE (1988) Biological nitrification in covered reservoirs containing chloraminated water. J. AWWA 80: 109–114

Yamanaka T & Fukumori Y (1988) The nitrite oxidizing system of *Nitrobacter winogradskyi*. FEMS Microbiol. Revs. 54: 259–271

Yamanaka T & Sakano Y (1980) Oxidation of hydroxylamine to nitrite catalyzed by hydroxylamine oxidoreductase purified from *Nitrosomonas europaea*. Curr. Microbiol. 4: 239–244

Yamanaka T, Shimura M, Takahashi K & Shibasaka M (1979) Highly purified hydroxylamine oxidoreductase derived from *Nitrosomonas europaea*. J. Biochem. 86: 1101–1108

Yoshida N (1988) ^{15}N depleted N_2O as product of nitrification. Nature 335: 528–529

Yoshioka T & Saijo Y (1984) Photoinhibition and recovery of NH_4^+ oxidizing bacteria and NO_2^- oxidizing bacteria. J. Gen. Appl. Microbiol. 30: 151–166

Biodegradation **3**: 265–284, 1992.

Versatility of soil column experiments to study biodegradation of halogenated compounds under environmental conditions

Jan Roelof van der Meer,[1] Tom N.P. Bosma, Wil P. de Bruin, Hauke Harms, Christof Holliger,[2] Huub H.M. Rijnaarts, Marijke E. Tros, Gosse Schraa & Alexander J.B. Zehnder[1]
Department of Microbiology, Wageningen Agricultural University, Hesselink van Suchtelenweg 4, 6703 CT Wageningen, The Netherlands (present addresses: [1] EAWAG/ETH, CH-8600 Dübendorf & [2] EAWAG/ETH, CH-6047 Kastanienbaum, Switzerland)

Key words: xenobiotic compounds, chlorobenzenes, tetrachloroethylene, reductive dehalogenation, residual concentration, desorption, genetic adaptation, *Pseudomonas* sp. strain P51, adhesion properties

Abstract

Soil column experiments were performed to obtain insight in the different biological and physico-chemical processes affecting biodegradation of halogenated compounds under natural conditions in a water infiltration site. Lower chlorinated aromatic compounds could be degraded under aerobic conditions, whereas highly chlorinated compounds and chlorinated aliphatic compounds were mainly transformed under anaerobic conditions. Microorganisms which derive energy from reductive dechlorination were enriched and characterized. It was found that microbes could adapt to using chlorinated benzenes by evolution of new enzyme specificities and by exchange of genetic material. For halogenated pollutants, which are generally hydrophobic, sorption processes control the concentration available for biodegradation. The effects of very low concentrations of halogenated compounds on their biodegradability are described. The use of isolated bacterial strains to enhance biodegradation was evaluated with respect to their temperature-related activity and to their adhesion properties.

Abbreviations: 3-CB - 3-chlorobenzoate, DCB - dichlorobenzene, HCH - hexachlorocyclohexane, IS - insertion sequence, PER - tetrachloroethylene, S_{min} - minimal substrate concentration for growth, TCB - trichlorobenzene, TRI - trichloroethylene, λ - filtration coefficient

Introduction

The widespread use of halogenated organic compounds for more than fifty years, has raised considerable concern about their persistence in the environment (Alexander 1985). These compounds are applied as solvents, pesticides, intermediates, or preservatives in industry, agriculture, or households (Pearson 1982). As a result, halogenated compounds can be found almost in any natural habitat. For instance, tetrachloroethylene, trichloroethylene, methylenechloride, or 1,2-dichlorobenzene, are used as solvents in dry cleaning and degreasing of electronic or metal parts, and enter the natural environment through spills, evaporation, or via wastewater discharges. Pesticides are usually directly introduced into the environment of their purpose, and in the case of lindane or DDT can end up in higher organisms via food webs. An important route via which halogenated compounds find their way to the groundwater is leaching from waste disposal sites. Higher chlorinated compounds can be formed during the incineration of domestic waste and be spread via the atmosphere (Pearson 1982).

Halogenated compounds are generally considered as being xenobiotic, i.e. foreign to existing life forms. This terminology refers to their uncommon chemical structures, or presence of uncommon side groups or atoms (such as halogens) on the molecules, which renders them more difficult for enzymatic attack. Xenobiotic compounds or their metabolites are often toxic to the biological world and may therefore persist in the environment for long times. Not all halogenated compounds, however, are to be considered 'xenobiotics', since more than 200 halogenated compounds are known to be produced by bacteria, algae and sponges (Faulkner 1980).

The fate of xenobiotic substances in the environment is governed by degradation processes among which biodegradation and by their distribution in the different environmental compartments, soil, water and air. The distribution is determined by physico-chemical properties such as hydrophobicity, solubility, or volatility. The distribution parameters influence largely the concentration of the compounds in the water phase and thus their mobility and transport, either solubilized or particle-bound, and their accessibility for biodegradation. Microorganisms can play an important role in the degradation of halogenated organic compounds. However, the persistence and accumulation of halogenated compounds in certain environmental compartments indicate their recalcitrance to biodegradation. The possibilities for biodegradation are dependent on the metabolic potential of the microorganisms and on physico-chemical factors which determine microbial activity in the environment, such as physical (temperature, water potential, accessibility of the substrate), chemical (pH, redox state, concentration of the compounds, presence of additional secondary substrates, salinity), and biological conditions (presence of predators, adsorption of microbes to surfaces) (Goldstein et al. 1985; Klein & Alexander 1986; Swindoll et al. 1988). Together all these factors determine whether or not a halogenated compound will disappear readily from the environment, or will persist and inflict long-term damage on the quality of our environment.

Aim of this research

Groundwater is an important source for drinking water in The Netherlands. In heavily populated areas the amount of groundwater is not sufficient to cover all needs, and surface waters are used as source for drinking water, either directly, or after infiltration into the soil. The quality of both groundwater and of surface water, such as river water, is increasingly threathened by the presence of organic micropollutants. Pesticide residues, solvent spills, or wastewater discharges account for the pollution of ground- and surface water, and cause the quality decline of this important resource. Increasing efforts are needed to purify the water before it can be used as drinking water.

To investigate the fate of organic compounds during infiltration of river water into the soil, collaborate projects were started between our laboratory and the Amsterdam Water Works in 1984. Soil column experiments were set up to mimic the infiltration conditions of water into the soil, and to study transport, distribution, and biodegradability of a defined set of mainly halogenated compounds (Fig. 1). These served as model compounds for the great variety of micropollutants in river water (Bosma et al. 1991; Bosma et al., in preparation). In this article an overview is given of a number of our studies which address various aspects of biodegradation in a soil-water system. The topics include, (i) the possibilities for biodegradation of organohalogen compounds under different redox conditions, (ii) biodegradation at very low substrate concentrations and the effect of adsorption and desorption on the biodegradability, (iii) the metabolic potential of the microorganisms present in the natural environment and the capacity to adapt to xenobiotic substrates, and (iv) the use of purified or enriched strains to enhance biodegradation in soil-type reactors. In this last part special emphasis is given to the temperature-related activity of microorganisms and the adhesion properties. The adhesion properties are important because they

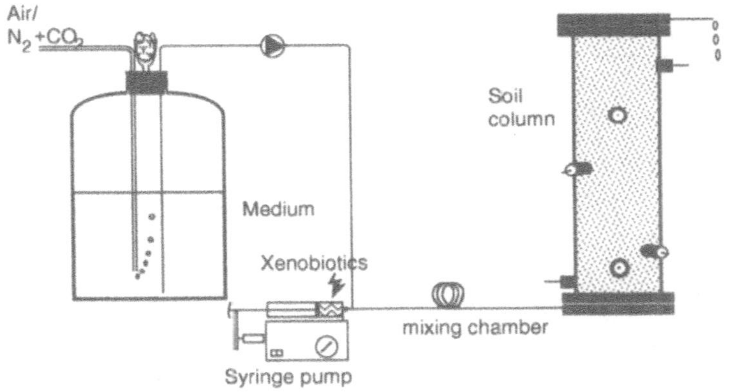

Fig. 1. The soil column system used in the different studies. A mineral medium, which can be aerated or purged with N_2/CO_2 during anaerobic studies, is pumped through a water-saturated soil column. Xenobiotic compounds are added to the mineral medium via a syringe pump, which allows accurate dosing. Stainless steel tubes at various positions pointing into the middle of the column are used to sample the liquid phase.

determine whether introduced microorganisms are retained or transported through the porous medium.

Biodegradation of organohalogen compounds under different redox conditions

Perhaps most information exists on the possibilities for degradation of halogenated compounds under different redox conditions. Many halogenated aromatic compounds can be degraded under aerobic conditions by microorganisms which either use these compounds as growth and energy source or co-metabolize them. The different pathways and reactions that microorganisms use to attack and metabolize halogenated compounds have been reviewed extensively (Chaudry & Chapalamadugu 1991; Commandeur & Parsons 1990; Haggblom 1990; Neilson 1990; Reineke & Knackmuss 1988). Of the halogenated aromatic compounds, the lower chlorinated aromatics are degraded generally more readily than higher chlorinated ones. Aerobic microorganisms can also degrade halogenated aliphatic compounds, such as haloalkanes and haloacids. Higher chlorinated aliphatic compounds (e.g. tetrachloroethylene) still remain problematic. The main mechanisms by which aerobic microorganisms remove halogen atoms include *substitution* (e.g. by the hydroxyl ion OH^-, a reaction catalyzed by hydroxylases), *oxidation* (the incorporation of oxygen, catalyzed by mono- and dioxygenases), or fortuitously (e.g. during lactonization of chloromuconic acid) (Chaudry & Chapalamadugu 1991; Commandeur & Parsons 1990; Haggblom 1990; Neilson 1990; Reineke & Knackmuss 1988). Less is known about the conversion reactions that can take place with halogenated compounds under anaerobic conditions (i.e. denitrifying, sulfate-reducing, or methanogenic), and the specific microorganisms which are involved in these reactions. The main mechanism which has been described is the *reductive dehalogenation* (Suflita et al. 1982), which includes processes like hydrogenolysis, hydrolytic reduction, or dihalo-elimination (Holliger 1992).

We used soil columns filled with sediment of the river Rhine or of the Dune infiltration area near Amsterdam in The Netherlands to study the possibilities for biodegradation of halogenated compounds under different redox conditions (Bosma et al., in preparation). The columns were percolated under water-saturated conditions with a mineral medium to which very low concentrations of different chlorinated compounds could be added (Fig. 1). Parallel experiments were performed in the laboratory of the Amsterdam Water Works, which made use of larger soil columns and a wider range of different organic micropollutants. The com-

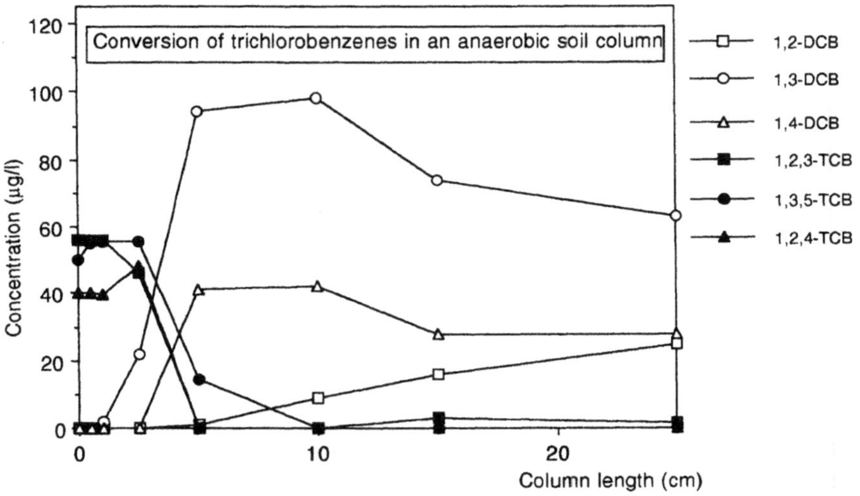

Fig. 2. Conversion of trichlorobenzenes in an anaerobic soil column after 300 days of continuous operation. Profile measurements of chlorobenzene concentrations along the length of the column show the disappearance of trichlorobenzene isomers and the formation of dichlorobenzene isomers. Figure redrawn from Bosma et al. (1988c).

pounds which were tested in these systems included chlorinated benzenes, non-chlorinated aromatics (benzene, toluene, naphthalene, di- and trimethylbenzenes), volatile chlorinated aliphatic compounds (chloroform, trichloroethylene, tetrachloroethylene, hexachloro-1,3-butadiene), and a number of pesticides (atrazine, bentazon). Our results (Table 1) showed that under aerobic conditions mainly transformations of lower chlorinated aromatic compounds are observed, such as dichlorobenzenes, 1,2,4-trichlorobenzene, or 2-chlorotoluene (not shown). All non-chlorinated aromatic compounds disappeared rapidly from the soil columns to undetectable levels. Under anaerobic (sulfate-reducing) conditions most of the higher chlorinated benzenes (Bosma et al. 1991; Bosma et al., in preparation; Bosma et al. 1988c) and most of the halogenated aliphatic compounds were transformed. In the dune infiltration areas and in most soils, both aerobic/anaerobic conditions may exist, thus promoting sequential transformation reactions which could favour a complete transformation of the higher chlorinated compounds. Two transformation processes were studied in greater detail and will be discussed further in this section, (i) anaerobic conversions of higher chlorinated benzenes, and (ii) anaerobic conversions of tetrachloroethylene.

Table 1. Observed degradation of halogenated compounds in Rhine and Dune sediments under different redox conditions.

Compound	Degradation	
	Aerobic	Sulfate reducing
Benzene	++	−
Monochlorobenzene	++	−
1,2-dichlorobenzene	+	+
1,3-dichlorobenzene	+/−	+
1,4-dichlorobenzene	+	+
1,2,3-trichlorobenzene	−	++
1,2,4-trichlorobenzene	+	++
1,3,5-trichlorobenzene	−	+
1,2,3,4-tetrachlorobenzene	−	++
1,2,4,5-tetrachlorobenzene	+	+
Pentachlorobenzene	−	++
Hexachlorobenzene	−	++
Tetrachloroethylene	−	++
Trichloroethylene	−	++
Chloroform	−	++

−: no removal observed.
+/−: removal less than 25%.
+: removal between 25% and 90%.
++: removal more than 90%.

Biodegradation of trichlorobenzene (TCB) isomers (1,2,4-TCB, 1,2,3-TCB, and 1,3,5-TCB)

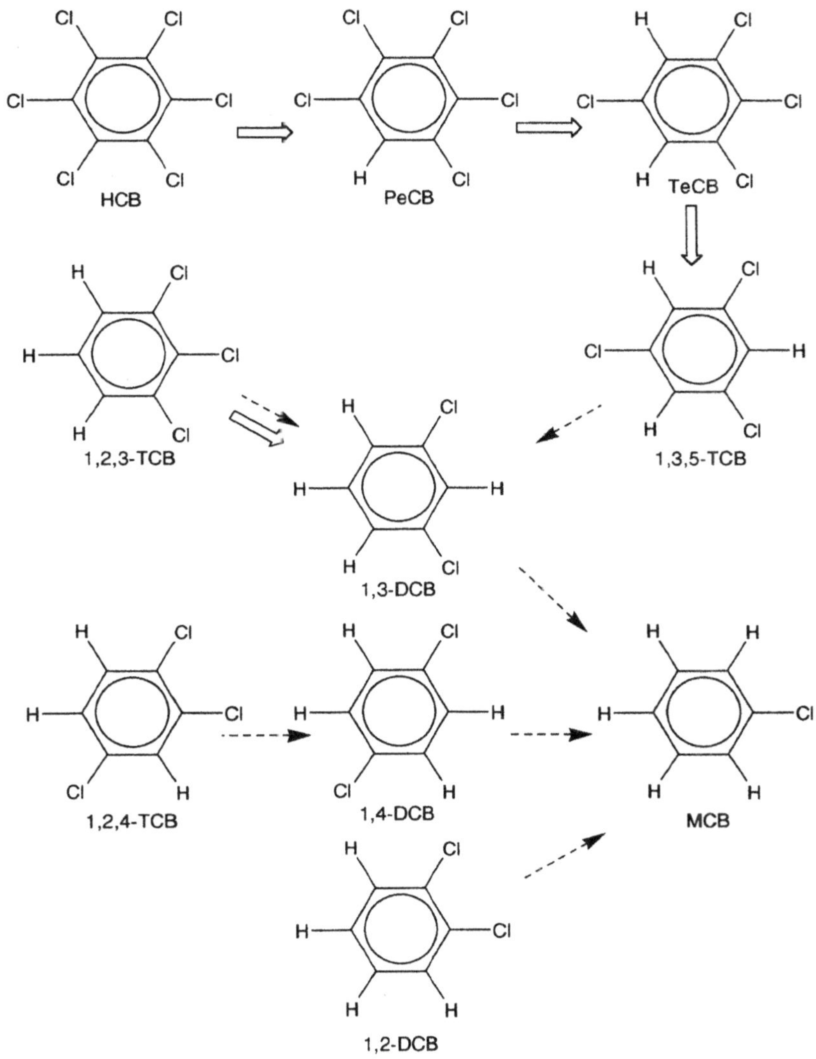

Fig. 3. Reductive dechlorination of highly chlorinated benzenes to monochlorobenzene, as observed in anaerobic soil columns (dotted arrows), and by a mixed culture which was enriched for reductive dechlorination of 1,2,3-TCB to 1,3-DCB (open arrows). Abbreviations: HCB, hexachlorobenzene; PeCB, pentachlorobenzene; TeCB, 1,2,3,5-tetrachlorobenzene; TCB, trichlorobenzene; DCB, dichlorobenzene; MCB, monochlorobenzene. Figure redrawn from Bosma et al. (1988c) and Holliger et al. (1992).

was studied in an anaerobic soil column for more than 450 days of continuous operation (Bosma et al. 1988c). During this period we detected transformations of all TCB-isomers in the column and a concomitant appearance of dichlorobenzenes (DCB) (Fig. 2). Applying one of each TCB-isomer at the time to this column revealed that the TCBs were transformed to a corresponding DCB by stepwise removal of one chlorine atom, suggesting a process of reductive dechlorination (Fig. 3). The DCB-isomers on their turn were transformed to monochlorobenzene, which was not further dechlorinated in the column. Subsequently, enrichments were started with material from this soil column to isolate microorganisms responsible for these conversion reactions. This resulted in an anaerobic mixed culture which could reductively dechlorinate 1,2,3-TCB to 1,3-DCB (Holliger et al.

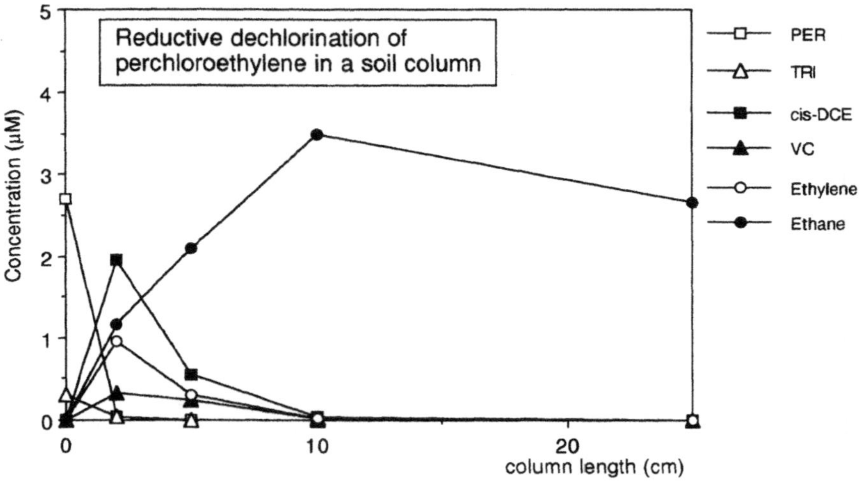

Fig. 4. Transformation reactions of tetrachloroethylene (PER) in an anaerobic soil column, indicating successive steps of dechlorination to trichloroethylene (TRI), *cis*-dichloroethylene (*cis*-DCE), vinylchloride (VC), ethylene and ethane.

1992). This mixed culture was also able to stepwise dechlorinate hexachlorobenzene, pentachlorobenzene and 1,2,3,5-tetrachlorobenzene to 1,3,5-TCB (Fig. 3). The dechlorinating activity in the mixed culture was most likely not caused by sulfate-reducing or methanogenic organisms, but appeared to be a specific reaction of bacteria which used 1,2,3-TCB as terminal electron acceptor and H_2 as electron donor. Interestingly, the mixed culture was not able to dechlorinate 1,3,5-TCB or 1,2,4-TCB which were being transformed in the original soil column (Bosma et al. 1988c; Holliger 1992). This indicated that in the original soil column other microorganisms must have been present with other specific dechlorinating activities.

In similar soil columns operated under sulfate-reducing conditions we detected transformation of tetrachloroethylene (PER) and trichloroethylene (TRI), resulting in a stoichiometric formation of *cis*-1,2-dichloroethylene (Bosma et al. 1988a). Subsequent studies in a packed-bed column which was filled with anaerobic granular sludge and sediment of the river Rhine near Wageningen, NL, clearly showed complete step-wise reductive dechlorination of PER to TRI, *cis*-dichloroethylene, vinylchloride, ethylene, and finally even to ethane (Figs. 4 and 5) (de Bruin et al. 1992). A coculture was enriched from this soil-slurry column by using a two-liquid-phase system which could dechlorinate PER and TRI (Holliger 1992). This enrichment was able to derive energy from the dechlorination of PER or TRI, used as electron acceptor, and H_2 or formate as electron donor. The reduction of PER to *cis*-1,2-dichloroethylene yields 377 kJ, which is by far more energy than can be obtained from sulfate reduction or methanogenesis (Holliger 1992). This raises the intriguing hypothesis that microorganisms in this enrichment found a new ecological niche by exploiting the energy content of a very specific xenobiotic compound.

Biodegradation at very low substrate concentrations

The actual concentrations of halogenated pollutants in the natural environment are mostly in the range of nM to μM. Although the compounds may be present in much higher amounts in the bulk, the concentration in the water phase is often rather low. These extremely low concentrations are due to the poor water solubility of most halogenated compounds, and/or their sorption to a solid phase, either to colloid particles (as in surface waters), or to the soil matrix or soil organic phase. The sorption behaviour of halogenated compounds to the organic phase in soils and sediments can be described by using empirical equations which are based on the octanol-water partition coefficients of the compounds (Schwarzenbach & Westall 1981). Lowering of the dissolved concentration as a result of sorption to a solid matrix can be beneficial in the case of toxic compounds, or it can prevent or limit biodegradation because of too low concentrations in the water phase (Rijnaarts et al. 1990). Sorption to the solid matrix initially acts as an effective sink for many hydrophobic compounds. However, the sorbed material may be released very slowly and can steadily pollute the environment for very long periods of time.

The effects of desorption rates on biodegradation were shown in studies performed in our laboratory with α-hexachlorocyclohexane (HCH). Biodegradation of α-HCH under aerobic conditions can proceed rapidly (Bachmann et al. 1988; Huntjens et al. 1988). However, soil samples contaminated with HCH-isomers taken from a waste-disposal site in The Netherlands, which were incubated as soil-slurries in end-over-end mixing systems, showed a much lower biodegradation rate (Rijnaarts et al. 1990). This could be attributed to the desorption kinetics of HCH from the soil matrix, which in this case was rate-limiting for bioconversion. Breakage of the soil aggregates by vigorous stirring, resulting in smaller average size of the soil particles, increased drastically the observed biodegradation rates, since more adsorbed HCH was exposed to the bulk liquid phase.

To determine whether very low concentrations of halogenated compounds can still be degraded by microorganisms or whether certain threshold levels exist for biodegradation (Alexander 1985), experiments were initiated to investigate the kinetics of substrate removal at very low concentrations in the nM to μM range. Inoculation into soil columns of a pure culture of *Pseudomonas* sp. strain P51, an organism able to use DCBs and 1,2,4-TCB as the sole carbon and energy sources (see below), exhibited a residual concentration for 1,2-DCB of approximately 67 nM, and for 1,2,4-TCB of 108 nM (van der Meer et al. 1987). At the flow rates operated in these columns (10.6 ml/h; concentrations of 1,2-DCB in the range of 134 nM to 6.7 μM), this

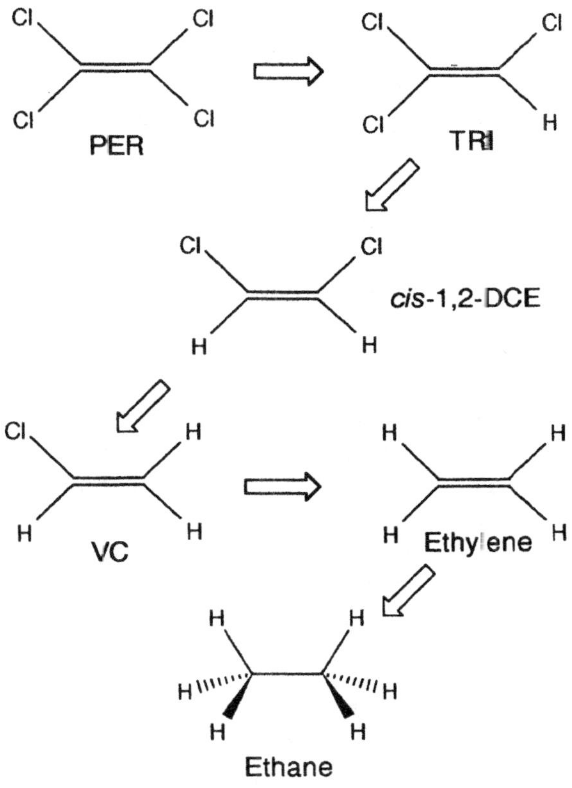

Fig. 5. Reductive dechlorination steps of tetrachloroethylene to ethane (see legends to Fig. 4). The transformation of PER to TRI and *cis*-1,2-DCE is carried out by the coculture described in the main text, which can derive energy from PER or TRI as sole electron acceptor and H_2 or formate as electron donor.

residual concentration could not be lowered. Incubation of the column effluent containing similarly low 1,2-DCB concentrations with *Pseudomonas* sp. strain P51 in batch experiments, resulted in a disappearance of 1,2-DCB to a value below the detection limit of 0.6 nM (van der Meer et al. 1987). Further experiments to study the nature of threshold levels were performed with *Pseudomonas* sp. strain B13 (Tros, unpublished). Hereto resting cell suspensions were prepared by growing the cells on 3-chlorobenzoate (3CB), washing them several times in buffered media without 3CB, and then starving them for substrate for about 20 h. Addition of [^{14}C]-labeled acetate or [^{14}C]-3CB to these cell suspensions caused uptake of the compounds resulting in a residual concentration which was below the detection limits, independent of the amount of substrate added (in a range of 10 nM to 1500 nM for 3CB, and 80 nm to 8000 nM for acetate). This indicated that no measurable threshold levels exist in batch incubations (Fig. 6A). In contrast, when strain B13 was cultivated in a recycling fermentor, residual concentrations were detected again (Fig. 6B). The use of a recycling fermentor enables the study of microbial growth at very low substrate concentrations (Pirt 1975). The biomass is completely retained in the fermentor (Fig. 6B), and a constant supply of fresh substrate in the medium results in growth of the biomass to a certain limit. Several phases can be observed from the start of a fermentation cycle (Fig. 6B), an exponential growth phase (I), two linear phases (II and III), in which the substrate addition becomes growth rate-limiting and the growth of biomass (dX/dt) becomes constant (Chesbro et al. 1979), and finally a phase (IV) where the substrate flow can only compensate the maintenance energy of the existing biomass, and growth becomes zero. In this phase a minimal substrate concentration (S_{min}) can be determined, which depends on the growth parameters of the specific strain on that particular substrate (μ_{max}, K_s, and maintenance rate (b)) (Chesbro et al. 1979). Using growth kinetic parameters which were independently derived for strain B13 on 3CB (Tros, unpublished) we could calculate an S_{min} of 8 μM at 20 °C, and 11 μM at 30 °C (see the equation of Fig. 6) (Pirt 1975). This value was in good agreement with the residual concentration of 10 μM observed in the recycling fermentor in phase IV (Fig. 6), suggesting that this residual concentration indicates a minimal substrate concentration for growth of *Pseudomonas* sp. strain B13 on 3-CB. These results showed that in a static system, such as a batch culture, a complete degradation of substrates is possible to unmeasurable low concentrations (<0.5 nM for 3-CB). However, in dynamic systems, such as a fermentor, a soil column, reactor, or perhaps the soil/water interphase, a residual concentration may occur, reflecting a minimal substrate concentration which depends on the growth kinetic parameters of bacterial strains with that specific metabolic substrate. Interestingly, we found no threshold levels for 1,2-DCB in naturally adapted soil columns (van der Meer et al. 1987), indicating that indigenous microorganisms may be present in the soil with different growth kinetic parameters, and perhaps specialized in low substrate level conversions.

Adaptation of microorganisms to xenobiotic compounds

Microorganisms are important for the breakdown of many halogenated compounds in the environment. However, since many of the chlorinated compounds have uncommon structures, the majority of the microorganisms present in the natural environment may not be able to deal effectively with these compounds, even if the environmental conditions are favourable (see above). The strong efforts of microbiologists in the last decades have resulted in the isolation of many new bacterial species, which are able to use a large variety of previously considered recalcitrant compounds as growth and energy substrates, such as chlorinated benzenes (Sander et al. 1991; Schraa et al. 1986; Spain & Nishino 1987; van der Meer et al. 1987; van der Meer et al. 1991b), chlorinated biphenyls (Furukawa et al. 1978; Mondello 1989), halogen-containing pesticides (Don & Pemberton 1981; Karns et al. 1983), or chloroanilines (Zeyer et al. 1985), suggesting that microbes adapted to xenobiotic compounds. However, this does not tell us what the catabolic potential of the indigenous microbes is, and if adaptation can occur under environmental conditions to compounds which were not previously recognized as metabolizable substrates. Adaptation may result from an induction of existing enzyme systems in microorganisms upon addition of a substrate, which may then be followed by growth of this sub-population of microorganisms owing to their activity on this substrate (Barkay & Pritchard 1988). Such an adaptation has been observed for growth on *p*-nitrophenol (Aelion et al. 1987). In the case of less easily degradable substrates such as chlorinated benzenes, adaptation may require genetic changes in the microorganisms or exchange of genetic material between different microorganisms before the suitable metabolic pathways arise (van

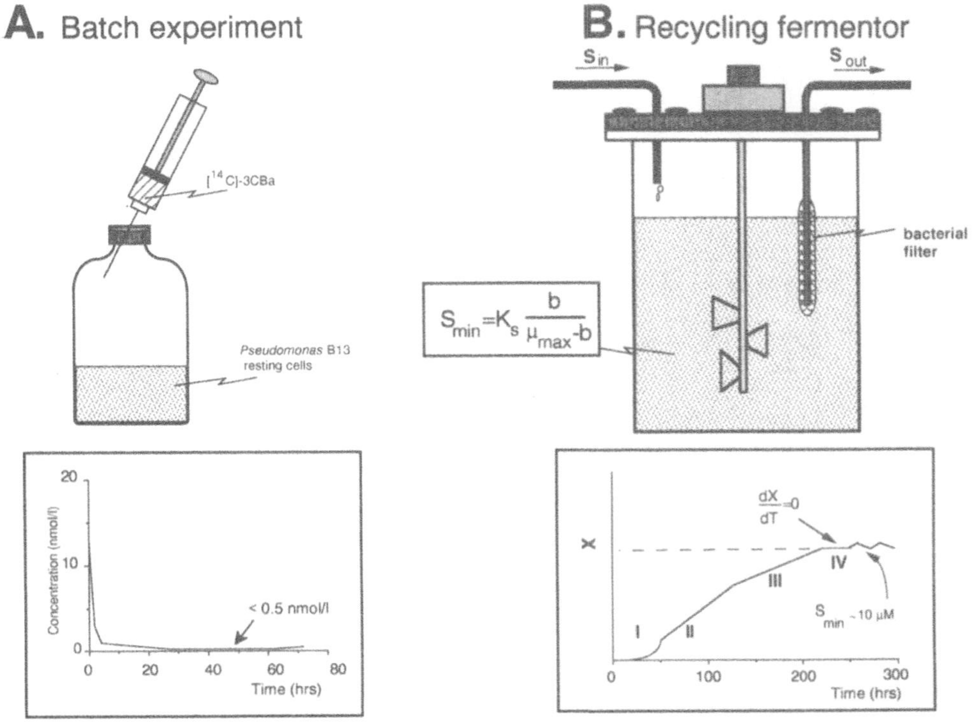

Fig. 6. Experimental set-up to detect the presence of threshold levels for microbial growth and activity. A). Batch experiments with resting cells of *Pseudomonas* sp. strain B13 to which [^{14}C]-3-chlorobenzoate (3CB) was added. Time measurements of the uptake of 3CB and the remaining concentration indicated that no threshold level can be detected (lower panel). B). The recycling fermentor which was used to study growth of strain B13 at very low substrate concentrations. A bacterial filter prevents outflow of the biomass. The biomass concentration in the fermentor as a function of time behaves schematically as drawn in the lower panel. For an explanation of the different phases, see the main text. In phase IV, the net growth of the biomass is zero, and in this phase the minimal substrate concentration (S_{min}) can be measured. S_{min} can be calculated as a function of growth kinetic parameters (K_s and μ_{max}), and of the maintenance rate, b (Chesbro et al. 1979; Pirt 1975).

der Meer et al. 1991c). Several genetic mechanisms may play a role in genetic adaptation, such as mutation, slippage-repair, recombination and transposition, or gene transfer (van der Meer et al. 1992). Here we will describe the adaptation to chlorinated benzenes in soil columns and the possible genetic changes which may have taken place upon adaptation to chlorinated benzenes in *Pseudomonas* sp. strain P51.

Only fairly recently microorganisms were isolated which were able to degrade chlorinated benzenes. The first such strain which could metabolize monochlorobenzene was obtained by Reineke & Knackmuss by using a chemostat for enrichment (Reineke & Knackmuss 1984). Several other strains were since then isolated, and the metabolic pathway for chlorobenzene degradation was largely unraveled (Sander et al. 1991; Schraa et al. 1986; Spain & Nishino 1987; van der Meer et al. 1991b). Aerobic metabolism of chlorinated benzenes proceeds similarly as that of other aromatic compounds, such as toluene, benzene, benzoate, or naphthalene. The initial attack of the aromatic ring is performed by an aromatic ring dioxygenase, a multi-enzyme complex which incorporates oxygen into the ring (Fig. 7). This intermediate is subsequently converted to a (chlorosubstituted) catechol, which is of a class of dihydroxylated compounds that are central intermediates in most aerobic aromatic pathways. Further ring-cleavage of the catecholic intermediates in bacteria is catalyzed by a class of enzymes, performing either intradiol

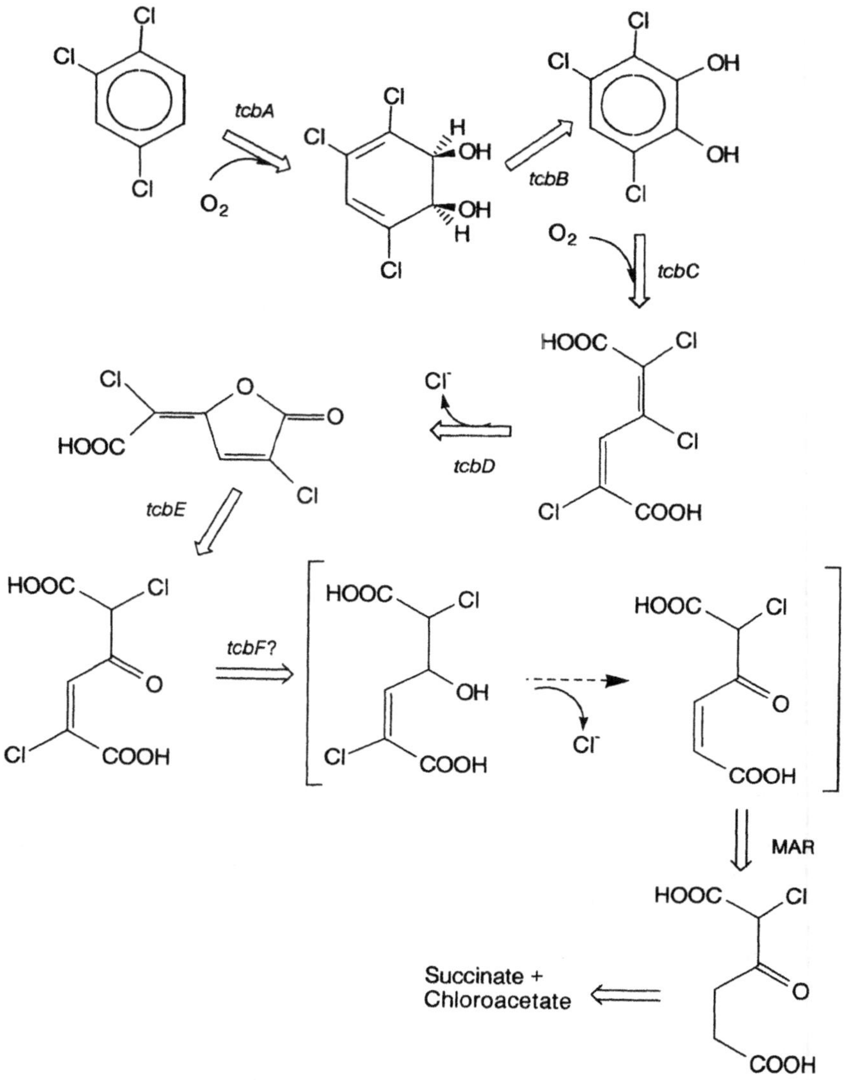

Fig. 7. Metabolic pathway of 1,2,4-trichlorobenzene in aerobic microorganisms proceeds via an aromatic ring dioxygenase, dehydrogenase, and a modified *ortho* cleavage pathway (Sander et al. 1991; van der Meer et al. 1991b). Two chloride atoms are released fortuitously during the conversion steps. The last part of the pathway shows the hypothetical function of the TcbF protein, the action of maleylacetate reductase (MAR) and finally the steps to succinate and chloroacetate. Hypothetical steps are indicated by open or dotted arrows; characterized enzymatic catalyzations are indicated by closed arrows.

(*ortho-*) cleavage (e.g. catechol 1,2-dioxygenase), or extradiol (*meta-*) cleavage (e.g. catechol 2,3-dioxygenase) (Harayama & Rekik 1989). Chloro-substituted catechols, which arise as intermediates in chlorobenzene degradation, are generally only successfully metabolized by the action of a series of enzymes, collectively referred to as the modified *ortho* cleavage pathway enzymes (Pieper et al. 1991; Reineke & Knackmuss 1988; Schlömann et al. 1990; Schmidt & Knackmuss 1980; Schmidt et al. 1980). This pathway leads probably to succinate and chloroacetate which are taken up in the central metabolic routes in aerobic microorganisms (Sander et al. 1991).

The metabolic details of the chlorobenzene pathway and that of other aromatic pathways suggested

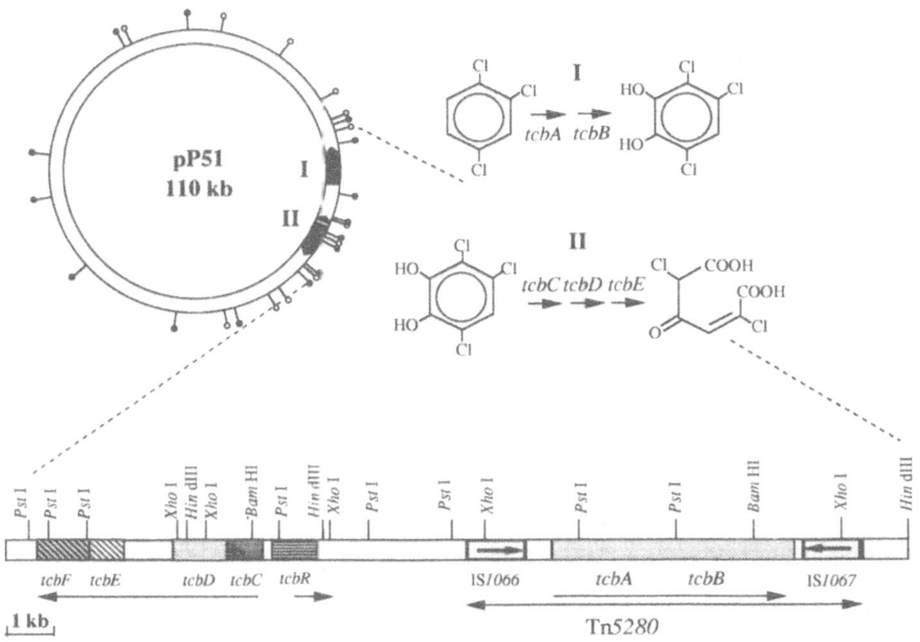

Fig. 8. Organization of chlorobenzene genes on plasmid pP51 of *Pseudomonas* sp. strain P51. The genes are located on the regions of the plasmid indicated by I and II, encoding the respective conversions shown in the figure. The directions of transcription are indicated by the arrows below the physical map. For gene assignments, see main text (van der Meer et al. 1991a–d).

the steps which would be necessary for a microorganism to achieve full chlorobenzene metabolism. Firstly, microorganisms would need to express the enzymes of the modified *ortho* cleavage pathway. However, strains such as *Pseudomonas* sp. strain B13 which degrades 3-chlorobenzoate by modified *ortho* cleavage, lack an appropriate aromatic ring dioxygenase to oxygenate chlorinated benzenes effectively to chlorinated catechols (Oltmanns et al. 1988). Therefore, the desired strain needs an effective aromatic ring dioxygenase and a *cis*-benzeneglycol dehydrogenase. Microorganisms which grow on toluene, such as *P. putida* F1, express a toluene dioxygenase which has a wide substrate range and is able to oxidize halogen-substituted benzenes (Gibson et al. 1990). However, in *P. putida* F1 catechol intermediates are cleaved by a catechol 2,3-dioxygenase and further degraded by the *meta* cleavage pathway enzymes. For most chlorosubstituted catechols this is an unproductive route and can lead to the formation of toxic intermediates (Bartels et al. 1984). Thus, a combination of a wide-substrate aromatic ring dioxygenase and benzeneglycol dehydrogenase in conjunction with a modified *ortho* cleavage pathway could be a successful way for microorganisms to metabolize chlorinated benzenes. Such a combination was achieved in the laboratory by mating two strains, one degrading toluene and the other degrading 3-chlorobenzoate (Kröckel & Focht 1987; Oltmanns et al. 1988), and it was shown that in these cases the transfer of a plasmid containing the genetic information for one of the determinants (toluene degradation or 3-chlorobenzoate degradation) had occurred (Carney et al. 1989).

We decided that metabolism of chlorinated benzenes would be a good case to study genetic adaptation of microorganisms in the natural environment. In soil columns which were percolated with mixtures of DCBs and TCBs (see above) we found that adaptation to 1,2-DCB could occur, after an initial period of up to 100 days without measurable degradation (van der Meer et al. 1987). From enrichments on 1,2-DCB, 1,4-DCB, and 1,2,4-TCB different microorganisms could be isolated which could use these compounds as the sole carbon and energy source (Schraa et al. 1986; van der Meer et al. 1991b). One of these strains, *Pseudomonas* sp. strain P51, was used for further characterization. We found that this strain induced modified *ortho* cleavage pathway enzymes when cultivated on 1,2,4-TCB or 1,2-DCB (van der Meer et al. 1991b). Bacteria which had been grown on chlorinated benzenes contained a large plasmid designated pP51, with a size of 110 kb (Fig. 8). This plasmid carried the genetic determinants for chlorobenzene degradation in strain P51. A genetic characterization of pP51 revealed two responsible regions: one for a modified *ortho* cleavage pathway and the other containing the genes for a chlorobenzene dioxygenase and a *cis*-benzeneglycol dehydrogenase (van der Meer et al. 1991b). Subcloning experiments and expression studies in *E. coli* identified most of the genes for the modified *ortho* pathway enzymes. Further DNA sequence analysis revealed that this region contains five contiguous open reading frames of which three could be assigned to an enzymatic function: *tcbC*, encoding chlorocatechol 1,2-dioxygenase, *tcbD*, encoding chloromuconate cycloisomerase, and *tcbE*, encoding dienelactone hydrolase (van der Meer et al. 1991a). Another open reading frame (*tcbF*) could be expressed in *E. coli*, but no enzymatic activity could be attributed to it. The primary sequence of the encoded TcbF polypeptide showed 30% identity with iron-containing alcohol dehydrogenases (van der Meer, unpublished), which suggests a possible function for TcbF (Fig. 7). A fifth open reading frame was found between the *tcbD* and *tcbE* gene, but this was not transcribed in *E. coli*. The DNA sequences of the *tcbCDEF* genes were between 50% and 75% identical to gene clusters for other modified *ortho* pathways (van der Meer et al. 1991a), such as *clcABD* of *P. putida* (pAC27) (Frantz &Chakrabarty 1987) and *tfdCDEF* of *A. eutrophus* JMP134 (pJP4) (Perkins et al. 1990). Furthermore, the genes of the modified *ortho* cleavage pathway were distantly related to those of the normal *ortho* cleavage pathway, which is found in many organisms growing on benzoate or hydroxybenzoate, suggesting an evolutionary relation and divergence of the two gene clusters from a common ancestor (van der Meer et al. 1991a). The enzymes of the modified *ortho* cleavage pathway catalyze conversion reactions that are analogous to those of the normal *ortho* cleavage pathway (Ornston et al. 1990; Reineke & Knackmuss 1988). However, in the modified *ortho* cleavage route enzymes occur which are not found in the normal *ortho* cleavage route and vice versa, such as dienelactone hydrolase (Fig. 7), whereas others, such as chlorocatechol 1,2-dioxygenase and chloromuconate cycloisomerase have relatives in the normal *ortho* cleavage pathway. The enzymes of the modified *ortho* cleavage pathway were in all cases described so far, less substrate-specific than their counterparts in the normal *ortho* cleavage pathway and could accomodate the chlorinated intermediates better than the non-chlorinated ones (Broderick & O'Halloran 1991; Kuhm et al. 1990; Ngai &Ornston 1988; Pieper et al. 1991). Analysis of catalytic activities of chlorocatechol 1,2-dioxygenases and of chloromuconate cycloisomerase activities indicated that there was a further subtle specialization of these enzymes to specific substrates (Kuhm et al. 1990; van der Meer et al. 1991a). For instance, chlorocatechol 1,2-dioxygenase of *Pseudomonas* sp. strain P51 showed a higher relative activity towards the substrate 3,4-dichlorocatechol than to the other dichlorinated catechols, whereas the chlorocatechol 1,2-dioxygenase of *A. eutrophus* JMP134 exhibited a higher relative activity to 3,5-dichlorocatechol. These data suggest that the chlorocatechol 1,2-dioxygenases of the different strains became evolutionary optimized for specific substrates, e.g. the intermediate 3,4-dichlorocatechol in 1,2-DCB degradation, and 3,5-dichlorocatechol in the degradation of 2,4-dichlorophenoxyacetic acid. Analysis of DNA sequence information suggests that the intradiol dioxygenases may have diverged via a mechanism of slippage-repair rather than via additive single-site mutations (Hartnett et al. 1990; Neidle et al. 1988; Ornston et al. 1990).

Genetic analysis of the genes encoding the chlorobenzene dioxygenase and the chlorobenzeneglycol dehydrogenase (tentatively named *tcbA* and *tcbB*) of *Pseudomonas* sp. strain P51 revealed another interesting aspect (van der Meer et al. 1991c). Southern hybridizations and subsequent DNA sequence information showed that a duplicated and

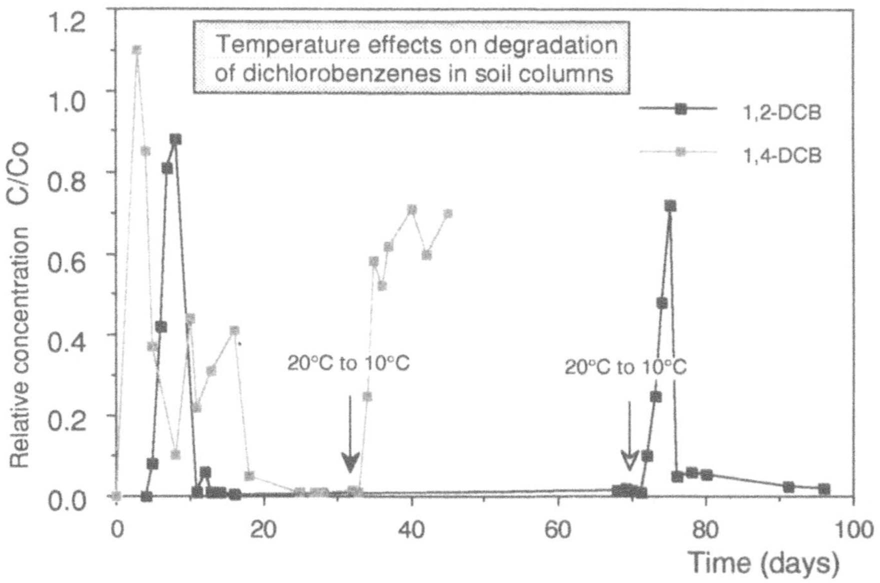

Fig. 9. Temperature effects on biodegradation of 1,2-DCB and 1,4-DCB in a naturally adapted column and an inoculated column, respectively. The experiment shown here started after the naturally adapted column to 1,2-DCB had been operated for 200 days continuously, and then for a period of 28 days had received no 1,2-DCB. After the addition of 1,2-DCB at a concentration of 25 μg/l to the column, the biological activity was restored within ten days operation. The column receiving 1,4-DCB (at a concentration of 25 μg/l) was inoculated with *Alcaligenes* sp. strain A175 (Schraa et al. 1986). At the time points indicated, the columns were changed from an operating temperature of 20 °C to 10 °C. Concentration measurements of dichlorobenzenes were done in the effluent of the columns, and are represented in the figure as the relative concentration measured in the effluent compared to the influent concentrations (C/C_0). Redrawn after (Bosma et al., in preparation).

inverted discrete DNA fragment was flanking the *tcbAB* gene loci. This fragment had a size of 1.1-kb and both copies were identical except for one base-pair. The sequence showed one open reading frame encoding a putative polypeptide with homology to polypeptides encoded by insertion elements (van der Meer et al. 1991c). Therefore, we designated these elements IS*1066* and IS*1067*. This suggested that the DNA fragment containing *tcbAB* plus the putative IS elements were a composite bacterial transposon. We were able to demonstrate that this transposable element, Tn*5280*, was functional and transposed at random and in single copy to the genome of *P. putida* (van der Meer et al. 1991c). These results suggested that perhaps the genes encoding the chlorobenzene dioxygenase and chlorobenzeneglycol dehydrogenase were recruited through the mediation of IS-elements by an ancestor plasmid of pP51, which already contained the genes encoding a modified *ortho* pathway. At present, we have not analyzed the origin of the IS elements IS*1066* and IS*1067* or the dioxygenase gene fragment in other bacterial species. However, preliminary studies which applied the technique of DNA-DNA colony hybridization (Sayler et al. 1985) with genetic markers from chlorobenzene degradation genes indicated the presence of other microorganisms in the original soils used for the column experiments, which carried genetic material homologous to the *tcb* genes.

The use of pure cultures or enrichment cultures to enhance biodegradation

Many attempts are made to evaluate the possibilities of applying pure cultures or enriched bacterial strains to enhance biodegradation of recalcitrant compounds in the environment. On the level of the bacterial strains, such efforts are directed to either isolate by enrichment techniques specific strains that will perform the desired metabolic conversions, or to construct strains by genetic means that will allow degradation of a previously considered unmetabolizable substrate (Latorre et al. 1984; Lehrbach et al. 1984; Rojo et al. 1987). However, as we have seen above, the success (or failure) to enhance biodegradation of specific compounds in the environment, does not depend solely on the catabolic potential of the microorganisms used. This section will deal with the effects of temperature-related activity and adhesion properties of microorganisms on the use and application of pure cultures in biodegradation.

We evaluated the use of two different pure cultures capable of metabolizing chlorinated benzenes, *Pseudomonas* sp. strain P51 and *Alcaligenes* sp. strain A175, to decontaminate low concentrations of DCBs and TCBs from percolated water in a soil column. Strain *Pseudomonas* sp. P51 was pregrown on 1,2,4-TCB and introduced into the soil column and was able to degrade different concentrations of DCBs and 1,2,4-TCB to a low, albeit detectable, level (see above). This microorganism remained active in degrading chlorinated benzenes as long as these compounds were present. However, a 28-day period without addition of fresh substrates resulted in a loss of the degradative capacity of the strain in the column. The activity of strain P51 in the column gave further evidence for chemotactic migration of the strain towards the substrates added in the influent (Bosma et al. 1988b). A soil column in which degradation of 1,2-DCB occurred after adaptation of the natural population, was able to restore the 1,2-DCB degradation activity after a period without substrate added (Fig. 9). In a new soil column which was inoculated with another chlorobenzene degrading strain, *Alcaligenes* sp. strain A175, we also found that the cells were able to retain their activity in the soil column and could degrade 1,4-DCB almost completely (Fig. 9). The effect of temperature on the biodegradation of 1,2-DCB and 1,4-DCB in the naturally adapted and this inoculated column was then examined by changing the operating temperature of the columns from 20 °C to 10 °C (Bosma et al., in preparation). The column which was inoculated with *Alcaligenes* sp. strain A175 was immediately disturbed in the degradation of 1,4-DCB, and after a 10-day period a new equilibrium was reached (Fig. 9). However, the total removal of 1,4-DCB after this change was only about 40% of the influent concentration applied to the column. The response of the naturally adapted column was initially similar to the inoculated column. A rapid decrease in 1,2-DCB removal was observed, and after a 10-day period the previous steady state concentration of 1,2-DCB was restored in the effluent of the column (Fig. 9). This demonstrated that the bacteria in this column were better suited to be active at lower temperatures. Changing the temperature of this column to an even lower value of 4 °C did not alter the steady state concentrations measured for 1,2-DCB (not shown).

A similar temperature response was found for soil columns which were inoculated with *Rhodococcus* sp. strain C125, a strain which can use toluene and 1,3-dimethylbenzene as the sole carbon and energy substrates (Schraa et al. 1987). Changing the operating temperature from 20 °C to 10 °C again resulted in disturbance of the degradation of toluene and 1,3-dimethylbenzene in the column. Whereas at 20 °C more than 90% of the influent concentrations of these compounds could be degraded (around 500 μg/l), only 20 to 30% removal was observed at 10 °C (Fig. 10). Enrichment cultures were started at a temperature of 10 °C, using contaminated soil from a petrol station as source for microorganisms (de Bruin, unpublished). Within three weeks, growth of the enrichments on toluene, 1,3- and 1,2-dimethylbenzene was observed, and the mixed cultures were added to the soil columns. Toluene and 1,3-dimethylbenzene were then easily degraded in these columns at a temperature of 10 °C to levels as low as those at 20 °C (Fig. 10). Temperatures of 10 °C are more often encountered in ground water than temperatures of 20 °C, and our results demonstrate that use of pure cultures which have been isolated on the basis of their faster growth rate at higher temperatures, can thus be limited in on-site applications.

Application of microorganisms in bioreactors or in groundwater systems will also require information about the adhesion of bacteria to solid particles and deposition in porous media. If, for instance, the introduced microorganisms do not ad-

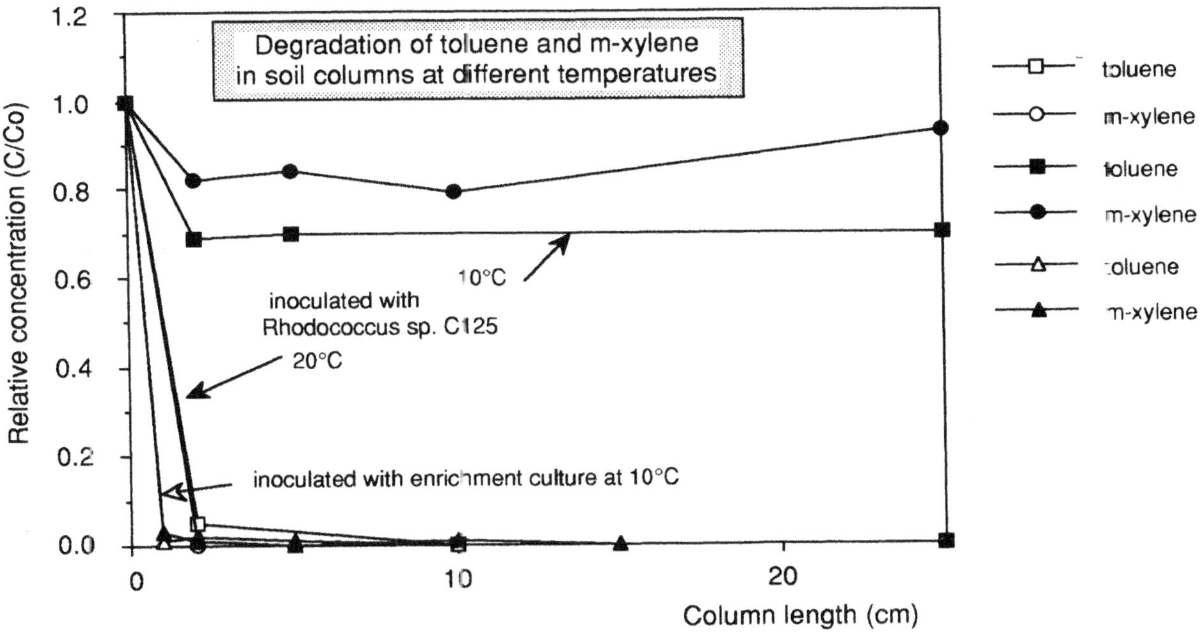

Fig. 10. Temperature effects on toluene and 1,3-dimethylbenzene degradation in soil-slurry columns, either inoculated with *Rhodococcus* sp. strain C125 (Schraa et al., 1987), or inoculated with a low temperature adapted enrichment. Influent concentrations of toluene and xylene measured approximately 500 μg/l.

here to the solid phase, they are not retained in a reactor or are transported in the soil with the water flow, and this may diminish the necessary desired biodegradation. Fundamental information about adhesion of microorganisms, which are regarded in this respect as colloidal particles, can be derived from batch and column experiments. The adhesion properties of microorganisms to well-defined surfaces such as glass, polystyrene, or teflon in batch systems was studied previously and shown to be controlled by repulsive electrostatic and attractive van der Waals forces, as described by the DLVO theory for colloidal particles (van Loosdrecht et al. 1989). The physico-chemical properties of the microorganisms can be experimentally studied by measuring their electrophoretic mobility (which characterizes the surface charge of the microorganism, van Loosdrecht et al. 1987a), and the contact angle of a drop of water on dried bacterial lawns (which provides information on the hydrophobicity of the organism, van Loosdrecht et al. 1987b). In dynamic systems, such as columns, adhesion of microorganisms is not only affected by surface characteristics but also by the hydrodynamic factors controlling transport of the particle from the bulk liquid phase to the solid surface (Rijnaarts et al., in preparation).

The adhesion of a number of bacterial strains was studied in more detail in column experiments, which contained the surface materials glass (a hydrophilic material), and teflon (PFA-teflon, an extremely hydrophobic material). Both surfaces are negatively charged and possess an electrokinetic potential of -42 ± 2 mV at an ionic strength of 0.01 M (Rijnaarts, unpublished). Three bacterial strains were tested, *Pseudomonas* sp. strain B13, *P. putida* mt-2 and *Sphingomonas* sp. strain HH19K (which degrades dibenzofuran) (Fortnagel et al. 1990), each having different cell surface characteristics. The contact angles of water observed for the three species varied from 32° for strain B13, 40° for strain mt-2, to 60° for *Sphingomonas*, indicating that this last strain was the most hydrophobic of the three. The electrophoretic mobilities at an ionic strength of 0.01 M measured $-2.1 \cdot 10^{-8}$, $-1.1 \cdot 10^{-8}$, and $-1.6 \cdot 10^{-8}$ $m^2 \cdot V^{-1} \cdot s^{-1}$, respectively.

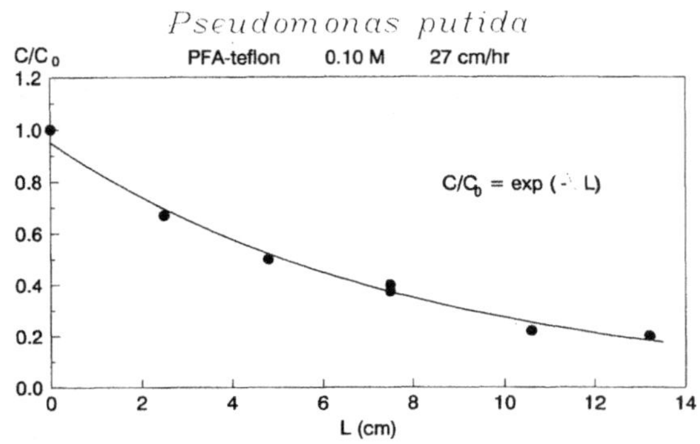

Fig. 11. Determination of the retention of *P. putida* in a PFA-teflon bead column. The decrease of cell numbers present in the liquid phase is indicated as a decrease in cell numbers per ml at a certain point in the column (C) compared to the numbers per ml in the influent (C_0). This decrease can be described using the exponential decay equation indicated, whereby L is the length of the column, and λ the filtration coefficient. Ionic strength of the liquid phase was 0.1 M, and flow rate imposed on the face of the column 27 cm/h.

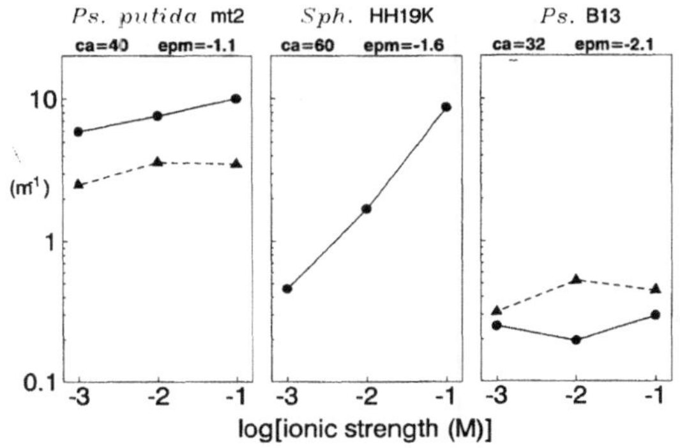

Fig. 12. Filtration coefficients determined from column adhesion experiments for three different microorganisms, *P. putida* mt-2, *Sphingomonas* sp. strain HH19K, and *Pseudomonas* sp. strain B13 under different ionic strengths and on two different surfaces, PFA-teflon (circles) and glass (triangles). The figure shows further cell surface characteristics of the microorganisms: the contact angle (ca, expressed in degrees), and the electrophoretic mobility (epm, $*10^{-8}\ m^2 \cdot V^{-1} \cdot s^{-1}$).

This indicated that strain B13 beared the highest surface charge. The columns were packed with spherical collectors of teflon and glass with an average radius of 0.2 mm. Bacterial suspensions of 10^8 cells per ml were fed to these columns and the cell numbers in the influent and the effluent of the columns were measured. After a number of pore volumes a steady state was reached resulting in a constant aqueous cell concentration at a specific location in the column. This normalized 'concentration' of cells (which is the ratio of the effluent concentration over the influent concentration, C/C_0) decreases exponentially with the length of the column (Fig. 11), with a decay constant termed the filtration coefficient λ (m^{-1}). The obtained filtration coefficients for the three strains at different ionic strengths and with two different surfaces, glass and teflon, showed interesting differences and have important implications for the behaviour of these strains (Fig. 12). *Pseudomonas* sp. strain B13 appeared to be a very poor adhering microorganism. Due to its hydrophilic nature and higher surface charge, the organism adheres better on hydrophilic surfaces (such as glass) than on hydrophobic ones (such as PFA-teflon). Changing the ionic strength of the liquid phase does not dramatically alter this adherence (Fig. 12). The organism can hardly be retained in columns such as described in these adhesion experiments. Similar observations were made when this strain was applied in soil columns. Wash-out of the strain from the column resulted in a very poor degradation of 3CB. In contrast, *P. putida* mt-2 which is a moderately hydrophilic, low-charged bacterium, behaved as a very good adhering organism both to hydrophilic and hydrophobic column material (Fig. 12). Results with *Sphingomonas* sp. strain HH19K demonstrated that ionic strength can determine the adherence of a strain to the solid matrix. This organism adhered poorly at low ionic strengths, but retained better at higher ionic strength (0.1 M).

Concluding remarks

Our studies have demonstrated the usefulness of soil column experiments to evaluate various aspects of biodegradation of halogenated compounds in an infiltration site. The system can be operated easily under variable conditions found in the environment (temperature, redox state, concentration of the compounds, etc.), and once biodegradation is initiated, the column can be used to enrich for microorganisms with the desired characteristics. Many interesting aspects were discovered and investigated further, such as (i) the isolation of anaerobic microorganisms which derive energy from reductive dechlorination of tetrachloroethylene or 1,2,3-trichlorobenzene, (ii) the occurrence and nature of residual concentrations and their effect on biodegradability, (iii) genetic adaptation to chlorinated benzenes, and (iv) the strong temperature effects on biodegradability of inoculated strains, and the effects of adherence properties of bacterial strains on their retainment.

Considering the possibilities to use naturally adapted or genetically engineered strains to combat pollution, we have shown that the success (or failure) to enhance biodegradation of specific compounds depends on various biological and physicochemical factors. In natural systems the environmental conditions will determine to a large extent the activities that microorganisms can display. Suitable microorganisms must be present to carry out the complete degradation, but different redox conditions (anaerobic, aerobic) will influence what type of biodegradation prevails, and this may favour particular transformation reactions (e.g. reductive dechlorination of polychlorinated benzenes and biphenyls in sediments), whereas others are repressed (e.g. aerobic conversion of lower chlorinated benzenes or biphenyls). We have shown that microorganisms in the natural environment can adapt to novel substrates, such as chlorinated benzenes. Therefore, to stimulate biodegradation in the natural environment it may be a good strategy to create optimal conditions for the activity of indigenous microorganisms. This may also enhance adaptation of suitable microorganisms, however, it is not thoroughly investigated what the full catabolic potential of microorganisms in the environment is (van der Meer 1992).

Under specific circumstances, such as in treatment systems, it may be more effective to apply specifically engineered or enriched strains. However, several aspects must be evaluated to obtain successful biodegradation. In on-site applications the desorption rates of the pollutants from the solid matrix may be rate-determining for biodegradation. Furthermore, biodegradation may be hampered in specific treatment systems due to growth kinetic parameters of the bacterial strains used. When the available concentration of the substrates becomes too low to sustain growth, a residual concentration of the pollutant may remain. The best strains to use in treatment systems are not necessarily those that have been isolated for their fastest growth rate at higher temperatures (20 °C to 30 °C). Depending on the temperature that is needed for the application, it is often better to enrich for low-temperature adapted strains or populations from the environment. Finally, the adhesion properties of the isolated microorganisms should be investigated to allow optimal retention of the microbial strains in the system.

Acknowledgements

This work was financed by grants from The Netherlands Integrated Soil Research Programme and the Innovative Research Program on Environmental Biotechnology.

References

Aelion CM, Swindoll CM & Pfaender FK (1987) Adaptation to and biodegradation of xenobiotic compounds by microbial communities from a pristine aquifer. Appl. Environ. Microbiol. 53: 2212–2217

Alexander M (1985) Biodegradation of organic chemicals. Environ. Sci. Technol. 18: 106–111

Bachmann A, de Bruin WP, Jumelet JC, Rijnaarts HHM & Zehnder AJB (1988) Aerobic mineralization of alpha-hexachlorocyclohexane in contaminated soil. Appl. Environ. Microbiol. 54: 548–554

Barkay T & Pritchard H (1988) Adaptation of aquatic microbial communities to pollutant stress. Microbiol. Sci. 5: 165–169

Bartels I, Knackmuss H-J & Reineke W (1984) Suicide inactivation of catechol 2,3-dioxygenase from *Pseudomonas putida* mt-2 by 3-halocatechols. Appl. Environ. Microbiol. 47: 500–505

Bosma TNP, Holliger C, van Neerven ARW, Schraa G & Zehnder AJB (1988a) Reductive dechlorination of chlorinated hydrocarbons in anaerobic sediment columns. In: Wolf K, van den Brink WJ & Colon FJ (Eds) Contaminated Soil '88 (pp 731–732). Kluwer Academic Publishers, Dordrecht

Bosma TNP, Schnoor JL, Schraa G & Zehnder AJB (1988b) Simulation model for biotransformation of xenobiotics and chemotaxis in soil columns. J. Contam. Hydrol. 2: 225–236

Bosma TNP, van der Meer JR, Schraa G, Tros ME & Zehnder

AJB (1988c) Reductive dechlorination of all trichloro- and dichlorobenzene isomers. FEMS Microbiol. Ecol. 53: 223–229
Bosma TNP, te Welscher RAG, Schraa G, Smeenk JGMM & Zehnder AJB (1991) Microbial aspects of the behaviour of chlorinated compounds during soil passage. In: Angeletti G and Bjorseth A (Eds) Organic Micropollutants in the Aquatic Environment (pp 184-192). Kluwer Academic Publishers, Dordrecht
Bosma TNP, te Welscher RAG, Smeenk JGMM, Ballemans EMW, Hoekstra NK, Schraa G & Zehnder AJB (in preparation) Biotransformation of organic contaminants in a dune infiltration area and in sediment columns.
Broderick JB & O'Halloran TV (1991) Overproduction, purification, and characterization of chlorocatechol dioxygenase, a non-heme iron dioxygenase with broad substrate tolerance. Biochemistry 30: 734–7358
Carney BF, Kröckel L, Leary JV & Focht DD (1989) Identification of *Pseudomonas alcaligenes* chromosomal DNA in the plasmid DNA of the chlorobenzene-degrading recombinant *Pseudomonas putida* strain CB1-9. Appl. Environ. Microbiol. 55: 1037–1039
Chaudry GR & Chapalamadugu S (1991) Biodegradation of halogenated organic compounds. Microbiol. Rev. 55: 59–79
Chesbro W, Evans T & Eifert R (1979) Very slow growth of *Escherichia coli*. J. Bacteriol. 139: 625–638
Commandeur LCM & Parsons JR (1990) Degradation of halogenated aromatic compounds. Biodegradation 1: 207–220
de Bruin WP, Kottermann MJJ, Posthumus MA, Schraa G & Zehnder AJB (1992) Complete biological reductive transformation of tetrachloroethene to ethane. Appl. Environ. Microbiol. 58: 1996–2000
Don RH & Pemberton JM (1981) Properties of six pesticide degradation plasmids isolated from *Alcaligenes paradoxus* and *Alcaligenes eutrophus*. J. Bacteriol. 145: 681–686
Faulkner DJ (1980) Natural organohalogen compounds. In: Hutzinger O (Ed) The Handbook of Environmental Chemistry, 1A. The Natural Environment and the Biogeochemical Cycles. Springer Verlag, New York
Fortnagel P, Harms H, Wittich R-M, Krohn S, Meyer H, Sinnwell V, Wilkes H & Francke W (1990) Metabolism of dibenzofuran by *Pseudomonas* sp. strain HH69 and the mized culture HH27. Appl. Environ. Microbiol. 56: 1148–1156
Frantz B & Chakrabarty AM (1987) Organization and nucleotide sequence determination of a gene cluster involved in 3-chlorocatechol degradation. Proc. Natl. Acad. Sci. USA. 84: 4460–4464
Furukawa K, Matsumura F & Tonomura K (1978) *Alcaligenes* and *Acinetobacter* strains capable of degrading polychlorinated biphenyls. Agric. Biol. Chem. 42: 543–548
Gibson DT, Zylstra GJ & Chauhan SS (1990) Biotransformations catalyzed by toluene dioxygenase from *Pseudomonas putida* F1. In: Silver S, Chakrabarty AM, Iglewski B & Kaplan S (Eds) *Pseudomonas*: Biotransformations, Pathogenesis and Evolving Biotechnology (pp 121–133). American Society for Microbiology, Washington, DC
Goldstein RM, Mallory LM & Alexander M (1985) Reasons for possible failure of inoculation to enhance biodegradation. Appl. Environ. Microbiol. 50: 977–983
Haggblom M (1990) Mechanisms of bacterial degradation and transformation of chlorinated monoaromatic compounds. J. Basic. Microbiol. 30: 115–141
Harayama S & Rekik M (1989) Bacterial aromatic ring-cleavage enzymes are classified into two different gene families. J. Biol. Chem. 264: 15328–15333
Hartnett C, Neidle EL, Ngai K-L & Ornston LN (1990) DNA sequences of genes encoding *Acinetobacter calcoaceticus* protocatechuate 3,4-dioxygenase: evidence indicating shuffling of genes and of DNA sequences within genes during their evolutionary divergence. J. Bacteriol. 172: 95–966
Holliger C (1992) Reductive dehalogenation by anaerobic bacteria. Ph.D. Thesis, Wageningen Agricultural University, Wageningen, The Netherlands
Holliger C, Schraa G, Stams AJM &Zehnder AJB (1992) Enrichment and properties of an anaerobic mixed culture reductively dechlorinating 1,2,3-trichlorobenzene to 1,3-dichlorobenzene. Appl. Environ. Microbiol. 58: 1636–1644
Huntjens JM, Brouwer W, Grobben K, Jansma O, Scheffer F & Zehnder AJB (1988) Biodegradation of α-hexachlorocyclohexane by a bacterium isolated from polluted soil. In: Wolf K, van den Brink W J & Colon FJ (Eds) Contaminated Soil '88. Kluwer Academic Publishers, Dordrecht
Karns JS, Kilbane JJ, Duttagupta S & Chakrabarty AM (1983) Metabolism of halophenols by 2,4,5-trichlorophenoxyacetic acid-degrading *Pseudomonas cepacia*. Appl. Environ. Microbiol. 46: 1176–1181
Klein TM & Alexander M (1986) Bacterial inhibitors in lake water. Appl. Environ. Microbiol. 52: 114–118
Kröckel L & Focht DD (1987) Construction of chlorobenzene-utilizing recombinants by progenitive manifestation of a rare event. Appl. Environ. Microbiol. 53: 2470–2475
Kuhm AE, Schlömann M, Knackmuss H-J & Pieper DH (1990) Purification and characterization of dichloromuconate cycloisomerase from *Alcaligenes eutrophus* JMP134. Biochem. J. 266: 877–883
Latorre J, Reineke W & Knackmuss H-J (1984) Microbial metabolism of chloroanilines: enhanced evolution by natural genetic exchange. Arch. Microbiol. 140: 159–165
Lehrbach PR, Zeyer J, Reineke W, Knackmuss H-J & Timmis KN (1984) Enzyme recruitment in vitro: use of cloned genes to extend the range of haloaromatics degraded by *Pseudomonas* sp. strain B13. J. Bacteriol. 158: 1025–1032
Mondello FJ (1989) Cloning and expression in *Escherichia coli* of *Pseudomonas* strain LB400 genes encoding polychlorinated biphenyl degradation. J. Bacteriol. 171: 1725–1732
Neidle EL, Hartnett C, Bonitz S & Ornston LN (1988) DNA sequence of the *Acinetobacter calcoaceticus* catechol 1,2-dioxygenase I structural gene *catA*: evidence for evolutionary divergence of intradiol dioxygenases by acquisition of DNA sequence repetitions. J. Bacteriol. 170: 4874–4880
Neilson AH (1990) The biodegradation of halogenated organic compounds - a review. J. Appl. Bacteriol. 69: 445–470
Ngai K-L & Ornston LN (1988) Abundant expression of *Pseu-*

domonas genes for chlorocatechol metabolism. J. Bacteriol. 170: 2412–2413

Oltmanns RH, Rast HG & Reineke W (1988) Degradation of 1,4-dichlorobenzene by constructed and enriched strains. Appl. Microbiol. Biotechnol. 28: 609–616

Ornston LN, Houghton J, Neidle EL & Gregg LA (1990) Subtle selection and novel mutation during evolutionary divergence of the β-ketoadipate pathway. In: Silver S, Chakrabarty AM, Iglewski B & Kaplan S (Eds) *Pseudomonas*: Biotransformations, Pathogenesis and Evolving Biotechnology (pp 207–225). American Society for Microbiology, Washington, DC

Pearson CR (1982) C1 and C2 halocarbons. In: Hutzinger O (Ed) The Handbook of Environmental Chemistry, 3B (pp 69–88). Springer Verlag, New York

Perkins EJ, Gordon MP, Caceres O & Lurquin PF (1990) Organization and sequence analysis of the 2,4-dichlorophenol hydroxylase and dichlorocatechol oxidative operons of plasmid pJP4. J. Bacteriol. 172: 2351–2359

Pieper DH, Kuhm AE, Stadler-Fritzsche K, Fischer P & Knackmuss H-J (1991) Metabolization of 3,5-dichlorocatechol by *Alcaligenes eutrophus* JMP 134. Arch. Microbiol. 156: 218–222

Pirt SJ (1975) Principles of Microbe and Cell Cultivation. Blackwell Scientifc Publications, Oxford

Reineke W & Knackmuss H-J (1984) Microbial metabolism of haloaromatics: isolation and properties of a chlorobenzene-degrading bacterium. Appl. Environ. Microbiol. 47: 395–402

Reineke W & Knackmuss H-J (1988) Microbial degradation of haloaromatics. Annu. Rev. Microbiol. 42: 263–287

Rijnaarts H, Norde W, Bouwer E, Lyklema J & Zehnder AJB (in preparation) Reversibility of bacterial adhesion: studies with various strains of *Pseudomonas* and coryneform bacteria

Rijnaarts HHM, Bachmann A, Jumelet JC & Zehnder AJB (1990) Effect of the desorption and intraparticle mass transfer on the aerobic biomineralization of α-hexachlorocyclohexane in a contaminated calcareous soil. Environ. Sci. Technol. 24: 1349–1354

Rojo F, Pieper DH, Engesser K-H, Knackmuss H-J & Timmis KN (1987) Assemblage of ortho cleavage route for simultaneous degradation of chloro- and methylaromatics. Science 238: 1395–1398

Sander P, Wittich R-M, Fortnagel P, Wilkes H & Francke W (1991) Degradation of 1,2,4-trichloro- and 1,2,4,5-tetrachlorobenzene by *Pseudomonas* strains. Appl. Environ. Microbiol. 57: 1430–1440

Sayler GS, Shields MS, Tedford ET, Breen A, Hooper SW, Sirotkin KM & Davis JW (1985) Application of DNA-DNA colony hybridization to the detection of catabolic genotypes in environmental samples. Appl. Environ. Microbiol. 49: 1295–1303

Schlömann M, Pieper DH & Knackmuss H-J (1990) Enzymes of haloaromatics degradation: variations of *Alcaligenes* on a theme by *Pseudomonas*. In: Silver S, Chakrabarty AM, Iglewski B & Kaplan S (Eds) *Pseudomonas*: Biotransformations, Pathogenesis, and Evolving Biotechnology (pp 185-197). American Society for Microbiology, Washington, DC

Schmidt E & Knackmuss H-J (1980) Chemical structure and biodegradability of halogenated aromatic compounds. Conversion of chlorinated muconic acids into maleoylacetic acid. Biochem. J. 192: 339–347

Schmidt E, Remberg G & Knackmuss H-J (1980) Chemical structure and biodegradability of halogenated aromatic compounds. Halogenated muconic acids as intermediates. Biochem. J. 192: 331–337

Schraa G, Bethe B, van Neerven ARW, van den Tweel WJJ, van der Wende E & Zehnder AJB (1987) Degradation of 1,2-dimethylbenzene by *Corynebacterium* strain C125. Anthonie van Leeuwenhoek 53: 159–170

Schraa G, Boone ML, Jetten MSM, van Neerven ARW, Colberg PJ & Zehnder AJB (1986) Degradation of 1,4-dichlorobenzene by *Alcaligenes* sp. strain A175. Appl. Environ. Microbiol. 52: 1374–1381

Schwarzenbach RP & Westall J (1981) Transport of non-polar organic compounds from surface water to groundwater: laboratory sorption studies. Environ. Sci. Technol. 15: 1360–1367

Spain JC & Nishino SF (1987) Degradation of 1,4-dichlorobenzene by a *Pseudomonas* sp. Appl. Environ. Microbiol. 53: 1010–1019

Suflita JM, Horowitz A, Shelton DR & Tiedje JM (1982) Dehalogenation: a novel pathway for the anaerobic biodegradation of haloaromatic compounds. Science 218: 1115–1117

Swindoll CM, Aelion CM & Pfaender FK (1988) Influence of inorganic and organic nutrients on aerobic biodegradation and on the adaptation response of subsurface microbial communities. Appl. Environ. Microbiol. 54: 212–217

van der Meer JR (1992) Molecular mechanisms of adaptation of soil bacteria to chlorinated benzenes. Ph.D. Thesis, Wageningen Agricultural University, Wageningen, The Netherlands

van der Meer JR, Roelofsen W, Schraa G & Zehnder AJB (1987) Degradation of low concentrations of dichlorobenzenes and 1,2,4-trichlorobenzene by *Pseudomonas* sp. strain P51 in nonsterile soil columns. FEMS Microbiol. Ecol. 45: 333–341

van der Meer JR, Eggen RIL, Zehnder AJB & de Vos WM (1991a) Sequence analysis of the *Pseudomonas* sp. strain P51 *tcb* gene cluster, which encodes metabolism of chlorinated catechols: evidence for specialization of catechol 1,2-dioxygenases for chlorinated substrates. J. Bacteriol. 173: 2425–2434

van der Meer JR, van Neerven ARW, de Vries EJ, de Vos WM & Zehnder AJB (1991b) Cloning and characterization of plasmid-encoded genes for the degradation of 1,2-dichloro-, 1,4-dichloro-, and 1,2,4-trichlorobenzene of *Pseudomonas* sp. strain P51. J. Bacteriol. 173: 6–15

van der Meer JR, Zehnder AJB & de Vos WM (1991c) Identification of a novel composite transposable element, Tn*5280*, carrying chlorobenzene dioxygenase genes of *Pseudomonas* sp. strain P51. J. Bacteriol. 173: 7077–7083

van der Meer JR, de Vos WM, Harayama S & Zehnder AJB (1992) Molecular mechanisms of genetic adaptation to xenobiotic compounds. Microbiol. Rev. 56: 677–694

van Loosdrecht MCM, Lyklema J, Norde W, Schraa G & Zehnder AJB (1987a) Electrophoretic mobility and hydrophobicity as a measure to predict the initial steps of bacterial adhesion. Appl. Environ. Microbiol. 53: 1898–1901

van Loosdrecht MCM, Lyklema J, Norde W, Schraa G & Zehnder AJB (1987b) The role of bacterial cell wall hydrophobicity in adhesion. Appl. Environ. Microbiol. 53: 1893–1897

van Loosdrecht MCM, Lyklema J, Norde W & Zehnder AJB (1989) Bacterial adhesion: a physicochemical approach. Microb. Ecol. 1: 1–15

Zeyer J, Wasserfallen A & Timmis KN (1985) Microbial mineralization of ring-substituted anilines through an *ortho*-cleavage pathway. Appl. Environ. Microbiol. 50: 447–453

Biodegradation 3: 285–298, 1992.

Bacterial PCB biodegradation

Alfred W. Boyle, Christopher J. Silvin, John P. Hassett, James P. Nakas & S.W. Tanenbaum
State University of New York, College of Environmental Science & Forestry, Syracuse, NY 13210, USA

Key words: PCB biodegradation, bacterial metabolism, *Comamonas*, *Pseudomonas*, *Rhodococcus* spp., anaerobic/anaerobic degradative pathways, environmental considerations

Abstract

The environment has become polluted with a variety of xenobiotics, including PCBs, as a result of the industrial development of useful halogenated compounds. While the PCBs may not exhibit the acute toxicity originally ascribed to them, they and their attendant byproducts remain as significant factors for adverse effects in the ecological food-chain. The use of microorganisms for bioremediation of PCBs is reviewed. This paper further details three new isolates obtained by conventional enrichment technics which show significant degradation capabilities for Aroclor 1242. These were identified by morphology, staining, and fatty acid analysis as *Comamonas testosteroni*, *Rhodococcus rhodochrous*, and a strain of *Pseudomonas putida*. These isolates demonstrated somewhat selective degradations of the congeners within Aroclor 1242; comprising total losses of 13.8, 19.1, and 24.6%, respectively. Each organism can attack dichloro-through tetrachloro-biphenyls. Analysis of chromatographic patterns from anaerobically digested Aroclor 1242 samples treated by these bacteria demonstrated decreases in di- through penta-substituted biphenyls. Each of these isolates, with discrete specificities, showed preferences for 'open' 2,3-sites, indicative of the action of 2,3-dioxygenase enzymes. The identification of many intermediates in the foregoing transformations was established by GC-MS analyses. Several variations in metabolic pathways, centering on the *meta* cleavage product 2-hydroxy-6-oxo-6-phenylhexa-2,4-dienoic acid (HOPDA) were suggested from these data. It is concluded that the described strains may be of future bioremediation use in processes which have an initial anaerobic dechlorination stage.

Introduction

Of a number of 'recalcitrant' organic halogen compounds in the environment that may pose significant threats to the global ecosystem, one such group is the polychlorinated biphenyls (PCBs). The parental backbone of the PCB molecule is biphenyl and PCBs are synthesized by its direct chlorination with anhydrous chlorine and iron filings or ferric chloride as a catalyst. This process creates a crude product which is then purified by treatment with alkali and distilled resulting in a mixture of chlorobiphenyls with different numbers of chlorine atoms per molecule. In theory, the biphenyl molecule can be substituted at any and all of its unlinked 10 carbons; this allows for 209 different congeners. However, mechanistic and statistical evidence indicate that all possible congeners are not formed (Huntzinger et al. 1974). Typically, a synthetic PCB mixture will contain between 60 and 80 different chlorinated biphenyls.

Such PCB mixtures have been synthesized since 1929 in a number of different countries (USA, Great Britain, France, Japan). In the United States and Great Britain, PCBs were produced by Monsanto and marketed under the trade name Aroclor. Different mixtures of Aroclor are referred to by use of a four-digit numbering system. The first two digits, 12, represent the biphenyl parent structure, and the last two, -21,-42, for example, represent the weight percent of chlorine.

The physical and chemical properties of PCB mixtures made them industrially useful. Aroclor mixtures are extremely thermostabile (up to 350 °C), resistant to oxidation, acids, bases, and have excellent electrical insulating as well as dielectric characteristics. It has been estimated in the United States alone, that 57×10^7 kilograms of PCBs were produced between the early 1920s and 1978 (Brown et al. 1984), and it is assumed that an equivalent quantity was manufactured in Europe, Russia and Japan. Dependent on composition, Aroclor mixtures have been used as industrial fluids, i.e., in hydraulic systems, gas turbines, and vacuum pumps; as dielectric fluids (capacitors, transformers); as plasticizers (adhesives, textiles, surface coatings, sealants, printing, copy paper); and as heat exchangers. These properties contribute, in parallel, to their relative metabolic inertness in the environment. It has been estimated that some 10×10^7 kilograms reside in the biosphere as contaminants of soil, the air, in (and under) rivers and waste streams, and as components of the lipoidal compartments of plant and animal wildlife. Because of this lipophilicity, the PCBs became bioaccumulated in ecosystem food chains. These topics and related facets have been comprehensively reviewed in a three-volume work edited by Waid (1986).

Table 1. Effects of PCBs on cellular physiological responses.

- → Various (−) effects on aquatic organism reproduction. [photosynthetic bact; phytoplankton; protozoa, *Daphnia*, salmon eggs; sea lions; seals, beluga whales, etc.]
- → Various (−) effects on bird reproduction.
- → Various (−) effects on terrestrial mammal reproduction.
- → Chromosome damage, teratogenicity, substandard juvenile growth, embryonic mortality.
- → Various (+) or (−) effects on mammalian enzymes.
- → Various (+) or (−) effects on lipid metabolism.
- → Various (+) or (−) hormonal or pseudohormonal manifestations, *via* binding to specific receptors.
- → Various (−) effects on immunologic defense system.
- → Various dermatological effects ['chloroacne'].
- → Degenerative neurological symptoms.
- → Experimentally induced tumorigenic or carcinogenic effects.

The health effects of halogenated aromatic hydrocarbons of technical use, including PCBs were initially intensively delineated in a monograph (Nicholson & Moore 1979) published under the auspices of the New York Academy of Sciences. More recent reviews of human health effects (Kimbrough 1987) and of mutagenicity and carcinogenicity of the PCBs (Safe 1989) are also available. The documented effects of PCBs at the organismal, cellular, or molecular biological levels are summarized in Table 1. Additionally, the specific binding of PCBs to rat liver cytosol proteins has been documented (Buff & Brundl 1992). These materials have thus been a cause for concern and have indeed been implicated in a number of well-publicized catastrophic events (Kimbrough 1987). As a result, the U.S. Congress banned their manufacture in 1976; but these long-lived chemicals are still found in service around the world and continue to be suspect in various environmental intoxication situations. This is especially so for marine organism die-offs (Hileman 1992). Yet, PCBs may not be the distinct causal factors in such syndromes, and a number of experts in this area attribute acute toxicities seen in humans and other animals to contaminating manufacturing byproducts and/or to abiotic transformation products which have accrued during PCB usage (Stone 1992). Identification of such putative 'supertoxic' species among PCB congeners remains as a daunting analytical task (cf. Erickson 1985; Storr-Hansen 1991). Despite such uncertainties, it is clear that PCBs can covalently bind to DNA *in vivo* and *in vitro*, that PCB mixtures and single congeners exhibit discernible mutagenic activity, and that more highly chlorinated PCB mixtures (>50% Cl by weight) are hepatocarcinogens in rodents. While lower chlorinated mixtures are not strongly carcinogenic, equivocal results from occupational studies suggest that humans exposed to PCBs may have an excess rate of cancer at some sites.

Insofar as PCB activities in enzyme inductions and repressions, as competitors for hormone receptor sites, as teratogens, and as diverse metabolic agonists or antagonists are concerned, it would appear (Table 1), besides total percent of chlorination, that substitution position for halogen is also critical. Thus, in terms of structure-function relationships, the possibility for the two phenyl rings to assume co-planarity is considered to be a parameter which influences the foregoing metabolic phenomena (Brunstrom 1989; Borlakaglu et al. 1990; Huang & Gibson 1992). It therefore cannot be gainsaid that the removal of PCBs from the environment represents a high priority challenge. Where these substances have accumulated in highly concentrated and accessible locales, direct incineration is the most likely practical approach. Additionally, solvent extraction processes have been patented and specific removal chemical interactions, for example, in transformer oil have been explored (Brunelle et al. 1985). However, for PCB depositions in soil, waste streams, and in or under lakes and rivers, alternative methodologies such as microbiol detoxification have also received considerable attention.

Microbial transformations of PCBs – descriptive overview

Studies dealing with the microbial breakdown of PCBs have been reported as early as 1973 (Ahmed & Focht 1973; Omori et al. 1973). Since that time, numerous experiments have been conducted to determine the nature of diverse microbial transformations of polychlorinated biphenyls. These transformations fall under two general categories; those which occur under anaerobic conditions and those which take place aerobically. In theory, the optimal biological process should mineralize PCBs to carbon dioxide, water and chlorine. In order for this to occur, chlorine must be removed from the biphenyl ring and resultant intermediates must be cleaved and oxidized. Under anaerobic conditions, one stage by which chlorine atoms can be removed from a chlorinated biphenyl is reductive dechlorination. Under aerobic conditions, oxidative dechlorination or hydrolytic dehalogenation, especially by members of the genus *Pseudomonas*, have been described (Scholten et al. 1991; Markus et al. 1986). Such biotransformations of halogenated organic compounds and their putative reaction mechanisms have been assessed by Hardman (1991) and by Chaudry & Chapalamadugu (1991). The aerobic and anaerobic biodegradations of PCBs have specifically been reviewed by Abramowicz (1990), Bedard (1990), and Mohn & Tiedje (1992).

Anaerobic biotransformations – field studies

Rhee and co-workers (1989) provided some initial evidence for the anaerobic degradation of polychlorinated biphenyls. Their studies (Chen et al. 1988) indicated the production of $^{14}CO_2$ when Hudson River sediments were incubated anaerobically with ^{14}C-labeled monochlorobiphenyls. Amending these sediments with biphenyl enhanced the degradation of the higher chlorinated congeners but not the lower ones (mono- and dichloro). Other important observations were that PCB degradation was seen only under a headspace of pure nitrogen, with no evidence for dechlorination or methanogenesis. However, at around the same time Brown et al. (1987) demonstrated in PCB contaminated sediments from the Hudson River that several dechlorination processes, presumably mediated by different populations of anaerobic microorganisms, appeared to be in operation. While none of the gas chromatographic distribution patterns showed the removal of *ortho*- substituted chlorines, they differed in the removal of *meta*- and *para*-substituted PCBs, and showed the accumulation of PCBs that were *ortho*-substituted. The authors hypothesized that the relevant microorganisms were selectively removing chlorines from PCB *meta* and *para* positions and that observed inconsistencies resulted from different microbial consortia with selective congener specificities.

Further studies of 'Silver Lake' sediments indicated the dechlorination of Aroclor 1260. This mixture consists primarily of hexa- and heptachlorobiphenyls, while tri- and tetrachlorobiphenyls account for less than 1%. Analysis of these sediments indicated 90 to 98% loss of the hexa- and heptachlorobiphenyls, and an increase in the amount of tri- and tetrachlorobiphenyls, which account for 57 to 82% of the PCBs in the dechlorinated sediment.

Laboratory studies of dechlorination

Evidence up to this point indicates that an undefined microbial metabolism was occurring in such sediments. In extension, researchers attempted to reproduce dechlorination in the laboratory to have a better understanding of this phenomenon. Three different groups have now shown the dechlorination of polychlorinated biphenyls using sediments under anaerobic, methanogenic conditions (Quensen et al. 1988; Abramowicz 1990; Vogel et al. 1989). The Quensen group has demonstrated dechlorination of Aroclor 1242, 1248, 1254 and 1260 by Hudson River sediment. Tiedje et al. (1989) showed the dechlorination of Aroclor 1242 and 1248 occurring as early as 8 weeks, with 80 to 90% loss of tetra- and pentachlorobiphenyls after 12 weeks. In following up these probes in order to identify responsible physiological groups of microorganisms Ye et al. (1992) found that heat- and ethanol-treated bacteria eluted from Hudson River sediments are indeed able to dechlorinate Aroclor 1242 under anaerobic conditions, as evidenced by a decrease in the highly chlorinated PCB congeners, with concurrent accumulations of the lesser chlorinated species. It was determined in this study that the cultures resistant to adverse experimental conditions, presumably anaerobic spore formers, gave PCB dechlorination patterns characterized by the removal of meta-chlorines and an increase in ortho- and para-substituted biphenyls. It has further been demonstrated by Van Dort & Bedard (1991) that reductive dechlorination of PCBs is effected by microbial anaerobes. Such findings are of significance because, by the application of numerous biological criteria (Table 1), the less substituted PCB congeners are usually less toxic; and furthermore, are more amenable (as seen below) to aerobic biotransformations. In exemplary studies, Pettigrew et al. (1990) demonstrated the mineralization of 4-chlorobiphenyl by a freshwater consortium of bacteria which included *Pseudomonas testosteroni, Ps. putida,* and an *Arthrobacter* isolate; and Kimbara et al. (1988) detailed a two-component *Pseudomonas* mixture which transformed highly chlorinated PCBs.

Aerobic catabolism

The aerobic biotransformations of PCBs essentially follow the steps which have been elucidated for biphenyl metabolism. The first step in this process is the conversion of biphenyl to cis 2,3-dihydro-2,3-dihydroxybiphenyl via a 2,3-dioxygenase. Then a NAD^+ dependent dehydrogenase converts this compound to 2,3-biphenyldiol. A dioxygenase enzyme causes meta cleavage to form 2-hydroxy-6-oxo-6-phenylhexa 2,4 dienoic acid (yellow compound) or ortho cleavage yielding 2-hydroxy-3-phenyl-6-oxohexa-2,4-dienoate and then phenylpyruvic acid. Further ring separation of the meta cleavage compound can occur by base hydrolysis to form acetophenones (Bedard et al. 1987a, 1987b) and a four carbon unit or benzoic acid and 2-oxo-penta-4-enoate. The first product, benzoate, then is converted by benzoate oxidase to catechol. Catechol is in turn acted on by a 1,2 oxygenase producing cis,cis muconate. Following the well known β-ketoadipate pathway, this compound is further converted to β-ketoadipate-enol-lactone by a muconolactonizing enzyme. This product is then hydrolyzed to β-ketoadipate. The final step is the cleavage of this structure by a thiolase yielding succinyl Co A and acetyl Co A, which are then shuttled off into the TCA cycle. The second product of the ring separation, 2-oxo-penta-4-enoate, forms the intermediates pyruvate and acetaldehyde. Alternatively, a 3,4-dioxygenase enzyme attack, to form a 3,4-dihydrodiol intermediate, has also been postulated (see below).

Early work by Catelani et al. (1973) identified this general pathway (Fig. 1) and since then, numerous attempts to isolate organisms and to identify the responsible enzymes for each step in this degradation scheme have been carried out. In addition, the genetic basis for possession and expression of the relevant enzymes have been extensively explored. The bulk of this research has been conducted over the past 25 years by a group led by

Furukawa in Japan. This work began with studies on the biodegradability of different PCBs by *Alcaligenes* sp. and then examined the effect of chlorine substitution on bacterial metabolism of PCBs (Furukawa et al. 1978, 1979). Furukawa & Chakrabarty (1982) then identified the genetic elements responsible for chlorobiphenyl degradation by the isolation of plasmids from *Acinetobacter* sp. P6, a strain previously reported to dissimilate 33 pure congeners of polychlorinated biphenyls and from *Arthrobacter* M5, a contaminant of the former isolate grown on 4-chlorobiphenyl. This early analysis of plasmids from chlorobiphenyl degrading organisms then led to the cloning of the chromosomal gene which encodes for chlorobiphenyl degradation in *Pseudomonas pseudoalcaligenes* and *Ps. aeruginosa*, along with purification of the translated, linked oxygenases (cf. Taira et al. 1992; and references therein). Biochemical and genetic characteristics of gram-positive bacteria (*Arthrobacter* sp., *Corynebacterium* sp., *Brevibacterium* sp., *Bacillus* sp.) and of gram-negative bacteria (*Pseudomonas, Alcaligenes, Achromobacter, Acinetobacter, Agrobacterium, Enterobacter* and *Flavobacterium* spp.) have also been delineated by Walia et al. (1988) and by Khan & Walia (1991). Experiments carried out by the Silvestre group in Canada (Sondossi et al. 1991) showed moreover, that a *Ps. testosteroni* isolate metabolized biphenyl, hydroxybiphenyl, chlorobiphenyl, and chloro-hydroxybiphenyl by a common metabolic pathway; and that this enzymatic route is specified by a conserved transmissible gene cluster. It was suggested that this metabolic pathway may be yet another manifestation of those bacterial genomes which are concerned with phenylpropanoid catabolism; and indeed, evidence for a super-family of proteins comes from the fact that there is a sequence homology between the *Pseudomonas* dihydrodiol dehydrogenase which metabolizes PCBs, and 'normal' dehydrogenases involved in polyol and sterol transformations (Baker 1990).

While there thus appears to be a conservation of the genes responsible for PCB degradation, different strains of bacteria show greatly different activity in regard to the congeners of PCBs which they degrade. Bedard (1990) isolated 25 different bacteria of which only three exhibited identical abilities to degrade mixtures of PCBs. From this study, Bedard and co-workers isolated two bacteria, *Pseudomonas putida* LB400 and *Alcaligenes eutrophus* H850, that exhibited the ability to degrade a large number or PCB congeners, including those substituted at the 2 and 3 carbons of both rings, a pattern which confers resistance to enzymatic degradation via 2,3-hydroxylation. Further analysis of these organisms indicated that the 2,3-dioxygenase in these organisms can attack *ortho*-chlorinated carbon (Bedard 1990). This capability obviously contributes to the significant degradative capabilities of these organisms. Another unique feature in these organisms and which may help to explain their ability to degrade such highly substituted PCB congeners is the additional presence of 3,4-dioxygenase enzyme activity.

Microcosms of microorganisms for the purpose of PCB mineralization thus continue to be sought. It has become apparent that there exists a gene pool in nature which may be exploited for such purpose *via* the current tools of genetic manipulation (Chakrabarty 1986; Furukawa & Suzuki 1988; Adams et al. 1992), by co-cultivation (Kimbara et al. 1988; Adriaens & Focht 1990), or by sequential applications to the substratum under appropriate physiological conditions (cf. Fathepure & Vogel 1991).

Current work in this laboratory

In keeping with the above considerations, we initiated a program to selectively obtain additional bacterial isolates which might contain elements both of anaerobic and aerobic PCB dissimilatory pathways, since it was shown that degradation of these compounds would probably greatly be improved by initial anaerobic dechlorination (see section *Laboratory studies of dechlorination*). We therefore proceeded to attempt to obtain facultative nitrate-respiring microorganisms (cf. Criddle et al. 1991) from biphenyl, biphenyl/PCB mixtures or from Aroclors. The technics (see below) essentially paralleled those detailed for ferulate catabolism as studied by Grbic-Galic (1985). While a number of the isolates we obtained exhibited distinct nitrate-reducing capabilities, it was not possible to link this physiological attribute to PCB biotransformation. Along this line, it was also noted in a study on the dechlorination of substituted benzenes using anaerobic biotic columns that there was no acceleration of these reactions under denitrifying conditions (Bosma et al. 1988). However from our experiments, a number of new and potentially useful bacterial strains and consortia for aerobic catabolism of PCBs were obtained. Several of these have, to date, been explored in some detail.

Materials and methods

Enrichment technics and culture isolation

Enrichments were set up either with PAS medium (Bedard et al. 1987) or B + M medium and various sludges as innocula (10% v/v). B + M medium consisted of (mg/l): $NaNO_3$ (450.0), K_2HPO_4 (100.0), $MgSO_4 \cdot 7H_2O$ (20.0), $FeSO_4 \cdot 7H_2O$ (5.0), $CaCl_2$ (2.0), $MnCl_2 \cdot 4H_2O$ (0.2), $NaMoO_4$ (0.1). Two sewage sludges, obtained from the Metropolitan Syracuse Treatment Plant (Syracuse, New York) and the Tully Municipal Sewage Treatment Plant (Tully, New York) and a paper mill sludge, obtained from the Department of Paper Science and Engineering, SUNY College of ESF (Syracuse, New York) were used. Enrichments were amended with 100 to 2,000 ppm biphenyl (Sigma Chemical, St. Louis, MO) and incubated at 23 °C in the dark. After two weeks of incubation, the enrichments were screened using a method similar to that of Sylvestre (1980). The inoculum was first streaked onto a PAS agar plate and then a biphenyl/ether solution was applied to the surface using a sterile Pasteur pipet in a fume hood. After 10–15 seconds, the ether evaporated off, leaving a thin film of biphenyl on the plate. This method was also used with 2-chloro- and 4-chlorobiphenyl (Lancaster Synthesis, Windham, NH). These plates were incubated at 23 °C and observed daily. Colonies showing a zone of clearing, indicative of substrate degradation, were selected for further study.

Culture maintenance and identification

Isolates were maintained in PAS broth with 500 ppm biphenyl on PAS/biphenyl plates, or on nutrient agar (Difco, Detroit, MI) supplemented with 0.1% yeast extract (Difco). Mixed cultures were separated using nutrient agar/yeast extract plates. Once pure cultures were obtained, PAS/biphenyl plates were used for strain maintenance. Selected isolates were identified by cellular fatty acid analysis (Microbial ID, Inc. Newark, DE).

Analysis of growth on aroclor 1242

Comamonas testosteroni was cultured in PAS medium containing 200 ppm Aroclor 1242 (Monsanto, St. Louis, MO) at 23 °C. Samples were taken for dilution plating on tryptic soy agar (Difco) to determine colony forming units (CFU)/ml and to record absorbance at 396 nm (Perkin Elmer UV/VIS spectrophotometer, model 4B) to determine the accumulation of degradative metabolites, specifically 2-hydroxy-6-oxo-6-phenylhexa-2,4-dienoic acid (HOPDA IV, Fig. 1) and its chlorinated analogs.

Degradative competence studies

Select strains were cultured in PAS with 500 ppm biphenyl and incubated at 30 °C on a rotary shaker (150 rpm). After 2–4 days of incubation, the cultures were filtered through sterile glass wool and centrifuged. The supernatants were decanted and the cells were washed with PAS buffer (PAS medium without yeast extract and trace metals) and recentrifuged. The cells were resuspended in PAS medium and distributed to sterile screw cap glass vials with teflon-lined septa (Supelco, Bellefonte, PA). Aroclor 1242 and biphenyl were added from concentrated acetone stocks to a final concentration of 10 ppm. After 2–4 days of incubation at 30 °C, perchloric acid was added (0.7% final concentration) to kill the culture. Controls had perchloric acid added at time zero. The cultures were extracted with 4 volumes of methylene chloride and subjected to gas chromatographic analysis.

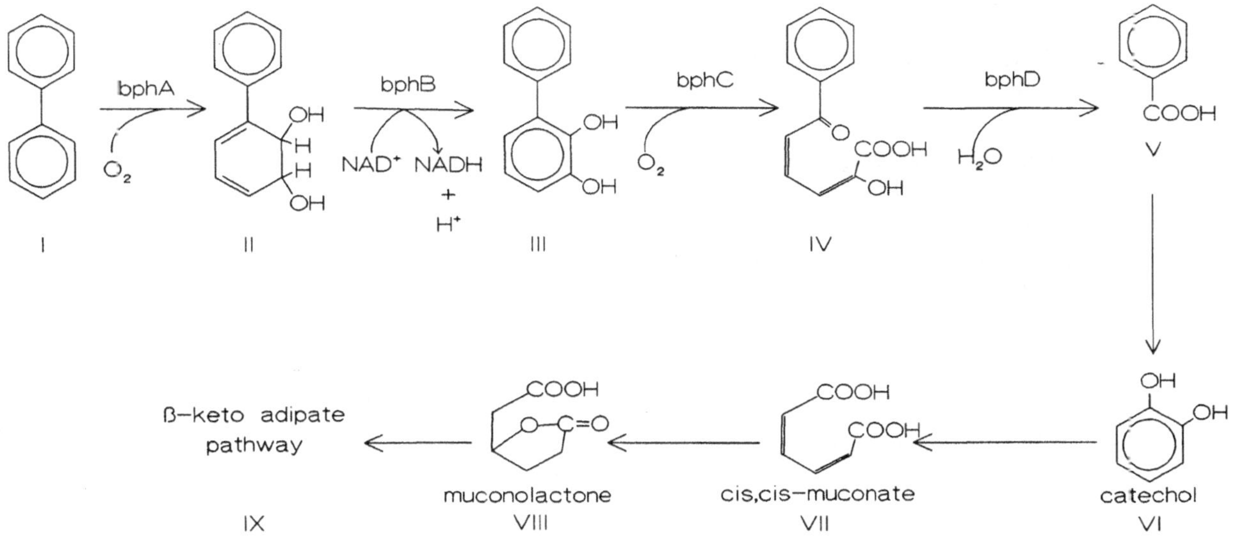

Fig. 1. Aerobic catabolic pathway for degradation of biphenyl and chlorinated biphenyls. The bph operon consists of at least seven genes and four expressed enzyme activities (Taira et al. 1992).

Metabolite production and recovery

Strains were inoculated into 2 l of PAS with 500 ppm biphenyl and 500 ppm Aroclor 1242 and incubated for 24–48 hours at 30 °C. The culture fluid was filtered through glass wool and centrifuged to pellet the cells. The supernatant was then acidified with concentrated H_2SO_4 to a pH of 1–2. The culture filtrate was then saturated with NaCl and extracted twice with ethyl acetate. The organic phase was collected, kept over anhydrous $NaSO_4$, and concentrated by flash evaporation. These preparations were then derivatized with methanolic HCl (Supelco, Inc.) to form the appropriate methyl esters for analysis by GC-MS.

Analytical methods

(i) Gas chromatography
Samples from degradation studies were analyzed on a Hewlett Packard 5890A GC equipped with SPB-5 fused silica capillary column (60 m, 0.2 mm I.D., Supelco, Inc.). The injector was set at 250 °C with a He carrier gas flow of 1.0 ml/min. Nitrogen was used as the make up gas at 30 ml/min. The electron capture detector was set at 300 °C. The temperature program was as follows: 150 °C held for 5 minutes then increased 2 °C/min to 250 °C and held for 10 minutes. Congener assignments for subsequent data given under 'Results and discussion' are tabulated in Table 2.

(ii) Gas chromatography-mass spectrometry
Samples for metabolite identification were analyzed using a Finnigan 9610 automated gas chromatography/EI-CI mass spectrometer system interfaced with a Varian 3400 GC equipped with a SPB-1 fused silica capillary column (30 m, 0.25 mm I.D.). The temperature program consisted of a 5 minute hold at 50 °C, then increased by 10 °C/min to 270 °C, with a 20 minute hold. Electron ionization was performed at 70 EV. Methane was the reagent gas for chemical ionization.

Results and discussion

Environmental bacterial isolates obtained by elective culturing of samples from contaminated locales are listed in Table 3. While practically all of these consortia (or single strains resolved therefrom) are nitrate reducers, it was not possible to propagate them anaerobically with nitrate on biphenyl, Arochlor 1242, or their admixtures. However, as detailed (Table 3), a number of the isolates did possess the ability to degrade these substrates aerobically.

From a five-membered bacterial consortium designated TS-1 obtained from sewage sludge, *Comamonas testosteroni* was isolated in pure culture. This organism when grown in media supplemented with yeast extract and Aroclor 1242, biphenyl, or mixtures of these reached populations approaching 10^8 CFU (Fig. 2) within 47 hrs. Scans of supernatants from *C. testosteroni* grown on biphenyl, biphenyl/Aroclor 1242, and Aroclor 1242 exhibited UV wavelength maxima at 396 nm characteristic of the biphenyl-derived meta-cleavage product (HOPDA; IV, Fig. 1). These supernatants also exhibited characteristic keto-enol tautaumerism at pH values between 4 and 7. HPLC analyses of extracts from cultures grown on biphenyl, biphenyl/Aroclor 1242, and Aroclor 1242 revealed the presence of benzoic acid and chlorinated benzoic

Table 2. Congener assignments for chromatographic peak numbers.

Peak number	Congeners in peak		
1	2,6	2,2′	
2	2,4	2,5	
3	2,3′,		
4	2,4	2,3	
5	2,2′,6		
6	2,2′,5	2,2′,4	4,4′
7	2,3,6	2,3′,6	
8	2,2′,3	2,4′,6	
10	2,4,5		
11	2,3′,5		
12	2,3′,4		
13	2,4′,5	2,4,4′	
14	2,3,3′	2′,3,4	2,2′,5,6′
15	2,2′,4,6′	2,3,4′	
16	2,2′,3,6		
17	2,2′,3,6′		
19	2,2′,5,5′		
20	2,2′,4,5′		
21	2,2′,4,4′	2,2′,4,5	2,4,4′,6
23	2,2′,3,5′		
24	3,4,4′	2,3,3′,6	2,2′,3,4′
25	2,2′,3,4	2,3,4′,6	
27	2,2′,3,3′		
28	2,2′,4,4′,6	2,3′,4,5	
29	2,3,4′,5		
30	2,4,4′,5		
31	2,3′,4′,5		
32	2,3′,4,4′	2,2′,3,5′,6	
34	2,2′,3,4′,6		
35	2,3,4,4′	2,3,3′,4′	
38	2,2′,3,4′,5	2,2′,4,5,5′	
39	2,2′4,4′,5		
42	2,2′,3′,4,5		
43	2,2′,3,4,5′	2,3,4,4′,6	
44	2,2′,3,4,4′		
46	3,3′,4,4′	2,3,3′,4′,6	
47	2,2′,3,3′,4	2,2′,3,4′,5′,6	
50	2′,3,4,4′,5	2,2′,3,4′,5′,6	2,3′,4,4′,5

Table 3. Environmental isolates with the ability to degrade chlorinated biphenyls.

Isolates	Source[1]	Bacterial composition[2]	Activity[3]
H-1	Hudson River	2 Gram (−) rods	Bp, 2-Cl, 4-Cl
H-2	Hudson River	2 Gram (−) rods	Bp, 2-Cl
X*	Hudson River	1 Gram (−) rod	Bp, 2-Cl, 4-Cl
S-2	sewage sludge #1	*Erwinia herbicola* *Enterobacter aerogenes* and *Corynebacterium sp.*	Bp, 2-Cl, 4-Cl, 1242
S-4	sewage sludge #1	*Aureobacterium saperdae* and *Bacillus cereus*	Bp, 2-Cl
S-29	sewage sludge #1	*Xanthomonas campestris*, *Citrobacter freundii*, *Pseudomonas fluorescens*, *P. putida* and *Comamonas testosteroni*	Bp, 2-Cl, 4-Cl, 1242
TS-1			(see text)
NB-1	sewage sludge #2	*P. putida*	Bp, 1242
OB-1	paper sludge	*Rhodococcus rhodochrous*	Bp, 1242

[1] Samples from these sources were used as inocula for various enrichments (see 'Materials and methods')
[2] Determined as outlined in 'Materials and methods'.
[3] Based on colony formation on overlay plates with indicated single congeners (Bp = biphenyl, 2-Cl = 2-chlorobiphenyl, 4-Cl = 4-chlorobiphenyl) or GC analyses of cell assays for Aroclor 1242.

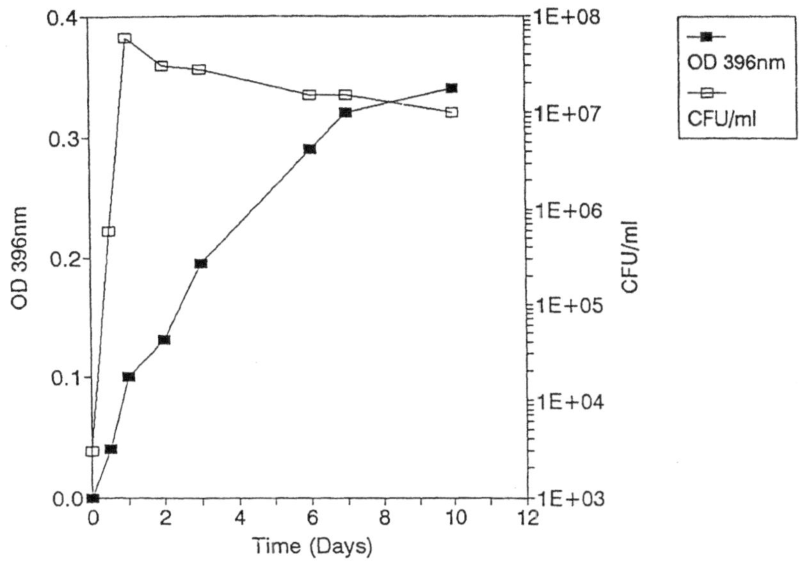

Fig. 2. Growth and meta-cleavage product formation by *C. testosteroni* on Aroclor 1242.

acid derivatives, respectively. Quantitation of products formed indicated that 190 mg/L and 138 mg/l of benzoic acid and chlorinated benzoic acid were formed within 5 days from an initial concentration of biphenyl and Aroclor 1242 at 1 g/l each. Analyses of individually resolved HPLC peaks by GC-MS indicated the presence of several expected metabolites including: 2-hydroxy-6-oxo-6-phenylhexa-2,4-dienoic acid (HOPDA), 2,6-dioxo-6-phenylhexanoic acid, γ-benzoyl-butyric acid, γ-benzoyl-propionic acid, phenylacetic acid, benzoic acid, and 2-oxo-penta-4-enoate. Thus, metabolism of biphenyl by *C. testosteroni* resulted in the formation of several metabolites which have been previously described in the literature. Moreover (see below) growth of this organism on chlorinated biphenyls resulted in the production of chlorinated benzoic acids indicating metabolic attack on the less chlorinated ring.

Aroclor 1242 was tested at concentrations between 5 ppm and 50 ppm as a substrate. As summarized in Table 4, treatment of 5 ppm and 10 ppm samples with the pure cultures isolated in this study resulted in a 13–24% loss in total congener concentration. Minor changes (< 1%) were observed at 25 ppm and no changes were detected in the 50 ppm sample. In the lower concentration (5 ppm and 10 ppm) experiments, the congener selectivity of *C. testosteroni* was identical, with 100% loss of 2,5-dichlorobiphenyl, 2,4-dichlorobiphenyl, 2,3′-dichlorobiphenyl, 2,3′,4-trichlorobiphenyl (possibly 2,3′,5-trichlorobiphenyl and 2′,3,5-trichlorobiphenyl); and an 87% loss in a peak representing 2,4′-dichlorobiphenyl and 2,3-dichlorobiphenyl. *C. testosteroni* was further grown in the presence of several congeners known to comprise Aroclor 1242. These included: 2-chlorobiphenyl, 4-chlorobiphenyl, 2,4-dichlorobiphenyl, and 4,4′-dichlorobiphenyl. In the presence of these Aroclor congeners, chlorinated benzoic acids were formed. From the dichlorobiphenyls, either 2- or 4-chlorobenzoic acid was identified as a product (indistinguishable by GC-MS).

Of some interest was the finding that this isolate could also grow on the following substrates as sole carbon sources: nitrobenzene, 4-nitrocatechol, 2-chloro-5-nitrobenzoic acid, p-nitrophenol, and 2,4,6-trinitrobenzene sulfonic acid. From these observations (Table 5) it can be concluded that our isolate of *C. testosteroni* contains dioxygenase(s) analogous to those demonstrated in *C. testosteroni* 'T-2' which catalyzes the conversion of p-sulfobenzoate into protocatechoic acid (Locher et al. 1991), or in a pseudomonad grown on 2,4-dinitrotoluene which catalyzed the displacement of the 4-nitro group by molecular oxygen (Spanggord et al. 1991). The genetic interrelationships, functional group specificities, annnd positional points of attack among these oxygenases are, as yet, unknown.

Workup of consortia S-29 or NB-1 provided isolates of *Pseudomonas putida* as the active component. Besides 'HOPDA' (2-hydroxy-6-oxo-6-phenylhexa-2,4-dienoic acid), there was identified in the growth filtrate of this microorganism mono-, di-, and trichlorobenzoic acids (Table 6). In like fashion, from paper sludge OB-1, a gram-positive microorganism identified as *Rhodococcus rhodochrous* was obtained. The identifiable metabolites formed by this isolate are given in Table 7. These include, besides HOPDA, benzoylbutanoic acid, dichlorobenzoylpropionic acid, phenylacetic acid, hydroxybenzoic acid, benzoic acid, and mono- and dichlorobenzoic acids. The foregoing breakdown distributions are in keeping with those found earlier by Yagi & Sudo (1980) who studied PCB metabolism by an *Alkaligenes* strain which oxidized high concentrations of dichlorobiphenyls.

To examine the effects of sequential anaerobic/aerobic treatments, 'site H' Hudson River samples (courtesy of D. Abramowicz, General Electricc Co.) were spiked with 300 ppm Aroclor, amended with RAMM medium (Shelton & Tiedje 1984), and incubated for six weeks under an anaerobic headspace. Replicate samples were then subjected to *R. rhodochrous, P. putida* NB1, or a mixture of the two inocula and kept under aerobic conditions. This process led to the loss off 4.9, 18.1 and 18.8%, respectively, in PCB content. Analysis of sequential treatments with *R. rhodochrous* for example, showed decreases in the tri-, tetra-, and pentachlorobiphenyls below those found in the parental Aroclor 1242 mix. Despite this stepwise processing it is conceivable that biphenyl supplementation might be needed in order to fully induce degradative enzymes for further aerobic PCB breakdown by thhe microorganisms at hand.

However, isolation of the actinomycete *R. rhodococcus* amplifies the available spectrum of PCB oxidizing/dechlorinating microorganisms, which typically to date, have mainly included closely phylogenetically-related gram-negative rods. In this context, it should be noted that Smith & Ratledge (1988) showed that a *Nocardia* species dissimilated biphenyl, *via* HOPDA andd benzoate to catechol and then to cis, cis-muconate. Additionally, an isolate also designated as *R. rhodococcus* was obtained by Fuchs et al. (1991) which could co-metabolize with ethanol 2-methylaniline or 4-chloro-2-methylaniline into the TCA cycle. Both catechol-1,2 and catcheol 2,3-dioxygenases were inducible enzymes in this strain, but prolonged culture gave rise to mutants which had lost the gene for the latter. Degradation of 2-methylaniline proceeded by way of *meta*-cleavage, whereas 4-chloro-2-methylaniline was split by *ortho*-cleavage. A

Table 4. Degradation of Aroclor 1242 by *Comamonas testosteroni, Pseudomonas putida* NB 1 and *Rhodococcus rhodochrous.*

Strain	Peaks degraded[a]	Total % loss[b]
C. testosteroni	2,3,4,12,15	13.8
C. putida NB1	1,2,3,4,10,11,12,13,14,15,28	19.12
R. rhodochrous	1,2,3,4,10,11,12,13,14,15, 28,35	24.59

[a] See Table 2 for congener assignments.
[b] Total percent loss of Aroclor 1242 as calculated from controls.

Table 5. Growth of *Comamonas testosteroni* on various aromatic substrates.

Substrate	Qualitative growth
p-Nitroaniline	(−)
4-Nitroguaiacol	(+/−)
3,5-Dinitrosalicyclicacid	(+/−)
5-Nitrovanillin	(+)
Nitrobenzene	(++)
Picrysulfonic acid	(+/−)
4-Nitrocatechol	(++)
p-Toluenesulfonate	(−)
2,4,6-Trinitrobenzenesulfonate	(+++)
p-Nitrophenol	(++)
2-Chloro-5-nitrobenzoic acid	(+++)
2,4-Dinitrotoluene	(−)
Phenylpyruvic acid	(+++)
Phenylacetic acid	(+++)
Sulfanilic acid	(+++)

Microorganism tested with 450 ppm of substrate in PAS salts buffer. (−), no detectable growth; (+), (++), (+++), One, two, and three log growth patterns.

Table 6. Mass spectrometric data of metabolites produced by *Pseudomonas putida* NB1 grown on biphenyl and Aroclor 1242.

Metabolite	E.I.ions[a] (m/z)	C.I. ions[b] (m/z)	Proposed structure
1	218, 133, 120, 105, 77, 55	221, 249, 261	2-hydroxy-6-oxo-6-phenylhexa-2,4-dienoic acid
2	238, 168, 139, 105, 77	239, 267, 279	trichlorobenzoic acid[d]
4	204, 173, 145, 109, 74	N.D.[c]	dichlorobenzoic acid[d]
5	170, 139, 111, 75	N.D.	monochlorobenzoic acid[d]

[a] M^+ (first value) and diagnostic ions identified from electron impact mass spectra.
[b] M^+ +1, M^+ +29 and M^+ +41 ions from chemical ionization mass spectra.
[c] Not determined.
[d] Identified as the methyl ester of indicated compound.

composite of the identification of aerobic PCB breakdown intermediates formed by the several isolates obtained here is given in Fig. 3.

Significance

Since the turn of this century, the environment has become polluted with a variety of man-made xenobiotics resulting from the development of herbicides, pesticides, munitions, refrigerants, and other industrially useful compounds. Those materials which are chlorinated or nitrated hydrocarbons, their manufacturing byproducts, and their partial metabolic transformants are often most toxic to plant and animal life at very low concentrations; and moreover, are quite persistent within given ecosystems. These factors are especially true for polychlorinated biphenyls (PCBs). Thus, the work reported here, as well as efforts from other laboratories, has been directed toward selection of microorganisms, which in concert or under cometabolic conditions, may be capable to a greater or lesser degree of mineralizing these compounds. Our research to date has revealed in nature the existence of a number of new bacterial isolates which selectively remove specific polychlorinated congeners from among that mix known as PCBs. These isolates could form the basis as agents for future bioremediation process development. Furthermore, the next obvious goal for application of such biocatalysts in cost-effective and environmentally compatible bioremediations would be to enlarge and redirect their metabolic potential. Such goals could be realized by the complementation or addition of specific heterologous enzyme-forming genes, so as to produce derivative constructs which would more completely mineralize most of the PCB congeneric substances. Provided that such genetically-engineered strains could thrive effec-

Table 7. Mass spectrometric data of metabolites produced by *Rhodococcus rhodochrous* grown on biphenyl and Aroclor 1242.

Metabolite	E.I.ions[a] (m/z)	C.I. ions[b] (m/z)	Proposed structure
1	218, 133, 120, 105, 77, 55	N.D.[c]	2-hydroxy-6-oxo-6-phenylhexa-2,4-dienoic acid
2	206, 175, 147, 105, 77, 55	207, 235, 247	benzoylbutanoic acid
3	260, 229, 201, 173, 145	261, 289, 301	dichlorobenzoyl-propionic acid
4	204, 177, 145, 109, 74	N.D.	dichlorobenzoic acid[d]
5	170, 139, 111, 75	N.D.	monochlorobenzoic acid[d]
6	136, 105, 77, 51	N.D.	benzoic acid[d]
7	152, 120, 92	153, 181, 193	hydroxy benzoic acid[d]
8	150, 91, 65	151, 179, 191	phenylacetic acid[d]

[a] M^+ (first value) and diagnostic ions identified from electron impact mass spectra.
[b] M^+ +1, M^+ +29 and M^+ +41 ions from chemical ionization mass spectra.
[c] Not determined.
[d] Identified as the methyl ester of indicated compound.

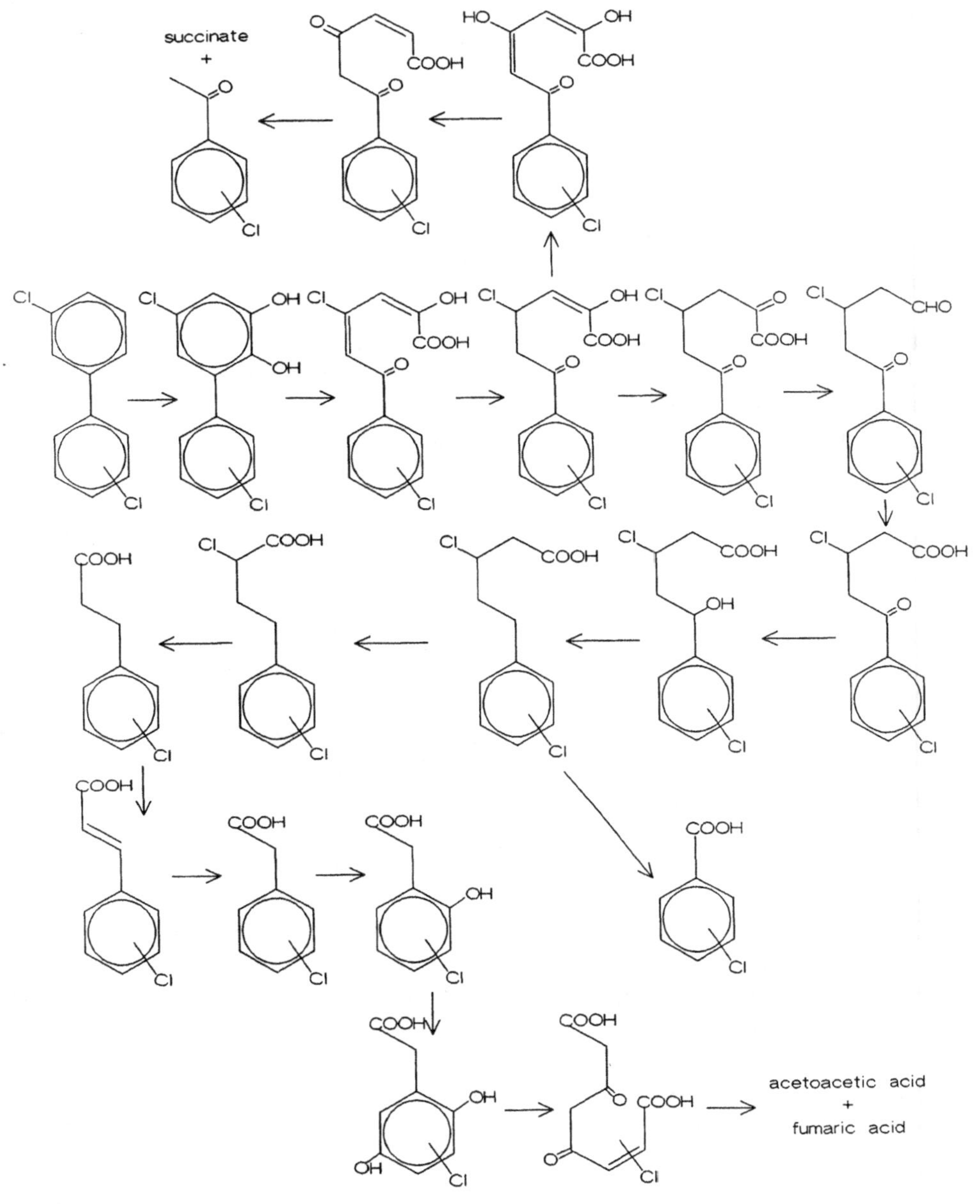

Fig. 3. Proposed composite reaction sequence to account for the metabolites identified in the oxidation of biphenyl or of chlorinated biphenyly by the isolates obtained in this study.

tively under field conditions and also overcome the current political barriers towards their use, it should, in theory, be possible to foresee their use in meeting legislative regulations regarding the clea-nup of such hazardous wastes.

Acknowledgements

This work was supported by grant awards from the N.Y. State Center for Hazardous Waste Management, SUNY at Buffalo; and the General Electric Co., Corporate Research and Development Biological Sciences Branch, Schenectady, NY. We thank Mee Kim for her excellent technical assistance.

References

Abramowicz DA (1990) Aerobic and anaerobic biodegradation of PCBs: a review. Crit. Revs. Biotechnol. 10: 241–251

Adams RH, Huang CM, Higson FK, Brenner V & Focht DD (1992) Construction of a 3-chlorobiphenyl-utilizing recombinant from an intergeneric mating. Appl. Environm. Microbiol. 58: 647–654

Adriaens P & Focht DD (1990) Continuous coculture degradation of selected PCB congeners by *Acinetobacter* spp. in an aerobic reactor system. Environ. Sci. Technol. 24: 1042–1049

Ahmed M, Focht DD (1973) Degradation of polychlorinated biphenyls by two species of *Achromobacter*. Can. J. Microbiol. 19: 47–52

Bedard DL (1990) Bacterial transformations of polychlorinated biphenyls. In: Kamely D (Ed) Biotechnology and Biodegradation (pp 369–391). Gulf Publ. Co., Houston

Bedard DL, Haberl ML, May RJ & Brennan MJ (1987a) Extensive degradation of Aroclors and environmentally transformed polychlorinated biphenyls by *Alcaligenes eutrophus* H850. Appl. Environ. Microbiol. 53: 1094–1102

Bedard DL, Haberl ML, May RJ & Brennan MJ (1987b) Evidence for novel mechanism of polychlorinated biphenyl metabolism in *Alcaligenes eutrophus* H850. Appl. Environ. Microbiol. 53: 1103–1112

Baker ME (1990) Sequence similarity between *Pseudomonas* dihydrodiol dehydrogenase, part of the gene cluster that metabolizes PCBs and dehydrogenases involved in the metabolism of ribitol, glucitol, and the synthesis of antibiotics and 17-β-estradiol, testosterone and corticosterone. Biochem. J. 267: 839–841

Borlakoglu JT, Welch VA, Wilkins JPG & Dils RR (1990) Transport and cellular uptake of PCBs I, II. Biochem. Pharmacol. 40: 265–272; 273–281

Bosma TNP, Van der Meer JR, Schraa G, Tros ME & Zehnder AJB (1988) Reductive dechlorination of all trichloro- and dichlorobenzene isomers. FEMS Microbiol. Ecol. 53: 223–229

Brown Jr. JF, Wagner RE, Bedard DL, Brennan MJ, Carnahan JC, May RJ & Tofflemire TJ (1984) PCB transformations in the upper Hudson sediments. Northeast Environ. Sci. 3: 167–179

Brown Jr. JF, Wagner RE, Feng H, Bedard DL, Brennan MJ, Carnahan JC & May RJ (1987) Environmental dechlorination of PCBs. Environ. Toxicol. Chem. 6: 579–593

Brunelle DJ, Mendiratta AK & Singleton DA (1985) Reaction/removal of PCBs from transformer oil: treatment of contaminated oil with polyethylene glycol/KOH. Environ. Sci. Technol. 19: 740–745

Brunstrom B (1989) Toxicity of coplanar polychlorinated biphenyls in avian embryos. Chemosphere 19: 765–768

Buff K & Brundl A (1992) Specific binding of PCBs to rat liver cytosol protein. Biochem. Pharmacol. 43: 965–970

Catelani D, Colombi A, Sorlini C & Treccani V (1973) Metabolism of biphenyl-2-hydroxy-6-oxo-6-phenylhexa-2,4-dienoate: the meta-cleavage product from 2,3-dihydroxybiphenyl by *Pseudomonas putida*. Biochem. J. 134: 1063–1066

Chakrabarty AM (1986) Genetic engineering and problems of environmental pollution. In: Rehn HJ (Ed) Biotechnology, Vol 8 (pp 516–530). VCH Publishers, New York

Chaudhry GR & Chapalamadugu S (1991) Biodegradation of halogenated organic compounds. Microbiol. Revs 55: 59–79

Chen M, Hong CS, Bush B & Rhee G-Y (1988) Anaerobic degradation of polychlorinated biphenyls by bacteria from Hudson River sediments. Ecotoxicol. Environ. Safety 16: 95–105

Criddle CS, Dewitt JR, Grbic-Galic D & McCarty PL (1990) Transformation of carbon tetrachloride by *Pseudomonas* sp. KC under denitrification conditions. Appl. Environ. Microbiol. 56: 3240–3246

Erickson MD (1985) Analytical Chemistry of PCBs. Butterworths, Stoneham, Massachusetts

Fathepure BZ & Vogel TM (1991) Complete degradation of polychlorinated hydrocarbons by a two-stage biofilm reactor. Appl Environ. Microbiol. 57: 3418–3422

Fuchs K, Schreiner A & Lingens F (1991) Degradation of 2-methylaniline and chlorinated isomers of 2-methylaniline by *Rhodococcus rhodochrous* strain CTM. J. Gen'l. Microbiol. 137: 2033–2039

Furukawa K & Chakrabarty AM (1982) Involvement of plasmids in total degradation of chlorinated biphenyls. Appl Environ Microbiol. 44: 619–626

Furukawa K & Suzuki H (1988) Gene manipulation of catabolic activities for production of intermediates of various biphenyl compounds. Appl. Microbiol. Biotechnol. 29: 363–369

Furukawa K, Tonomura K & Kamibayashi A (1978) Effect of chlorine substitution on the biodegradability of polychlorinated biphenyls. Appl. Environ. Microbiol. 35: 223–227

Furukawa K, Tonomura K, & Kamibayashi A (1979) Effect of chlorine substitution on the bacterial metabolism of various polychlorinated biphenyls. Appl. Environ. Microbiol. 38: 301–310

Grbic-Galic D (1985) Fermentative and oxidative transformation of ferulate from a facultative anaerobic bacterium isolated from sewage sludge. Appl. Environ. Microbiol. 50: 1052–1057

Hardman, DJ (1991) Biotransformation of halogenated compounds. Crit. Revs. Biotechnol. 11: 1–40

Hileman B (1992) Effects of organohalogens on marine animals to be investigated. Chem. Eng. News 3/9: 23–24

Huang S & Gibson GG (1992) Species and congener specific induction of hepatic cytochrome P4504A by PCBs. Biochem. Pharmacol. 43: 637–639

Huntzinger O, Safe S, Zitko V (1974) The Chemistry of PCBs. CRC Press Inc., Cleveland

Khan AA & Walia SK (1991) Expression, localization and functional analysis of PCB degradation genes cbp ABCD of *Ps. putida*. Appl. Environ. Microbiol. 57: 1325–1332

Kimbara K, Hashimoto T, Fukuda M, Koana T, Takagi M, Oishi M & Yano K (1988) Isolation and characterization of a mixed culture that degrades PCBs. Agric. Biol. Chem. 52: 2885–2891

Kimbrough, RD (1987) Human health effects of PCBs and PBBs. Ann. Rev. Pharmacol. Toxicol. 27: 87–111

Locher HH, Leisinger T & Cook AM (1991) 4-Sulphobenzoate 3,4-dioxygenase. Purification and properties of a desulphonative two-component enzyme system from *C. testosteroni* T-2. Biochem. J. 274: 833–842

Markus A, Krekel D & Lingens F (1986) Purification and some properties of component A of the 4-chlorophenylacetate 3,4-dioxygenase from *Pseudomonas* sp. J. Biol. Chem. 261: 12883–12888

Mohn WW & Tiedje JM (1992) Microbial reductive dehalogenation. Microb. Revs. 56: 482–507

Nicholson WJ & Moore JA (1979) Health effects of halogenated aromatic hydrocarbons. Ann. N.Y. Acad. Sci. 320, New York

Omori T, Ikai T, Minoda Y and Yamada K (1973) Utilization of polyphenyl and polyphenyl-related compounds by microorganisms. Agric. Biol. Chem. 37: 1599–1605

Pettigrew CA, Breen A, Corcoran C & Saylor GS (1990) Chlorinated biphenyl mineralization by individual populations and consortia of freshwater bacteria. Appl. Environ. Microbiol. 56: 2036-2045

Quensen III JF, Tiedje JM, Boyd SA (1988) Reductive dechlorination of polychlorinated biphenyls by anaerobic microorganisms from sediment. Science 242: 752

Rhee G-Y, Bush B, Brown MP, Kane M & Shane L (1989) Anaerobic biodegradation of polychlorinated biphenyls in Hudson River sediments and dredged sediments in clay encapsulation. Wat. Res. 23: 957–964

Safe S (1989) Polychlorinated biphenyls (PCBs): mutagenicity and carcinogenicity. Mutation. Res. 220: 31–47

Scholten JD, Chang KH, Babbitt HC, Sylvestre M & Dunaway-Mariano D (1991) Novel enzymic dehalogenation of a chlorinated aromatic. Science 253: 182–185

Shelton DR & Tiedje TM (1984) Isolation and partial characterization of an anaerobic consortium that mineralizes 3-chlorobenzoate. Appl. Microbiol. Biotechnol. 48: 840–848

Smith MR & Ratledge C (1989) Catabolism of biphenyl by *Pseudomonas* NCIB 10643 and *Nocardia* NCIB 10503. Appl. Microbiol. Biotechnol. 30: 395–401

Sondossi M, Sylvestre M, Ahmad D & Masse R (1991) Metabolism of hydroxybiphenyl and chlorobiphenyl by biphenyl/chlorobiphenyl degrading *Ps. testosteroni* strain B-356. J. Industr. Microbiol. 7: 77–88

Spanggord RJ, Spain JC, Nishino SF & Mortelmans KE (1991) Biodegradation of 2,4-dinitrotoluene by a *Pseudomonas* sp. Appl. Environ. Microbiol. 57: 3200–3205

Stone R (1992) Swimming against the PCB tide. Science 255: 798–799

Storr-Hansen E (1991) Comparative analysis of thirty PCB congeners on two capillary columns of different polarity with non-linear multi-level calibration. J. Chromatog. 558: 375–391

Sylvestre M (1980) Isolation method for bacterial isolates capable of growth on *p*-chlorobiphenyl. Appl. Environ. Microbiol. 39: 1223–1224

Taira K, Hirose J, Hayashida S & Furukawa K (1992) Analysis of bph operon from PCB-degrading strain of *Pseudomonas pseudoalcaligenes* KF-707. J. Biol. Chem. 267: 4844–4853

Tiedje NM, Boyd SA & Quensen JF (1989) Reductive dechlorination of PCB's by anaerobic microorganisms. In: Research and Development Program for the Destruction of PCBs, 8th Progress Report, pp 37–47. GE Corporate Research and Development Center, Schenectady, NY

Van Dort HM & Bedard DL (1991) Reductive *ortho* and *meta* dechlorination of a polychlorinated biphenyl congener by anaerobic microorganisms. Appl. Environ. Microbiol. 57: 1576–1578

Vogel TM, Nies L & Anid PJ (1989) Long term anaerobic-aerobic degradation of PCB's. In: Research and Development Program for the Destruction of PCBs, 8th Progress Report, pp 71–80. GE Corporate Research and Development Center, Schenectady, NY

Waid JF (1986) PCBs and the Environment. Vols 1, 2 & 3. CRC Press Inc., Boca Raton

Walia S, Tewari R, Brieger G, Thimm V & McGuire T (1988) Biochemical and genetic characterization of soil bacteria degrading PCB. In: Abbon R (Ed) Hazardous Waste: Detection, Control, Treatment. Elsevier Publishers, Amsterdam

Yagi O & Sudo R (1980) Degradation of polychlorinated biphenyls by microorganisms. Water. Polln. Control Fed. J. 52: 1035–1043

Ye D, Quenson JF, Tiedje JM & Boyd SA (1992) Anaerobic dechlorination of PCBs (Aroclor 1242) by pasteurized and ethanol-treated microorganisms from sediments. Appl. Environ. Microbiol. 58: 1110–1114

Biodegradation **3**: 299–313, 1992.

Biodegradation of haloalkanes

Shimshon Belkin
Environmental Microbiology, Ben Gurion University of the Negev,
The Jacob Blaustein Institute for Desert Research, Sede Boker Campus 84990, Israel

Key words: biodegradation, bromoalkanes, dehalogenase, environmental pollution, haloalkanes, *Pseudomonas* sp.

Abstract

Halogenated alkanes constitute a significant group among the organic pollutants of environmental concern. Their industrial and agricultural uses are extensive, but until 1978 they were considered to be non-biodegradable. In recent years, microorganisms were described that could degrade, partially or fully, singly or in consortia, many of the compounds tested. The first step in haloalkane degradation appears to be universal: removal of the halogen atom(s). This is mediated by a group of enzymes, generally known as dehalogenases, acting in most cases either as halidohydrolases or oxygenases. Nevertheless, information is still severely lacking regarding the biochemical pathways involved in these processes, as well as their genetic control.

A recently isolated *Pseudomonas* strain, named ES-2, was shown to possess a very wide degradative spectrum, and to contain at least one hydrolytic dehalogenase. The utilization by this organism of water-insoluble haloalkanes, such as 1-bromooctane, appears to consist of three phases: extracellular emulsification by a constitutively excreted surface active agent, periplasmic dehalogenation by an inducible dehalogenase, and intracellular degradation of the residual carbon skeleton.

Introduction

Of the man-made chemicals currently in use in industry and agriculture, halogenated organic compounds have probably received the widest scientific attention. Used as herbicides, pesticides, refrigerants, fire retardants, solvents, degreasers and as various intermediates in organic synthesis, these compounds are often of environmental importance due to their toxicity and often limited biodegradability. Both of these characteristics are, in fact, often conveyed by the halogen(s) covalently bound to the carbon skeleton.

Of the numerous halogenated organics, the degradation of which has been studied, attention has been largely focused on chlorinated aromatics. Two other partially overlapping sub-groups have been relatively ignored: halogenated aliphatics and brominated compounds; their common component, the brominated aliphatics, have probably drawn minimal attention.

Of the various halogenated aliphatics in widespread use, the simplest configuration is that of the halogenated alkanes. These straight carbon chains offer simplified biochemical analysis to the scientist, and possibly facilitated biochemical solutions to the microorganisms involved in their biodegradation. They are therefore an ideal choice for studying and modeling microbial haloorganic metabolism.

While several excellent reviews have been published in recent years concerning the biodegradation of halogenated organic compounds (Knackmuss 1981; Lal & Saxener 1982; Slater & Bull 1982;

Leisinger 1983; Motosugi & Soda 1983; Ghosal et al. 1985; Alexander 1981, 1985; Reineke & Knackmuss 1988; Morgan & Watkinson 1989; Chaudhry & Chapalamadugu 1991), most of them mention haloalkanes only briefly or not at all. The purpose of the present review is to try and collate the available information concerning the biodegradation of haloalkanes, and amend it by new data from our laboratory concerning mostly bromoalkane metabolism.

Occurrence and environmental significance of haloalkanes

In natural environments, the occurrence of haloaliphatic compounds may be of both biogenic and anthropogenic origin, with the latter category in most cases greatly outweighing the former. An exception may be the case of chloromethane; sources such as forest fires and decomposition of seaweeds appear to provide its major environmental input (Edwards et al. 1982; Lovelock 1975). Marine algae have also been shown to release other haloalkanes, among them 1-bromopentane (Gschwend et al. 1985).

Other known sources of haloalkanes are all of human origin; these compounds find their way into the environment mostly due to either agricultural uses or to improperly disposed industrial (and to some extent municipal) wastes. A third major anthropogenic source, mainly of trihalomethanes, is water chlorination. A number of organic precursors present in water, many of them humic in nature, are transformed to trihalomethanes in the presence of free chlorine (Pearson 1982). Both chloro- and bromohalomethanes are thus formed; the bromine is believed to originate from the bromide naturally present in the water, liberated as molecular bromine in the presence of excess chlorine. When effluents of sewage treatment are chlorinated, or when chlorine is used for the bleaching of textiles or paper pulp, the potential also exists for the formation of higher molecular weight halogenated compounds (Pearson 1982).

Various halogenated alkanes are used in agriculture, as fumigants, herbicides and pesticides, and in industry as solvents, intermediates in various synthetic processes, flame retardants and more. A list of some commonly used haloalkanes is presented in Table 1.

In view of these extensive uses, it is not surprising that haloalkanes, especially the short chain ones, are often found in all components of the biosphere – soil, water and air (Leisinger 1983). The toxicity and potential genotoxicity of many of these compounds (Veissman & Hammer 1985) turn this observation into an acute environmental problem. Consequently, haloaliphatics constitute a massive group in the EPA priority pollutants list (31 out of 114 organics). Of these, 19 are haloalkanes.

Since many of these compounds do not decompose readily in nature, understanding their biodegradation pathways is of utmost practical importance. While in most developed countries legislative measures now exist to curtail their use and release, remediation steps are needed in order to rehabilitate environments polluted over years of ignorance and/or neglect. Bioremediation procedures, increasingly implemented in many areas, are often hampered by a lack of basic understanding of many of the microbial processes involved. The present review, aimed at integrating much of what is known about haloalkane degradation, also serves to point out how much still remains unknown.

Haloalkane metabolism

Reports on microbial degradation of haloalkanes may be roughly divided into three broad categories, often intermixed: (1) utilization of the organic substrate for growth, as a sole carbon and energy source; (2) various co-metabolic degradative pathways, and (3) partial degradation, not supporting growth, which often involves only dehalogenation.

Utilization of haloalkanes as single carbon and energy source

Halogenated alkanes were considered to be non-biodegradable, until microbial utilization was demonstrated by Omori & Alexander (1978a, 1978b). Since then the list of haloalkanes capable of supporting bacterial growth has grown; Table 2 lists these compounds, arranged by length of carbon chain.

Utilization of the two compounds at the end of the list, 1-chlorohexadecane and 1-chlorooctadecane, is different from those preceding them in that they were shown to be incorporated, intact, into the membrane lipids of the *Mycobacterium* species studied (Murphy & Perry 1983). However, since these compounds appeared to promote growth in the absence of other organics, they must have also been degraded to allow carbon utilization by the cells. Except for the compounds degraded by *Arthrobacter* HA1 (Scholtz et al. 1987a, 1987b, 1988) and for one other mention of 1,10-dichlorododecane (Yokota et al. 1986), all compounds in Table 2 are 1 to 9 carbon atoms long, and appear to be aerobically utilized by a limited number of microbial isolates. The capabilities of *Arthrobacter* HA1 (Scholtz et al. 1987a, 1987b, 1988) were subjected to the most methodical screening. This isolate was originally shown to grow on C_4 to C_8 1-chlorosubstituted, C_2 to C_8 1-bromosubstituted and C_2 to C_7 1-iodosubstituted alkanes. Later reports (Scholtz et al. 1988) expanded its biodegradative spectrum to C_{12}, C_{14} and C_{16} haloalkanes, though the growth rates were very slow. None of the C_9 alkanes tested was degraded by this organism. The highest specific growth rate measured was 0.14 h^{-1} for 1-chloropentane and 1-chlorohexane; the corresponding growth yields were 5.2 and 5.5 g protein/mol C. For other isolates growing on various haloalkanes, the values, when reported, were similar. In no case was growth on haloalkanes observed at a generation time shorter than 4 hours. Interestingly, in some cases (Scholtz et al. 1987a), growth on haloalkanes was not paralleled by growth on the equivalent non-substituted alkanes. The hypothesis proposed by Omori & Alexander (1978b) that the capacity for dehalogenation of chloroalkanes is closely related to the process of metabolizing structurally similar alkanes, is therefore not necessarily universally true.

Table 1. Industrial and agricultural uses of various haloalkanes.

Compound (synonym)	Uses
Chloromethane (methyl chloride)	Intermediate, refrigerant
Dichloromethane (methyl chloride)	Solvent
Trichloromethane (chloroform)	Intermediate, solvent
Chlorobromomethane	Intermediate (agrochemicals), fire extinguisher, solvent
Bromomethane (methyl bromide)	Soil and space fumigant, intermediate
Dibromomethane (methylene bromide)	Intermediate (agrochemicals and biocides)
Chloroethane (ethyl chloride)	Intermediate, solvent, refrigerant, alkylating agent
1,2-Dichloroethane (ethylene dichloride)	Solvent, intermediate (dyes, perfumes, pharmaceuticals, lacquers)
Chloroethane (vinyl chloride)	Intermediate (PVC production)
1,1,1-Trichloroethane (chlorothene)	Solvent
1,1,2-Trichloroethane (vinyl trichloride)	Solvent
1,1,2,2-Tetrachloroethane (cellon)	Solvent
Hexachloroethane (perchloroethane)	Solvent, explosives, rubber vulcanizing
Bromoethane (ethyl bromide)	Ethylating agent, refrigerant
Dibromoethane (ethylene dibromide, EDB)	Anti-knock additive, nematicidic soil fumigant, solvent
1-Bromopropane (propyl bromide)	Intermediate (pharmaceuticals, insecticides, quarternary ammonium compounds, flavors and fragrances)
1,2-Dichloropropane (propyl bromide)	Nematicidic soil fumigant, solvent
1,2 Dibromo-3-chloropropane (DBCP)	Nematicidic soil fumigant
1-Bromobutane (butyl bromide)	Intermediate (pharmaceuticals, quarternary ammonium compounds)
1-Bromooctane (octyl bromide)	Intermediate (UV absorbers, quarternary ammonium compounds, encapsulated dyes)

Table 2. Microbial utilization of haloalkanes as sole carbon and energy sources.

No. of carbon atoms	Compound	Organism	Specific growth rate (h^{-1})	Growth yield[a]	Reference
1	Chloromethane	*Hyphomicrobium* sp.	0.09	5.0	Hartmans et al. (1986)
	Dichloromethane	*Pseudomonas* DM1	0.11	15.7[b]	Brunner et al. (1980)
		Hyphomicrobium DM2	0.07	–	Stucki et al. (1981)
2	1,2-Dichloroethane	*Xanthobacter autotrophicus*	0.11	–	Janssen et al. (1985)
		Pseudomonas fluorescens	–	–	Vandenberg & Kunka (1988)
		unidentified, gram-negative rod, DE2	0.08	–	Stucki et al. (1983)
	1-Bromoethane	*X. autotrophicus* GJ10	0.03	–	Janssen et al. (1985)
		Acinetobacter sp. GJ70	0.06	–	Janssen et al. (1987)
		Arthrobacter sp. HA1	–	–	Scholtz et al. (1987a)
	1-Iodoethane	"	–	–	"
3	1-Chloropropane	*X. autotrophicus* GJ10	0.12	–	Janssen et al. (1985)
	1,3-Dichloropropane	"	0.09	–	"
	1,2-Dichloropropane	*P. fluorescens*	–	–	Vandenberg & Kunka (1988)
	2,2-Dichloropropane	"	–	–	"
	1-Bromopropane	*Arthrobacter* sp. HA1	0.12	4.0	Scholtz et al. (1987a)
		Acinetobacter sp. GJ70	–	–	Janssen et al. (1987)
	1-Iodopropane	*Arthrobacter* sp. HA1	0.09	4.2	Scholtz et al. (1987a)
		Acinetobacter sp. GJ70	–	–	Janssen et al. (1987)
4	1-Chlorobutane	*Corynebacterium* m15-3	–	–	Yokota et al. (1986, 1987)
		Corynebacterium m2c-32			
		X. autotrophicus GJ10	0.10	–	Janssen et al. (1985)
		Arthrobacter sp. HA-1	0.11	5.6	Scholtz et al. (1987a, 1988)
		Acinetobacter sp. GJ 70	0.15	–	Janssen et al. (1987)
	1,2-Dichlorobutane	*Corynebacterium* m15-3	–	–	Yokota et al. (1986, 1987)
		Corynebacterium m2c-32			
	1-Bromobutane	*Arthrobacter* sp. HA-1	0.11	4.1	Scholtz et al. (1987a)
	1-Iodobutane	"	–	–	"
5	1-Chloropentane	*Corynebacterium* sp.	–	–	Yokota et al. (1986, 1987)
		Arthrobacter sp. HA1	0.14	5.2	Scholtz et al. (1987a)
		Acinetobacter sp. GJ70	0.21	–	Janssen et al. (1987)
	1,5-Dichloropentane	*Pseudomonas* sp.	–	–	Omori & Alexander (1978)
	1-Bromopentane	*Arthrobacter* sp. HA1	–	–	Scholtz et al. (1987a)
	1-Iodopentane	"	–	–	"
6	1-Chlorohexane	*Arthrobacter* sp. HA1	0.14	5.5	Scholtz et al. (1987a)
	1,6-Dichlorohexane	*Acinetobacter* sp. GJ70	–	–	Janssen et al. (1987)
		Pseudomonas sp.	–	–	Omori & Alexander (1978)
	1-Bromohexane	*Arthrobacter* sp. HA1	–	–	Scholtz et al. (1987a)
	1-Iodohexane	"	–	–	"
7	1-Chloroheptane	*Arthrobacter* sp. HA1	–	–	Scholtz et al. (1987a)
		Pseudomonas sp.	–	–	Omori & Alesander (1978)
	1-Bromoheptane	*Arthrobacter* sp. HA1	–	–	Scholtz et al. (1987a)
		Pseudomonas sp.	–	–	Omori & Alexander (1978)
	1-Iodoheptane	*Arthrobacter* sp. HA1	–	–	Scholtz et al. (1987a)
		Pseudomonas sp.	–	–	Omori & Alexander (1978)
8	1-Chlorooctane	*Arthrobacter* sp. HA1	–	–	Scholtz et al. (1987a)
	1-Bromooctane	"	–	–	"
9	1,9-Dichlorononane	"1,9-dichlorononane-utilizing" H10-2, H15-1	–	–	Yokota et al. (1986)

Table 2. Continued.

No. of carbon atoms	Compound	Organism	Specific growth rate (h^{-1})	Growth yield[a]	Reference
	1,9-Dichlorononane	*Acinoetobacter* sp. GJ70	–	–	Janssen et al. (1987)
		Pseudomonas sp.	–	–	Omori & Alexander (1978)
10	1-Chlorodecane	*Arthrobacter* sp. HA1	slow[c]	–	Scholtz et al. (1988)
	1-Bromodecane	"	very slow[d]	–	"
12	1-Chlorododecane	*Arthrobacter* sp. HA1	slow	–	Scholtz et al. (1988)
	1-Bromododecane	"	very slow	–	"
	1,10-Dichlorododecane	"1,9-dichlorononane-utilizing" H10-2, H15-1	–	–	Yokota et al. (1986)
14	1-Bromotetradecane	*Arthrobacter* sp. HA1	very slow	–	Scholtz et al. (1988)
16	1-Chlorohexadecane	*Arthrobacter* sp. HA1	slow	–	Scholtz et al. (1988)
	1-Bromohexadecane	"	slow	–	"
	1-Chlorohexadecane	*Mycobacterium vaccae (M. convolutum)*	–	–	Murphy & Perry (1983)
18	1-Chlorooctadecane	*Mycobacterium vaccae (M. convolutum)*	–	–	"

[a] g protein·mol^{-1} C, except where mentioned.
[b] g dry weight·mol^{-1} substrate.
[c] Growth period – 7 days.
[d] Growth period – 14 days.

Cometabolism of haloalkanes

Out of 500 soil enrichments carried out by Omori & Alexander (1978a), only three have yielded microorganisms capable of growth on 1,9-dichlorononane as a single carbon source, while many of the isolates dehalogenated this molecule. One of their conclusions (Omori & Alexander 1978b) was that halogenated compounds may be acted on in nature by cometabolism. Indeed, many examples were reported in which halogenated alkanes, as well as other haloorganics, underwent transformations in the presence of a growth substrate. Many of these instances involve cooxidations by monoxigenases of methane-oxidizing (Perry 1979; Dalton & Stirling 1982; Little et al. 1988; Oldenhuis et al. 1989) or nitrifying (Rasche et al. 1990; Hyman et al. 1988) bacteria. These reports deal almost exclusively with cooxidations of short (1–3 carbon atoms) haloalkanes; however, indications for cometabolism of longer haloalkanes also exist. For instance, Scholtz et al. (1988) present a long list of substrates up to 7 carbon atoms in length dehalogenated by *Arthrobacter* HA1 in the presence of butanol. Another example of cometabolism is a bromoethanol-resistant mutant of *Acinetobacter* sp. GJ70, possibly with impaired alcohol dehydrogenase activity (Janssen et al. 1987). This mutant lost the ability to grow on bromoethane, 1-bromopropane, 1-chlorobutane or 1-chloropentane as a sole carbon source as well as the equivalent alcohols. However, it could well cometabolize 1,2 dibromoethane in the presence of citrate. Worthy of special mention, though not directly involving haloalkanes, is the description of codegradation of chlorinated paraffins with n-hexadecane by bacterial isolates capable of dehalogenating several haloalkanes (Omori et al. 1987). Other interesting examples are the anaerobic transformations of carbon tetrachloride and brominated trihalomethanes under methanogenic (Bouwer & McCarty 1983a) and denitrifying (Bouwer & McCarty 1983b) conditions.

As shown in many of these examples, the cometabolic requirements may be very specific in nature. This may be of special significance when bioremediation measures are considered, for instance, for contaminated soils and aquifers. Indeed, it has been shown that addition of non-selective carbon sources, such as glucose or amino acids, inhibited mineralization in subsurface samples of xenobiotic substrates, among them dibromoethane (Swindoll et al. 1988). Apparently, the easily degradable organic amendments were preferentially utilized by indigenous bacterial populations. Nevertheless, many reports exist (see Morgan & Watkinson 1989, for review) describing microbially mediated removal of haloalkanes, as well as other contaminants, from environments such as soils (Castro & Belser 1968; Pignatello 1986), aquifers (Baldauf 1981; Werner 1989) and wastewater (Omori & Alexander 1978b). It is impossible to determine whether in these cases degradation was carried out cometabolically, but it is probably safe to assume that it involved the synergistic activity of mixed bacterial consortia.

Dehalogenation

An essential step in any microbial transformation of halogenated compounds is the removal of the halogen from the organic molecule, a step known as dehalogenation. The process may occur abiotically, as demonstrated by halide release into the medium under sterile conditions (Omori & Alexander 1978a; Yokota et al. 1986; Janssen et al. 1987); however, in most cases dehalogenation is enzymatically driven. The enzymes catalyzing the process are generally known as 'dehalogenases' (Jensen 1960). As will be described below, these include several enzymes highly varied in their mode of action.

The number of halogenated compounds which were shown to be microbially dehalogenated far exceeds the number of such compounds actually supporting growth. In many cases dehalogenations were observed and the reaction products were identified, but there were no indications of further uses the microorganisms involved may make of those compounds. Dehalogenases, therefore, possess a wider specificity towards their substrates than the subsequent metabolic machinery needed to utilize the residual carbon skeleton. One of the possible reasons for this may be that for the longer alkanes, dehalogenation takes place on the outside of the cell membrane; the end product may still be too large to penetrate into the cell.

The environmental importance of the degradation of haloorganics, as well as other potential biotechnological applications of this process, have led to extensive studies of dehalogenation mechanisms. While aromatic chlorides were the major group investigated, substantial information concerning haloalkanes has also accumulated. Table 3 lists some of the relevant findings concerning intact cell and cell-free activities of various dehalogenases. In order to allow some generalizations to be made, only data from the better characterized systems, using pure bacterial strains only, were selected.

As can be seen from the data in Table 3, as well as from additional information not included therein, haloalkane dehalogenases may be generally grouped into the following three major categories:

1. Oxygenases, leading to the formation of the corresponding aldehyde. Further dehydrogenation yields the acid species (Yokota et al. 1986):

$$
\begin{aligned}
&{}^{1}/_{2}O_2 + \text{R-CH}_2\text{X} \xrightarrow[\text{oxygenase}]{} \text{R-CHOHX} \\
&\xrightarrow[\text{spontaneous}]{} \text{HX} + \text{R-CHO} \xrightarrow[\text{dehydrogenase}]{} \text{R-COOH}
\end{aligned}
\quad (1)
$$

This type of sequence was reported for the 1-chlorobutane grown *Corynebacterium* m15-3 (Yokota et al. 1986, 1987). Similar activities were described for the methane- and ammonia-monoxygenases (Oldenhuis et al. 1989; Rasche et al. 1990, respectively).

2. Halidohydrolases, substituting the halide with a hydroxyl:

$$
H_2O + \text{R-CH}_2\text{X} \xrightarrow[\text{hydrolase}]{} \text{R-CH}_2\text{OH} + \text{HX} \quad (2)
$$

This reaction, invariably leading to medium acidification, appears to be shared by many of the organisms in Table 2, although this does not neces-

Table 3. Microbial dehalogenations of haloalkanes.

Organism	Mode of enzyme activity	Induction[a]	Location[b]	Substrate[c]	Product	Growth[d]	Comments	Reference
Methylotroph DM1 (*Pseudomonas* sp.)	hydrolytic (?)	+	s	Dichloromethane	formaldehyde	+	monoxygenase not likely	Brunner et al. (1980)
Xanthobacter autotrophicus GJ10	hydrolytic	–	s (?)	bromoethane	ethanol	+	a different enzyme exists for alkanoic acids	Janssen et al. (1985) Kuening et al. (1985)
				1-chlorobutane	butanol	+		
				1,2 dichloroethane	2-chloroethanol	+		
				1,3 dichloropropane	3-chloropropanol	–	–	
				1,2 dibromoethane	2-bromoethanol			
Acinetobacter sp. GJ70	hydrolytic	?	?	1-chlorobutane	butanol	+		Janssen et al. (1987)
				1-iodopropane	propanol	+		
				1-chloropropane	?	–		
				1,2 dibromoethane	2-bromoethanol	–		
Hyphomicrobium sp. DM2	[e]	+	s	dichloromethane	formaldehyde	+		Kohler-Staub & Leisinger (1985) Stucki et al. (1981)
Methylosinus trichosporium OB3b	methane monooxygenase	+	s	trichloroethylene	trichloroethylene oxide (?)	+/–	induction by copper starvation; other C_1-C_3 haloalkanes at least partially degraded	Oldhenius et al. (1989)
Nitrosomonas europaea	ammonia monooxygenase	?	m	1,2dichloropropane	?	+/–	other C_1-C_3 fumigants degraded	Rasche et al. (1990)
				bromoethane	formaldehyde			
Arthrobacter sp. HA1	1) hydrolytic	+	s	C_2-C_{10} 1-chloroalkanes	alcohols[f]	+	other haloalkanes including many dihaloalkanes, also dehalogenated; induced by 1-chlorobutane	Scholtz et al. (1987a,b)
				C_2-C_9 1-bromoalkanes	alcohols[f]	+		
				C_1-C_7 1-iodoalkanes	alcohols[f]	+		
	2) hydrolytic	+	s	$C_{10,12,16}$ 1-chloroalkanes	prob. alcohols[f]	+ (slow)	induced by 1-bromohexadecane	Scholtz et al. (1988)
				$C_{10,12,14,16}$ 1-bromoalkanes	prob. alcohols[f]	+ (slow)		
	3) hydrolytic	+	s	1-bromoalkanes	?	+	unstable enzyme, induced by 1-chlorobutane	Scholtz et al. (1987b)
Corynebacterium m-15-3	1)hydrolytic	+	s	1-chlorobutane	n-butanol	+	anaerobic	Yokota et al. (1986, 1987)
	2) oxygenase	?	?	1-chlorobutane	butyric acid	+	aerobic	Yokota et al. (1986)
				1,3 dichloropropane	3-chloropropinic acid	?	aerobic	Yokota et al. (1986)
Methylotroph H-2	oxygenase	?	?	1,2 dichloroethane	2-chloroacetic acid			Yokota et al. (1986)

[a] (+), an inducible enzyme; (–), a constitutive enzyme; (?), data not available.
[b] (s), soluble; (m), membrane bound; (?), data not available.
[c] For some broad-spectrum enzymes of microorganisms, only representative substrates are presented.
[d] (+), growth substrate, (–), non-growth substrate; (+/–), cosubstrate.
[e] Glutathione-dependent nucleophilic substitution, leading to a non-enzymatic hydrolysis.
[f] When determined.

sarily imply a uniformity in enzyme structure and function. In fact, in several cases, the presence of more than one enzyme form has been reported within a single microbial species. In *Arthrobacter* HA1, for instance, the existence of four different halidohydrolases was postulated (Scholtz et al. 1988), three of them haloalkane dehalogenating. Selective induction of the different enzymes was shown to be controlled by the haloalkane serving as a carbon source for growth. This organism was also the first to be reported to hydrolyze the carbon-halogen bond in long-chain 1-haloalkanes. *Xantobacter autotrophicus* GJ10 is another organism containing more than one dehalogenase. In addition to the haloalkanes-specific enzyme mentioned in Table 2, it also contains an alkanoic acid-specific halidohydrolase (Janssen et al. 1985; Kuening et al. 1985). These two enzymes appear to have a non-overlapping activity spectrum. In all tested cases, the haloalkane halidohydrolases did not dehalogenate the haloacids, even those with the same number of carbon atoms.

When the halogen position on the alkane chain is not terminal, dehalogenation appears to be preceded by oxidation of the terminal methyl group to 2- or 3-haloaliphatic acids (Yokota et al. 1986). In the case of a halogen substituted in the second position, it is then removed from the halo-acid by a 2-haloacid-type dehalogenase (Motosugi et al. 1982):

$$\text{R-CHX-CH}_3 \longrightarrow \text{R-CHX-COOH} \xrightarrow{H_2O} \text{R-CHOH-COOH} + \text{HX} \quad (3)$$

For the 3-halogenated acid formed by oxidation of the 3-haloalkanes, halogen removal appears to be mediated by a membrane-bound enzyme system, such as the one involved in β-oxidation of fatty acids:

$$\text{R-CHX-CH}_2\text{-CH}_3 \rightarrow \text{R-CHX-CH}_2\text{-COOH} \rightarrow \text{R-CH}_2\text{-CH}_2\text{-COOH} + X^- \quad (4)$$

While all of these reactions are aerobic in nature, only those mediated by the oxygenase (see Eq. 1) appear to be obligately aerobic. The halidohydrolase reaction was shown to proceed also anaerobically, although sometimes at slower rates.

3. Reductases, leading directly to the formation of the corresponding alkane:

$$H_2 + RX \rightarrow RH + HX \quad (5)$$

This type of mechanism has been described for various haloaromatic compounds (Kobayashi & Rittman 1982; Lal & Saxena 1982; Reineke & Knackmuss 1988); to my knowledge, however, it has not been described for haloalkanes except for an unconfirmed report by Lang et al. (Ann. Meet. Am. Water Works Assoc. 1981). There are, however, some indications of reductive dehalogenations of C1 and C2 halogenated aliphatics under methanogenic and nitrifying conditions (Bouwer & McCarty 1983a, 1983b). The actual metabolic pathways in these two cases have not been resolved.

Finally, a dehalogenation reaction which deserves special mention is that reported for *Hyphomicrobium* DM-2 (Kohler-Staub & Leisinger 1985; Stucki et al. 1981). Dichloromethane dehalogenation by this organism was demonstrated to be a glutathione-dependent nucleophilic substitution. It was proposed that an S-chloromethyl glutathione conjugate is formed enzymatically, which is then hydrolyzed non-enzymatically to an unstable 2-hydroxymethyl glutathione; the latter compound decomposes to formaldehyde, thus regenerating reduced glutathione.

Genetics of haloalkane degradation

As pointed out in the introduction, most of the research into the biodegradation of halogenated organics has been focused on aromatic chlorides. It is not surprising, therefore, that available information concerning the genetics of haloalkane degradation is scant. This is in sharp contrast to the constantly increasing genetic information regarding the degradation of halogenated aromatics (see Chaudhry & Chapalamadugu 1991, for a recent review). In fact, in all the literature cited to this point, including the articles from which the data for the compilation of Tables 2 and 3 were derived, the subject is hardly mentioned. It is even unclear, for instance, whether the genetic information for dehalogenation and degradation of the haloalkanes mentioned in these tables is plasmid- or chromosome- encoded. In the *Pseudomonas* strain described later in this review (ES-2), which is capable of degrading a wide spectrum of haloalkanes, repeated attempts to show the presence of plasmids have failed.

In contrast to haloalkanes, several reports describe plasmids involved in the degradation of haloalkanoic acids. Hardman et al. (1986) have examined four *Pseudomonas* and two *Alcaligenes* species, capable of growth on 2-chloropropionic acid and chloroacetic acid. All their isolates contained a single plasmid (pUU204). When the *Pseudomonas* isolates were cured of their plasmids, dehalogenase activities were lost. However, attempts to reintroduce the plasmid to the cured original strains or to new hosts have failed. Two different plasmids, involved in the degradation of haloacetate, pU01 and pU011, were isolated by Kawasaki et al. (1981) from *Moraxella* strain B. Both plasmids also code for mercury resistance, but no further genetic information is available.

It should be remembered that, as mentioned above, in all cases examined the dehalogenase acting on haloacids was different from that involved in dehalogenating haloalkanes. The information presented above, in addition to being very preliminary, is thus also only partially relevant to haloalkane degradation. It was included in this review since, to my knowledge, this is the only genetic information concerning the biodegradation of halogenated aliphatics in general.

Recent advances

We have recently studied several bacterial isolates from an experimental reactor system for the treatment of chemical industrial wastewater, rich in haloorganics, including many brominated com-

Table 4. Degradation[a] of nonsubstituted and substituted alkanes by strain ES-2.

Chain length (No of C atoms)	Non-substituted	1-chloro	1-bromo	1-iodo	1-fluoro	1-alcohol	1-acid
1						–	–
2						+	+
3			+/–[b]			+	+
4		–	+/–[b]			+	+
6	–	–	+		–	–	–
7			+				
8	–	–	+	+		+	+
9			+		–		
10	–	+	+			+	+
11							+
12			+				+
13			+				
14			+				
16	–	+	+			+	+
17							+
18	–	+	+			+	+
20							–
22							–
26							–

[a] Utilization of the substrate as a single carbon and energy source.
[b] Dehalogenation but no growth.

pounds. One of these isolates, a gram negative motile rod named ES-2 (Shochat et al., in preparation) is probably a *Pseudomonas* sp., and will be briefly described here. Under optimal growth conditions, with 1-bromooctane as a sole carbon and energy source, its generation time is 113 min. Its dehalogenase activity is inducible, and it is capable of growth on a variety of haloalkanes.

Biodegradation and dehalogenation

The biodegradation spectrum of this strain appears to be wider than any previously described for a haloalkane degrading organism (Table 4), except possibly *Arthrobacter* HA1 (Scholtz et al. 1987a, 1987b, 1988). Originally isolated on 1-bromooctane as a sole carbon and energy source, ES-2 is capable of growth on all the 1-bromoalkanes tested in the C_6–C_{18} range. In addition, it can debrominate 1-bromopropane and 1-bromobutane, but it could not utilize the latter two compounds for growth. It grows also on 1-chloroalkanes, though only at the C_{10}–C_{18} range. C_2–C_{18} fatty acids and alcohols also support growth, with the exception of the C_6 species - caproic acid and hexanol. It grew on a variety of sugars, tricarboxylic and amino acids (not shown), but *not* on the non-substituted alkanes (C_6–C_{18}), chloroacetate and aromatics (benzene, phenol, bromophenol) tested.

One of the interesting observations regarding the ES-2 degradation spectrum (Table 4) is that the organism exhibited a wider specificity for bromides than for chlorides. It dehalogenated C_3 to C_{18} bromoalkanes, utilizing all of them but bromopropane and bromobutane for growth, while only chloroalkanes of 10 carbon atoms or longer were metabolized. The dehalogenase, acting on bromopropane through bromooctane, was incapable of removing the chloride from the parallel chloroalkanes. For alkanes C_{10} and longer, both halogens were removed; but it is unclear whether this latter activity was mediated by the same enzyme dehalogenating the shorter bromoalkanes. It is difficult to envisage how a dehalogenase would lose its bromide-specificity with longer substrates; it is therefore likely that there are at least two dehalogenases involved. There are several other indications for the presence of more than one dehalogenase in ES-2 cells. One of them is the puzzling observation that ES-2 can utilize bromooctane, bromohexane, and octanol for growth, but not hexanol. The first identified product of bromooctane degradation was octanol (Table 5), indicating that dehalogenation is carried out hydrolytically. If the same enzyme would act on bromohexane, the hexanol formed would not be able to support growth. Thus, bromohexane (and possibly lower bromoalkanes) is probably dehalogenated by a different enzyme. Nevertheless, hexanol is produced from 1-bromohexane by resting cells of ES-2 (Table 5). However, unlike the octanol produced from bromooctane, which is found only inside the cells, the hexanol produced from bromohexane is detectable only in the medium. It

Table 5. ES-2 degradation products of some haloalkanes.

Substrate	Growth[a]	Location[b]	Product[c]	Growth[d]	Location[b]
1-Bromobutane	–	EC	?	–[e]	EC
1-Bromohexane	+	EC	1-hexanol	–	EC
1-Bromooctane	+	EC	1-octanol	+	IC
1-Bromodecane	+	EC	1-decanol	+	IC
1-Chlorodecane	+	EC	1-decanol	+	IC

[a] Use of substrate as a single carbon and energy source for growth.
[b] Cells were separated from the medium and disrupted by vigorous shaking with glass beads. Both medium and the cleared supernatant of the broken cells were extracted with ethyl acetate, and assayed by gas chromatography. EC – extracellular; IC – intracellular.
[c] Identified by gas chromatography, using an SPB-1 column and a flame ionization detector.
[d] Known use of product as a single carbon and energy source for growth.
[e] Since ES-2 is incapable of growth on 1-bromobutane, the same is likely to be true for its yet unidentified degradation product.

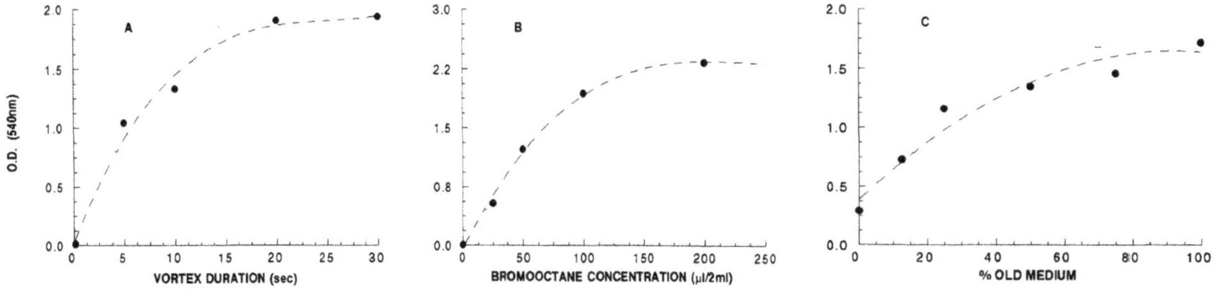

Fig. 1. A rapid method for determination of emulsification activity. A 24-hour cell-free growth medium was used. (A), 2 ml medium and 100 μl 1-bromooctane vortexed for different durations; (B), different concentrations of 1-bromooctane in 2 ml medium, vortexed for 30 sec; (C), different ratios of 24-hour to fresh medium (2 ml total volume), vortexed for 30 sec with 100 μl 1-bromooctane. In all cases, optical density at 540 nm was determined immediately following the mixing.

is therefore unclear at present how bromohexane is metabolized and utilized for growth. A degradation product of 1 bromobutane was also found extracellularly; it has not been identified yet, but it was shown not to be butanol (Table 5).

Emulsification

One of the reasons contributing to the fact that most biodegradation studies involving haloalkanes have been restricted to short chain compounds may be attributed to the insolubility in water of most longer haloalkanes. However, even the reports on the dehalogenation or degradation of clearly insoluble halogenated aliphatics generally do not mention this obviously significant factor. For instance, Omori & Alexander (1978a) in their pioneering work on haloalkane degradation mention that the substrate, 1,9-dichlorononane, was separated from the medium by centrifugation; however, no mention is made of the obvious accessibility problems of the substrate to the bacteria. Similarly, Janssen et al. (1987) mention that some substrates were poorly soluble, thus explaining low utilization efficiency. Conversely, Scholtz et al. (1987b) presumed that the rate of dissolution of the sparingly soluble chloroalkanes (C_6 and longer) was higher than the rate of hydrolysis, and therefore not rate limiting for dehalogenation by the cell-free enzyme of Arthrobacter HA1. Such an assumption, however, cannot be justified for intact cells. In all cases where water-insoluble haloalkanes were partially or fully utilized (see Tables 2 and 3), a solubilization step must have taken place before any metabolic activity.

Emulsification of insoluble organic compounds has been investigated in detail in the past (see Rosenberg 1986, for review), mostly with regard to petroleum hydrocarbons. Many microorganisms produce surface active materials, ranging in structure from low molecular weight glycolipids, via polyanionic heteropolysaccharides containing covalently linked hydrophobic side chains, to polysaccharide-protein complexes. In most cases, the biosurfactants are extracellular; they were shown to be involved in cell adhesion, emulsification, dispersion, flocculation, and aggregation phenomena (Rosenberg 1986). All of these activities are significant for the physiology and ecology of the biosurfactant-producing strains; in addition, they present the potential for various industrial applications. Both of these aspects – the natural and the industrial – are manifested in the case of the emulsifiers produced by haloalkane-degrading microorganisms. For both reasons, therefore, more effort should be directed towards their research.

During growth, strain ES-2 produces an extracellular surface-active compound which leads to emulsification of the substrate. When bromooctane is added to a culture of this bacterium, it initially rests as a single large droplet at the bottom of the incubation vessel. After agitation and growth, at least part of it is transformed into an emulsion of fine droplets. Excretion of the emulsifying agent is observed immediately upon inoc-

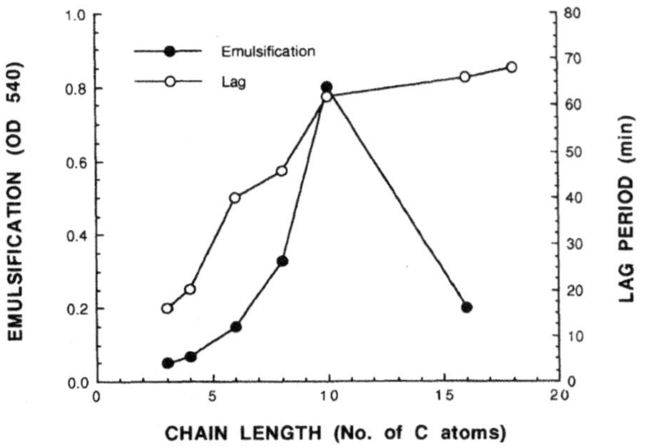

Fig. 2. Emulsification of 1-bromoalkanes by strain ES-2. (•), emulsifying activity determined with 2 ml of a 24-hour medium of a 1-bromooctane grown culture and 100 μl of the tested 1-bromoalkane; (○), length of lag period preceding debromination in a concentrated cell suspension, with 20 mM of the tested 1-bromoalkane.

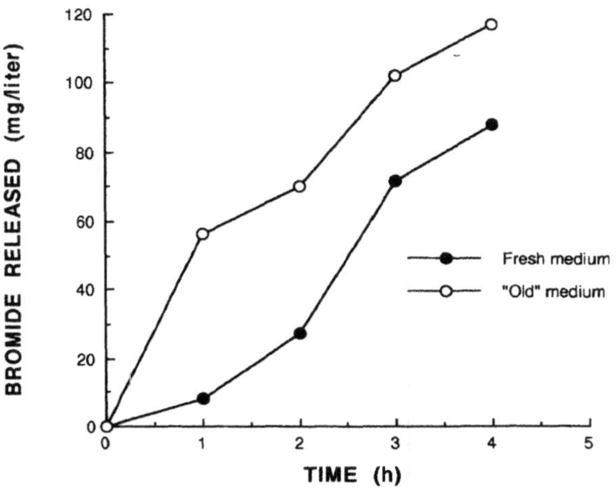

Fig. 3. Importance of emulsifier for bromooctane dehalogenation. Late exponential phase cells were harvested by centrifugation, washed with fresh medium, and resuspended in either fresh (•) or "old" (○) medium. 1-bromooctane (20 mM) was added, and bromide release was followed for 4 hours.

ulation, and proceeds throughout the lag and exponential growth phases. Emulsifier production was observed irrespectively of the solubility of the growth substrate. For example, emulsification activities determined in a growth medium of acetate-grown or bromooctane-grown cells were similar. This may mean that the emulsifier also plays a yet undetermined role beyond accessing water-insoluble compounds.

The chemical nature of the ES-2 emulsifier has not been determined. Either due to a different mode of action or to a relatively low potency, its activity could not be measured by the methods described by Rosenberg et al. (1979) and by Sar & Rosenberg (1983, 1989). A different methodology was therefore developed which allowed a rapid assessment of relative emulsification activity, as described in Fig. 1. The method involved a 30-sec intensive mixing (by vortex) of a 2-ml sample with 100 μl of substrate, and an immediate determination of the optical density at 540 nm. Using this approach, the specificity of the ES-2 emulsifier to various bromoalkanes was determined. The results are shown in Fig. 2, along with another parameter affected by carbon chain length – the length of the lag period preceding bromide release. While maximal emulsification values were obtained for 1-bromodecane, the length of the lag period preceding bromide release by resting cells increased with increasing chain length; up to C_{10}, at least, the correlation to emulsifier activity was actually negative. Debromination, therefore, is apparently also restricted by other parameters, possibly steric in nature, on top of the insolubility of the substrate. However, as demonstrated in Fig. 3, the presence of the emulsifier was essential for dehalogenation. When cells were suspended in a 24-h old medium supplemented with fresh bromooctane, bromide release proceeded from time zero, as compared to nearly a 1-h lag in fresh medium.

Location of bromooctane degradation

As mentioned above, extracellular emulsification was the first step in bromooctane utilization. The following stage, dehalogenation, appears to take place in the periplasmic space. This assumption is based on the observation that bromooctane could not be found intracellularly, whereas the product, octanol, was not found in the medium (Table 5). The subsequent degradation of octanol, possibly by dehydrogenations to octyl aldehyde and octanoic acid, followed by β-oxidation, is thus believed to be intracellular.

Utilization of 1-bromooctane is thus a three-step process, taking place in separate locales:

1) extracellular emulsification by a constitutively excreted emulsifier
2) periplasmic dehalogenation by an inducible halidohydrolase
3) intracellular degradation

It remains to be determined how universal this sequence of events is for other substrates.

Concluding remarks

Two main conclusions may be drawn from the data presented in this review. The first is that, contrary to earlier beliefs, many haloalkanes are biodegradable, either partially or fully, by a variety of microorganisms. The second conclusion is that the study of this industrially, agriculturally and – primarily – environmentally important group of organic chemicals is still only in the preliminary stages. While an appreciable amount of data has accumulated concerning their environmental distribution, available information regarding degradative biochemical pathways is lacking, and that concerning genetic control is practically nonexistent. Thus, based upon these two conclusions, it may be stated that in many cases, environmental hazards involving haloalkanes may be potentially biologically solved or at least neutralized. On the other hand, further work is needed in order to allow a better understanding of the processes involved, and hence facilitate the implementation of such measures.

Acknowledgements

The project was supported by a grant from the Ramat Hovav industries. Special thanks for their cooperation are conveyed to the management and staff of Bromine Compounds, Ltd, Machteshim Chemical Works, Ltd, and the Ramat Hovav Local Industrial Council. The original data reported here form part of the M.Sc. thesis of Eyal Shochat.

References

Alexander M (1981) Biodegradation of chemicals of environmental concern. Science 211: 132–38

Alexander M (1985) Biodegradation of organic chemicals. Environ. Sci. Technol. 19: 106–111

Baldauf G (1981) Der Fall Grenzach – Beispiel einer Grundwasserverschmutzung mit umweltrelevanten Stoffen. In: Halogenkohlenwasserstoffe in Grundwassern. DVGW-Schriftenreihe. Wasser Nr. 29, Frankfurt.

Bouwer EJ & McCarty PL (1983a) Transformations of 1- and 2-carbon halogenated aliphatic organic compounds under methanogenic conditions. Appl. Environ. Microbiol. 4: 1286–1294

Bouwer EJ & McCarty PL (1983b) Transformations of halogenated organic compounds under denitrification conditions. Appl. Environ. Microbiol. 45: 1295–1299

Brunner W, Staub D & Leisinger T (1980) Bacterial degradation of dichloromethane. Appl. Environ. Microbiol. 40: 950–958

Castro CE & Belser NO (1968) Biodehalogenation. Reductive dehalogenation of the biocides ethylene dibromide, 1,2-dibromo-3-chloropropane, and 2,3-dibromobutane in soil. Environ. Sci. Technol. 2: 779–783

Chaudhry GR & Chapalamadugu S (1991) Biodegradation of halogenated organic compounds. Microbiol. Rev. 55: 59–79

Dalton H & Stirling DI (1982) Co-metabolism. Phil. Trans. R. Soc. Lond. B. 297: 481–496

Edwards PR, Campbell I & Milne GS (1982) The impact of chloromethanes on the environment. Part 2: methyl chloride and methylene chloride. (pp 619–622) Chem. Ind. Lond.

Ghosal D, You I-S, Chatterjee DK& Chakrabarty AM (1985) Microbial degradation of halogenated compounds. Science 228: 135–142

Gschwend PM, MacFarlane JK & Newman KA (1985) Volatile halogenated organic compounds released to seawater from temperate marine microalgae. Science 227: 1033–1035

Hardman DJ, Gowland PC & Slaater JH (1986)) Plasmids from soil bacteria enriched on halogenated alkanoic acids. Appl. Environ. Microbiol. 51: 44–51

f09Hartmans S, Schmuckle A, Cook AM & Leisinger T (1986) Methyl chloride: naturally occurring toxicant and C-1 growth substrate. J. Gen Microbiol. 132: 1139–1142

Janssen DB, Scheper A, Dijkhuizen L & Witholt B (1985) Degradation of halogenatedc aliphatic compounds by*Xanthobacter autotrophicus* GJ10. Appl. Enviro. Microbiol. 49: 673–677

Janssen DB, Jager D & Witholt B (1987) Degradation of n-haloalkanes and alpha, omega-dihaloalkanes by wild-type and mutants of *Acinetobacter* sp. strain GJ70. Appl. Environ. Microbiol. 539: 561–566

Jensen HL (1960) Decomposition of chloroacetate by bacteria. Acta. Agric. Scand. 10: 83–103

Kawasaki H, Yahara H & Tonomura K (1981) Isolation and characterization of plasmid PU-01 mediating dehalogenation of haloacetate and mercury resistance in *Morasella* sp. Agric Biol. Chem. 45: 1477–1482

Keuning S, Janssen DB & Witholt B (1985) Purification and characterization of hydrolytic haloalkane dehalogenase from *Xanthobacter autotrophicus* GJ10. J. Bacteriol. 163: 635–639

Knackmuss H-J (1981) Degradation of halogenated and sulfonated hydrocarbons. In: Leisinger, T, Cook, AM, Hutter R & Neusch J (Eds) Microbial Degradation of Xenobiotics and Recalcitrant Compounds (pp 189–212). Academic Press, London

Kobayashi H & Rittmann BE (1982) Microbial removal of hazardous organic compounds. Environ. Sci. Technol. 16: 170A–183A

Kohler-Stabu D & Leisinger T (1985) Dichloromethane dehalogenase of *Hyphomicrobium* sp. strain DM2. Bacteriol. 162: 676–681

Lal R & Saxena DM (1982) Accumulation, metabolism, and effects of organochlorine insecticides on microorganisms. Microbiol. Rev. 46: 95–127

Leisinger T (1983) Microorganisms and xenobiotic compounds. Experientia 39: 1183–1191

Little CD, Palumbo AV, Herbes SE, Lidstrom ME, Tyndall RL & Gilmer PJ (1988) Trichloroethylene biodegradation by a methane-oxidizing bacterium. Appl. Environ. Microbiol. 54: 951–956

Lovelock JE (1975) Natural halocarbons in the air and in the sea. Nature 256: 193–194

Morgan P & Watkinson RJ (1989) Microbiological methods for the cleanup of soil and ground water contaminated with halogenated organic compounds. FEMS Microbiol. Rev. 63: 277–300

Motosugi K & Soda K (1983) Microbial degradation of synthetic organochlorine compounds. Experientia 39: 1214–1220

Motosugi K, Esaki N & Soda K (1982) Purification and properties of a new enzyme, DL-2-haloacid dehalogenase, from *Pseudomonas* sp. J. Bacteriol. 150: 522–527

Murphy GL & Perry JJ (1983) Incorporation of chlorinated alkanes into fatty acids of hydrocarbon-utilizing mycobacteria. J. Bacteriol. 156: 1158–1164

Oldenhuis R, Vink RLJM, Janssen DB & Witholt B (1989) Degradation of chlorinated aliphatic hydrocarbons by *Methylosinus trichosporium* OB3b expressing soluble methane monooxygenase. Appl. Environm. Microbiol. 55: 2819–2826

Omori T & Alexander M (1978a) Bacterial and spontaneous dehalogenation of organic compounds. Appl. Environ. Microbiol. 35: 512–516

Omori T & Alexander M (1978b) Bacterial dehalogenation of halogenated alkanes and fatty acids. Appl. Environ. Microbiol. 35: 867–871

Omori T, Kimura T & Kodama T (1987) Bacterial cometabolic degradation of chlorinated paraffins. Appl. Microbiol. Biotechnol. 25: 553–557

Pearson CR (1982) C_1 and C_2 halocarbons. In: Hutzinger O (Ed) The Handbook of Environmental Chemistry, Vol 3, Part B (pp 69–88). Springer-Verlag, Berlin, Heidelberg, New York

Perry JJ (1979) Microbial cooxidations involving hydrocarbons. Microbiol. Rev. 43: 59–72

Pignatello JJ (1986) Ethylene dibromide mineralization in soils under aerobic conditions. Appl. Environ. Microbiol. 51: 588–592

Rasche ME, Hyman MR & Arp DJ (1990) Biodegradation of halogenated hydrocarbon fumigants by nitrifying bacteria. Appl. Environ. Microbiol. 56: 2568–2571

Reineke W & Knackmuss H-J (1988) Microbial degradation of haloaromatics. Annu. Rev. Microbiol. 42: 263–287

Rosenberg E (1986) Microbial surfactants. CRC Crit. Revs. Biotechnol. 3: 109–132

Rosenberg E, Zuckerberg A, Rubinovitz C & Gutnick DL (1979) Emulsifier of *Arthrobacter* RAG-1: isolation and emulsifying properties. Appl. Environ. Microbiol. 37: 402–408

Sar N & Rosenberg E (1983) Emulsifier production by *Acinetobacter calcoaceticus* strains. Curr. Microbiol. 9: 309–314

Sar N & Rosenberg E (1989) Colonial differentiation and hydrophobicity of a *Vibrio* sp. Curr. Microbiol. 18: 331–334

Scholtz R, Scmuckle A, Cook AM & Leisinger T (1987a) Degradation of eighteen 1-monohaloalkanes by *Arthrobacter* sp. strain HA1. J. Gen. Microbiol. 133: 267–273

Scholtz R, Leisinger T, Suter F & Cook AM (1987b) Characterization of 1-chlorohexane halidohydrolase,a dehalogenase of wide substrate range from an *Arthrobacter* sp. J. Bacteriol. 169: 5016–5021

Scholtz R, Messi F, Leisinger T & Cook AM (1988) Three dehalogenases and physiological restraints in the biodegradation of haloalkanes by *Arthrobacter* sp. strain HA1. Appl. Environm. Microbiol. 54: 3034–3038

Slater JH & Bull AT (1982) Environmental microbiology: biodegradation. Phil. Trans. R. Soc. Lond. B. 297: 575–597

Slater H, Lovatt D, Weightman AJ, Senior E & Bull AT (1979) The growth of *Pseudomonas putida* on chlorinated aliphatic acids and its dehalogenase activity. J. Gen. Microbiol. 114: 125–136

Stucki G, Galli R, Ebersold H-R & Leisinger T (1981) Dehalogenation of dichloromethane by cell extracts of *Hyophomicrobium* DM2. Arch. Microbiol. 130: 366–371

Stucki G, Krebser U & Leisinger T. (1983) Bacterial growth on 1,2-dichloroethane. Experientia 39: 1271–1273

Swindoll CM, Aelion CM & Pfaender FK (1988) Influence of inorganic and organic nutrients on aerobic biodegradation and on the adaptation response of subsurface microbial communities. Appl. Environ. Microbiol. 54: 212–217

Tsang JSH, Sallis PJ, Bull AT & Hardman DJ (1988) A monobromoacetate dehalogenase from *Pseudomonas cepacia* MBA4. Arch. Microbiol. 150: 441–446

Vandenbergh PA & Kunka BS (1988) Metabolism of volatile chlorinated aliphatic hydrocarbons by *Pseudomonas fluorescens*. Appl. Environ. Microbiol. 54: 2578–2579

Veissmann M & Hammer MJ (1985) Water Supply and Pollution Control, 4th edition. Harper & Row, New York.

Vogel TM & McCarty PL (1987) Abiotic and biotic transformations of 1,1,1-trichloroethane under methanogenic conditions. Environ. Sci. Technol. 21: 1208–1213

Vogel TM, Criddle CS & McCarty PL (1987) Transformations of halogenated aliphatic compounds. Environ. Sci. Technol. 21: 722–736

Werner P (1989) Experiences in the use of microorganisms in soil and aquifer decontamination. In: Kobus & Kinzelbach (Eds) Contaminant Transport in Groundwater (pp 59–63). Balkema, Rotterdam

Yokota T, Fuse H, Omori T & Minoda Y (1986) Microbial dehalogenation of haloalkanes mediated by oxygenase or halohydrolase. Agric. Biol. Chem. 5: 453–460

Yokota T, Omori T & Kodam T (1987) Purification and properties of haloalkane dehalogenase from *Corynebacterium* sp. strain m15-3. J. Bacteriol. 169: 4049–4054

Biodegradation 3: 315–335, 1992.

Oil spill bioremediation: experiences, lessons and results from the *Exxon Valdez* oil spill in Alaska

P.H. Pritchard,[1] J.G. Mueller,[2] J.C. Rogers,[3] F.V. Kremer[4] & J.A. Glaser[5]
[1] *Environmental Protection Agency, Environmental Research Laboratory, Gulf Breeze, FL 32561-5299, USA;* [2] *SBP Technologies, Inc., EPA Environmental Research Laboratory, Gulf Breeze, FL 32561-5299, USA;* [3] *Environmental Protection Agency, Environmental Research Laboratory, Athens, GA 30613, USA;* [4] *Environmental Protection Agency, Center for Environmental Research Information, Cincinnati, OH 45268, USA;* [5] *Environmental Protection Agency, Risk Reduction Engineering Laboratory, Cincinnati, OH 45268, USA*

Key words: oil mineralization, hydrocarbon degrading microorganisms, oleophilic fertilizers, n-C18/phytane ratios, aromatic hydrocarbons, oil biodegradation rates

Abstract

The use of bioremediation as a supplemental cleanup technology in the *Exxon Valdez* oil spill, in Prince William Sound, Alaska, has proven to be a good example of the problems and successes associated with the practical application of this technology. Field studies conducted by scientists from the U.S. Environmental Protection Agency have demonstrated that oil degradation by indigenous microflora on the beaches of Prince William Sound was accelerated by adding fertilizer directly to the surfaces of oil-contaminated beaches. Although several types of fertilizers were used in the studies, only the results from the application of an oleophilic fertilizer are presented. The fertilizer enhanced biodegradation of the oil, as measured by changes in hydrocarbon composition and bulk oil weight per unit of beach material, by approximately two-fold relative to untreated controls. Laboratory studies verified the usefulness of the oleophilic fertilizer as a nutrient source, but the contribution of its oleophilic components towards enhancing biodegradation is still unclear.

These studies supported bioremediation as a useful cleanup strategy that was subsequently used by Exxon on a large scale. The *Exxon Valdez* experience has also provided a number of informative lessons that have significant relevance to future oil bioremediation efforts. This paper discusses these lessons and the difficulties in assessing the effectiveness of bioremediation in the field.

Introduction

The potential usefulness of bioremediation to clean soils, sediments and aquifer materials contaminated with oil and petroleum hydrocarbons has been recognized for some time (Morgan & Watinkson 1989; Lee & Levy 1989; Bartha 1986). The recognition of oil as a complex but largely biodegradable mixture of hydrocarbons and the knowledge that hydrocarbon-degraders can be enriched in many, if not most, types of environments (Lee & Levy 1987; Atlas 1981), has contributed greatly to the development of oil bioremediation technologies. In addition, bioremediation is rapidly gaining acceptance as a viable technology and if used prudently, can provide efficient, inexpensive and environmentally safe cleanup of waste chemicals.

The feasibility of bioremediation as a supplemental cleanup tool in the *Exxon Valdez* oil spill, in Prince William Sound, Alaska was timely and scientifically reasonable (Pritchard & Costa 1991). Analysis of data collected from subsequent field studies to determine if oil degradation by indigenous microflora on the beaches of Prince William Sound could be significantly accelerated by fertilizer application (Pritchard et al. 1991) suggested a new and workable remedial action.

A number of useful lessons were learned that will have significant meaning to future oil bioremediation efforts. This paper summarizes some of these lessons and highlights several of the difficulties that arose during the use of bioremediation in Alaska. Results from the application of an oleophilic fertilizer as part of a field demonstration project conducted at Snug Harbor on Knight Island in Prince William Sound will be the primary focus of discussion. We hope this information will stimulate others to carefully consider the process by which bioremediation success is measured in the field and will further emphasize the need to continually expand our research data base, despite the quantity of background information that may already exist.

Background

Any bioremediation effort must invariably involve a scientifically valid demonstration of process effectiveness and environmental safety. Effectiveness, in the case of oil bioremediation, means establishing that: a) the removal or disappearance of the oil is primarily attributable to biodegradation and not other processes, and b) enhanced biodegradation rates of oil are sufficiently faster (with statistical verification, if possible) than natural rates to justify expenditure of effort to implement the bioremediation process on a large scale. Although environmental safety issues are not addressed in this paper, they too require considerable effort to verify that no adverse ecological effects occurred as a result of the fertilizer application.

For the *Exxon Valdez* oil spill, both aspects were crucial to the eventual acceptance of bioremediation by the public and state and federal regulatory agencies. While confidence provided by almost 20 years of accumulated research data on oil biodegradation laid the foundations for our approach in Alaska, the complexity of the filed application and the problems associated with demonstrating successful bioremediation on a large scale were substantial. These aspects are discussed below. Remember that the effectiveness issues addressed in Prince William Sound, where oil-contaminated gravel and cobblestone beaches were treated, will not be the same as those addressed for oil-contaminated sandy beaches, marshes and wetlands.

Enrichments of oil degrading microbial communities

Despite the cold water temperatures of Alaska, significant enrichment of oil degrading microorganisms was expected in the beach material following exposure to the oil. Research by Atlas and his colleagues supported this assumption (Atlas 1981). Approximately two months after the spill, concentrations of oil degraders averaged 10^6 per gram of oiled beach material, representing as much as a 10,000-fold increase in the number of oil degrading microorganisms relative to uncontaminated beaches (Table 1). Studies at a later date by Lindstrom et al. (1991) have showed similar differences. Enrichments of this magnitude suggested the oil was being degraded (previous laboratory studies have demonstrated that Prudhoe Bay crude oil is readily biodegradable), that nitrogen and phosphorus were available to support the growth of the hydrocarbon degraders, and that the cold temperatures (10–16 °C) were probably not overly restrictive to the degradative capabilities of the indigenous microflora. This high biomass, a great excess of degradable organic carbon from the oil, and a finite supply of nitrogen in the water, suggested the potential of nitrogen-limited biodegradation and also opened the possibility of accelerating oil biodegradation by overcoming this limitation.

In this context, it was unnecessary to experimentally reverify the stimulatory effect of nitrogen on oil biodegradation in the laboratory, even using oil-contaminated beach samples from Prince William Sound, since numerous published reports in the literature supported this as a sensible approach. Instead, because of the importance and magnitude of the problem, it was reasonable to go directly to the field and conduct a practical demonstration. Thus, in the very early stages of the oil spill cleanup program in Alaska, the simple measurement of the number of oil degrading microorganisms, along with a wealth of relevant supporting literature, provided an initial indicator of the feasibility of oil bioremediation in this situation.

Other indicators of oil degradation activity

There is a strong tendency, by most decision-making officials involved in oil spill cleanup, to generate 'site-specific' information about bioremediation potential; information from other sites or experiences is often not acceptable or applicable. Circumstances in other spills may dictate that information on the enrichment of oil degraders in the contaminated areas alone is not sufficient for initial decision-making on bioremediation. A further discussion of these indicators, even though they were not used initially in Alaska, is therefore relevant.

For example, mineralization studies involving measurements of total CO_2 production can provide excellent initial information. The approach provides rapid, relatively unequivocal, time-course data suitable for testing different biological treatment options (e.g., effects of adding nitrogen). If natural oil degradation is occurring in contaminated beach material, then considerable amounts of CO_2 should be produced from oil mineralization compared to a control containing uncontaminated beach material. Likewise, nitrogen or other additives should readily stimulate the biodegradation rate in oil contaminated beach material. Biometer flask systems (Bartha & Pramer 1965), designed to trap CO_2 in side-arms containing an alkaline solution (Fig. 1) can be adapted to measure oil mineralization rates from contaminated beach material (Mueller et al. 1992). The procedure entails placing oil contaminated beach material and its associated microbial community (mixed sand and gravel in the case of Prince William Sound) directly in the flasks and simulating tidal exchange by flushing fresh seawater in and out (Mueller et al. 1992). An example of the data generated from such a mineralization test is shown in Fig. 2. This experiment was performed with oil-contaminated Prince William Sound beach material taken after our bioremediation field demonstration had begun. The biometer flask method is indicative of the short term site-specific tests that can be initially conducted as a data gathering exercise to develop cleanup strategies for a particular oil spill. Note the enhancing effect of adding nitrogen fertilizer (Fig. 2). By comparing rates of CO_2 production, an estimation of the extent of enhancement in oil biodegradation can be obtained. The system can be easily adapted to test beach material from other oil spill sites.

Table 1. Relative concentration[a] (Log_{10} of the cell numbers/g of beach material and standard deviation) of oil-degrading microorganisms in samples from beaches exposed and unexposed to oil.

Unexposed sites	High tide	Mid tide	Low tide
Tatitlek	2.41 (±0.58)	4.31 (±1.14)	6.11 (±2.05)
Snug Corner Cove	2.31 (±0.54)	2.51 (±0.55)	<2.1
Hell's Hole	<2.1	2.51 (±0.89)	<2.1
Commander Cove	4.51 (±1.14)	<2.1	3.11 (±0.45)
Oil exposed site	**Eagle Beach**	**Otter Beach[b]**	**Seal Beach[b]**
Snug Harbor[c]	5.18 (±0.91)	5.75 (±1.29)	5.65 (±1.04)

[a] Based on most probable number determination.
[b] Prior to fertilizer application.
[c] Mid tide zone.

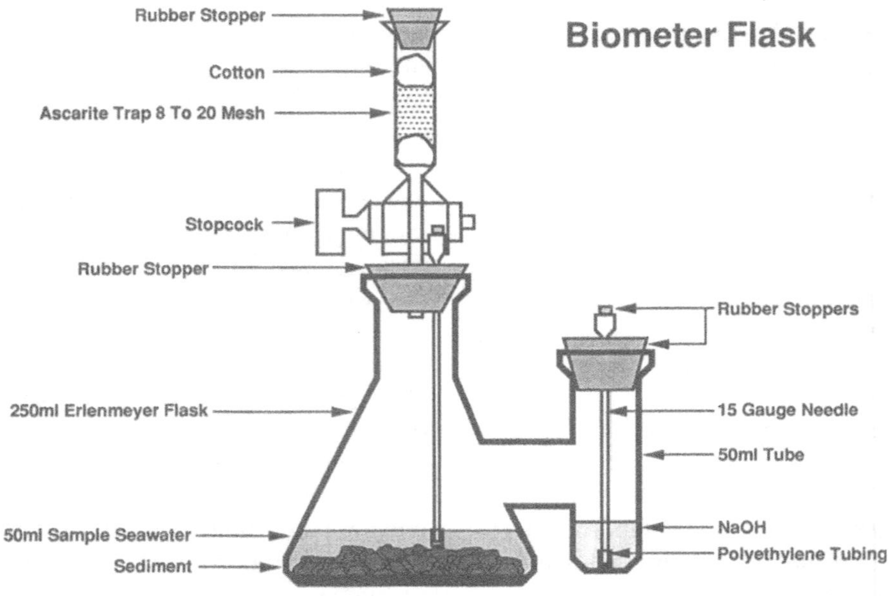

Fig. 1. Modified biometer flasks used to measure mineralization of oil from contaminated beach samples taken directly from the field.

Since any other forms of organic matter in the beach material can also produce CO_2, care must be taken to assure that one is measuring only oil mineralization. In addition to running control flasks with uncontaminated beach material as indicated above (with and without added nitrogenous nutrients) and ensuring that mineralization in the presence of the oil is considerably above background, one can also add a radiolabelled hydrocarbon. If oil is actively degraded (with or without nitrogenous nutrients added), the production of radiolabelled CO_2 should be extensive and immediate. Results have demonstrated that phenanthrene worked well as an indicator because it was rapidly and completely sorbed to the oil (Mueller et al. 1992; Pritchard et al. 1991). Since the tidal cycle washed out any residual aqueous-phase phenanthrene, we assume that CO_2 production was mostly from the bacterial communities associated with oiled surfaces.

Oil spill bioremediation as a finishing step

We assumed that bioremediation, for more heavily contaminated areas, would be most effective as a finishing step, following some physical measure to remove bulk oil. Clearly, the time required for cleanup depends on the mass of oil present; since oil degradation occurs primarily at the water interface, the more oil, the longer it will take to be degraded. In Alaska, where the treatment window for active oil biodegradation was limited to the summer months when highest water temperatures occurred, the time required for successful bioremediation was particularly critical. In other oil spill situations, more time may be available for cleanup and therefore, consideration of oil bioremediation as a primary cleanup method may be reasonable. Greater oil mass, however, will undoubtedly require some procedure for maximizing the interaction between the oil degrading microbial communities and the oil, such as by mechanical tilling.

The physical washing procedure employed by Exxon in the Alaskan spill removed bulk oil, but the beach material remained quite contaminated, aesthetically unpleasing and the oil residues ecologically available. As there were very few follow up treatment alternatives, bioremediation was appealing. However, without the initial removal of this bulk oil, bioremediation may not have been practical in Alaska because even if oil biodegradation was quite active, there was only a limited time

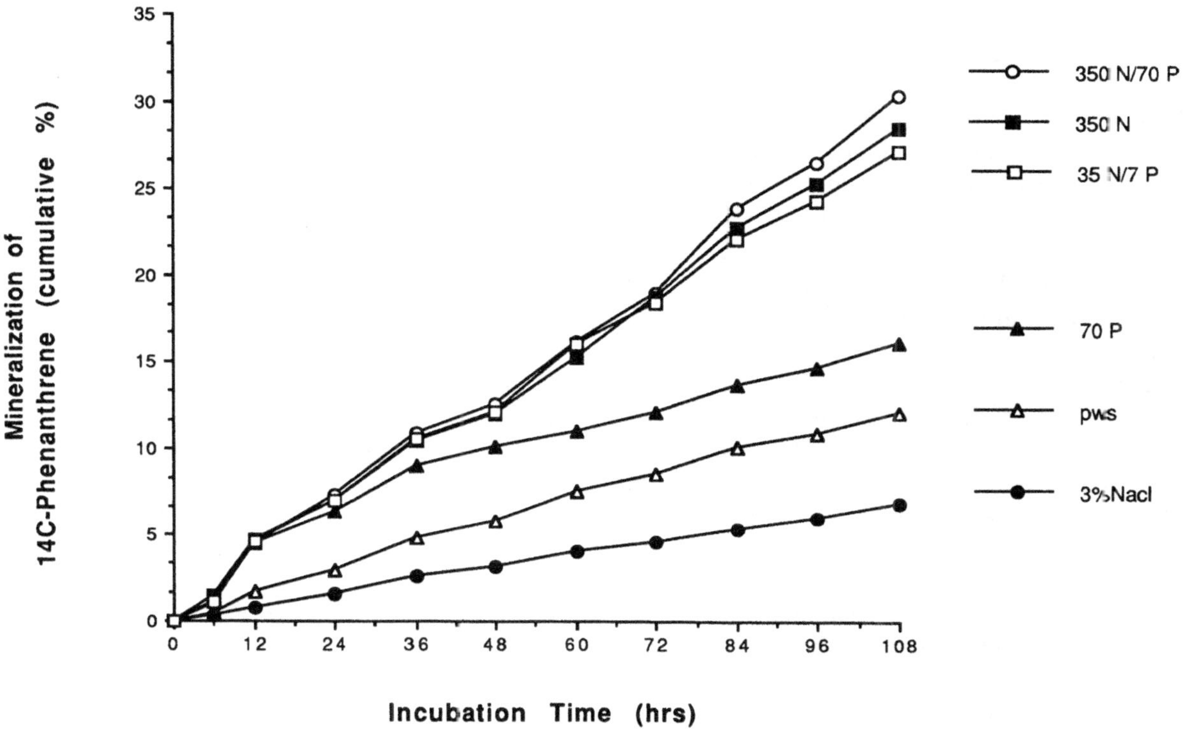

Fig. 2. Cumulative CO_2 production from ^{14}C-phenanthrene added to oil-contaminated beach samples taken from Prince William Sound and as affected by the addition of different concentrations of phosphorus (P, 70 ppm) and nitrogen (N, 35 and 350 ppm). Nutrients were added as inorganic salts (sodium phosphate and ammonium chloride) dissolved in Prince William Sound (PWS) seawater.

to degrade oil to any significant depth. Thus, for more short-term results, the oil must first be further distributed throughout the beach material to increase the surface to volume ratio. Since surface cultivation (i.e., tilling) was unreasonable on most of the cobblestone beaches in Alaska, effective bioremediation depended, in part, on bulk oil removal by the physical washing process and the consequential dispersion of residual oil into the beach material. However, the physical washing may have forced oil to a greater depth in the beach material and possibly affected its ultimate biodegradation or bioremediation.

The highly porous beaches of Prince William Sound were advantageous for bioremediation; oil that diffused into the gravel was spread over a much greater surface area and the oil degrading bacteria were continually bathed at each high tide with oxygenated water. On a sand or mud shoreline, where porosity is much lower, most of the oil will probably be located at the surface. However, the contaminated beach material can be physically removed and the remaining contaminated beach material can potentially be treated by bioremediation, as a finishing step. Tilling may be used to further disperse the remaining oil into the beach material, increasing the surface to volume ratio. Tilling also aerates beach material and helps disseminate added fertilizers.

Choice of fertilizer formulations

We assumed for Alaskan beaches that nitrogen (and, to a lesser degree, phosphorus) had to be applied in a manner that would passively expose the oil degrading microbial communities to the elevated nutrient concentrations over an extended time. Due to the large tidal volume and significant wave action, soluble or easily dissolved fertilizer materials placed on the beach surface would likely wash away in a few days. To avoid this problem, two types of slow release fertilizers were initially considered; solid pelletized formulations and liquid oleophilic formulations. Characteristics of each fertilizer selected are summarized in Table 2.

Although we will not present data here on the effectiveness of the solid fertilizer formulations in bioremediation, briefly summarizing some of the criteria used to make the fertilizer selections is instructive, as this could help in making fertilizer selections in the future. The three main criteria were a) ease of application and potential to retain position on the beaches, b) nutrient release characteristics, and c) physical durability over time. As it turned out, fertilizer granules seemed to best fit the criteria (Pritchard et al. 1991). They were easy to apply over a large surface area and were found to adhere tightly to the oiled beach material and worked their way down under the cobble where they were difficult to dislodge. Nutrient release characteristics, which were determined in simple laboratory test systems (Safferman 1992; Glaser et al. 1991), showed that much of the nitrogen (ammonia and nitrate) and phosphorus (phosphate) released from the granules occurred in the first 24–72 hours. However, sufficient quantities continued to be released steadily for a considerable time thereafter. Thus, as the tides washed the beach, nutrients were distributed to the microbial communities associated with the oil for a period of 2 weeks or longer. Although the physical condition of the granules slowly deteriorated on contact with seawater, they persisted on the beaches, with fertilizer inside, for 2–3 weeks after application. The fertilizer granules were ultimately used by Exxon in combination with the liquid oleophilic fertilizer for much of the large scale applications in Prince William Sound.

Fertilizer briquettes were also found to be satisfactory based on the above criteria (Safferman 1992). The briquettes are approximately the size of charcoal briquettes and contained organic sources of nitrogen and phosphorus that slowly hydrolyzed to release urea and phosphate. Nutrient release characteristics of the briquettes were very similar to the granules. Although the briquettes maintained their physical integrity, if broadcast over the

Table 2. Description of fertilizers tested.[a]

Commercial name	Woodace	Customblen	Inipol
Manufacturer	Vigoro Industries	Sierra Chemical Co.	Elf Aquataine
Form	Briquette	Granule	liquid
Size	5 × 5 × 5 cm	2–3 mm dia	–
N-source	Isobutyraldehyde diurea	Ammonium nitrate	Urea
N:P:K	14:3:3	28:8:0	7.3:2.8:0
Specific gravity	1.8	1.8	0.996 gm/ml
Viscosity	–	–	250 cst
Application rate in 12 × 35 m plots	986 gm/m^2	100 g/m^2	284 gm/m^2
Method of application	Net bags (11.8 kg ea.)	Fertilizer spreader	Back pack sprayer
Test areas	Snug Harbor	Snug Harbor Passage Cove	Snug Harbor Passage Cove

[a] Mention of trade names or commercial products does not constitute endorsement or recommendation for use.

beach surface, they were easily moved around by the tides and waves, resulting in heterogenous distributions. Consequently, they had to be packaged in net bags and secured to the beach with metal stakes. Although this was effective, it presented significant logistical problems precluding their consideration for large scale use. The briquettes were, however, used as part of our initial field demonstration in bioremediation and appeared effective in enhancing oil biodegradation (Pritchard et al. 1991).

In consideration of future applications, rather than broadcast fertilizer granules and briquettes on to the beach surface, they could also be buried in the beach material, e.g. in trenches running parallel to the water line in the contaminated intertidal zone. Depending on the porosity of the beaches, tides, and interstitial water movement, nutrients could be effectively distributed to the bacteria. Because of the physical integrity of the briquettes, they would be most suitable for this type of application. A burial approach was initially considered in Alaska but never tested, because of potential logistical restrictions and vigorous storm action. However, fertilizer granules were applied by Exxon and the State of New York in this manner to a sandy beach on Prall's Island (located in an estuary southwest of New York City) contaminated with diesel fuel. Initial reports (Madden 1991) suggested that the application was successful in enhancing diesel oil biodegradation. Distribution of nutrients will, of course, depend on the hydrodynamics of interstitial water in the beach and, in many cases, specific information will be lacking. Based on these limited successes, one could quite easily perform a pilot study, of several days duration, to actually measure movements of nutrients in interstitial water (Madden 1991).

Pragmatically, the best criteria for determining how much slow release fertilizer to place on a given beach, was to apply as much as possible without exceeding toxic concentrations of ammonia and/or nitrate. Ammonia and/or nitrate are toxic (96-hr LC50 = 10–15 ppm) to mysid shrimp, a good sentinel for most sensitive species (Pritchard et al. 1991). Given the pulsed nature of nutrient release from these fertilizer formulation, concentrations of applied ammonia/nitrate could exceed toxic concentrations and still be acceptable if dilution quickly reduced the concentrations. How much the toxic concentration can be exceeded will depend on the dilution capacity in the area of application and data from laboratory toxicity tests using pulsed exposure to the nutrients.

Alternatively, fertilizer application rates could be determined using information on the amount of nutrients required to metabolize a given quantity of oil. Since this approach will depend on the amount of oil treated and the C:N ratios for optimal oil biodegradation (Morgan & Watkinson 1989), it is likely to be a complicated assessment due to the heterogeneity of oil distribution invariably encountered in the field and the impact of tidal flux and dilution.

The concept of oleophilic fertilizers is based on the use of organic sources of nitrogen and phosphorus in a liquid carrier that is miscible with oil. In theory, when the liquid carrier is applied, the nutrients adhere to the oil, keeping them in contact for sustained periods, with the bacteria growing on the oil's surface. Several types of oleophilic fertilizers have been successively tested in both laboratory and small-scale field experiments. Most of these were designed and tested with the idea of treating oil on the surface of water rather than oil on beach material. Pioneering studies by Atlas and Bartha (Atlas & Bartha 1973) demonstrated that the addition of paraffinized urea and octylphosphate to Prudhoe Bay Crude oil on the water surface significantly enhanced the biodegradation of the oil. Similar success has been reported for a commercial product, Victawet 12 (2-ethylhexyl-dipolyethylene oxide phosphate) (Bergstein & Vestal 1978), for several natural sources of lipophilic nitrogen and phosphorous, such as soybean lecithin and ethyl allophanate (Olivieri et al. 1978), and for $MgNH_4PO_4$ incorporated into a paraffin support base (Olivieri et al. 1976). Uncertainties related to the specific mechanisms of enhanced oil degradation, the factors affecting physical release rates of the nutrients *in situ,* and the effect of adding large quantities of organic carbon to the oil, have perhaps limited the application of these products. In addition, several reports comparing enhanced oil degradation by oleophilic and regular fertilizer showed little difference (Lee & Levy 1987; Halmo 1985). Therefore, the advantage of using oleophilic fertilizers has not been clearly established. None the less, the concept of oleophilic fertilizers was intriguing to us given that it was the prevailing method for placing nutrients directly in contact with the microbial communities, and thus further field testing seemed justified.

The oleophilic fertilizer selected was Inipol EAP 22, produced by Elf Aquataine Company in France, since it was the only commercially available source with large production capability. This unique product is a stable micro-emulsion consisting of a 'core' of urea (the nitrogen source) surrounded by an oleic acid carrier. Tri(laureth-4) phosphate (a surfactant and the source of phosphorus) acts as an emulsion stabilizer and butoxy ethanol reduces viscosity. Inipol EAP 22 has shown promise in tests conducted in the laboratory and in large outdoor tanks using different types of oil-contaminated beach materials and different environmental locations (Lee & Levy 1987; Sveum & Ladousse 1989; Tramier & Sirvins 1983). Interesting results were obtained from a test conducted on an oil-contaminated beach in Spitsbergen, Norway (Sveum & Ladousse 1987). Inipol appeared to increase oil biodegradation rates when applied to coarse-grained, gravely beach material but not when applied to fine-grained sandy beach material. Based on these studies, Inipol appeared to have significant potential for bioremediation. When Inipol was applied to oil-contaminated mixed sand and gravel from Prince William Sound in laboratory studies designed to provide intermittent submersion with seawater, approximately 60% of the urea (measured as total Kjeldahl nitrogen, TKN) was released within the first few minutes after application (Safferman 1992). Measurements of total phosphorus showed a similar release pattern. A 6-hour wait before the first submersion produced the same results as a 5-minute wait, suggesting that the Inipol 'sets' quite quickly. However, following this initial burst of TKN and phosphorus, there was essentially no release of these materials thereafter. Presumably, the residual nitrogen and phosphorus, although tightly held to the oiled beach material, was available to the bacteria degrading the oil. Experimentally demonstrating this availability was difficult and is discussed below.

Inipol application in Prince William Sound was initially conducted with a backpack sprayer to give a thin coating over the oiled beach material (Pritchard et al. 1991). Coatings were controlled visually because the oil appeared wet and became a deeper black in color following coverage by the Inipol. This visual effect disappeared several hours after application. Manufacturers recommended application rate (5% of the oil) could not be followed because oil concentrations were highly variable. In the large scale use of Inipol by Exxon, an application rate of approximately 0.3 liter/m^2 was used, based, in large part, on simulating the surface coverage used in our field demonstration project.

Visual changes

Test beaches at Snug Harbor, where Inipol was applied as part of a field demonstration of its effectiveness in bioremediation, produced some surprising visual results (Pritchard & Costa 1991). These beaches were moderately contaminated with oil and had not been subjected to the physical washing process at the time of our test. They were selected because the concentration and distribution of oil was thought to simulate physically-washed beaches. Visually, the cobble areas had a thin coating of sticky oil covering the rock surfaces and mixed sand and gravel under the cobble. Oil penetration was < 0.3 cm below the surface. In some areas, small patches of thick oil and mousse could be found.

Approximately 10–14 days following oleophilic fertilizer application, reductions in the amount of oil on rock surfaces were visually apparent. It was particularly evident from aerial observations, where the contrast with oiled areas surrounding the plot was dramatic, etching a 'clean' rectangle (area where the Inipol was applied) on the beach surface (Fig. 3). The contrast was also impressive at ground level; there was a precise demarkation between fertilizer-treated and untreated areas. At this time, the untreated control plots appeared unaltered visually.

Fig. 3. Visual effect of Inipol fertilizer application on the removal of oil from test plots in Snug Harbor on Knight Island, Prince William Sound. Fertilizer treated area is shown on the right.

Close examination showed that much of the oil on the surface of the cobble was gone, yet considerable amounts of the oil remained under the cobble and in the mixed sand and gravel below. Remaining oil was not dry and dull as was the oil on the untreated control beach, but appeared softened and wetter. It was also very sticky to the touch, but with no tendency to come off the rocks. Concurrently monitoring of the shore line waters revealed that no oil slicks or oily materials were found leaving the beach during tidal flushing.

The visual disappearance of oil on the cobble surface 2–3 weeks following Inipol application was largely due to biodegradation and not a chemical washing phenomena. Supporting chemical data are presented below. When large concentrations of Inipol were applied repeatedly on several mini plots in Snug Harbor to maximize any potential chemical washing effects, oil removal was not affected; that is a period of at least 2–3 weeks was required to see any 'cleaning' effect regardless of the amount of Inipol applied. The application of aqueous fertilizer solutions (tested at a different beach), which contained only inorganic chemicals and no organic surfactant-like materials, also produced the 'cleaning' effect in about the same time period, further supporting the role of biodegradation rather than chemical washing (Pritchard et al. 1991). Finally, laboratory tests performed by Exxon researchers (R. Prince, S. Hinton, and J. Bragg, personal communications) have shown that Inipol, when compared to chemicals specifically designed as rock washers, was ineffective. Also, they observed that in microcosm studies Inipol caused the oil to become more tightly associated with the beach material; that is, in the absence of Inipol, the oil had much more of a tendency to move to the glass walls of the microcosms.

Six to eight weeks after fertilizer application, the contrast between the treated and untreated areas on the cobble beach had lessened. Reoiling of the Inipol-treated beach from oiled subsurface material and/or the concurrent slow removal of oil on the surface of the beach material surrounding the treated areas was probably responsible for this decrease in contrast. Toward the end of the summer season, the area used for the bioremediation studies became steadily cleaner, including the control plots. Several storms and more frequent rainfall, as well as natural biodegradation, contributed to these changes.

Since the disappearance of oil on the untreated control beaches eventually 'caught up' to that obtained on the treated beach, it can be argued that the cost and effort of bioremediation may not have been justified. However, rapid oil disappearance, caused by the application of the oleophilic fertilizer, potentially made these beaches safer for local wildlife (less tendency for fur and feathers to become oiled) in a shorter period of time, a highly desirable situation where breeding and migratory activities are prominent as they were in Alaska. Depending on the time of season, accommodation of these activities could not have been realized by 'doing nothing' or by just performing physical washing. On beaches more heavily oiled than our test plots in Snug Harbor, natural biodegradation would not likely remove all of the oil before the winter season curtailed this degradative activity. Accelerating the degradation through the application of fertilizer during the active summer months undoubtedly reduced the need for further cleanup and limited the potential of ecological exposure to the oil the following spring. Finally, the more quickly oil can be removed from a beach, the more rapidly recolonization of those animals and plants killed initially by the oil spill can occur. Consequently, bioremediation could help accelerate biological recovery of the intertidal zones during times when biological activities were at their peak.

Measures of oil spill bioremediation effectiveness

Obtaining definitive information on the role of biodegradation in the removal of oil residues from beach material, or from any complex environmental matrix, is a difficult task. In general, to assess the effectiveness of oil spill bioremediation, information must be produced on (1) qualitative changes in hydrocarbon composition that are indicative of biological processes and (2) decay rates of bulk oil that are also indicative of biological processes.

Qualitative information establishes the *extent* to which biodegradation has occurred. However, with a complex chemical mixture like Prudhoe Bay Crude oil, the removal of more than just a few short-chain hydrocarbons (representing only a small percentage of the oil) and removal of more than just the aliphatic fraction of the oil (i.e., leaving behind aromatic, heterocyclic, branched hydrocarbons, polar chemicals, etc.), is desirable.

Simultaneously, the quantitative information can establish that the enhancement of oil biodegradation by the fertilizer treatments was sufficient to merit full-scale operation; generally a 2- to 3-fold enhancement over the untreated controls might be acceptable scientifically and probably to many decision makers and regulatory groups, but this is not based on a comprehensive data base. Both types of information, however, were difficult to obtain in the field.

Any bioremediation testing program that is based on analytical techniques involving decreases of oil residues or hydrocarbons resolvable by gas chromatography is open to scientific criticism because other environmental fate processes (including photolysis, physical dissolution, chemical washing, volatility, etc.) can affect or contribute to the disappearance of oil. To confront these potential criticisms, we chose an approach that integrated several analytical procedures with several key assumptions. The first assumption was that the disappearance of several target hydrocarbon groups could be used as definitive indicators of biodegradation. Furthermore, strong indications of biodegradation would be associated with substantial changes in the composition of other fractions in the oil, particularly selected aromatic hydrocarbons.

The second key assumption was that if a correlation between changes in hydrocarbon composition and changes in residue weight of the oil could be established, disappearance rates of the residue weights could be used as the primary quantitative measure of fertilizer effect (i.e., significant differences between treated and control plots). The rate information could then be used to estimate cleanup effectiveness over extended time periods.

Some discussion as to why this approach was selected is in order because the criteria as to what constitutes biodegradation of oil is complicated and controversial. Many studies have considered measurements of reductions in aliphatic hydrocarbons concentrations as generally indicative of overall biodegradation but the value of such reductions is often questioned because these hydrocarbons are: a) frequently the most readily degradable fraction, b) the least toxic and c) often only a small percentage (by weight) of the oil. Measuring compositional changes in the aromatic fraction adds a further dimension to biodegradation, as these hydrocarbons are less readily degradable and potentially cause chronic toxicity. Common procedures for measuring the aromatic hydrocarbons are usually based on mass spectral analysis which concentrates on only 10–20 selected compounds and represents only a very small fraction of the total aromatics. Whether these selected aromatic hydrocarbons are reasonable surrogates for the degradation of the rest of the aromatic compounds has not been established. However, if the aromatic hydrocarbons are shown to degrade, aliphatic hydrocarbons will almost certainly be extensively degraded as well.

Regardless of compositional changes, it seems intuitive that bioremediation should effect the removal of bulk material as well; changes in hydrocarbon composition without much change in oil residue removal seems to present only part of the picture. Moreover, oil biodegradation under optimized conditions in the laboratory will result in as much as a 40–60% reduction in the total weight of oil (Atlas & Bartha 1973). Therefore, some reduction in oil residue weight should be reasonably expected, even under field conditions. The ultimate measure of biodegradation would be to fractionate the oil into aliphatic, aromatic, heterocyclic, polar and asphaltene fractions and determine weight loss for each of these fractions. These fractions can be further analyzed by gas chromatography to determine qualitative changes in composition. This analytical procedure has been used by Westlake and his colleagues in several studies (Jobson et al. 1972).

Changes in the normal alkane to branched alkane composition

Results with the oleophilic fertilizer provide perspective on measuring effectiveness in bioremediation at a field demonstration level. Chemical analysis of the oiled beach material, exposed and unexposed to the oleophilic fertilizer, was accomplished by collecting beach material according to a block design (21 samples taken at each sampling time, seven each in contiguous blocks along a line in the high, mid and low tide zones of the beach) and then extracting samples (with methylene chloride) from the cobble surface and from the mixed sand and gravel under the cobble (Pritchard et al. 1991). It is assumed that all areas of the treated beach were equally exposed to the fertilizer, although there was no certainty in this respect. The weight of oil recovered on a mg/g basis of beach material was determined gravimetrically and hydrocarbon composition was determined by gas chromatography of the extracts following standard analytical procedures.

Changes in composition were determined first by examining the resolvable alkanes. Historically, this is done by calculating the weight ratio between a hydrocarbon that is known to readily biodegrade (generally the C-17 and C-18 normal alkanes) and ones that are slower to biodegrade, such as the branched alkanes, pristane and phytane, which elute closely to the n-C17 and n-C18 alkanes, respectively (Kennicutt 1988). We focused on the n-C18/phytane ratio because pristane is sometimes found naturally in seawater (C. Henry, pers. comm.). The ratio concept is based on the idea that most nonbiological fate processes (physical weathering, volatilization, leaching, etc.) will not produce differential losses of normal and branched hydrocarbons that have similar gas chromatographic and correspondingly, chemical behavior. Support for this concept can be found in biogeochemical studies on oil (Kennicutt 1988).

The technique has two disadvantages. First, since the branched alkanes do, in fact, biodegrade (Prinik et al. 1977; Mueller et al. 1992), they need only degrade more slowly than the straight chain alkane to obfuscate the ratio method resulting in a very conservative measure of biodegradation. Second, octadecane is likely to be one of the first alkanes degraded, making the ratio technique useful only in the early stages of oil degradation. Also, because the alkanes may not make up a large percentage of the crude oil (6% in Prudhoe Bay Crude), the change in ratios may not be indicative of significant or longer-term changes in other components of the oil.

The effects of the oleophilic fertilizer, Inipol EAP 22, on changes in alkane and branched alkane composition in samples from the cobble surface, are shown in Fig. 4A and B. Extensive decay in the n-C18/phytane ratio occurred through time. The decay rate for the fertilizer-treated beach was slightly faster; the difference in the slope of linear regression lines was approximately 1.2 but was not statistically significant ($p = 0.05$). However, comparing the slopes of these decay curves is complicated by the great heterogeneity of the samples and because of the apparent lag observed on the Inipol-treated beach. The cause of this lag is not known but it does not appear to be Inipol specific, as a similar lag was seen on a beach treated with fertilizer briquettes (data not shown). If decay rates are compared following the lag (between the 6/25 and 8/26 sampling dates), the fertilizer treatment produced a decay rate approximately 2 times that seen on the control beach. Therefore, based on the assumptions described above regarding the meaning of the ratio changes, biodegradation was

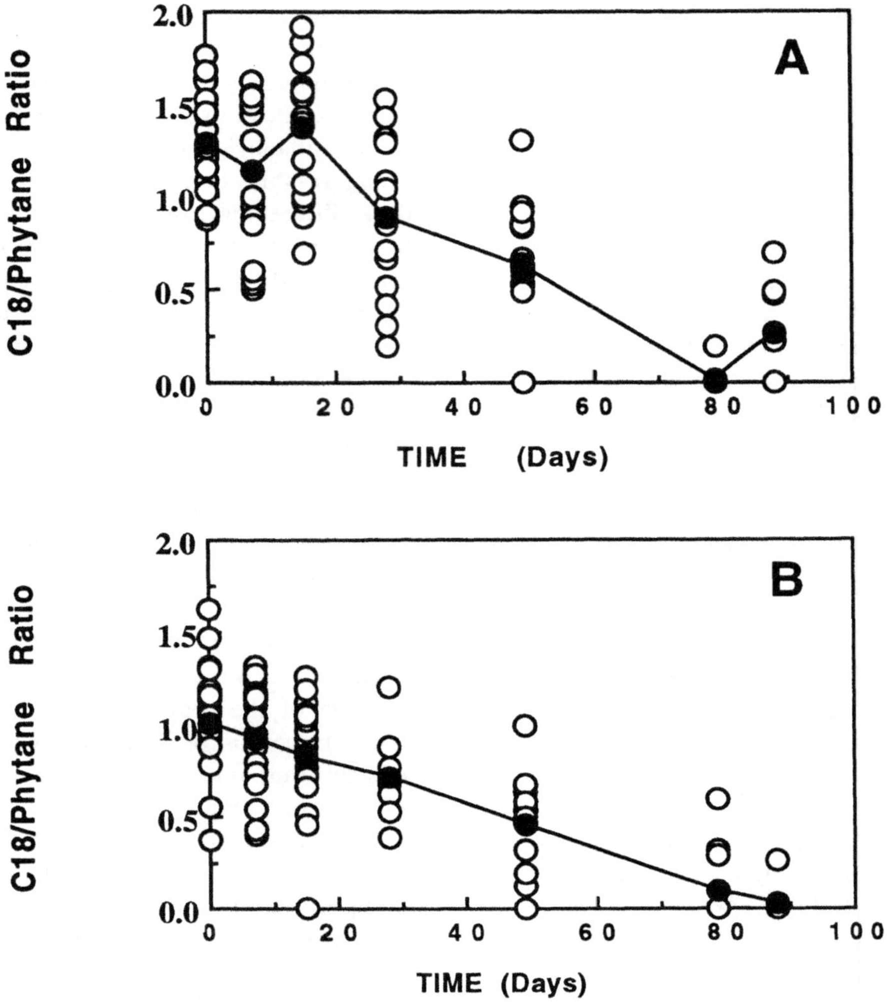

Fig. 4. Changes in the n-C18/phytane ratio through time in oil extracted from cobblestone samples from Inipol fertilizer-treated (A) and untreated (B) beach plots in Snug Harbor. Shaded symbols are means. Fertilizer was applied immediately following the day 0 sampling (6/8/89).

occurring on both beaches, but appeared to be enhanced by the application of the oleophilic fertilizer. This corresponds with the visual observations discussed above. The sustained nature of this enhancement is interesting given that the release of nutrients from the fertilizer probably occurred over a considerably shorter time period.

Note that oil had already undergone biodegradation prior to fertilizer application as the ratio for undegraded weathered Prudhoe Bay crude was around 2.0. Also, large decreases in the ratios were invariably linked to considerable reduction, if not complete removal, in the concentrations of the resolvable (by gas chromatography) alkanes, n-C17 to n-C30.

The large variability in the ratios shown in Fig. 4 (many samples showed evidence of extensive biodegradation while others showed very little) was a function of the highly heterogenous distribution of oil on the beach. Possibly the same amount of biodegradation was occurring in each sample. But since biodegradation takes place on the oil's surface, a grab sampling procedure (which was almost unavoidable in this case) necessarily encompasses sufficient quantities of undegraded oil from below that surface, thereby diluting the measure of biodegradation.

Much less change in the n-C18/phytane ratio, if any, occurred in the mixed sand and gravel under the cobble for the oleophilic fertilizer-treated beach and the untreated control (data not shown). The absence of a change in the n-C18/phytane ratios through time for the subsurface mixed sand and gravel samples suggested that oil biodegradation was not occurring. Examination of phytane itself in the mixed sand and gravel under the cobble on the Inipol-treated beach showed that its loss was as fast as that for n-C18 alkane. Biodegradation was either not occurring (i.e., some nonbiological process was removing both hydrocarbons simultaneously) or phytane was actually being degraded as fast as the n-C18. Phytane degradation is not common but it does occur (Prinik et al. 1977; Mueller et al. 1992) and we have readily isolated phytane-degrading microorganisms from the beach material in Prince William Sound. Thus, microbial communities on Alaskan beaches may have a very pronounced ability to degrade branched alkanes and the concept of using phytane as a internal biological marker becomes compromised. The cobble surface samples also showed significant decreases in phytane through time but at a slower rate than the n-C18, thus giving the observed decay in the ratio. Why phytane degradation was possibly much faster in mixed sand and gravel is not clear but it may be related to the greater surface area available. Again, if degraded oil was not removed physically as it was on the beach surface, continued degradation of the degraded oil would occur; bacteria on the beach surface, however, would continually be exposed to fresh undegraded oil as the degraded oil was removed. This could also, explain the visual observation of more oil loss on the beach surface.

Alternatively, results from the mixed sand and gravel samples under the cobble could suggest that Inipol was acting in a chemical manner (surfactant effect) to remove aliphatic and branched hydrocarbons simultaneously. However, n-C18 and phytane disappeared in essentially the same manner in mixed sand and gravel samples from the untreated control beach. Since there was no possibility of a chemical effect on the control plot, the conclusion is that phytane removal was primarily due to biodegradation, although this cannot be proven.

Compositional changes in aromatic hydrocarbons

At this point we can conclude that biodegradation of the aliphatic fractions of oil was occurring on the samples taken from the cobble surface and, quite possibly, in the mixed sand and gravel samples as well. As presented above, biodegradation of the aliphatic fraction was not sufficient in itself to establish the overall effectiveness of bioremediation. Thus, another approach for examining biodegradation effects was to perform selective mass ion spectrometry following gas chromatographic analysis. A variety of aromatic hydrocarbons, that are perhaps more difficult to degrade than the aliphatic hydrocarbons, can be examined with this method (Kennicutt 1988; Rowland 1986). This was important not only because it tracks another degradable fraction of the oil, but because certain polycyclic aromatic hydrocarbons (PAHs) are known to be potential pro-carcinogens under specific conditions. Observing their removal from the oil would therefore imply a reduction in potential adverse ecological effects. Whether this toxicity issue is really relevant due to improbable exposure scenarios, it was none the less an additional factor to be considered in effectiveness assessments.

Furthermore, as the low solubility of the PAHs makes them to difficult to degrade, they can be used as an indicator of the extent of oil degradation. Mass spectral analysis of a variety of aromatic and heterocyclic hydrocarbons in several samples of oil with greatly reduced n-C18/phytane ratios and aliphatic hydrocarbon concentrations is shown in Table 3. The selected aromatic hydrocarbons represent a group of methyl-substituted homologs that are found close to the mass number of each parent chemical structure (based on known standards). The values in the table are normalized to hopane (17 alpha, 21 beta), a multi-ring cyclic alkane (C30). Hopane and its homologs, which are quite resistant to biological attack, have been used for some time as conserved internal biomarkers in oil by the geochemists (Kennicutt 1988). However, unlike the n-C18/phytane ratio, the relative changes cannot be attributed to biodegradation with as much confidence; differential decay between a hydrocarbon and hopane could be due to nonbiological processes since there may be considerably less chemical similarity between the target hydrocarbon and hopane. None the less, hopane provides a consistent standard to normalize the concentrations of aromatic hydrocarbons.

From Table 3, the relative differences in the amount of each group of homologs between the beach samples and Prudhoe Bay crude oil indicates the degree of compositional change that occurred in the samples on the beach. Despite the limited number of analyses, samples that showed low n-C18/phytane ratios also showed large changes in many of the aromatic and heterocyclic hydrocarbons. Of the 58 PAHs monitored, only a few remained in samples with a low ratio. There are insufficient analyses at this time to demonstrate a further relationship between the amount of ratio reduction and the amount of aromatic hydrocarbons remaining. One cannot explicitly state that disappearance of the aromatic and heterocyclic hydrocarbons was due to biodegradation but the suggestion is strong. Many of the higher molecular weight PAHs are unlikely to be affected by chemical or physical processes because of their low solubility and volatility. Biodegradation may be the only mechanism that could explain their disappearance. In addition, these samples were taken from the beach approximately 50 days after the application of the Inipol fertilizer and any chemical effect from the fertilizer that could cause these changes in composition was equally unlikely.

Samples with reduced n-C18/phytane ratios also showed increased concentrations (percent of total oil residue weight) of the hexane insoluble (but methylene chloride soluble) fraction. Chemically, this fraction should consist primarily of asphaltene hydrocarbons (Speight et al. 1984) which are quite resistent to biodegradation. However, polar materials originating from the partial oxidation (most likely by bacteria) of high molecular weight hydrocarbons would also accumulate in this fraction. Samples with increased percentages of asphaltene material relative to Prudhoe Bay crude suggest that considerable removal of other components in the oil had occurred, possibly because of biodegrada-

Table 3. Relative concentrations of aromatic, hererocyclic and cyclic hydrocarbons normalized to hopane in oil from cobble surfaces in Snug Harbor.

	Oleophilic fertilizer treatment		
	Purdhoe Bay Crude[a]	Day 0	Day 50[b]
n-C18	32.28	22.05	0.30 (0.23)
Phytane	15.2	15.62	1.7 (1.16)
n-C18/Phytane ratio	2.08	1.41	0.46 (0.30)
C3/Naphthalenes	11.92	2.51	0.00
C3/Fluorenes	2.67	2.24	0.01
C3/Phenanthrenes	3.75	3.84	1.11 (0.39)
C2/Pyrenes	1.14	0.92	0.27 (0.16)
C3/Dibenzothlophene	3.97	4.14	1.28 (0.47)
C2/Chrysenes	1.15	0.98	0.62 (0.25)
Benzo (e) pyrene	0.18	0.15	0.10 (0.05)
Total PAHs[c]	58	29	3.8 (0.8)
% Hexane insoluble weight	6%	11%	36.2% (21.5)

[a] Artificially weathered to simulated oil found on Prince William Sound Beaches (Lindstrom et al. 1991).
[b] Mean (standard deviation) N = 5; randomly selected from the 8/26/89; all samples 50 mg of extracted oil residue per ml.
[c] Number of PAH peaks monitored in mass spectral analysis.

tion. However, this percentage increase should also occur with hopane, as it is also a conserved internal biological marker (Kennicut 1988; Lindstrom et al. 1991). Results showed hopane only increased slightly, whereas the hexane insoluble fraction increased (percent of total oil residue weight) almost six-fold. The lack of correspondence between hopane and the hexane insoluble fraction is difficult to explain in chemical terms. However, it does make sense biologically if we assume that biodegradation products are accumulating in the hexane insoluble fraction through time. Controlled laboratory experiments involving the generation of extensively biodegraded oil, support this interpretation (P. Chapman, pers. comm.). If this is the case in the field, clearly oil residues remaining on the beach experienced considerable biodegradation.

Relationship of compositional change to residue weight change

A relationship was then established between compositional changes and the loss of oil residues from beach material. Due to limitations of the analytical procedures routinely used to measure hydrocarbon composition, less than 10% of the oil by weight is represented, as was discussed above. Chemically analyzing the remaining 90% of the oil to improve on measures of effectiveness was not practical because of cost and analytical complexity. Thus, some bulk measurement of oil degradation was needed. Since bioremediation is designed to ultimately degrade as much oil as possible from contaminated beaches, monitoring changes in total oil residues seemed to provide the best measure of biodegradation extent. However, this is only reasonable if some correlation can be established between changes in oil residue weight and changes in hydrocarbon compositions, since it is the latter that can be linked directly to biodegradation.

Figure 5 shows a plot of the changes in residue weight against changes in the n-C18/phytane ratio. 'Change' in this case is expressed as the percent reduction of the median values (from 21 field samples at each sampling time) relative to the $t = 0$ sampling for each parameter. Median values were used in this case because of their independence from out-lying data points and because they are probably a good representative of the samples taken on a particular sampling date. The data points reflect the sampling times with the progression of time going from right to left in the figure. A positive correlation was apparent. The differences in slopes suggests that for Inipol-treated plots biodegradation (i.e., change in the n-C18/phytane ratio) were accompanied by a greater removal of oil residue. The results strongly suggest that reductions in the ratio consistently accompanied reductions in oil residue weight or vice versa. This was unexpected since the n-C18 alkane, being one of the most rapidly degraded hydrocarbons, could easily be reduced to the point of rendering the ratio technique ineffective, prior to extensive change in the oil residue weight. However, this was apparently not the case. The results can possibly be explained by assuming that oil biodegradation worked at the surface of an oil layer covering the cobble and that degraded oil residues continuously sloughed from this layer, leaving behind largely undegraded oil. When oil was extracted from the cobble surface during sampling, a mixture of degraded and undegraded oil was obtained, with the undegraded oil providing most of the measured amount of n-C18 alkane (and phytane). With time, the undegraded oil became less and less of the total oil residue recovered. Thus, in a sense, the degradation of normal and branched alkanes was internally compensated as a function of the rate at which bacteria at the oil–water interface were exposed to undegraded oil.

Based on these results, reductions in oil residue weight can be used as a direct measure of biodegradation and the effect of fertilizer addition on biodegradation rates assessed. The decrease in oil residue weight from cobble surfaces on Inipol-treated and untreated control beaches is shown in Fig. 6. Assuming that decay rates for the oil residues were first order, slopes and half lives of the oil can be calculated (Table 4). Application of oleophilic fertilizer caused an apparent greater than two-fold increase in the loss rate of oil residues. The effect, was not unfortunately, statistically significant (p = 0.05) due to the large variability in the data. Other statistical treatments are currently being examined in an attempt to significantly verify the effect.

We, therefore, attribute the greater rate of decay on the fertilizer-treated beach to an enhancement of biodegradation from the provision of nitrogen and phosphorus nutrients. Interestingly, the enhancement effect of the fertilizer appeared to be sustained for as long as 90 days. This time period was well beyond that in which nutrients would be released or in which the fertilizer might have a chemical washing effect. Thus 'priming' the biodegradation process with nutrients seemed to produce a sustained enhancement effect. One can generalize and say that over a 120-day period (i.e., the maximum window for Alaska in which water temperatures are > 10 °C and thereby adequate for oil biodegradation), bioremediation could remove (assuming linearity) approximately 2.5 times more oil from the cobble surface than on the untreated control beach. Coincidentally, respiratory activities by oil degrading microbial communities on fertilized beaches has been shown to increase an average of 3-fold (Lindstrom et al. 1991). Thus with an initial concentration of 1.0 mg of oil per gram of beach material (cobble surface), bioremediation can potentially remove most of the oil in a single summer season. This was consistent with our visual observations. The absence of any effect on oil residues in the mixed sand and gravel under the cobble may be related, as we discussed above, to their being partially trapped in the beach matrix. On the other hand, oil may not have been spread in a thin enough layer over the beach material to allow bioremediation to have an effect during this testing period, or possibly the Inipol was unable to provide nutrients to this area of the beach; i.e., it was primarily acting at the beach surface.

Nutrients and microbial biomass

Following the application of the oleophilic fertilizer, interstitial water samples were taken during several tidal cycles to determine if increased concentrations of nitrogen and phosphorus could be observed. Water samples were taken using a modified root feeder apparatus which sampled water 10–15 centimeters below the surface of the mixed sand and gravel. Sampling was conducted 2, 10, and 30 days after fertilizer application. Elevated nitrogen concentrations were seen only on the day 2 sampling (Table 5). However, the variability was quite large with somewhat of a bias toward one side of the treated area. If all of the nitrogen in the fertilizer was released at once into a hypothetical body of water overlying the beach test plot at high tide, one would expect concentrations of approximately 200–300 μm N. Obviously, these concentrations were reached in some areas of the beach. Given that three tidal cycles had occurred prior to this sampling, much of the nitrogen in the Inipol fertilizer was probably released in the first few days. This corresponds with the nutrient release data generated from laboratory studies described above. Thus, the enhancing effect of the oleophilic fertilizer on oil biodegradation may have been the result of an initial pulse of nutrients rather than a

Table 4. Rate analysis of natural Log-transformed oil residue weights (mg/g) in cobble surface samples (July 8, 1989 to July 29, 1989 only) for test beaches at Snug Harbor.

Beach	Slope (*Std Dev*)	Significance of slope[a] Greater than zero *N*	*T-value*	*P*	Half-life *days*	Time to remove *90% days*
Inipol treated	− 0.016 (0.007)	80	− 2.4	0.02	44	146
Untreated control	− 0.006 (0.010)	65	− 0.56	0.58	124	411

[a] Only the INIPOL rate is significantly different from zero (p = 0.05); slopes were significantly not different from each other (p = 0.05).

Table 5. Ammonia nitrogen (μM) in interstitial water samples taken on an incoming tide, 2 days following application of oleophilic fertilizer on a cobble beach in Snug Harbor.

	Block[a]			
	1	3	5	7
High tide zone	57	300	10	4
Mid tide zone	410	61	3	6
Low tide zone	190	3	2	3

[a] Blocks were 5 m long and 4 m wide running end-to-end parallel to the water line and covering 3 parallel zones, each 4 m wide running side-to-side up the beach. Blocks 2, 4, and 6 in each zone were not sampled.

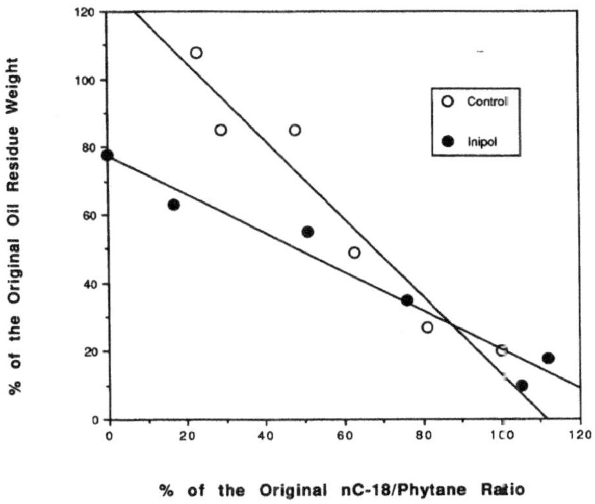

Fig. 5. Relationship of changes in oil residue weight to changes in the n-C18/phytane ratio (expressed as % reduction of the median at each sampling interval) from oil samples taken from cobble surfaces for both Inipol fertilizer-treated and untreated beach plots in Snug Harbor. Data points at the lower left corner of the graph represent samples taken later in the sampling period.

sustained concentration of nutrients over extended periods. Other laboratory and field data support this possibility (Pritchard et al. 1991).

Increases in oil biodegradation rates as a result of fertilizer application would also result in increases in the number or activity (growth rate) of oil-degrading bacteria. To determine if this was the case, beach samples (mixed sand and gravel) were analyzed using an MPN procedure (Pritchard et al. 1991) in which changes in the physical consistency of the oil were monitored as an indication of oil biodegradation. There was no significant difference between the control and treated beaches over the three months sampling period (data not shown). However, as indicated above, the concentrations of oil degraders was very high to start with and with the large variability observed in the data, increases of approximately two orders of magnitude were needed to be significant. In addition, increases in biomass could be obscured by sloughing of the cells or predation by protozoa. Field studies the following summer (1990) were finally able to demonstrate significant increases in hydrocarbon degraders but only in the beach subsurface (Lindstrom et al. 1991).

Mechanism of oleophilic fertilizer action

Despite the success of enhancing oil biodegradation rates through the application of fertilizers, the mechanism of action of the oleophilic fertilizer was still not clear. Was it enhancing oil biodegradation by the supply of nitrogen and phosphorus or by some other mechanism? Or was it causing removal of the oil by some chemically-enhanced process?

To answer these questions, a series of laboratory studies was conducted. All experiments used oiled beach material from Prince William Sound and the test system consisted of the biometer shake flasks described above. Incubations were carried out at 15 °C using a 12-hour tidal cycle: beach material was incubated without overlying sea water (low tide) for 12 hours then covered with seawater (high tide) every 12 hours and gently shaken at 75 rpm. Fertilizers were applied either to the overlying water (soluble nutrients) or directly to the oiled beach material at 'low tide' (Inipol EAP 22). Biodegradation activity was measured as either total CO_2 production or as ^{14}C-CO_2 production from radiolabelled phenanthrene. Carbon dioxide was trapped in dilute NaOH (0.1 N) contained in the flask side arm (Fig. 1). Phenanthrene, which was added by dissolving in the tidal water, completely sorbed to the oiled beach material and consequently its min-

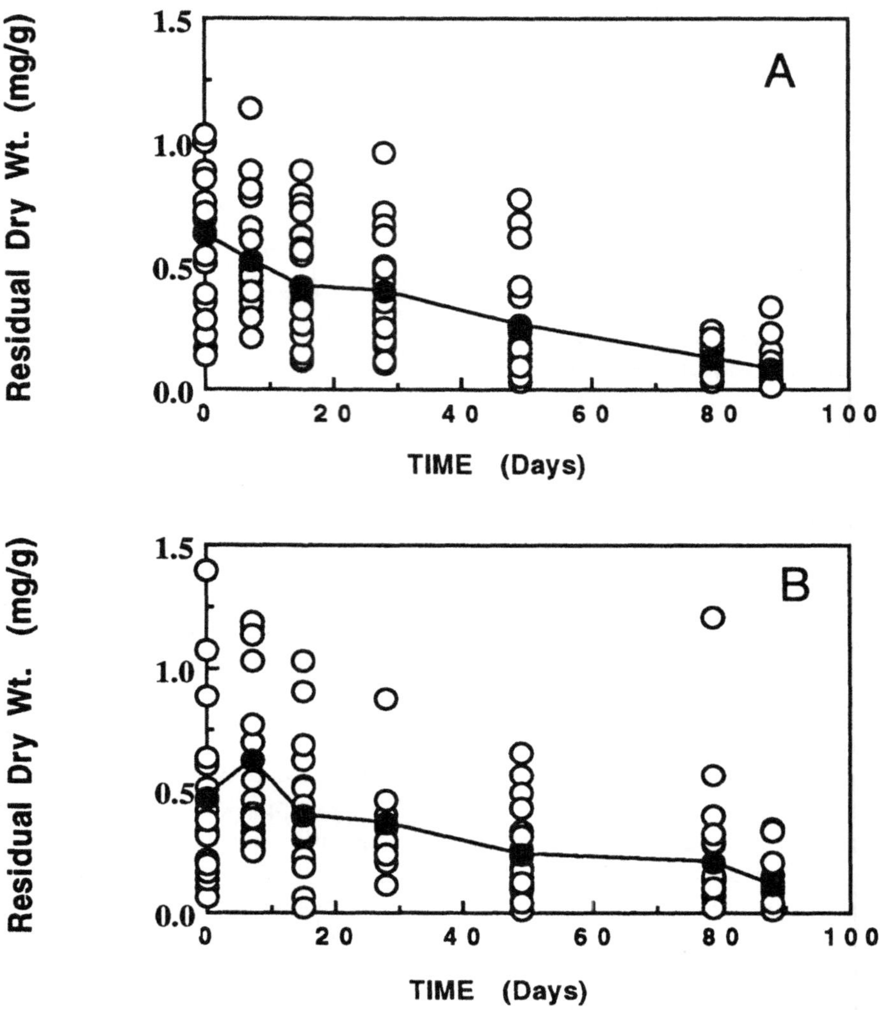

Fig. 6. Changes in oil residue weight through time extracted from cobble stone samples from Inipol fertilizer-treated (A) and untreated (B) beach plots in Snug Harbor. Shaded symbols are means. Fertilizer was applied immediately following the day 0 sampling (6/8/83).

eralization was probably catalyzed by the microbial communities on the surface of the oil.

To determine if Inipol was enhancing oil biodegradation by nutrient supply alone, flasks were set up to compare the effect of soluble nutrients with Inipol. Figure 7 shows that mineralization of the oil (oil was the dominant carbon source by far so we assumed that most of the CO_2 was from the oil and not other organic material) was enhanced by both the addition of soluble nutrients and Inipol. Where the total concentration of nitrogen was the same for the two different fertilizers (Inipol and low level soluble nutrients), mineralization enhancement was approximately the same. These results argue that the other organic materials in the Inipol have no effect on the mineralization and it was the concentration of nitrogen supplied that was important. Unfortunately, experiments of greater duration were not performed and thus the long term effect of Inipol carbon cannot be assessed. Similar laboratory experiments have shown that phosphorus stimulated mineralization only to a minor extent. Some of the CO_2 production in these experiments was undoubtedly due to the mineralization

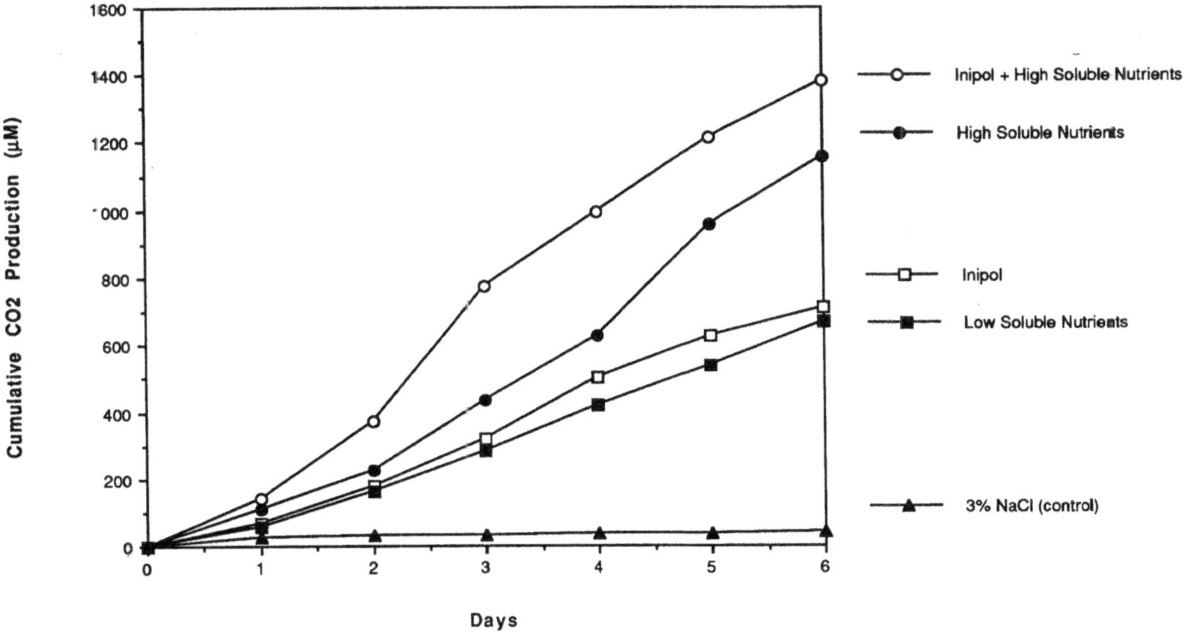

Fig. 7. Effect of Inipol fertilizer and soluble inorganic nutrients on the mineralization of oil in mixed sand and gravel samples from Prince William Sound.

of organic materials in the Inipol fertilizer but Inipol carbon accounted for < 1% of the total oil carbon in the flask.

Additionally, only the replacement water ('high tide') on the first day was supplemented with nutrients; all subsequent replacements had no added nutrients. Thus, the mineralization results reflected the effect of a single pulse exposure to the nutrients at high concentrations. If we assume that much of the Inipol nitrogen was released during the first exposure to water, then it too could represent a single pulse of nitrogen at high concentrations.

Where Inipol was combined with a high level of soluble nutrients (Fig. 7), more mineralization occurred than with the high level nutrients alone. Again, this extra mineralization can be accounted for by the additional nitrogen provided in the Inipol fertilizer.Thus the ability of the oleophilic fertilizer to 'hold' nitrogen close to the bacterial communities involved in oil biodegradation cannot be directly verified with these experiments, since a simple single exposure to nitrogen in any form may have been enough to enhance oil biodegradation to an equal degree.

An additional set of experiments was conducted to ascertain the importance of the individual chemical components in Inipol. Equivalent samples of oiled beach material where exposed to oleic acid and tri(laureth-4) phosphate at concentrations approximately equal to that from an Inipol application, and the flasks incubated in the presence of soluble nutrients in the 'tidal water'. There was no effect due to these organic materials, as the amount of mineralization was no greater than that with the soluble nutrients alone (Fig. 8). Soybean oil was also checked as it has high concentrations of oleic acid and nitrogen containing organic compounds (e.g., amino acids); i.e., like a natural oleophilic fertilizer. Unfortunately, soybean oil was essentially ineffective. Despite the conceptual appeal of soybean oil, either the nitrogen content was too low or it was not released as fast as nitrogen from the urea in Inipol.

Inipol, then, appeared to serve as an effective nutrient source for oil-degrading microbial com-

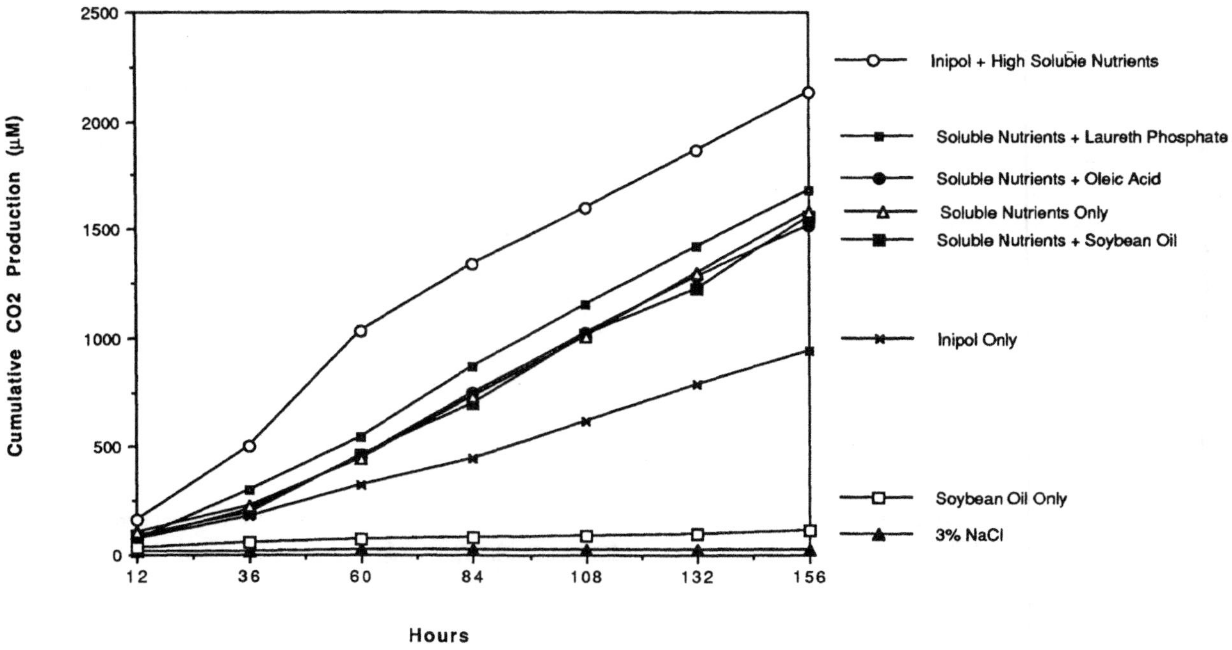

Fig. 8. Effect of Inipol fertilizer, soluble inorganic nutrients and other organic materials on the mineralization of oil in mixed sand and gravel samples from Prince William Sound.

munities but the contribution of its oleophilic nature for the bioremediation of contaminated beach material from Prince William Sound is still unclear.

Summary and conclusions

The results from our field demonstration of oil spill bioremediation in Prince William Sound has shown that the use of the oleophilic fertilizer, Inipol EAP 22, served as an effective nutrient source for oil-degrading microbial communities. It appeared to enhance oil biodegradation, as measured by both changes in oil composition and oil residue weights, by as much as two-fold compared to the untreated controls although this effect could not be statistically verified due to high variability in field samples. The Exxon Corporation and regulatory authorities believed this was enough of a response to merit incorporation of bioremediation on a large scale into the remedial action cleanup plan for oil contaminated beaches in Prince William Sound. Our studies support the belief that the visual observation of oil removal from the beaches 8–10 days following application of the Inipol was due largely to bioremediation and not to a chemical washing effect.

Overall, rapid oil disappearance brought on by the application of the oleophilic fertilizer potentially made these beaches safer for local wildlife (less tendency for fur and feathers to become oiled) and minimized the movement of undegraded oil from the beaches into the water column. These changes occurred in a shorter period of time then those limited changes observed in untreated control plots, and possibly helped accelerate biological recovery of the intertidal area.

We have also summarized some of the lessons associated with measuring bioremediation success in the field. Effective measures of biodegradation and interpretation of resulting data, given the highly variable nature of field studies, has been emphasized as a key element in this success. In addition, the difficulties that arose during our use of bioremediation efforts in Alaska were also discussed . Hopefully this will help guide similar applications of bioremediation at future spills.

Acknowledgements

The authors would like to recognize the scientific contributions of A. Venosa, S. Safferman, E. Opatken, J. Baker and R. Cripe. The technical assistance of D. Dalton, M. Shelton, D. Miller, L. Bosworth, S. Montgomery and C. Robinson is greatly appreciated. Data management and analysis were admirably handled by A. Neale, D. Heggem and R. Gerlach. Special thanks to C. Costa, R. Cripe and L. Suit for their technical coordination and scientific review.

References

Atlas R & Bartha R (1973) Stimulated biodegradation of oil slicks using oleophilic fertilizers. Environ. Sci. Technol. 7: 538–541

Atlas RM (1981) Microbial degradation of petroleum hydrocarbons: an environmental perspective. Microbiol. Rev. 45: 180–209

Bartha R (1981) Microbial degradation of petroleum hydrocarbons: an environmental perspective. Microbiol. Rev. 45: 180–209

Bartha R & Pramer D (1965) Features of a flask and method for measuring the persistence and biological effects of pesticides in soil. Soil Sci. 100: 68–70

Bergstein PE & Vestal JR (1978) Crude oil biodegradation in arctic tundra ponds. Arctic. 31: 159–169

Chianelli RR, Aczel T, Bare RE, George GN, Genowitz MW, Grossman MJ, Haith CE, Kaiser FJ, Lessard RR, Liotta R, Mastracchio RL, Minal-Bernero V, Prince RC, Robbins WK, Stiefel EI, Wilkinson JB, Hinton SM, Bragg JR, McMillem SJ & Atlas RM (1991) Bioremediation technology development and application to the Alaskan spill. In: Proceedings 1991 Oil Spill Conf. (pp 545–555). Am. Petroleum Inst., Washington, DC

Glaser JA, Venosa AD & Opatken EJ (1991) Development and evaluation of application techniques for the delivery of nutrients to contaminated shoreline in Prince William Sound. In: Proceedings 1991 Oil Spill Conf. (pp 556–562). Am. Petroleum Inst., Washington, DC

Halmo G (1985) Enhanced biodegradation of oil. In: Proceedings 1985 Oil Spill Conf. (pp 531–537). Am. Petroleum Inst., Washington, DC

Jobson AM, Cook FD & Westlake DWS (1972) Microbial utilization of crude oil. Appl. Microbiol. 23: 1082–1089

Kennicutt MC (1988) The effect of biodegradation on crude oil bulk and molecular composition. Oil Chem. Pollution. 4: 89–112

Lee K & Levy EM (1987) Enhanced biodegradation of light crude oil in sandy beaches. In: Proceedings, 1987 Oil Spill Conf. (pp 411–479). Am. Petroleum Inst. Washington, DC

Lee K & Levy EM (1989) Biodegradation of petroleum in the marine environment and its enhancement. In: Aquatic Toxicology and Water Quality Management (pp 217–243). New York

Lindstrom JE, Prince RC, Clark JC, Grossman MJ, Yeager TR & Brown JF and EJ (1991) Microbial populations and hydrocarbon biodegradation potential in fertilized shoreline sediments affected by the t/v *Exxon Valdez* oil spill. Appl. Environ. Microbiol. 57: 2514–2522

Madden PC (1991) Final report Prall's Island bioremediation project. Exxon Res. and Engineering, Florham Park, New Jersey, 82 pp

Morgan P & Watkinson PJ (1989) Hydrocarbon degradation in soils and methods for soil biotreatment. CRC Critical Rev. Biotech. 8: 305–353

Mueller JG, Resnick SM, Shelton ME & Pritchard PH (1992) Effect of inoculation on the biodegradation of weathered Prudhoe Bay crude oil. J. Ind. Microbiol. 10: 95–102

Olivieri R, Bacchin P, Robertiello A, Oddo N, Degen L & Tonolo A (1976) Microbial degradation of oil spills enhanced by a slow-release fertilizer. Appl. Environ. Microbiol. 31: 629–634

Olivieri R, Robertiello A & Degen L (1978) Enhancement of microbial degradation of oil pollutants using lipophilic fertilizers. Marine Pollut. Bull. 9: 217–220

Pirnik MP, Atlas RM & Bartha R (1977) Hydrocarbon metabolism by *Brevibacterium erythrogenes*: normal and branched alkanes. J. Bacteriol. 119: 868–878

Pritchard PH, Costa CF & Suit L (1991) Alaska oil spill bioremediation project. U.S. EPA, Office of Res. and Dev. Report. EPA/600/9–91/046a, 522 pp

Pritchard PH & Costa CF (1991) EPA's Alaskan oil spill bioremediation project. Environ. Sci. and Technol. 25: 372–379

Rowland SJ, Alexander R, Kazi RI, Jones DM & Douglas AG (1986) Microbial degradation of aromatic components of crude oils: a comparison of laboratory and field observations. Org. Geochem. 9: 153–161

Safferman I (1992) Selection of nutrients to enhance biodegradation for the remediation of oil spilled on beaches. In: Proceedings 1992 Oil Spill Conf. (pp 571–576). Am. Petroleum Inst., Washington, DC

Speight JG, Long RB & Trowbridge TD (1984) Factors influencing the separation of asphaltenes from heavy petroleum feedstocks. Fuel 63: 616–620

Sveum P & Ladousse A (1989) Biodegradation of oil in the Arctic: enhancement by oil-soluble fertilizer application. In: Proceedings 1989 Oil Spill Conf. (pp 439–446). Am. Petroleum Inst., Washington, DC

Tramier B & Sirvins A (1983) Enhanced oil biodegradation: a new operational tool to control oil spills. In: Proceedings 1983 Oil Spill Conf. (pp 155–119). Am. Petroleum Inst., Washington, DC

Biodegradation 3: 337–350, 1992.

Petroleum bioremediation – a multiphase problem

Eugene Rosenberg, Rachel Legmann, Ariel Kushmaro, Ran Taube, Ellik Adler[1] & Eliora Z. Ron
Department of Molecular Microbiology and Biotechnology, The George S. Wise Faculty of Life Sciences, Tel-Aviv University, Tel-Aviv, Israel; [1] *Marine and Coastal Environment Division, Office of the Environment, Jerusalem, Israel*

Key words: petroleum, bioremediation, hydrocarbon pollution, bioemulsifiers, petroleum microbiology

Abstract

Microbial degradation of hydrocarbons is a multiphase reaction, involving oxygen gas, water-insoluble hydrocarbons, water, dissolved salts and microorganisms. The fact that the first step in hydrocarbon catabolism involves a membrane-bound oxygenase makes it essential for microorganisms to come into direct contact with the hydrocarbon substrate. Growth then proceeds on the hydrocarbon/water interface. Bacteria have developed two general strategies for enhancing contact with water-insoluble hydrocarbons: specific adhesion mechanisms and production of extracellular emulsifying agents. Since petroleum is a complex mixture of many different classes of hydrocarbons, of which any particular microorganism has the potential to degrade only part, it follows that the microorganisms must also have a mechanism for desorbing from 'used' oil droplets.

The major limitations in bioremediation of hydrocarbon-contaminated water and soil is available sources of nitrogen and phosphorus. The usual sources of these materials, e.g. ammonium sulfate and phosphate salts, have a high water solubility which reduces their effectiveness in open systems because of rapid dilution. We have attempted to overcome this problem by the use of a new controlled-release, hydrophobic fertilizer, F-1, which is a modified urea-formaldehyde polymer containing 18% N and 10% P as P_2O_5. Microorganisms were obtained by enrichment culture that could grow on crude oil as the carbon and energy source and F-1 as the nitrogen and phosphorus source. The microorganisms and the F-1 adhered to the oil/water interface, as observed microscopically and by the fact that degradation proceeded even when the water phase was removed and replaced seven times with unsupplemented water – a simulated open system. Strains which can use F-1 contain a cell-bound, inducible enzyme which depolymerizes F-1.

After optimizing conditions in the laboratory for the use of F-1 and the selected bacteria for degrading crude oil, a field trial was performed on an oil contaminated sandy beach between Haifa and Acre, Israel, in the summer of 1992. The sand was treated with 5 g F-1 per kg sand and inoculated with the selected bacteria; the plot was watered with sea water and plowed daily. After 28 days the average hydrocarbon content of the sand decreased from 5.1 mg per g sand to 0.6 mg per g sand. Overall, there was an approx. 86% degradation of pentane extractables as demonstrated by dry weight, I.R. and GLC analyses. An untreated control plot showed only a 15% decrease in hydrocarbons. During the winter of 1992, the entire beach (approx. 200 tons of crude oil) was cleaned using the F-1 bacteria technology. The rate of degradation was 0.06 mg g^{-1} sand day^{-1} (10 °C) compared to 0.13 mg g^{-1} sand day^{-1} during the summer (25 °C).

Introduction

Hydrocarbons are a ubiquitous class of natural compounds. Not only are they found in petroleum-polluted areas, but small concentrations are present in most soils and sediments (e.g. Giger & Blumer 1974; Stephenson 1966). It is, therefore, not surprising that hydrocarbon-oxidizing bacteria are located in virtually all natural areas, although with large variations in cell concentration (Rosenberg 1991). Why then is petroleum pollution a major ecological problem? The reason is that for microorganisms to degrade hydrocarbons, they must have available sources of oxygen, nitrogen and phosphate, elements which are not present in sufficient quantities in crude oil and petroleum products.

The use of hydrocarbons as substrates for bacterial growth presents special problems to both the microorganisms using them as a source of carbon and energy (Table 1) and to investigators in the field of hydrocarbon microbiology.

There are two essential characteristics that define hydrocarbon-oxidizing microorganisms:

1. membrane-bound, group-specific oxygenases and
2. mechanisms for optimizing contact between the microorganisms and the water-insoluble hydrocarbon.

Several reviews have been published on the microbial metabolism of straight-chain and branched alkanes (e.g. Singer & Finnerty 1984), cyclic alkanes (e.g. Perry 1981) and aromatic hydrocarbons (e.g. Cerniglia 1984). It has been established that the first step in the degradation of hydrocarbons by bacteria is the introduction of both atoms of molecular oxygen into the hydrocarbon. In the case of aromatic hydrocarbons, ring fission requires a dihydroxylation reaction and the subsequent formation of a *cis*-dihydrodiol (Gibson 1968) and is carried out by a membrane-bound enzyme system (Crutcher & Geary 1979). Further oxidation leads to the formation of catechols that are substrates for another deoxygenase that catalyzes ring fission (Dagley 1971).

In general, alkanes are terminally oxidized to the corresponding alcohol, aldehyde and fatty acid (McKenna & Kallio 1965). Fatty acids derived from alkanes are then further oxidized to acetate and propionate (odd-chain alkanes) by inducible oxidation systems. Different microorganisms exhibit different group specificities. For example, some grow on alkanes of six to ten carbons in chain length, whereas others grow on long chain alkanes. Some of the oxygenases are encoded on plasmids and others on chromosomal genes. Subterminal oxidation apparently occurs in some bacterial species (Markovitz 1971).

The low solubility of hydrocarbons in water, coupled to the fact that the first step in hydrocarbon degradation involves a membrane-bound oxygenase, makes it essential for bacteria to come into direct contact with the hydrocarbon substrates. Two general biological strategies have been proposed for enhancing contact between bacteria and water-insoluble hydrocarbons:

1. specific adhesion mechanisms and
2. emulsification of the hydrocarbon.

In order to understand the special cell surface properties of bacteria that allow them to grow on hydrocarbons, it is necessary to consider the dynamics of petroleum degradation in natural environments. Following an oil spill in the sea, the hydrocarbons rise to the surface and come into contact with air. Some of the low molecular weight hydrocarbons volatilize; the remainder are metabolized relatively rapidly by microorganisms which take up soluble hydrocarbons. These bacteria do not adhere to oil and do not have a high cell-surface hydrophobicity (Rosenberg & Rosenberg 1985). The next stage of degradation involves microorganisms with high cell-surface hydrophobicity, which can adhere to the residual high molecular weight hydrocarbons.

Table 1. Requirements for biodegradation of petroleum.

A. Microorganisms with:
  1. Hydrocarbon-oxidizing enzymes
  2. Ability to adhere to hydrocarbons
  3. Emulsifier-producing potential
  4. Mechanisms for desorption from hydrocarbons

B. Water

C. Oxygen

D. Phosphorus

E. Utilizable nitrogen source

In the case of *A. calcoaceticus* RAG-1, this adherence is due to thin hydrophobic fimbriae (Rosenberg et al. 1982). Mutants lacking these fimbriae failed to adhere to hydrocarbons and were unable to grow on hexadecane. Other bacteria exhibit high cell surface hydrophobicity as a result of a variety of fimbriae and fibrils, outer-membrane and other surface proteins and lipids. Bacterial capsules and other anionic exopolysaccharides appear to inhibit adhesion to hydrocarbons (Rosenberg et al. 1983).

Desorption from the hydrocarbon is a critical part of the growth cycle of petroleum-degrading bacteria. Petroleum is a mixture of thousands of different hydrocarbon molecules. Any particular bacterium is only able to use a part of the petroleum. As the bacteria multiply at the hydrocarbon/water interface of a droplet, the relative amount of nonutilizable hydrocarbon continually increases, until the cells can no longer grow. For bacteria to continue to multiply, they must be able to move from the depleted droplet to another one. *A. calcoaceticus* RAG-1 has an interesting mechanism for desorption and for ensuring that it only reattaches to a droplet of fresh oil (Rosenberg et al. 1989). When cells become starved on the 'used' hydrocarbon droplet or tar ball, they release their capsule. The capsule is composed of an anionic heteropolysaccharide, with fatty acid side-chains, referred to as emulsan (Rosenberg 1986). The extracellular, amphipathic emulsan attaches avidly to the hydrocarbon/water interface, thereby displacing the cells to the aqueous phase. Each 'used' oil droplet or tar ball is then covered with a monomolecular film of emulsan. The hydrophilic outer surface of the emulsan-coated hydrocarbon prevents reattachment of the RAG-1 cells. The released capsule-deficient bacteria are hydrophobic and readily adhere to fresh hydrocarbon substrate.

Many hydrocarbon-degrading microorganisms produce extracellular emulsifying agents. In some cases, emulsifier production is induced by growth on hydrocarbons (Hisatsuka et al. 1971). Mutants which do not produce the emulsifier grow poorly on hydrocarbons (Itoh & Suzuki 1972). Pretreatment of oil with emulsifying agents can both inhibit and stimulate oil biodegradation (e.g. Foght et al. 1989; Nakahara et al. 1981). As discussed above, emulsification may be a by-product of a cell/hydrocarbon detachment process.

Acinetobacter sp. HO1-N accumulates extracellular membrane vesicles of 20–50 nm in diameter when grown on hexadecane (Kappeli & Finnerty 1980). The isolated vesicles partition exogenously supplied hydrocarbons in the form of a microemulsion. These vesicles appear to play a role in the uptake of alkanes. Recently, Miller & Bartha (1989) have been able to overcome the difficulties involved in the transport of water-insoluble, solid hydrocarbons by using unilamellar vesicles. A *Pseudomonas* isolate grew on octadecane (C_{18}) and hexatri-acontane (C_{36}) with K_s values of 2,450 and 2,700 mg/liter, compared to 60 and 41 mg/liter, respectively, when the hydrocarbon was presented in the form of liposomes. The data clearly demonstrate the importance of transport in the microbial metabolism of recalcitrant hydrocarbons.

The low solubility of hydrocarbons in water, coupled to the fact that the first step in hydrocarbon degradation involves a membrane-bound oxygenase, makes it essential for bacteria to come into direct contact with their hydrocarbon substrates.

As mentioned above, microorganisms which have the genetic potential to bind, emulsify, transport and degrade hydrocarbons are widely distributed in Nature. Thus, the rate limiting step in biodegradation of hydrocarbons is not the lack of appropriate microorganisms. Depending upon the particular environmental situation, the extent of degradation depends on the availability of moisture, oxygen and utilizable sources of nitrogen and phosphorus. Thus, effectiveness of a bioremediation program depends on defining the limitations and overcoming them – in a practical way.

The requirements for oxygen and moisture is not a problem in oil spills in aquatic environments. However, on land, oxygen and water are often rate-limiting. If oil has not penetrated too deeply into the ground, then watering and tilling are often advantageous. These problems have been discussed theoretically by Atlas (1991), and Bartha et al. (1990) have provided some recent useful data on the subject.

The major limitation in the biodegradation of hydrocarbons on land and water is an available source of nitrogen and phosphorus. In theory, approximately 150 mg of nitrogen and 30 mg of phosphorus are consumed in the conversion of 1 g of hydrocarbon to cell material. The nitrogen and phosphorus requirements for maximum growth of hydrocarbon oxidizers can generally be satisfied by ammonium phosphate. Alternatively, these requirements can be met with a mixture of other salts, such as ammonium sulfate, ammonium nitrate, ammonium chloride, potassium phosphate, sodium phosphate and calcium phosphate. When ammonium salts of strong acids are used, the pH of the medium generally decreases with growth. This problem can often be overcome by using urea as the nitrogen source. All of these compounds have a high water solubility which reduces their effectiveness in open systems because of rapid dilution. In principle, this problem can be solved by using oleophilic nitrogen and phosphorus compounds with low C:N and C:P ratios. For example, it was found that a combination of paraffinized urea and octyl phosphate was able to replace nitrate and inorganic phosphate, respectively (Atlas & Bartha 1973).

Recently, commercial nitrogen and phosphorus-containing fertilizers that have an affinity for hydrocarbons are being developed for treating oil pollution. The most extensively studied example is the oleophilic fertilizer Inipol EAP 22, an oil-in-water microemulsion, containing a N:P ratio of 7.3:2.8 (Glaser 1991). Inipol EAP 22 is composed of oleic acid, lauryl phosphate, 2-butoxyl-1-ethanol, urea and water. The fertilizer was used in large quantities to treat the oil-contaminated shoreline following the 1989 oil spill in Prince William Sound, Alaska. Although a number of articles (reviewed by Pritchard 1993) have reported that the fertilizer treatment was successful, several questions still remain, mainly because of a lack of appropriate controls. There are at least three problems with Inipool EAP 22. First, it contains large amounts of oleic acid, which serves as an alternative carbon source, thereby further increasing the C:N ratio in the environment. Second, it contains an emulsifier which could be harmful to the environment. Third, as soon as the fertilizer comes into contact with water, the emulsion breaks, releasing the urea to the water phase where it is not available for the microorganisms.

We report here, laboratory and field experiments using a new controlled-release, hydrophobic fertilizer, referred to as F-1.

Isolation of oil-degrading bacteria using fertilizer F-1 as the source of nitrogen and phosphorus

A mixed bacterial culture was obtained by enrichment culture procedure using crude oil as a carbon and energy source, F-1 as a nitrogen and phosphorus source and sea water as the source of other minerals. Fresh sea water and tar served as the enrichment inoculum. After three transfers of 1 ml into 20 ml fresh medium, turbid cultures were obtained during overnight incubation. Microscopic examination demonstrated a mixed bacterial population and emulsified oil droplets. When plated on marine agar, several different colony types appeared from which three different strains were isolated for further studies. Each of these pure cultures was able to grow on and emulsify oil utilizing F-1 as its sole nitrogen and phosphorus source. The bacteria were referred to as ER strains RL3 (ER-LR3), RL4 (ER-RL4) and RT (ER-RT).

Growth of mixed cultures

The kinetics of growth of the mixed culture on crude oil and acetate is shown in Figs. 1 and 2. The mixed culture had a doubling time of approximately 2 h, reaching stationary phase after 40 h at 2×10^8 cells/ml. The pH of the growth medium decreased from an initial value of 7.8 to a minimum of 6.8 at 24 h and then increased to a final value of 7–7.2. The turbidity of the culture rose sharply up to 32000 Klett units (K.U.), mostly because of emulsification of the oil. In order to estimate how much of the turbidity was contributed by the cells, the culture was grown on 0.5% acetate medium reaching 500 K.U. and 2×10^9 cells/ml. Assuming that the K.U./cell was the same in both media, then

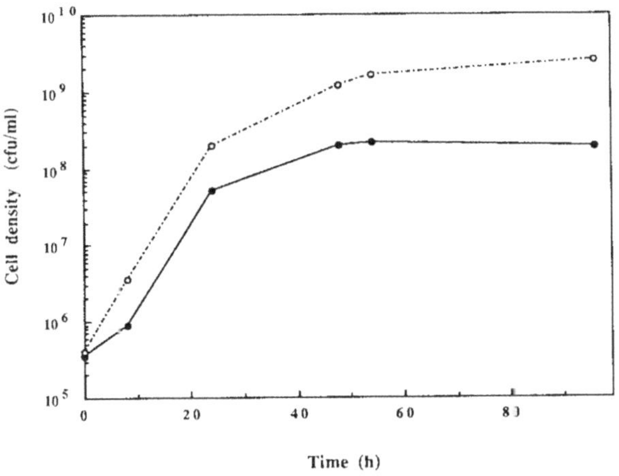

Fig. 1. Growth kinetics of the mixed culture on crude oil (●) and acetate (○) media, using F-1 as the source of nitrogen and phosphorus. To 125 ml flasks containing 10 ml sterile sea water, 0.05 ml crude oil and 0.01 ml of the enrichment culture (three days growth) were added 0.01 ml fertilizer F-1. After incubation with mixing at 25 °C, approximate dilutions were plated for viable cell density (CFU/ml) on Marine agar (18.7 g Marine broth MA-2216 [Difco Labs], 10g NaCl and 16 g agar per liter).

2×10^8 cells/ml in the oil medium would account for only 50 K.U. out of the 3200 K.U. measured. The remainder would be due to the emulsified oil.

The main interest of the mixed culture is in its ability to utilize fertilizer F-1 as the sole nitrogen and phosphorus source when grown on crude oil. Thus, it was essential to demonstrate that the

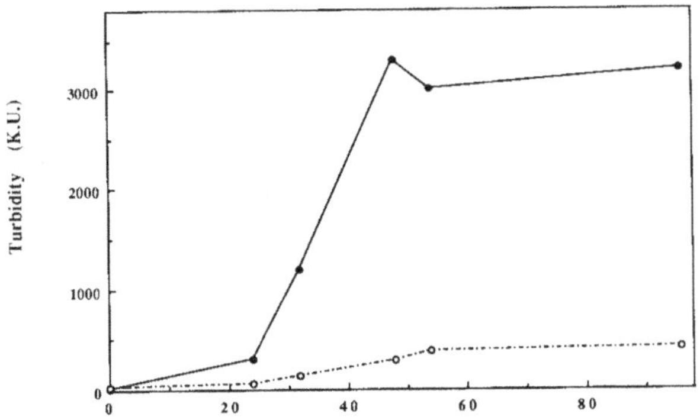

Fig. 2. Growth and emulsification of crude oil by the mixed culture, using F-1 as the source of nitrogen and phosphorus. The experiment was performed as described in Fig. 1, using 0.5% in crude oil (●) and 0.5% acetate (○) as the carbon source.

Table 2. Growth of the mixed culture as a function of fertilizer F-1 concentration.[a]

Fertilizer (%)	Turbidity (K.U.)	Cell density (c.f.u./ml)
0.0	22	3.0×10^6
0.05	800	1.4×10^8
0.1	790	1.6×10^8
0.2	1200	1.8×10^8
0.5	1250	2.1×10^8
1.0	2000	4.0×10^8
1.6	2650	4.0×10^8

[a] The experiment was carried out as described in Fig. 1 using different concentrations of F-1 and incubation for 72 h.

culture's growth was dependent on the fertilizer. Accordingly, the mixed culture was grown on different concentrations of fertilizer as is shown in Table 2. The data show that at the lowest level of fertilizer tested – 0.05%, the culture reached 1.4×10^8 cells/ml and 800 K.U., compared to 3.0×10^6 cells/ml and 22 K.U. without fertilizer. Further increase in the concentration of F-1 led to slight increases in both cell density and turbidity.

As demonstrated above, emulsification of the endogenous crude oil did not proceed in parallel with the viable cell numbers. Thus, it was interest-

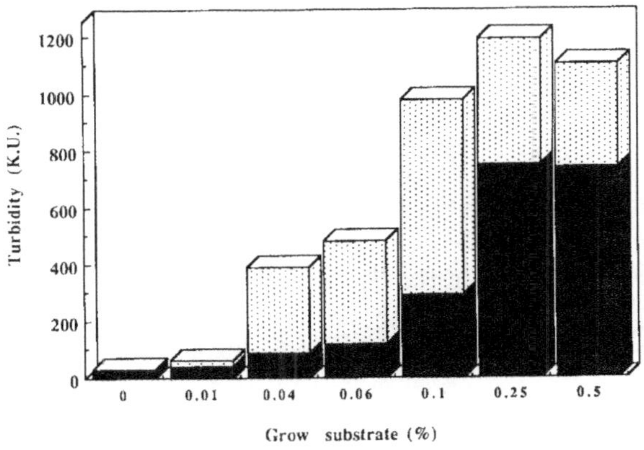

Fig. 3. Emulsifying potential of the mixed culture. To 125 ml flasks containing 10 ml sterile sea water and 0.1% F-1. Different concentrations of crude oil were added. After inoculation with 10 μl of the enrichment culture, the flasks were incubated with mixing at 25 °C for 72 h. Samples were removed for determination of turbidity (dark bars). Crude oil was then added to the flask to reach a final concentration of 1%, mixed at 25 °C for 1 h and then turbidity determined (spotted bars).

ing to determine how much oil could be emulsified when it was added at the end of the growth experiment (Fig. 3). The data demonstrate that the cultures have the potential to emulsify higher amounts of oil than were present in the growth media. This effect was especially pronounced when 0.1% crude oil was used as the substrate. The turbidity in that experiment rose to 980 K.U. with addition of 0.9% crude oil, compared to 290 K.U. at the end of the growth period. When higher concentrations of oil were used as growth substrate, e.g. 0.5%, then the increase in turbidity following addition of oil and shaking was reduced.

Table 3. Characterization of strains.

Property	Strain		
	RL3	RL4	RT
Colony on marine agar			
diameter	2 mm	<1 mm	1–2 mm
color	white	white	yellowish
Bacterial shape	rod	rod	rod
Dimensions (µm) of cells			
from marine agar (16 h)	0.3/1.5	0.6/1.2	0.6/1.0
from broth, on oil (72 h)	0.5/1.6	0.5/2.2	0.44/1.1
Motility	+	+	+
Flagellar arrangement			
from marine agar (16 h)	peritrichal	polar	none
from broth, on oil (72 h)	none	polar	none
Growth temperature, °C			
4 °C (100 h)	–	–	–
20–25 °C (24 h)	+	+	+
37 °C (24 h)	+	+	+
41 °C (100 h)	–	–	–
NaCl (6%) tolerance	+	+	+
NaCl requirement for growth	–	+	+
Lipase	+	+	+
Oxidase	+	+	+
Catalase	+	+	+
Starch hydrolysis	–	+	+
Plasmids	–	2 > 60 kb	2 > 60 kb 1–3 kb
Antibiotic sensitivity			
ampicillin	S[b]	S	S
tetracycline	R[b]	R	R
penicillin G	S	S	R
erythromycin	S	S	S
nalidixic acid	S	R	R

[a] Determined by electron microscopy.
[b] S is sensitive and R is resistant.

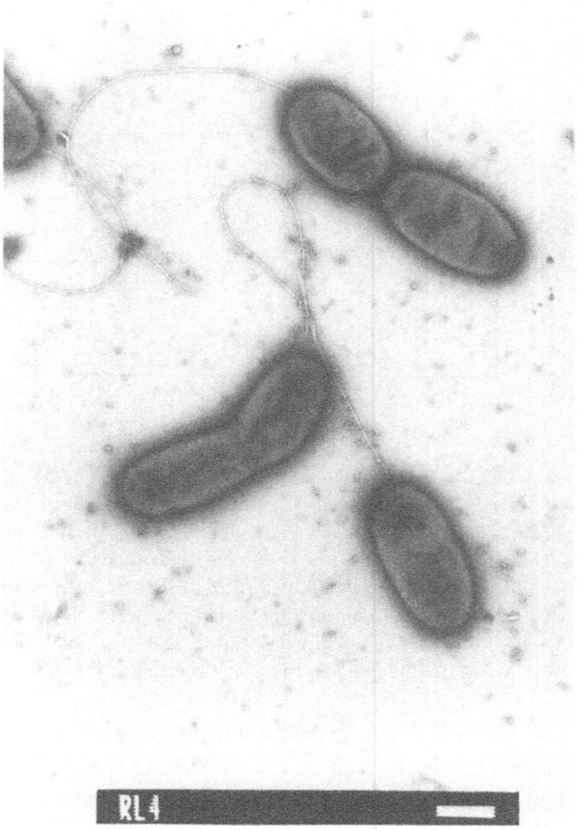

Fig. 4. Electron microscopy of the pure strains: RL3 (A), RL4 (B) and RT (C).

Characterization of the pure culture

The three pure bacterial strains that were isolated from the mixed culture were able to grow on and emulsify crude oil, utilizing fertilizer F-1 as a nitrogen and phosphorus source. The general characteristics of the three strains are summarized in Table 3. All of the strains are gram negative, aerobic, motile, oxidase positive, urease negative and rod shaped. Electronmicrographs of the strains are shown in Fig. 4. The strains have been tentatively identified as *Pseudomonasceae* genus, *Gluconbacter* (strain RT), *Pseudomonadaceae* genus *Pseudomonas* (strain RL4) and *Pseudomonas alcaligenes* (strain RL3)

The kinetics of growth and emulsification of the three strains are shown in Figs. 5–7. All the strains

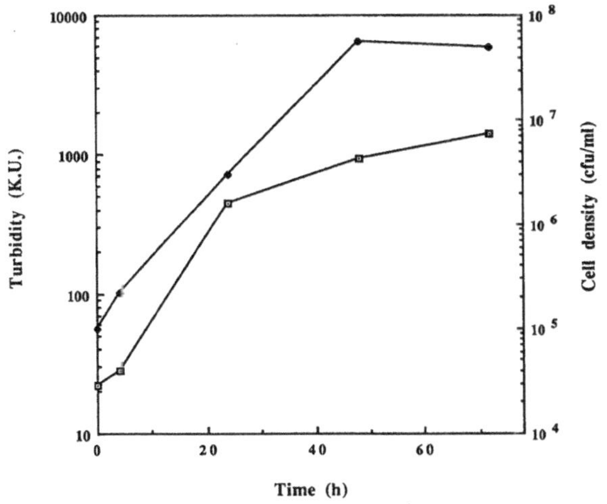

Fig. 5. Kinetics of growth and emulsification of crude oil by strain RL3. Sea water (20 ml) containing 0.5% crude oil and 20 mg fertilizer F-1 was inoculated with 20 µl of strain RL3 (three days growth). The culture was incubated at 25 °C with shaking; samples were removed for the determination of viable count (●) and turbidity (□).

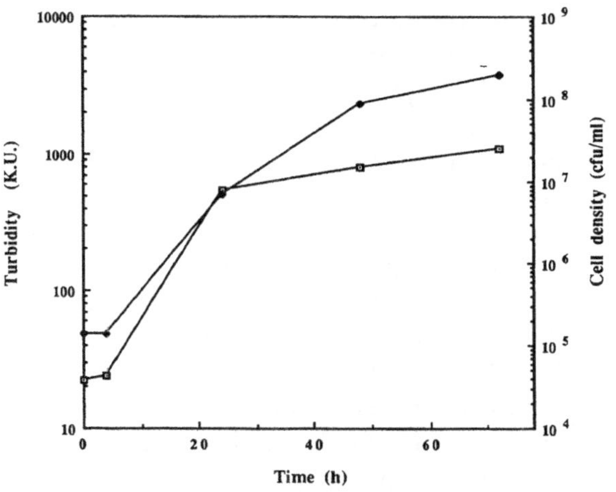

Fig. 6. Kinetics of growth and emulsification of crude oil by strain RT. The experiment was performed as described in Fig. 5.

have the ability to grow and emulsify oil, using F-1 as a nitrogen and phosphorus source. The strains showed similar doubling times, approximately 3h. Strain RT reached the highest growth density (2×10^8 c.f.u./ml), although its emulsification (1000 K.U.) ability was almost the same as the other two strains. The pH in the growth media (in all th strains) decreased from an initial value of 7.8 to about 7.0 at the end of the growth period. Here, as in the case of the mixed culture, the turbidity of the cultures came mostly from the emulsified oil.

Table 4 shows the ability of the pure cultures to grow on different hydrocarbons. The nitrogen and phosphorus source was fertilizer F-1. All of the strains grew on decane, hexane, hexadecane, kerosine, crude oil and gas oil and failed to grow on toluene, naphthalene and heptane. Only strain RL3 grew on xylene. Tetradecane, iso-octane and paraffin oil served as substrate for strains RL3 and RL4, but not for strain RT. The failure of the strains to grow on toluene, naphthalene and heptane could be the result of the toxicity of these hydrocarbons when present at 0.5%. Lower concentrations were not tested.

When grown on crude oil medium, all three pure cultures adhered to the oil. An example of strain RL3 adhering to an oil droplet is shown in Fig. 8.

Adhesion of fertilizer F-1 to crude oil

A key property of F-1 is that it is insoluble in water and binds to the hydrocarbon/water interface. This phenomenon is demonstrated qualitatively in Fig. 9 and quantitatively in Table 5. When the powdered F-1 was added to a suspension of crude oil in

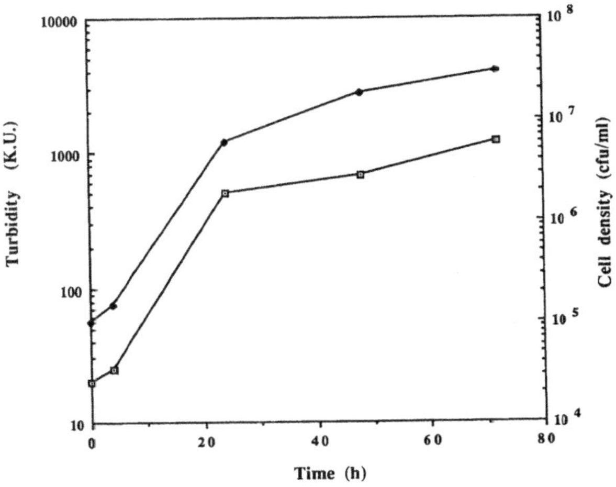

Fig. 7. Kinetics of growth and emulsification of crude oil by strain RL4. The experiment was performed as described in Fig. 5.

sea water, the fertilizer adhered to the oil and floated to the surface.

One of the major problems with supplementing a contaminated oil spill in the sea or in a moist environment with water soluble nitrogen and phosphorus compounds is that they will not be concentrated near the oil where growth must occur. Rather, the compounds will spread and dilute in the open system. It was therefore interesting to test whether or not this problem could be overcome by use of the fertilizer F-1. The procedure that was used compared growth and emulsification (turbidity) after the fertilizer or soluble nutrients and oil were mixed and the aqueous phase was removed and replaced with fresh sea water. The data (Table 5) clearly show that F-1 can support growth on crude oil even after the aqueous phase had been removed and replaced twice with fresh sea water containing no additions. None of the cultures that received ammonium sulfate and phosphate buffer, and treated in a similar manner, grew significantly. This demonstrates that the soluble nutrients were efficiently removed by the wash out process. Strain *A. calcoaceticus* RAG-1 (Rosenberg et al. 1982) which grew well on crude oil supplemented with ammonium sulfate and phosphate salts, failed to grow on F-1 as its course of nitrogen and phosphorus, no matter whether a wash out procedure took place or not. The pure strains RL3, RL4 and RT and the mixed culture grew after the dilution procedure, reaching more than 10^7 cells/ml and high values of turbidity. The mixed culture showed the highest ability to grow and reached 1×10^8 cells/ml and 900 K.U. Strain RL3 was almost as effective as the mixed culture in growth yield and emulsification of the crude oil.

Table 4. Hydrocarbon utilization by pure strains.

Hydrocarbon[a]	Strains		
	RL3	RL4	RT
Decane	+	+	+
n-hexane (vapor)	+	+	+
Toluene	–	–	–
Xylene	+	–	–
Naphthalene	–	–	–
Hexadecane	+	+	+
Tetradecane	+	++	–
Heptane	–	–	–
Kerosene	+	+	+
Crude oil	+	+	+
Gas-oil	+	+	+
Iso-octane	+	+	–
Paraffin oil	++	++	–

[a] Except for n-hexane, growth was tested in shake flasks containing 0.5% of the hydrocarbon and 0.1% F-1.

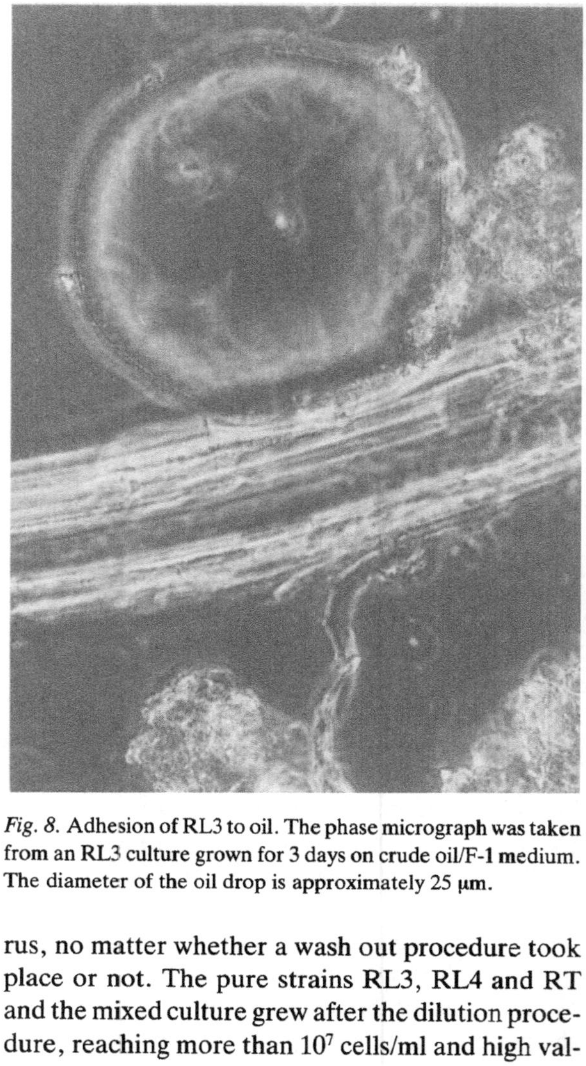

Fig. 8. Adhesion of RL3 to oil. The phase micrograph was taken from an RL3 culture grown for 3 days on crude oil/F-1 medium. The diameter of the oil drop is approximately 25 μm.

Isolation and characterization of the inducible cell-bound activity for F-1 utilization

Regarding fertilizer F-1, it is important to understand the mechanism by which the isolated culture utilized F-1 as its sole nitrogen and phosphorus source. Preliminary experiments indicated that the F-1-utilizing strains had the enzymatic capabilities

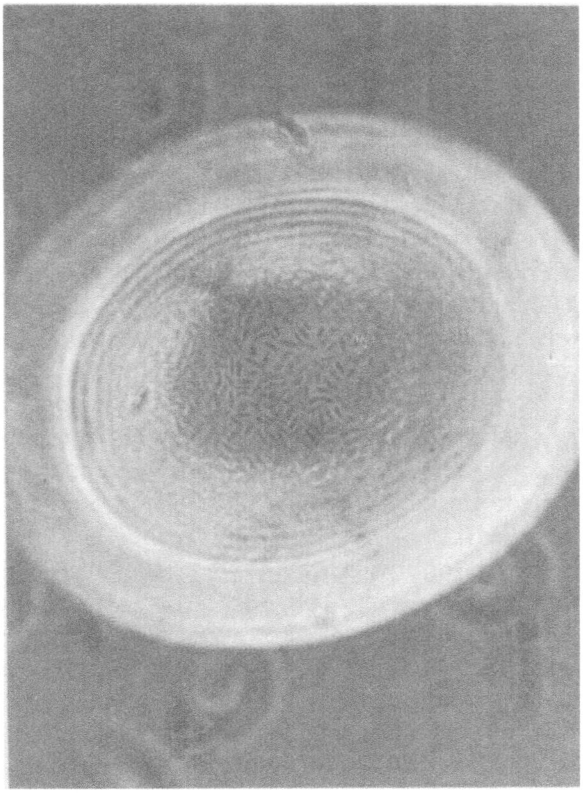

Fig. 9. Adhesion of F-1 to an oil droplet. The photograph was taken through a phase microscope from an RL3 culture grown on crude oil/F-1 medium.

of degrading it. The determination of 'F-1-ase' activity of the pure strains was based on a bioassay for degrading the F-1 to water soluble, low molecular weight compounds. The quantity of low molecular weight F-1 was estimated by a bioassay using strain RL3. The sample to be tested (whole bacteria or bacterial extracts) was initially dialyzed in order to remove low molecular weight nitrogen and phosphorus compounds. It was then incubated with the insoluble fraction of F-1 in a dialysis bag, which was put in water containing cells of an indicator strain (RL3 bacteria) but no nitrogen source. If the sample is active, the F-1 in the dialysis bag will be degraded to low molecular weight compounds that will dialyze out of the bag and enable the bacteria outside the bag to grow. One unit of F-1-ase activity was defined as the activity resulting in bacterial increase of 10^6 cells.

The data presented in Table 6 demonstrate the F-1-ase activity of strains RL3, RL4 and RT. The activity was mainly associated with the cells and not the extracellular fluid.

Table 7 shows that the F-1-ase activity of cells is influence by the source of nitrogen and phosphorus in the growth medium. Cells grown on media containing F-1 had higher activities of F-1-ase than cells grown on ammonium sulfate and phosphate buffer. Strain RL2 yielded the highest specific activity of F-1-ase in broken cell preparations.

Table 5. Growth of bacteria on crude oil and F-1 following removal of water soluble nutrients.[a]

Strain	Cell density (c.f.u./ml)				Turbidity (K.U.)	
	(time 0)		(time 120 h)		(time 120 h)	
	UF-1	AS	UF-1	AS	F-1	AS
RL4	5.4×10^5	6.5×10^5	9.0×10^7	4.0×10^3	550	22
RL3	8.0×10^5	4.0×10^5	9.0×10^7	2.0×10^6	860	25
RT	3.0×10^5	4.0×10^5	6.0×10^7	2.0×10^6	490	20
RAG-1	1.0×10^3	1.0×10^3	1.0×10^3	1.0×10^3	20	20
Mixed culture	3.0×10^5	5.0×10^5	1.0×10^8	1.0×10^6	900	23

[a] F-1 medium contained 0.2% fertilizer (equivalent to 0.05% nitrogen). AS medium contained 0.1% ammonium sulfate (equivalent to 0.2% nitrogen) and 10 mM phosphate buffer. Both media contained 0.5% crude oil as a carbon source. Each 20 ml medium was mixed well and then allowed to stand for 5 min allowing the crude oil and the aqueous phase to separate. The aqueous phase was then removed as completely as possible and replaced with 20 ml fresh sea water. The procedure of removal of the remaining water soluble nutrients was repeated once more. Each flask was then inoculated with 40 μl of a pure or a mixed culture and incubated for 120 h at 20 °C with shaking. Viable counts were determined after inoculation (time 0) and at 120 h. Turbidity was measured at 120 h.

Laboratory experiments using F-1 and the mixed culture to treat oil-contaminated beach sand

The effect of temperature on the biodegradation of crude oil-contaminated sand is shown in Fig. 10. After 30 days at 20 °C and 30 °C, the amount of remaining pentane-extractable hydrocarbon had decreased to 45% and 55%, respectively, of the initial value. At 60 °C, 85% of the hydrocarbon was still present, indicating that the evaporation of the heavy crude oil that was used in the experiment was not a major factor.

Experiments demonstrating the requirement for added bacteria and fertilizer F-1 on biodegradation of crude oil in beach sands is summarized in Table 8. When both bacteria and fertilizer were added, 45–54% of the pentane-extractable hydrocarbons were degraded in 10–15 days. Without addition of fertilizer or bacteria, only 9% was degraded in 15 days. The optimum amount of fertilizer was approximately 5–10 mg per g sand. The Ashdod sand showed a strong requirement for added bacteria, 45% degradation with bacteria, compared to 8% without added bacteria. It should be noted that soil chronically polluted with crude oil can have a relatively high indigenous concentration of useful bacteria, suggesting that the requirement for added bacteria will vary, depending on the condition of the soil.

Table 6. F-1-ase activity of pure cultures.[a]

Fraction assayed	F-1(ase) activity (units)
Control	190
RL3 broth	1,000
RL3 cells	16,000
RL4 broth	800
RL4 cells	17,000
RT broth	1,200
RT cells	18,000

[a] The control was either sea water or a culture of RAG-1. For strains RL3, RL4 and RT, cultures were prepared on Acetate medium containing 0.1% F-1. After 3 days incubation, the cells were separated from the culture broth by centrifugation at 10,000 × g for 15 min. The cells were then resuspended in sea water to 10^9 per ml and assayed as described in the text. The cell-free broth was concentrated 150-fold prior to the assay.

Field experiment using F-1 and the mixed culture to treat oil-contaminated Haifa beach-sand

In August 1991, approximately 100 tons of heavy crude oil was accidentally spilled about 3 km north of Zvulon Beach between Haifa and Akko in Israel. The oil-contaminated sand was collected into piles and subsequently spread over the beach.

Prior to the field trial, a simulation trial for bioremediation of the Haifa contaminated beach sand was performed in the laboratory, The experiment was performed on one kilogram samples of the hydrocarbon-contaminated sand from Haifa beach in plastic containers. The results (Fig. 11) indicated that in a period of about 5 weeks the concentration of hydrocarbons was reduced by 10% without any treatment. Plowing and wetting resulted in a 20% reduction of hydrocarbon, while the microbial treatment reduced the hydrocarbons by about 70%. Initially, two plots of 50 m^2 were selected, one for the experiment and the second to serve as a control.

In the field trial the experimental plot was inoculated with 20 l of the mixed culture containing strains RL3, RL4 and RT. Fertilizer F-1 (38 kg, in the form of a fine powder) was then added as the source of nitrogen and phosphorus. The experiment plot was watered from the adjacent sea (water temperature was 27 °C). The plot was then tilled by hand with the help of a simple rake to the depth of about 5 cm. The control plot was left undis-

Table 7. Induction of F-1-ase activity by F-1.

Strain	N,P Source[a]	Protein (mg/ml)	F-1(ase) activity	
			U/ml	U/mg protein
RL2	F-1	0.33	1,400	4,200
RL3	AS	0.04	40	1,000
RL4	F-1	0.23	290	1,230
RL4	AS	0.02	5	230
RT	F-1	0.30	1,000	3,300
RT	AS	ND	ND	ND

[a] The strains were grown in Acetate medium containing either 0.1% F-1 or 0.2% ammonium sulfate and 20 mM phosphate buffer (AS). After 3 days of growth, the cells were harvested, sonicated and assayed as described in the text.

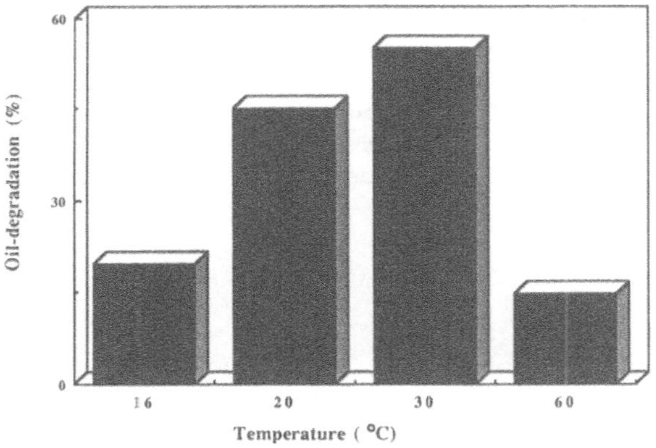

Fig. 10. Temperature dependence of enhanced biodegradation of hydrocarbon-contaminated sand from Haifa beach. To 300 g of hydrocarbon-contaminated sand (7.15 mg/g sand) from Haifa beach in 100 cm^2 plastic containers were added F-1 (10 mg/g sand), followed by addition of 2 ml of the 'mixed culture' (1×10^8). The containers were incubated for 30 days at various temperatures. The sand was plowed and kept moist by addition of sea water daily. The amount of residual hydrocarbon remaining in the sand was determined by dry weight following pentane extraction.

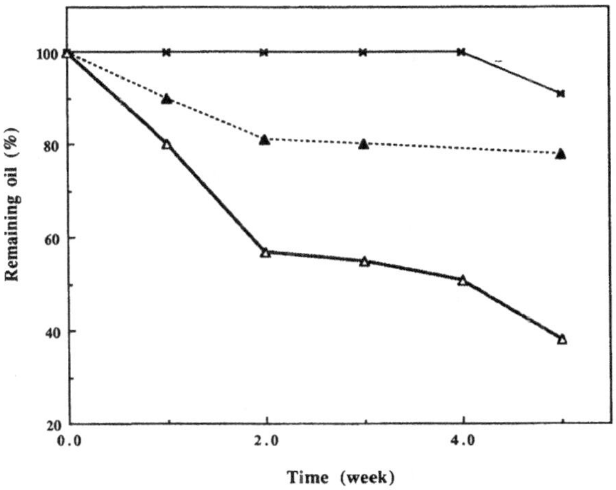

Fig. 11. Trial for bioremediation technology and application to the Haifa contaminated beach sand. One kilogram samples of hydrocarbon-contaminated sand (4.7 mg/g sand) from Haifa beach in 100 cm^2 plastic containers were treated as follows: no treatment (×), plowed and kept moist (▲) and addition of fertilizer F-1 (5 mg/g sand) and 2 ml bacterial 'mixed culture' 5×10^8 cells/ml (△). The amount of residual hydrocarbon remaining in the sand was determined by dry weight following pentane extraction (as described in Materials and methods) from removed samples of 50 g at various times. The containers were incubated for 5.5 weeks outdoors in temperatures varying from 20–35 °C. The sand was plowed and kept moist by addition of sea water daily.

turbed. For each time point, triplicate 50 g samples were extracted and the results presented are the average of the three values. Each 50 g sample was placed in a 500 ml bottle that contained 50 ml n-pentane. After shaking vigorously for 5 min, 25 ml of the pentane extract were transferred to a 100 ml flask containing anhydrous $CaCl_2$. After standing for 5 min with occasional stirring, the dried pentane extract was filtered through a Whatman 2V paper and then evaporated in vacuo (water aspirator) at 30 °C. The residual hydrocarbon was dissolved in 3 ml n-pentane and placed in small glass vials in a chemical hood. The weight of the pentane-extractable hydrocarbon was determined by weighing to constant weight at room temperature on an analytical balance. A control of n-pentane taken through the same procedure contained 0.54 mg residues per 100 ml pentane. Thus, the contribution of the nonvolatiles in the pentane to the values reported, was 0.006 mg/g sand, which is negligible. The values reported were, however, corrected for the moisture content of the specific sand sample.

The experimental data are presented in Tables 9

Table 8. Biodegradation of oil-contaminated beach sand.[a]

Sand source	Addition (per g sand)		Time (days)	Degradation[b] (%)
	F-1 (mg/g)	Bacteria		
Ashdod	6.7	0	10	8
"	6.7	+	10	48
Tel Baruch	0	+	12	5
"	3.3	+	12	35
"	6.7	+	12	51
"	10	+	12	54
"	20	+	12	46

[a] To 500 g of sand from the beach in 100 cm^2 plastic containers were added 2 ml of heavy crude oil (6.9 mg/g sand) that were applied over the surface with a pasteur pipet. The containers were then allowed to remain at room temperature (20–32 °C) for 1 day (to let the volatile fractions evaporate). Fertilizer F-1 was then spread on the surface, followed by addition of 2 ml of the 'mixed culture' (8.2×10^8 cells/ml). The sand was kept moist by addition of sea water daily.

[b] The amount of hydrocarbon remaining was determined by dry weight following pentane extraction.

and 10. On day 0 (September 1, 1991), the core samples were taken prior to any treatment. At the beginning of the experiment, the experimental plot contained significantly more hydrocarbon (3.80 mg/g sand) than the control plot (2.30 mg/g sand). There was only a small decrease in the first day. However, by the fourth day, 30% of the hydrocarbon had been degraded. The biodegradation continued, reaching 50% on day 9 and 84.5% on day 25, when the experiment was concluded. The control plot showed a 18% degradation by day 9 which remained relatively constant until day 25.

The relatively small variations from day to day were not due to the pentane extraction procedure, because the average standard deviations were very low (0.03 mg/g sand, corresponding to less than 2%). Therefore, we assume that the core sampling was probably the source of the day to day variation.

After the first day, the experimental plot was watered every afternoon with 1.5 m^2 sea water and then raked to a depth of 5 cm. Ten random core samples (5 cm diameter × 10 cm depth) were taken, prior to the daily watering, from both the experimental and control plots. The samples were mixed thoroughly in the field and brought to the laboratory for analysis.

Gas chromatographic analyses of the alkane fraction of the extracted sand from the experimental plot at day 0 and at day 25 are summarized in Table 11. Extractions with pentane and CCl_4 gave similar results, i.e. approximately 85% of the hydrocarbon was removed from the sand in 25 days. The C_{20}–C_{32} fraction was the most abundant at day 0, 221 mg per kg sand and fell to 13 mg/kg sand by day 25, a 94% removal. The C_{14}–C_{18} and C_{36}–C_{40}

Table 9. Enhanced biodegradation of hydrocarbon-contaminated sand at Haifa beach during September 1991.[a]

Day	Pentane-extractables[b] (mg/g sand)	Degradation (%)
0	3.80	0
1	3.70	2.6
4	2.76	30
9	1.89	50
14	0.88	77
21	1.40	63
25	0.59	84.5

[a] An experimental plot of 50 m^2 was selected on the Haifa contaminated beach. The initial concentration of hydrocarbon-contamination in the sand was 3.8 mg/g sand in the upper 10 cm. The plot was treated with 38 kg (~ 15 mg/g sand) fertilizer F-1 in the form of a fine powder and inoculated with 20 l of the 'mixed culture' (5 × 10^8 cells/ml). The plot was watered with sea water and tilled by hand with a rake to the depth of about 5 cm daily.
[b] For each determination, ten random core samples were taken and mixed thoroughly in the field. For each time point triplicates of 50 g samples were extracted in pentane and determined by dry weight. The results presented are the average of three values (the average standard deviation was 0.1 mg/g sand).

Table 10. Natural biodegradation (no treatment) of hydrocarbon-contaminated sand at the Haifa beach, September 1991.[a]

Day	Pentane-extractable[b] (mg/g sand)	Degradation (%)
0	2.30	0
4	2.53	0
9	1.88	18
14	1.70	26
21	1.94	15.6
25	1.95	15.6

[a] A 50 m^2 control plot was selected on the Haifa contaminated beach. The initial concentration of hydrocarbon-contamination in the sand was 2.3 mg/g sand in the upper 10 cm. The plot was left undisturbed.
[b] See Table 9; the average standard deviation was 0.15 mg/g sand.

Table 11. Gas chromatographic analysis of the extracted sand from experimental plot.[a]

	Extracted hydrocarbon (mg/kg sand)		Residual hydrocarbon (%)
	Day-0	Day-25	
Total CCl_4 extractable[a]	3138	474	15.1
Total pentane extractable	3800	590	15.5
C.G. analysis[b]			
C_{14}–C_{18}	23.7	4.7	19.8
C_{20}–C_{32}	221	13	5.9
C_{36}–C_{40}	132	35	26.5
C_{14}–C_{40}	377	53	14.1

[a] See Table 9.
[b] Samples of pentane extracted residual hydrocarbon (day-0 and day-25) were analyzed by gas chromatography at the Israel Institute of Petroleum and Energy.

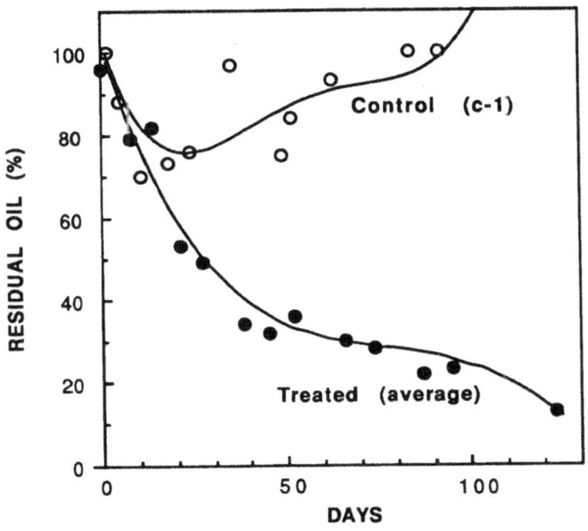

Fig. 12. Microbial clean-up of the Zvulon Beach. The Zvulon Beach in Haifa was treated with bacteria and fertilizer (●) or untreated (○). The treatment and analyses was carried out essentially as described for Fig. 11 and in the text.

fractions were degraded to lesser degrees, 80% and 74%, respectively. Overall, there was an 86% degradation of the alkanes, similar to the data obtained by dry weight analyses.

Visual examination of the beach sand following the treatment, in addition to the analytical data described above, indicated that the F-1/bacteria technology was applicable for bioremediation of the remaining sand.

Microbial clean-up of the Zvulon Beach

The clean-up was of 30,000 m^2 which were treated essentially as in the field trial. As the clean-up was in winter time, it was not necessary to water. It should be noted that the winter of 1992 was unusually hard, and temperatures were around 5–10 °C for a couple of months. The results of the experiment (a brief summary presented in Fig. 12) indicated that even under these conditions 88% of the oil was degraded after 4 months.

Acknowledgements

We wish to thank Naomi Kayam for technical assistance, Nimrod Otiz, Chaim Abir and Motti Ben-Yehuda for help with the field experiments, and Makhteshim Chemical Works Ltd for financial support.

References

Atlas RM (1991) Microbial hydrocarbon degradation-bioremediation of oil spills. J. Chem. Tech. Biotechn. 52: 149–156

Atlas RM & Bartha R (1973) Stimulated biodegradation of oil slicks using oleophilic fertilizers. Environ. Sci. & Technol. 7: 538–541

Bartha R (1990) Bioremediation potential of terrestrial fuel spills. AEM 56: 652–656

Cerniglia CE (1984) Microbial transformation of aromatic hydrocarbons. In: Atlas RM (Ed) Petroleum Microbiology (pp 98–128). Macmillan Publishing Co., New York

Crutcher SE & Geary PJ (1979) Properties of the iron-sulphur proteins of the benzene dioxygenase system from *Pseudomonas putida*. Biochem. J. 177: 393–400

Dagley S (1971) Catabolism of aromatic compounds by microorganisms. Adv. Microbiol. Physiol. 6: 1–46

Foght JM, Gutnick DL & Westlake DWS (1989) Effect of emulsan on biodegradation of crude oil by pure and mixed bacterial cultures. Appl. Environ. Microbiol. 55: 36–42

Gibson DT (1968) Microbial degradation of aromatic compounds. Science 161: 1093–1097

Giger W & Blumer M (1974) Polycyclic aromatic hydrocarbons in the environment: Isolation and characterization by chromatography, visible, ultraviolet and mass spectrometry. Analytical Chemistry 46: 1663–1671

Glaser JA (1991) Nutrient-enhanced bioremediation of oil-contaminated shoreline: The Valdez experience. In: Hinchee RE & Olfenbuttel RF (Eds) On Site Bioreclamation (pp 336–384). Butterworth-Heinemann, Stoneham, MA

Hisatsuka K, Nakahara T, Sano N & Yamada K (1971) Formation of rhamnolipid by *Pseudomonas aeruginosa* and its function in hydrocarbon fermentation. Agric. Biol. Chem. 35 686–692

Itoh S & Suzuki T (1972) Effect of rhamnolipids on growth of *Pseudomonas aeruginosa* mutant deficient in n-paraffin utilizing ability. Agric. Biol. Chem. 36: 2233–2235

Kappeli O & Finnerty WR (1980) Characteristics of hexadecane partition by the growth medium of *Acinetobacter* sp. Biotech. Bioeng. 22: 495–503

Markovetz AJ (1971) Subterminal oxidation of aliphatic hydrocarbons by microorganisms. CRC Crit. Rev. Microbiol. 1: 225–238

McKenna EJ & Kallio RE (1965) The biology of hydrocarbons. Ann. Rev. Microbiol. 19: 183–208

Miller RM & Bartha R (1989) Evidence from liposome encapsulation for transport-limited microbioal metabolism of solid alkanes. Appl. Environ. Microbiol. 55: 269–274

Nakahara T, Hisatsuka K & Minoda Y (1981) Effect of hydrocarbon emulsification on growth and respiration of microorganisms in hydrocarbon media. J. Ferm. Technol. 59: 415–418

Perry JJ (1984) Microbial metabolism of cyclic alkanes. In: Atlas RM (Ed) Petroleum Microbiology (pp 61–98). Macmillan Publishing Co., New York

Pritchard PH, Mueller JG, Rogers JC, Kremer FV & Glaser JA (1992) Oil spill bioremediation: experiences, lessons and results from the *Exxon Valdez* oil spill in Alaska. Biodegradation 3: 315–335 (this issue)

Rosenberg E (1986) Microbial surfactants. CRC Crit. Rev. in Biotechnol. 3: 109–132

Rosenberg E (1991) Hydrocarbon-oxidizing bacteria. In: Ballows, A (Ed) The Prokaryotes (pp 441–459). Springer-Verlag, Berlin

Rosenberg M, Hayer EA, Delaria J & Rosenberg E (1982) Role of thin fimbriae in adherence and growth of *Acinetobacter calcoaceticus* RAG-1 on hexadecane. Appl. Environ. Microbiol. 44: 929–937

Rosenberg E, Kaplan N, Pines O, Rosenberg M & Gutnick D (1983) Capsular polysaccharides interfere with adherence of *Acinetobacter*. FEMS Microbiol. Lett. 17: 157–161

Rosenberg M & Rosenberg E (1985) Bacterial adherence at the hydrocarbon-water interface. Oil & Petrochem. Pollution 2: 155–162

Rosenberg E, Rosenberg M, Shoham Y, Kaplan N & Sar N (1989) Adhesion and desorption during growth of *Acinetobacter calcoaceticus* on hydrocarbons. In: Cohen Y & Rosenberg E (Eds) Microbial Mats (pp 218–226). ASM Publications, Washington, DC

Singer ME & Finnerty WR (1984) Microbial metabolism of straight-chain and branched alkanes. In: Atlas RM (Ed) Petroleum Microbiology (pp 1–60). Macmillan Publishing Co., New York

Stevenson FJ (1966) Lipids in soil. J. Am. Oil Chem. Soc. 43: 203–210

Biodegradation 3: 351–368, 1992.

Biodegradation of polycyclic aromatic hydrocarbons

Carl E. Cerniglia
Microbiology Division, National Center for Toxicological Research, Jefferson, AR 72079, USA

Key words: biodegradation, degradation, detoxification, dioxygenase, hydroxylation, monooxygenase, polycyclic aromatic hydrocarbons, ring cleavage

Abstract

The intent of this review is to provide an outline of the microbial degradation of polycyclic aromatic hydrocarbons. A catabolically diverse microbial community, consisting of bacteria, fungi and algae, metabolizes aromatic compounds. Molecular oxygen is essential for the initial hydroxylation of polycyclic aromatic hydrocarbons by microorganisms. In contrast to bacteria, filamentous fungi use hydroxylation as a prelude to detoxification rather than to catabolism and assimilation. The biochemical principles underlying the degradation of polycyclic aromatic hydrocarbons are examined in some detail. The pathways of polycyclic aromatic hydrocarbon catabolism are discussed. Studies are presented on the relationship between the chemical structure of the polycyclic aromatic hydrocarbon and the rate of polycyclic aromatic hydrocarbon biodegradation in aquatic and terrestrial ecosystems.

Introduction

Polycyclic aromatic hydrocarbons (PAHs) constitute a class of hazardous organic chemicals consisting of three or more fused benzene rings in linear, angular and cluster arrangements (Fig. 1). PAHs mostly occur as a result of fossil fuel combustion, as by-products of industrial processing, and during the cooking of foods (Lijinsky 1991) (Table 1). PAHs enter the environment from a multiplicity of sources which include: direct aerial fallout, chronic leakage of industrial or sewage effluents, accidental discharges during the transport, use and disposal of petroleum products, or from natural sources such as oil seeps and surface water run-off from forest and prairie fire sites (Giger & Blumer 1974; LaFlamme & Hites 1978). More specifically, industrial effluents from coal gasification and liquefaction processes, waste incineration, coke, carbon black, and other petroleum-derived products release high quantities of PAHs into the environment (Table 1). Exposure to PAHs constitutes a significant health risk for people living in industrialized areas of the world (Thakker et al. 1985; Dipple et al. 1990). PAHs are on the United States Environmental Protection Agency's priority pollutant list since some are known carcinogens and mutagens (Keith & Telliard 1979) (Fig. 1).

PAHs are hydrophobic compounds and their persistence within ecosystems is due chiefly to their low water solubility. PAHs rapidly become associated with sediments (Hites et al. 1977, 1980; Gschwend & Hites 1981; Means et al. 1980), where they may become buried and persist until degraded, resuspended, bioaccumulated, or removed by dredging. The lipophilicity, environmental persistence, and genotoxicity increase as the molecular size of PAHs increases up to 4 or 5 fused benzene rings (Fig. 1) and toxicological concern shifts towards chronic toxicity, primarily carcinogenesis (Jacob et al. 1986; Miller & Miller 1981). The possible fates of PAHs in the environment include volatilization, photooxidation, chemical oxidation, bioaccumulation, adsorption to soil particles, leaching and microbial degradation (Fig. 2).

The concentration of PAHs in the environment

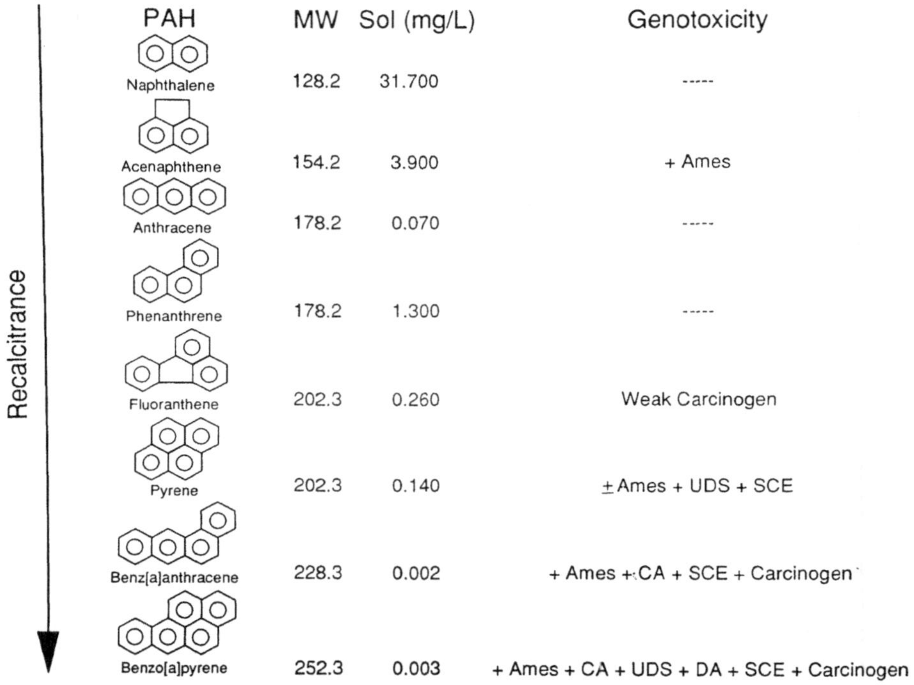

Fig. 1. Chemical structures, physical and toxicological characteristics of polycyclic aromatic hydrocarbons. The symbols are: (DA) DNA adducts, (SCE) sister chromatid exchange, (CA) chromosomal aberrations, (Ames) *Salmonella typhimurium* reversion assay, (UDS) unscheduled DNA synthesis, (–) not genotoxic.

varies widely, depending on the level of industrial development and contamination with petroleum products (Jones et al. 1989; Wild et al. 1991a, 1991b). For example, PAH contamination ranges from a low of 5 ng/g of soil in an undeveloped area to 1.79×10^6 ng/g at an oil refinery. Additionally, the concentration of PAHs in marine sediments can exceed 10^5 ng/g in urban estuaries (Johnson & Larsen 1985). A major source of PAH contamination is creosote because PAHs represent approximately 85–90% of creosote constituents. A representative example of a capillary column gas chromatographic analysis of an organic solvent extract of a PAH-contaminated soil from a creosote waste site is presented in Fig. 3.

Recent studies have shown that microbiological degradation of PAHs is the major process that results in the decontamination of sediment and surface soil (Sims et al. 1990). These compounds can be totally degraded (mineralized) or be partially transformed (Fig. 2) by either a community of microorganisms or by a single microorganism (Gib-

Table 1. Major sources and transport mechanisms of PAHs in the environment.

Natural oil seeps
Refinery and oil storage wastes
Accidental spills from oil tankers and other ships
Municipal and urban wastewater discharge runoff
River-borne pollution
Atmospheric fallout of fly ash particulates
Petrochemical industrial effluents
Coal tar and other coal processing wastes
Automobile engine exhausts
Combustion of fossil fuels (gasoline, kerosene, coal, diesel fuel)
Smoked, charcoal broiled, or pan fried foods
Tobacco and cigarette smoke
Forest and prairie fires
Rural and urban sewage sludge
Refuse and waste incineration
Coal gasification and liquefaction processes
Creosote and other wood preservative wastes
Commercial and pleasure boating activities

Reference: IARC (1972–1990) Monographs on the Evaluation of Carcinogenic Risks to Humans, Vol 1–49, International Agency for Research in Cancer, Lyons, France.

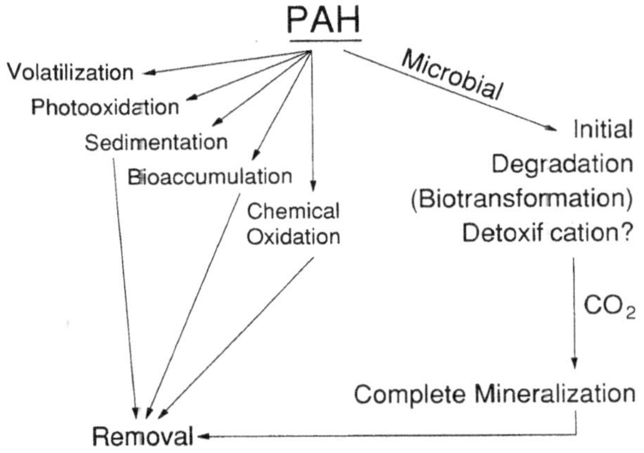

Fig. 2. Schematic representation of the environmental fate of polycyclic aromatic hydrocarbons.

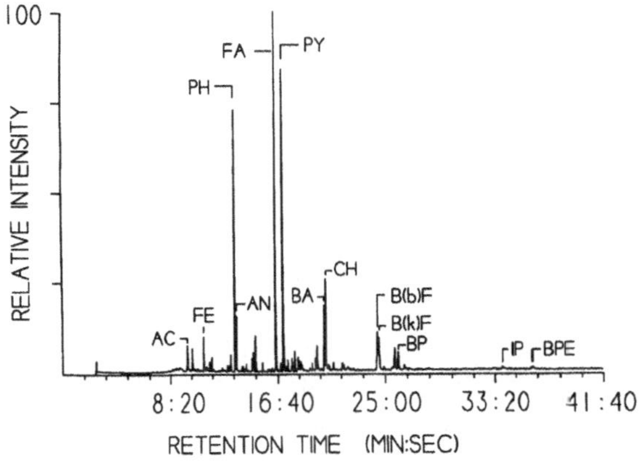

Fig. 3. Capillary gas chromatographic trace of a polycyclic aromatic hydrocarbon contaminated soil sample. PAHs and their symbol abbreviations are: (AC) Acenaphthene, (ACY) Acenaphthylene, (AN) Anthracene, (BPE) Benzo[ghi]perylene, (BP) Benzo[a]pyrene, (B(b)F) Benzo[b]fluoranthene, (B(k)F) Benzo[k]fluoranthene, (BA) Benz[a]anthracene, (CH) Chrysene, (DBA) Dibenz[a,h]anthracene, (FA) Fluoranthene, (FE) Fluorene, (IP) Indeno[1,2,3-cd]pyrene, (NA) Naphthalene, (PH) Phenanthrene, (PY) Pyrene.

son & Subramanian 1984; Cerniglia 1984; Cerniglia & Heitkamp 1989). Bioremediation of soils contaminated with PAHs is an alternative means of detoxifying hazardous waste, which may replace traditional methods such as incineration (Aronstein et al. 1991; Ellis et al. 1991). This review provides recent information on the general principles of the microbial degradation of PAHs, with special emphasis on the biochemical mechanisms for the degradation.

General principles of polycyclic aromatic hydrocarbon metabolism

The metabolism of PAHs by pure cultures of microorganisms and cometabolic transformations by mixed microbial communities have been studied for almost 80 years. However, in order for bioremediation to be considered a viable technology for the decontamination of PAH-contaminated sites, we need to understand more about the microorganisms, enzymatic processes and the environmental conditions necessary to optimize the degradation of these compounds. The known principles in the metabolism of PAHs are summarized below:

1. A wide variety of bacteria, fungi and algae have the ability to metabolize PAHs (Tables 2–4). Generally, the rate of degradation of PAHs is inversely proportional to the number of rings in the PAH molecule. Thus, the lower weight PAHs are biodegraded more rapidly than the higher weight compounds (Cerniglia & Heitkamp 1989).
2. The microbial degradation of PAHs such as naphthalene, phenanthrene, anthracene and acenaphthene has been well documented and the biochemical pathways have been elucidated (Cerniglia 1984; Cerniglia et al. 1992; Gibson & Subramanian 1984; Schocken & Gibson 1984; Pothuluri et al. 1992b). Less is known about the microorganisms capable of utilizing high molecular weight PAHs as sole sources of carbon and energy (Barnsley 1975; Heitkamp & Cerniglia 1988, 1989; Heitkamp et al. 1988; Kelley & Cerniglia 1991; Mueller et al. 1990; Walter et al. 1991; Weissenfels et al. 1990, 1991).
3. Biodegradation mechanisms, both procaryotic and eucaryotic, require the presence of molecular oxygen to initiate enzymatic attack on the PAH rings (Gibson et al. 1968; Chapman 1979; Dagley 1971, 1975; Gibson & Subramanian 1984). Other enzyme systems, such as methane monooxygenases (Colby et al. 1977, 1978; Dalton et al. 1981) and lignin peroxidases (Bumpus 1989; Bumpus et al. 1985; Gold et al. 1989;

Haemmerli et al. 1986; Hammel 1989; Aust 1990) may also be important in the catabolism of PAHs. There is also increasing interest and speculation concerning anaerobic decomposition of PAHs (Vogel & Grbic-Galic 1986; Grbic-Galic & Vogel 1987; Mihelcic & Luthy 1987, 1988).

4. The involvement of plasmids in the degradation of toluene, naphthalene, biphenyl and phenanthrene has been reported (Williams 1981; Zylstra & Gibson 1991; Zylstra et al. 1990). Some catabolic genes, including regulatory genes, have been extensively analyzed (Zylstra et al. 1991). However, little is known about the genetic and regulatory mechanisms involved in the bacterial catabolism of high molecular weight PAHs.
5. Lower molecular weight PAHs, such as naphthalene and phenanthrene, are degraded rapidly in sediments, whereas higher molecular weight PAHs, such as benz[a]anthracene, chrysene or benzo[a]pyrene, are quite resistant to microbial attack. The potential biodegradation rates for PAHs are higher in PAH-contaminated sediments than in pristine sediments (Bauer & Capone 1985, 1988; Heitkamp & Cerniglia

Table 2. Polycyclic aromatic hydrocarbons oxidized by different species of bacteria.

Compound	Organisms	References
Naphthalene	*Acinetobacter calcoaceticus, Alcaligenes denitrificans, Mycobacterium* sp., *Pseudomonas* sp., *Pseudomonas putida, Pseudomonas fluorescens, Pseudomonas paucimobilis, Pseudomonas vesicularis, Pseudomonas cepacia, Pseudomonas testosteroni, Rhodococcus* sp., *Corynebacterium renale, Moraxella* sp., *Streptomyces* sp., *Bacillus cereus*	Ryu et al. 1989; Weissenfels et al. 1990, 1991; Kelley et al. 1991a; Dunn & Gunsalus 1973; Davies & Evans 1964; Foght & Westlake 1988; Jeffrey et al. 1975; Mueller et al. 1990; Kuhm et al. 1991; Walter et al. 1991; Dua & Meera 1981; Tagger et al. 1990; Garcia-Valdes et al. 1988; Trower et al. 1988; Grund et al. 1992; Cerniglia et al. 1984; Barnsley 1983
Acenaphthene	*Beijerinckia* sp., *Pseudomonas putida, Pseudomonas fluorescens, Pseudomonas cepacia, Pseudomonas* sp.	Chapman 1979; Schocken & Gibson 1984; Ellis et al. 1991
Anthracene	*Beijerinckia* sp., *Mycobacterium* sp., *Pseudomonas putida, Pseudomonas paucimobilis, Pseudomonas cepacia, Rhodococcus* sp., *Flavobacterium* sp., *Arthrobacter* sp.	Colla et al. 1959; Akhtar et al. 1975; Jerina et al. 1976; Evans et al. 1965; Ellis et al. 1991; Weissenfels et al. 1991; Foght & Westlake 1988; Walter et al. 1991; Mueller et al. 1990; Savino & Lollini 1977
Phenanthrene	*Aeromonas* sp., *Alcaligenes faecalis, Alcaligenes denitrificans, Arthrobacter polychromogenes, Beijerinckia* sp., *Micrococcus* sp., *Mycobacterium* sp., *Pseudomonas putida, Pseudomonas paucimobilis, Rhodococcus* sp., *Vibrio* sp., *Nocardia* sp., *Flavobacterium* sp., *Streptomyces* sp., *Streptomyces griseus, Acinetobacter* sp.	Kiyohara et al. 1976, 1982, 1990; Weissenfels et al. 1990, 1991; Keuth & Rehm 1991; Jerina et al. 1976; Colla et al. 1959; West et al. 1984; Kiyohara & Nagao 1978; Heitkamp & Cerniglia 1988; Guerin & Jones 1988, 1989; Treccani et al. 1954; Evans et al. 1965; Foght & Westlake 1988; Mueller et al. 1990; Sutherland et al. 1990; Ghosh & Mishra 1983; Savino & Lollini 1977; Trower et al. 1988; Barnsley 1983
Fluoranthene	*Alcaligenes denitrificans, Mycobacterium* sp., *Pseudomonas putida, Pseudomonas paucimobilis, Pseudomonas cepacia, Rhodococcus* sp., *Pseudomonas* sp.	Kelley & Cerniglia 1991; Walter et al. 1991; Weissenfels et al. 1991; Foght & Westlake 1988; Mueller et al. 1989, 1990
Pyrene	*Alcaligenes denitrificans, Mycobacterium* sp., *Rhodococcus* sp.	Heitkamp et al. 1988; Walter et al. 1991; Weissenfels et al. 1991; Grosser et al. 1991
Chrysene	*Rhodococcus* sp.	Walter et al. 1991
Benz[a]anthracene	*Alcaligenes denitrificans, Beijerinckia* sp., *Pseudomonas putida*	Gibson et al. 1975; Mahaffey et al. 1988; Weissenfels et al. 1991
Benzo[a]pyrene	*Beijerinckia* sp., *Mycobacterium* sp.	Gibson et al. 1975; Heitkamp & Cerniglia 1988; Grosser et al. 1991

Table 3. Polycyclic aromatic hydrocarbons oxidized by different species of fungi.

Compound	Organisms	References
Naphthalene	*Absidia glauca, Aspergillus niger, Basidiobolus ranarum, Candida utilis, Choanephora campincta, Circinella* sp., *Claviceps paspali, Cokeromyces poitrassi, Conidiobolus gonimodes, Cunninghamella bainieri, Cunninghamella elegans, Cunninghamella japonica, Emericellopsis* sp., *Epicoccum nigrum, Gilbertella persicaria, Gliocladium* sp., *Helicostylum piriforme, Hyphochytrium catenoides, Linderina pennispora, Mucor hiemalis, Neurospora crassa, Panaeolus cambodginensis, Panaeolus subbalteatus, Penicillium chrysogenum, Pestalotia* sp., *Phlyctochytrium reinboldtae, Phycomyces blakesleeanus, Phytophthora cinnamomi, Psilocybe cubensis, Psilocybe strictipes, Psilocybe stuntzii, Psilocybe subaeruginascens, Rhizophlyctis harderi, Rhizophlyctis rosea, Rhizopus oryzae, Rhizopus stolonifer, Saccharomyces cerevisiae, Saprolegnia parasitica, Smittium culicis, Smittium culisetae, Smittium simulii, Sordaria fimicola, Syncephalastrum racemosum, Thamnidium anomalum, Zygorhynchus moelleri*	Cerniglia & Gibson 1977; Cerniglia et al. 1978; Smith & Rosazza 1974; Cerniglia & Crow 1981; Ferris et al. 1973
Anthracene	*Bjerkandera* sp., *Cunninghamella elegans, Phanerochaete chrysosporium, Ramaria* sp., *Rhizoctonia solani, Trametes versicolor*	Cerniglia 1982; Cerniglia & Yang 1984; Hammel et al. 1991; Sutherland et al. 1992; Field et al. 1992
Acenaphthene	*Cunninghamella elegans*	Pothuluri et al. 1992b
Phenanthrene	*Cunninghamella elegans, Phanerochaete chrysosporium, Trametes versicolor*	Cerniglia & Yang 1984; Cerniglia et al. 1989; Morgan et al. 1991; Sutherland et al. 1991; Bumpus 1989; Hammel et al. 1992
Fluoranthene	*Cunninghamella elegans*	Pothuluri et al. 1990, 1992a
Pyrene	*Cunninghamella elegans, Phanerochaete chrysosporium*	Cerniglia et al. 1986; Hammel et al. 1986
Benz[a]anthracene	*Cunninghamella elegans*	Cerniglia et al. 1980d
Benzo[a]pyrene	*Aspergillus ochraceus, Bjerkandera adusta, Bjerkandera* sp., *Candida maltosa, Candida tropicalis, Chrysosporium pannorum, Cunninghamella elegans, Mortierella verrucosa, Neurospora crassa, Penicillium* sp., *Phanerochaete chrysosporium, Ramaria* sp., *Saccharomyces cerevisiae, Trametes versicolor, Trichoderma viride*	Cerniglia & Gibson 1979, 1980a, 1980b; Ghosh et al. 1983; Cerniglia & Crow 1981; Lin & Kapoor 1979; Bumpus et al. 1985; Wiseman & Woods 1979; Field et al. 1992

Table 4. Polycyclic aromatic hydrocarbons oxidized by different species of cyanobacteria and algae.

Compound	Organisms	References
Naphthalene	*Oscillatoria* sp. (strain JCM), *Oscillatoria* sp. (strain MEV), *Microcoleus chthonoplastes, Nostoc* sp., *Anabaena* sp. (strain CA), *Anabaena* sp. (strain 1F), *Agmenellum quadruplicatum, Coccochloris elabens, Aphanocapsa* sp., *Chlorella sorokiniana, Chlorella autotrophica, Dunaliella tertiolecta, Chlamydomonas angulosa, Ulva fasciata, Cylindrotheca* sp., *Amphora* sp., *Nitzschia* sp., *Synedra* sp., *Navicula* sp., *Porphyridium cruentum*	Cerniglia et al. 1979, 1980a, 1980b, 1982; Narro et al. 1992a
Phenanthrene	*Oscillatoria* sp. strain JCM, *Agmenellum quadruplicatum*	Narro et al. 1992b
Benzo[a]pyrene	*Selenastrum capricornutum*	Warshawsky et al. 1988, 1990

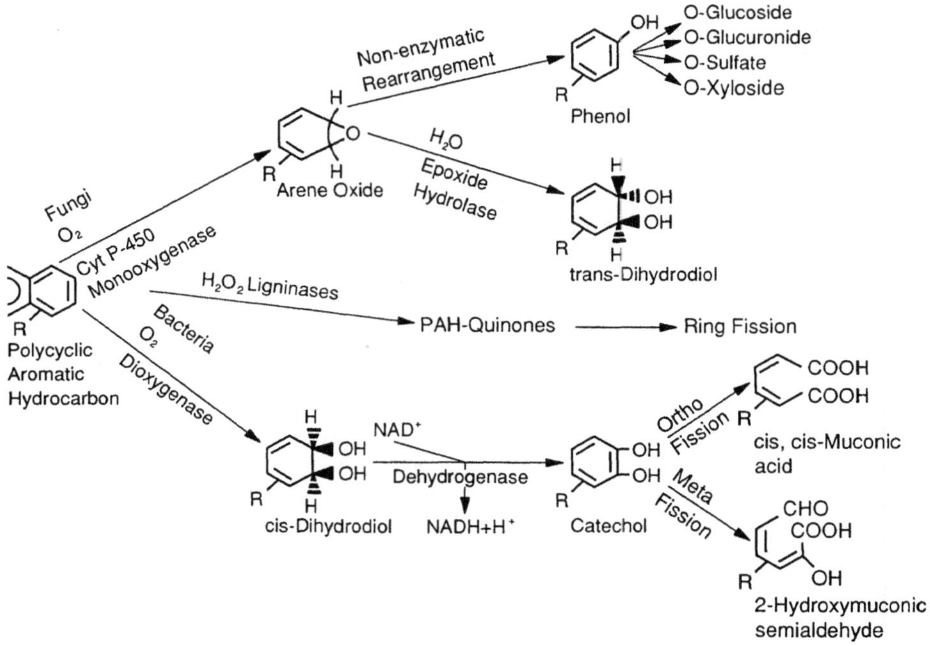

Fig. 4. Pathways for the microbial catabolism of polycyclic aromatic hydrocarbons.

1987; Herbes & Schwall 1978; Foght & Westlake 1988; Keck et al. 1989; Wang et al. 1990).

6. Filamentous fungi hydroxylate PAHs as a prelude to detoxification, whereas most bacteria oxidize PAHs as a prelude to ring fission and carbon assimilation (Cerniglia 1984; Cerniglia et al. 1992; Sutherland 1992). Similar to bacteria, fungal metabolism of PAHs is also highly regio-and stereoselective (McMillan et al. 1987, 1988; Cerniglia 1984; Cerniglia & Gibson 1980a, 1980b; Cerniglia et al. 1983a, 1983b, 1984, 1990).
7. Microbial degradation of PAHs in aquatic and terrestrial ecosystems is influenced strongly by a wide variety of abiotic and biotic factors which include: temperature, pH, soil type, aeration, nutrients, depth, diffusion, microbial adaptations, bioavailability, previous chemical exposure, water availability, sediment toxicity, physico-chemical properties of the PAH, concentration of the PAH and seasonal factors (Manilal & Alexander 1991; Atlas 1991; Wang et al. 1990; Shiaris 1989; Cerniglia & Heitkamp 1989; Wild et al. 1991a, 1991b).

Metabolism of polycyclic aromatic hydrocarbons

Hydroxylation of benzenoid aromatic compounds always involves the incorporation of molecular oxygen (Gibson & Subramanian 1984). Procaryotic microorganisms utilize dioxygenase enzymes to incorporate both atoms of molecular oxygen into the aromatic nucleus to form *cis*-dihydrodiols (Gibson et al. 1990) (Fig. 4). *Cis*-dihydrodiols are stereoselectively dehydrogenated by *cis*-dihydrodiol dehydrogenases (Patel & Gibson 1974), which rearomatize the benzene nucleus to form dihydroxylated intermediates. Subsequent enzymatic fission of the aromatic ring by procaryotic organisms is also catalyzed by highly regio- and stereoselective dioxygenases (Gibson & Subramanian 1984). As a prerequisite for this reaction, the aromatic ring must contain two hydroxyl groups placed either *ortho* or *para* to each other (Evans et al. 1965; Gibson et al. 1968; Barnsley 1976a, 1976b; Chapman 1979). If the hydroxyl moieties are *ortho* to each other, oxygenolytic ring cleavage can occur either between the two hydroxyl groups, by *ortho* or intradiol cleaving dioxygenases, or adjacent to the two hydroxyl groups, by *meta* or extradiol cleaving dioxygenases (Fig. 5).

A considerable amount of research on the microbial degradation of naphthalene, including the elucidation of the metabolic pathways, enzymatic mechanisms and genetic regulation, has been conducted (Davies & Evans 1964; Tattersfield 1927; Tausson 1927; Treccani et al. 1954; Catterall et al. 1971; Dunn & Gunsalus 1973; Barnsley 1976a, 1976b, 1983; Boronin et al. 1980; Cane & Williams 1982; Connors & Barnsley 1982; Ensley & Gibson 1983; Garcia-Valdes et al. 1988; Jeffrey et al. 1975; Kiyohara & Nagao 1978; Patel & Gibson 1974; Schell 1983; Tagger et al. 1990; Utkin et al. 1990). Despite the results on naphthalene degradation, there is little information on the genetic and physiological basis for the microbial degradation of higher-molecular weight PAHs. Thus, naphthalene was chosen as an example to illustrate the general principles outlined above.

Pseudomonas putida PpG7 (Dunn & Gunsalus 1973) and *Pseudomonas putida* strain NCIB 9816 (Davies & Evans 1964; Yen & Serdar 1988) metabolize naphthalene through the pathway shown in Fig. 6. The first step in the pathway involves the formation of *cis*-(1*R*, 2*S*)-dihydroxy-1,2-dihydronaphthalene (naphthalene *cis*-1,2-dihydrodiol) (Jeffrey et al. 1975). This reaction is catalyzed by a multicomponent enzyme system, designated naphthalene dioxygenase (Ensley et al. 1982) in Fig. 6. An iron-sulfur flavoprotein, reductase (Haigler & Gibson 1990a), initially accepts electrons from NADH or NADPH and then transfers them to a 2Fe.2S ferredoxin, designated ferredoxin NAP (Haigler & Gibson 1990b). Ferredoxin NAP in turn reduces the terminal iron-sulfur protein, designated ISP NAP (Ensley & Gibson 1983), which catalyzes the oxidation of naphthalene to naphthalene *cis*-1,2-dihydrodiol (Fig. 6). The genes for the individual components of naphthalene dioxygenase in *Pseudomonas putida* PpG7 and NCIB 9816 are encoded on the NAH7 plasmid (Yen & Gunsalus 1982; Serdar & Gibson 1989) and have been designated nahAa, nahAb, and nahAcAd (Fig. 6). Ensley et al. (1987) have cloned the nahAcAd gene in *E. coli* and shown the expression of naphthalene dioxygenase activity. The nucleotide sequence for

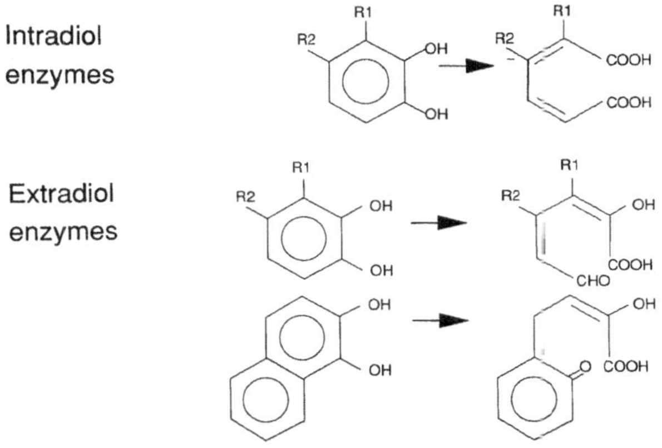

Fig. 5. Ortho-(intradiol) and *meta* (extradiol)-ring fission pathways for the catabolism of polycyclic aromatic hydrocarbons.

the naphthalene dioxygenase genes from *P. putida* strains has recently been reported (Simon et al. 1992).

Figure 4 shows the different routes of initial oxidative attack of PAHs by eucaryotic and procaryotic microorganisms. A diverse group of ligninolytic and non-ligninolytic fungi have the ability to oxidize PAHs (Table 3). The sites of enzymatic attack, mechanisms, chemical intermediates and pathways for the metabolism of several PAHs by eucaryotic microorganisms have been extensively reviewed (Cerniglia 1984; Holland et al. 1986; Cerniglia et al. 1992; Sutherland 1992). The best studied non-ligninolytic fungus, *Cunninghamella elegans*, utilizes cytochrome P-450 monooxygenase enzymes to incorporate one atom of molecular oxygen into the aromatic nucleus and reduce the remaining atom to water (Cerniglia 1984). The resulting arene oxide intermediate can undergo further metabolism by epoxide hydrolase to form a dihydrodiol with the *trans*-configuration or can undergo nonenzymatic rearrangement to form a phenol, which can be conjugated with sulfate, glucuronic acid, or glucose (Fig. 4). *C. elegans* has been shown to oxidize numerous PAHs and their methyl, nitro, and fluoro-substituted derivatives to *trans*-dihydrodiols, phenols, quinones, tetralones and conjugates of these primary metabolites (Table 5). Although small amounts of mutagenic and carcinogenic metabolites are found during the metabolism of PAHs, most fungal transformation

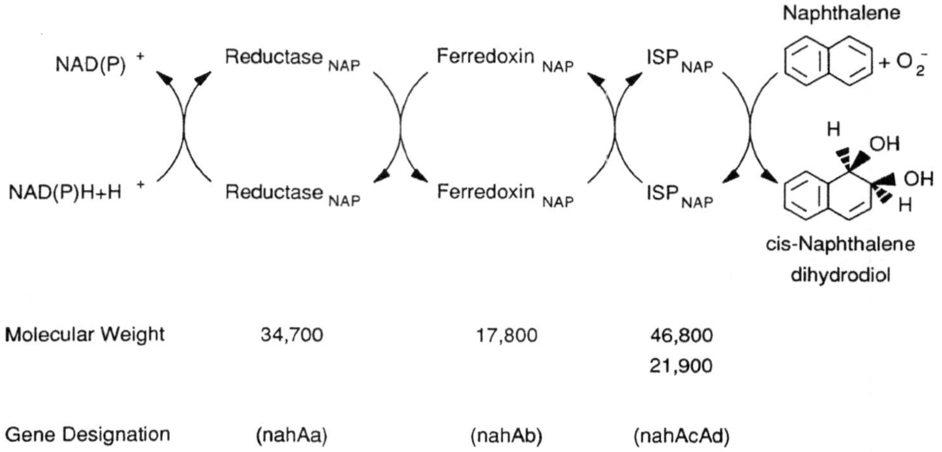

Fig. 6. Oxidation of naphthalene to naphthalene *cis*-1,2-dihydrodiol by naphthalene dioxygenase. (ISP = iron-sulfur protein).

products formed by *C. elegans* are less mutagenic than the parent compound and thus detoxify the PAHs (Cerniglia et al. 1985a, b).

Table 5. Metabolites produced from polycyclic aromatic hydrocarbons by fungi.

Compound	Metabolites	References
Naphthalene	1-Naphthol, 2-naphthol, naphthalene *trans*-1,2-dihydrodiol, 4-hydroxy-1-tetralone, 1,2-naphthoquinone, 1,4-naphthoquinone, sulfate and glucuronide conjugates	Ferris et al. 1973; Smith & Rosazza 1974; Cerniglia & Gibson 1977, 1978; Cerniglia et al. 1978, 1983a, 1985a
Acenaphthene	1-Acenaphthenone, 1,2-acenaphthenedione, *cis*-1,2-dihydroxyacenaphthene, *trans*-1,2-dihydroxyacenaphthene, 1,5-dihydroxyacenaphthene, 1-acenaphthenol, 6-hydroxyacenaphthenone	Pothuluri et al. 1992b
Anthracene	Anthracene *trans*-1,2-dihydrodiol, 1-anthrol, anthraquinone, phthalate, glucuronide, sulfate and xyloside conjugates	Cerniglia 1982; Cerniglia & Yang 1984; Sutherland et al. 1992; Field et al. 1992; Hammel et al. 1991
Phenanthrene	Phenanthrene *trans*-1,2-dihydrodiol, phenanthrene *trans*-3,4-dihydrodiol, phenanthrene *trans*-9,10-dihydrodiol, glucoside conjugate of 1-phenanthrol, 3-,4- and 9-hydroxyphenanthrene, 2,2-diphenic acid	Cerniglia & Yang 1984; Cerniglia et al. 1985a, 1989; Sutherland et al. 1991; Hammel et al. 1992
Fluoranthene	Fluoranthene *trans*-2,3-dihydrodiol, 8 and 9-hydroxyfluoranthene *trans*-2,3-dihydrodiols, glucoside conjugates	Pothuluri et al. 1990
Benz[a]anthracene	Benz[a]anthracene *trans*-3,4-dihydrodiol, benz[a]anthracene *trans*-8,9-dihydrodiol, benz[a]anthracene *trans*-10,11-dihydrodiol, phenolic and tetrahydroxy derivatives of benz[a]anthracene, glucuronide and sulfate conjugates	Cerniglia et al. 1980d, 1985a
Pyrene	1-Hydroxypyrene, 1,6-pyrenequinone, 1,8-pyrenequinone, glucoside conjugates	Hammel et al. 1986; Cerniglia et al. 1986
Benzo[a]pyrene	Benzo[a]pyrene *trans*-4,5-dihydrodiol, benzo[a]pyrene *trans*-7,8-dihydrodiol, benzo[a]pyrene *trans*-9,10-dihydrodiol, benzo[a]pyrene-1,6-quinone, benzo[a]pyrene-3,6-quinone, benzo[a]pyrene-6,12-quinone, 3-hydroxybenzo[a]pyrene, 9-hydroxybenzo[a]pyrene, 7β, 8α, 9α, 10β-tetrahydrobenzo[a]pyrene, 7β, 8α, 9β, 10α-tetrahydroxy-7,8,9,10-tetrahydrobenzo[a]pyrene, benzo[a]pyrene 7,8-dihydrodiol-9,10-epoxide, glucuronide and sulfate conjugates	Haemmerli et al. 1986; Cerniglia & Gibson 1979, 1980a, 1980b; Lin & Kapoor 1979; Wiseman & Woods 1979; Cerniglia et al. 1980c

Recently, Sutherland et al. (1992) reported that another fungus, *Rhizoctonia solani*, metabolizes anthracene to *trans*-1,2-dihydroxy-1,2-dihydroanthracene (anthracene *trans*-1,2-dihydrodiol) and three xyloside conjugates. The formation of xyloside conjugates suggests that *R. solani* may be as useful as *C. elegans* in detoxifying PAHs.

Several research groups have demonstrated that white rot fungi are able to degrade PAHs and have potential for the bioremediation of PAH-polluted soils and sediments (Aust 1990; Bumpus 1989; Gold et al. 1989; Hammel 1989; Haemmerli et al. 1986; Sanglard et al. 1986). The best studied strain, *Phanerochaete chrysosporium*, produces extracellular lignin peroxidases that not only degrade lignin-related compounds but also catalyze the one-electron oxidation of PAHs to quinones (Fig. 4). Degradation of ^{14}C-labeled PAHs to minor amounts of $^{14}CO_2$ and polar metabolites has also been demonstrated (Bumpus et al. 1985). The lignin peroxidases have been shown to oxidize PAHs that have ionization potentials less than about 7.6 eV (Haemmerli et al. 1986).

Hammel et al. (1992) reported that the ligninolytic fungus, *P. chrysosporium*, oxidizes phenanthrene at its C-9 and C-10 positions to give 2,2′-diphenic acid as a ring cleavage product. Their results indicate that *P. chrysosporium* utilizes a different pathway from *C. elegans* and bacteria, since the major site of enzymatic attack by most bacteria is at the C-3 and C-4 positions of phenanthrene. Recently, Sutherland et al. (1991) reported that *P. chrysosporium* under nonligninolytic conditions metabolizes phenanthrene to phenols and *trans*-dihydrodiols, similarly to *C. elegans*. This observation indicates that several enzymatic mechanisms may occur in *P. chrysosporium* for the initial oxidative attack on PAHs (Hammel et al. 1992).

Procaryotic and eucaryotic photoautotrophs (Table 4) may be an exception to the oxidation mechanisms illustrated in Fig. 4, since both *cis*- and *trans*-dihydrodiols have been detected as transformation products formed from PAHs by cyanobacteria and green algae (Cody et al. 1984; Lindquist & Warshawsky 1985; Warshawsky et al. 1988; Narro et al. 1992a, 1992b) (Table 6). Cerniglia et al. (1979, 1980a, 1980b, 1980c) have shown that the marine cyanobacterium *Oscillatoria* sp. strain JCM oxidizes naphthalene under photoautotrophic conditions, predominantly to 1-naphthol. Deuterated naphthalene-labeling experiments indicate that 1-naphthol is formed through a naphthalene 1,2-oxide intermediate rather than by direct hydroxylation (Narro et al. 1992a). Another marine cyanobacterium, *Agmenellum quadruplicatum* PR-6, has also been shown to oxidize phenanthrene via a monooxygenase to form *trans*-9,10-dihydroxy-9,10-dihydrophenanthrene (phenanthrene *trans*-9,10-dihydrodiol) and 1-methoxyphenanthrene as major metabolites. The metabolism of naphthalene and phenanthrene by cyanobacteria is similar to that reported for mammalian and fungal enzymes. The freshwater green alga, *Selenastrum capricornutum*, grown under photoautotrophic conditions, oxidizes benzo[a]pyrene to *cis*-dihydrodiols (Table 6) with the predominant isomer being *cis*-11,12-dihydroxy-11,12-dihydrobenzo[a]pyrene (benzo[a]pyrene *cis*-11,12-dihydrodiol). The formation of *cis*-dihydrodiols suggests a dioxygenase similar to that found in heterotrophic procaryotes rather than the monooxygenase found in cyanobacteria and eucaryotes (Warshawsky et al. 1988, 1990). Further studies are necessary to characterize the enzymes involved in PAH oxidation by algae.

Methanotrophs have been shown to transform a variety of xenobiotics, including alkanes, alkenes and aromatic compounds (Colby & Dalton 1976; Colby et al. 1977, 1978; Dalton et al. 1981). Recently, Brusseau et al. (1990) reported that *Methylococcus trichosporium* OB3b oxidizes naphthalene to 1- and 2-naphthol. The exact mechanism of PAH degradation in methanotrophs has not been determined.

Biodegradation of high molecular weight PAHs

The persistence and bioavailability of PAHs in the environment depend on the physical and chemical characteristics of the PAH as well as the composition and chemical characteristics of the sediment (Wang et al. 1990; Park et al. 1990; Sims et al. 1990). The recalcitrance of PAHs for microbial

Table 6. Metabolites produced from PAHs by different species of cyanobacteria and algae.

Compound	Metabolites	References
Naphthalene	1-Naphthol, 4-hydroxy-1-tetralone, naphthalene *cis*-1,2-dihydrodiol	Cerniglia et al. 1979, 1980a, 1980b; Narro et al. 1992a
Phenanthrene	Phenanthrene *trans*-9,10-dihydrodiol, 1-methoxyphenanthrene	Narro et al. 1992b
Benzo[a]pyrene	Benzo[a]pyrene *cis*-4,5-dihydrodiol, benzo[a]pyrene *cis*-7,8-dihydrodiol, benzo[a]pyrene *cis*-9,10-dihydrodiol, benzo[a]pyrene *cis*-11,12-dihydrodiol	Cody et al. 1984; Lindquist & Warshawsky 1985; Warshawsky et al. 1990

degradation increases directly with the molecular weight and the octanol: water partition coefficient (log Kow) and inversely with aqueous water solubility (Fig. 1) since high molecular weight PAHs are more slowly desorbed and therefore less available for biological uptake (Aronstein et al. 1991). Calculated half-lives of PAHs in sediments vary between laboratory and field studies (Wild et al. 1991a, 1991b). This variation is probably a consequence of differences in the environmental parameters of the site, experimental design and methodology, biological components, and season of sampling. For example, PAH degradation is influenced by environmental factors, including temperature, pH, nutrients, oxygen and water availability, salinity and soil type (Atlas 1991). In general, low molecular-weight PAHs, such as naphthalene (Heitkamp et al. 1987, 1991), have short half-lives, whereas the high molecular weight PAHs, such as benzo[a]pyrene, are recalcitrant and persist indefinitely (Fig. 7). Also, half-lives tend to be shorter in PAH-contaminated sediments than in pristine sediments, which is probably due to the elevated populations of PAH-degrading bacteria in the contaminated sediments (Herbes & Schwall 1978; Shiaris 1989; Guerin & Jones 1989).

Although microbial metabolic pathways for degradation of PAHs containing up to three rings have been proposed (Gibson & Subramanian 1984), until recently little information has been available concerning the ability of microorganisms to metabolize the larger, more recalcitrant, high molecular weight PAHs. Biotransformation of the tetracyclic PAHs fluoranthene and pyrene by eukaryotes has been reported (Table 3) and bacteria that are able to oxidize or co-oxidize fluoranthene and pyrene have been isolated (Table 2). Information on the biochemical pathways for microbial catabolism of higher molecular weight PAHs is scarce.

Research in this laboratory has demonstrated the ability of *Mycobacterium* sp. PYR-1 to substantially mineralize fluoranthene, naphthalene and pyrene in pure culture (Kelley & Cerniglia 1991; Kelley et al. 1991a, 1991b; Heitkamp et al. 1988) and to significantly enhance mineralization of fluoranthene and pyrene in sediments containing indigenous microorganisms (Kelley & Cerniglia 1991; Heitkamp & Cerniglia 1989). Figures 8 and 9 show the proposed pathways for the catabolism of fluoranthene and pyrene by *Mycobacterium* sp. PYR-1. These reaction sequences show that fluoranthene and pyrene biodegradation can range from initial oxygenation and the formation of dead-end products to complete mineralization. From a toxicological standpoint, the question which needs to be addressed is whether high molecular weight PAH biodegradation intermediates in the environ-

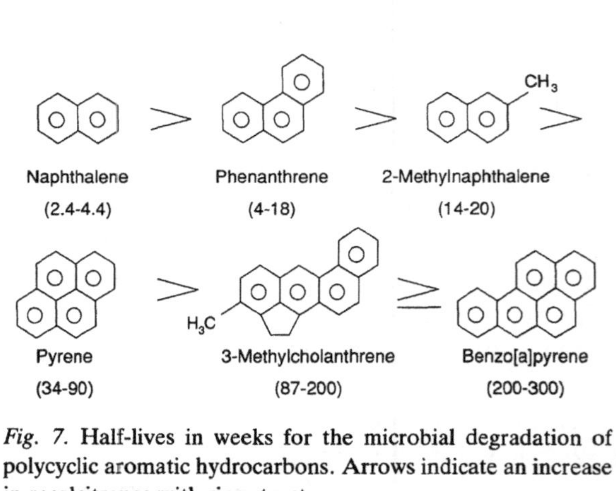

Fig. 7. Half-lives in weeks for the microbial degradation of polycyclic aromatic hydrocarbons. Arrows indicate an increase in recalcitrance with ring structure.

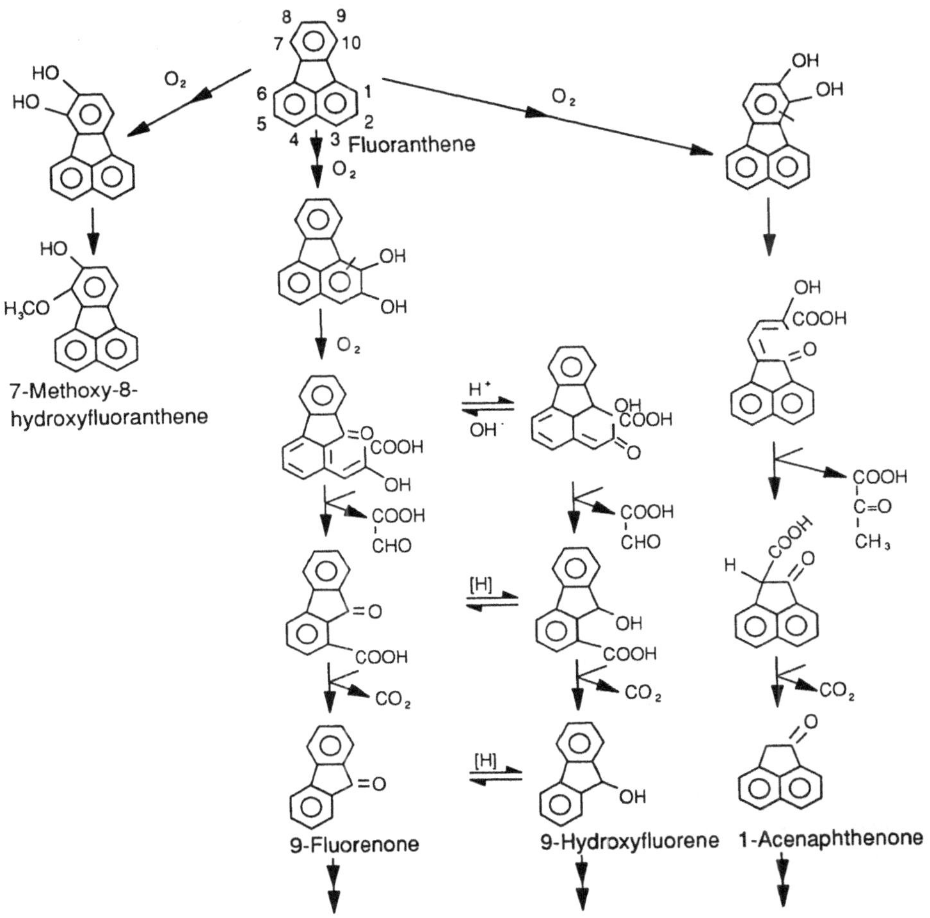

Fig. 8. Pathways proposed for the metabolism of fluoranthene by *Mycobacterium* sp. PYR-1.

ment represent a human health risk and safety problem.

Conclusion

Microbial degradation of PAHs is thought to be the major process involved in effective site bioremediation. The number of PAH compounds now known to be biodegraded has increased; however, the number of isolates of microorganisms that can completely mineralize high molecular weight PAHs, such as benzo[a]pyrene, is limited. Four and five-ring PAHs are very recalcitrant due to their low water solubilities and the resonance energies of their structures.

Microorganisms capable of rapidly mineralizing the more recalcitrant and potentially carcinogenic PAHs need to be isolated and characterized. Although the degradative pathways for PAH catabolism have been widely studied, there have been relatively few investigations on the regulatory mechanisms involved in controlling the microbial degradation of these ubiquitous pollutants.

Most studies regarding the toxicity and biodegradability of PAHs have been conducted on single compounds to minimize the number of variables. PAHs exist in environmental samples as complex mixtures, the concentration of each PAH at polluted sites depends upon the source of the sample, susceptibility to microbial degradation, and abiotic and environmental factors. Generally, the rates of

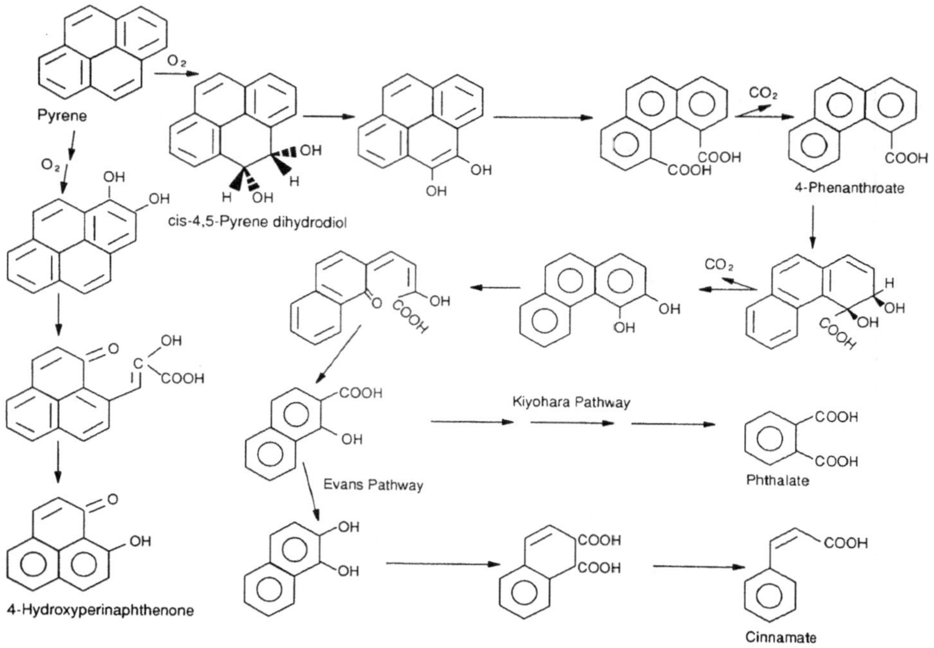

Fig. 9. Pathways proposed for the metabolism of pyrene by *Mycobacterium* sp. PYR-1.

biodegradation of individual hydrocarbons have been found to be related to their chemical structure, degree of aromaticity, concentration and physicochemical properties. Biodegradation of a component in an industrial mixture also depends to a large extent on the qualitative hydrocarbon content of the mixture, preexposure conditions and concentration. Little is known about the biodegradation of complex mixtures of PAHs, especially the effect of one PAH component on the biodegradability of another. Further experiments are required with mixtures of PAHs to determine the full degradative potential of microorganisms and the mechanisms by which they metabolize PAHs.

Acknowledgements

I gratefully acknowledge John B. Sutherland, Kay Shuttleworth, Warren Campbell, Melissa Morehead, Jairaj V. Pothuluri, Ashraf A. Khan, and Mohamed Nawaz for critically reading the manuscript; Kat Wheeler for her assistance in preparing the figures and Donna Fowler and Gail Shelton for assistance in preparing the manuscript.

References

Akhtar MN, Boyd DR, Thomas NJ, Koreeda M, Gibson DT, Mahadevan V & Jerina DM (1975) Absolute stereochemistry of the dihydroanthracene *cis*- and *trans*-1,2-diols produced from anthracene by mammals and bacteria. J. Chem. Soc. Perkin Trans. I: 2506–2511

Aronstein BN, Calvillo YM & Alexander M (1991) Effect of surfactants at low concentrations on the desorption and biodegradation of sorbed aromatic compounds in soil. Environ. Sci. Technol. 25: 1728–1731

Atlas RM (1991) Microbial hydrocarbon degradation – Bioremediation of oil spills. J. Chem. Technol. Biotechnol. 52: 149–156

Aust SD (1990) Degradation of environmental pollutants by *Phanerochaete chrysosporium*. Microb. Ecol. 20: 197–209

Barnsley EA (1975) Bacterial degradation of fluoranthene and benzo[a]pyrene. Can. J. Microbiol. 21: 1004–1008

Barnsley EA (1976a) Role and regulation of the *ortho* and *meta* pathways of catechol metabolism in pseudomonads metabolizing naphthalene and salicylate. J. Bacteriol. 125: 404–408

Barnsley EA (1976b) Naphthalene metabolism by *Pseudomonas*: the oxidation of 1,2-dihydroxynaphthalene to 2-hydroxychromene-2-carboxylic acid and the formation of 2-hydroxybenzalpyruvate. Biochem. Biophys. Res. Commun. 72: 1116–1121

Barnsley EA (1983) Bacterial oxidation of naphthalene and phenanthrene. J. Bacteriol. 153: 1069–1071

Bauer JE & Capone DG (1985) Degradation and mineralization

of the polycyclic aromatic hydrocarbons anthracene and naphthalene in intertidal marine sediments. Appl. Environ. Microbiol. 50: 80–90

Bauer JE & Capone DG (1988) Effects of co-occurring aromatic hydrocarbons on the degradation of individual polycyclic aromatic hydrocarbons in marine sediment slurries. Appl. Environ. Microbiol. 54: 1649–1655

Boronin AM, Kochetkov VV & Skryabin GK (1980) Incompatibility groups of naphthalene degradative plasmids in *Pseudomonas*. FEMS Microbiol. Lett. 7: 249–252

Brusseau GA, Hsien-Chyang T, Hanson RS & Wackett LP (1990) Optimization of trichloroethylene oxidation by methanotrophs and the use of a colorimetric assay to detect soluble methane monooxygenase activity. Biodegradation 1: 19–29

Bumpus JA (1989) Biodegradation of polycyclic aromatic hydrocarbons by *Phanerochaete chrysosporium*. Appl. Environ. Microbiol. 55: 154–158

Bumpus JA, Tien M, Wright D & Aust SD (1985) Oxidation of persistent environmental pollutants by a white rot fungus. Science 228: 1434–1436

Cane PA & Williams PA (1982) The plasmid-coded metabolism of naphthalene and 2-methylnaphthalene in *Pseudomonas* strains. Phenotypic changes correlated with structural modification of the plasmid PWW60-1. J. Gen. Microbiol. 128: 2281–2290

Catterall FA, Murray K & Williams PA (1971) The configuration of the 1,2-dihydroxy-1,2-dihydronaphthalene formed by the bacterial metabolism of naphthalene. Biochem. Biophys. Acta 237: 361–364

Cerniglia CE (1982) Initial reactions in the oxidation of anthracene by *Cunninghamella elegans*. J. Gen. Microbiol. 128: 2055–2061

Cerniglia CE (1984) Microbial metabolism of polycyclic aromatic hydrocarbons. In: Laskin A (Ed) Advances in Applied Microbiology, Vol 30 (pp 31–71). Academic Press, New York

Cerniglia CE & Crow SA (1981) Metabolism of aromatic hydrocarbons by yeasts. Arch. Microbiol. 129: 9–13

Cerniglia CE & Gibson DT (1977) Metabolism of naphthalene by *Cunninghamella elegans*. Appl. Environ. Microbiol. 34: 363–370

Cerniglia CE & Gibson DT (1978) Metabolism of naphthalene by cell extracts of *Cunninghamella elegans*. Arch. Biochem. Biophys. 186: 121–127

Cerniglia CE & Gibson DT (1979) Oxidation of benzo(a)pyrene by the filamentous fungus *Cunninghamella elegans*. J. Biol. Chem. 254: 12174–12180

Cerniglia CE & Gibson DT (1980a) Fungal oxidation of benzo (a)pyrene and (±)-trans-7,8-dihydroxy-7,8-dihydrobenzo(a) pyrene: evidence for the formation of a benzo(a)pyrene 7,8-diol-9,10-epoxide. J. Biol. Chem. 255: 5159–5163

Cerniglia CE & Gibson DT (1980b) Fungal oxidation of (±)-9,10-dihydroxy-9,10-dihydrobenzo(a)pyrene: formation of diastereomeric benzo(a)pyrene 9,10-diol 7,8-epoxides. Proc. Natl. Acad. Sci. U.S.A. 77: 4554–4558

Cerniglia CE & Heitkamp MA (1989) Microbial degradation of polycyclic aromatic hydrocarbons in the aquatic environment. In: Varanasi U (Ed) Metabolism of Polycyclic Aromatic Hydrocarbons in the Aquatic Environment (pp 41–68). CRC Press, Boca Raton, FL

Cerniglia CE & Yang SK (1984) Stereoselective metabolism of anthracene and phenanthrene by the fungus *Cunninghamella elegans*. Appl. Environ. Microbiol. 47: 119–124

Cerniglia CE, Hebert RL, Dodge RH, Szaniszlo PJ & Gibson DT (1978) Fungal transformation of naphthalene. Arch. Microbiol. 117: 135–143

Cerniglia CE, Gibson DT & Van Baalen C (1979) Algal oxidation of aromatic hydrocarbons: formation of 1-naphthol from naphthalene by *Agmenellum quadruplicatum* strain PR-6. Biochem. Biophys. Res. Commun. 88: 50–58

Cerniglia CE, Van Baalen C & Gibson DT (1980a) Metabolism of naphthalene by the blue-green alga, *Oscillatoria* sp., strain JCM. J. Gen. Microbiol. 116: 485–495

Cerniglia CE, Gibson DT & Van Baalen C (1980b) Algal oxidation of naphthalene. J. Gen. Microbiol. 116: 495–500

Cerniglia CE, Mahaffey W & Gibson DT (1980c) Fungal oxidation of benzo(a)pyrene: formation of (−)-*trans*-7,8-dihydroxy-7,8-dihydrobenzo(a)pyrene by *Cunninghamella elegans*. Biochem. Biophys. Res. Commun. 94: 226–232

Cerniglia CE, Dodge RH & Gibson DT (1980d) Studies on the fungal oxidation of polycyclic aromatic hydrocarbons. Bot. Mar. 23: 121–124

Cerniglia CE, Gibson DT & Van Baalen C (1982) Aromatic hydrocarbon oxidation by diatoms isolated from the Kachemak Bay region of Alaska. J. Gen. Microbiol. 128: 987–990

Cerniglia CE, Althaus JR, Evans FE, Freeman JP, Mitchum RK & Yang SK (1983a) Stereochemistry and evidence for an arene-oxide-NIH shift pathway in the fungal metabolism of naphthalene. Chem.-Biol. Interact. 44: 119–132

Cerniglia CE, Fu PP & Yang SK (1983b) Regio- and stereoselective metabolism of 4-methylbenz[a]anthracene by the fungus *Cunninghamella elegans*. Biochem. J. 216: 377–384

Cerniglia CE, Freeman JP & Evans FE (1984) Evidence for an arene oxide-NIH shift pathway in the transformation of naphthalene to 1-naphthol in *Bacillus cereus*. Arch. Microbiol. 138: 283–286

Cerniglia CE, White GL & Heflich RH (1985a) Fungal metabolism and detoxification of polycyclic aromatic hydrocarbons. Arch. Microbiol. 143: 105–110

Cerniglia CE, Freeman JP, White GL, Heflich RH & Miller DW (1985b) Fungal metabolism and detoxification of the nitropolycyclic aromatic hydrocarbon, 1-nitropyrene. Appl. Environ. Microbiol. 50: 649–655

Cerniglia CE, Kelly DW, Freeman JP & Miller DW (1986) Microbial metabolism of pyrene. Chem.-Biol. Interact. 57: 203–216

Cerniglia CE, Campbell WL, Freeman JP & Evans FE (1989) Metabolism of phenanthrene by the fungus *Cunninghamella elegans*: identification of a novel metabolite. Appl. Environ. Microbiol. 55: 2275–2279

Cerniglia CE, Campbell WL, Fu PP, Freeman JP & Evans FE (1990) Stereoselective fungal metabolism of methylated anthracenes. Appl. Environ. Microbiol. 56: 661–668

Cerniglia CE, Sutherland JB & Crow SA (1992) Fungal metabolism of aromatic hydrocarbons. In: Winkelmann G (Ed) Microbial Degradation of Natural Products (pp 193–217). VCH Press, Weinheim

Chapman PJ (1979) Degradation mechanisms. In: Bourquin AW & Pritchard PH (Eds) Proceedings of the workshop: Microbial Degradation of Pollutants in Marine Environments (pp 28–66). U.S. Environmental Protection Agency, Gulf Breeze, FL

Cody TE, Radike MJ & Warshawsky D (1984) The phytotoxicity of benzo[a]pyrene in the green alga, *Selenastrum capricornutum*. Environ. Res. 35: 122–131

Colby J & Dalton H (1976) Some properties of a soluble methane monooxygenase from *Methylococcus capsulatus* strain Bath. Biochem. J. 157: 495–497

Colby J, Stirling DI & Dalton H (1977) The soluble methane monooxygenase of *Methylococcus capsulatus* (Bath): its ability to oxygenate *n*-alkanes, n-alkenes, ethers and alicyclic, aromatic and heterocyclic compounds. Biochem. J. 165: 395–402

Colby J, Stirling DI & Dalton H (1978) Resolution of the methane monooxygenase of *Methylococcus capsulatus* (Bath) into three components. Purification and properties of component C, a flavoprotein. Biochem. J. 171: 461–468

Colla A, Fiecchi A & Treccani V (1959) Ricerche sul metabolismo ossidativo microbico dell'antracene e del fenantrene. Ann. Microbiol. 9: 87–91

Connors MA & Barnsley EA (1982) Naphthalene plasmids in pseudomonads. J. Bacteriol. 149: 1096–1101

Dagley S (1971) Catabolism of aromatic compounds by microorganisms. Adv. Microb. Physiol. 6: 1–46

Dagley S (1975) A biochemical approach to some problems of environmental pollution. Essays Biochem. 11: 81–138

Dalton H, Golding BT, Waters BW, Higgins R & Taylor JA (1981) Oxidations of cyclopropane, methylcyclopropane, and arenes with the mono-oxygenase system from *Methylococcus capsulatus*. J. Chem. Soc. Chem. Commun. 1981: 482–483

Davies JI & Evans WC (1964) Oxidative metabolism of naphthalene by soil pseudomonads. Biochem. J. 91: 251–261

Dipple A, Cheng SC & Bigger CAH (1990) Polycyclic aromatic hydrocarbon carcinogens. In: Pariza MW, Aeschbacher HU, Felton JS & Sato S (Eds) Mutagens and Carcinogens in the Diet (pp 109–127). Wiley-Liss, New York

Dua RD & Meera S (1981) Purification and characterization of naphthalene oxygenase from *Corynebacterium renale*. Eur. J. Biochem. 120: 461–465

Dunn NW & Gunsalus IC (1973) Transmissible plasmid coding early enzymes of naphthalene oxidation in *Pseudomonas putida*. J. Bacteriol. 114: 974–979

Ellis B, Harold P & Kronberg H (1991) Bioremediation of a creosote contaminated site. Environ. Technol. 12: 447–459

Ensley BD & Gibson DT (1983) Oxidation of naphthalene by a multicomponent enzyme system from *Pseudomonas* sp. strain NCIB 9816. J. Bacteriol. 155: 505–511

Ensley BD, Gibson DT & LaBorde LA (1982) Naphthalene dioxygenase: purification and properties of a terminal oxygen component. J. Bacteriol. 149: 948–954

Ensley BD, Osslund TD, Joyce M & Simon MJ (1987) Expression and complementation of naphthalene dioxygenase activity in *Escherichia coli*. In: Hagedorn SR, Hanson RS & Kunz DA (Eds) Microbial Metabolism and the Carbon Cycle (pp 437–455). Harwood Academic Publishers, New York

Evans WC, Fernley HN & Griffiths E (1965) Oxidative metabolism of phenanthrene and anthracene by soil pseudomonads; the ring fission mechanism. Biochem. J. 95: 819–821

Ferris JP, Fasco MJ, Stylianopoulou FL, Jerina DM, Daly JW & Jeffrey AM (1973) Mono-oxygenase activity in *Cunninghamella bainieri*: Evidence for a fungal system similar to liver microsomes. Arch. Biochem. Biophys. 156: 97–103

Field JA, DeJong E, Costa GF & DeBont JAM (1992) Biodegradation of polycyclic aromatic hydrocarbons by new isolates of white rot fungi. Appl. Environ. Microbiol. 58: 2219–2226

Foght JM & Westlake DWS (1988) Degradation of polycyclic aromatic hydrocarbons and aromatic heterocycles by a *Pseudomonas* species. Can. J. Microbiol. 34: 1135–1141

Garcia-Valdes E, Cozar E, Rotger R, Lalucat J & Ursing J (1988) New naphthalene-degrading marine *Pseudomonas* strains. Appl. Environ. Microbiol. 54: 2478–2485

Ghosh DK & Mishra AK (1983) Oxidation of phenanthrene by a strain of *Micrococcus*: evidence of protocatechuate pathway. Curr. Microbiol. 9: 219–224

Ghosh DK, Dutta D, Samanta TB & Mishra AK (1983) Microsomal benzo[a]pyrene hydroxylase in *Aspergillus ochraceus* TS: Assay and characterization of the enzyme system. Biochem. Biophys. Res. Commun. 113: 497–505

Gibson DT & Subramanian V (1984) Microbial degradation of aromatic hydrocarbons. In: Gibson DT (Ed) Microbial Degradation of Organic Compounds (pp 181–252). Marcel Dekker, New York

Gibson DT, Koch JR & Kallio RE (1968) Oxidative degradation of aromatic hydrocarbons by microorganisms. I. Enzymatic formation of catechol from benzene. Biochemistry 7: 2653–2661

Gibson DT, Mahadevan V, Jerina DM, Yagi H & Yeh HJC (1975) Oxidation of the carcinogens benzo[a]pyrene and benz[a]anthracene to dihydrodiols by a bacterium. Science 189: 295–297

Gibson DT, Zylstra GJ & Chauhan S (1990) Biotransformations catalyzed by toluene dioxygenase from *Pseudomonas putida* F1. In: Silver S, Chakrabarty AM, Iglewski B & Kaplan S (Eds) *Pseudomonas*: Biotransformations, Pathogenesis, and Evolving Biotechnology (pp 121–132). American Society for Microbiology, Washington, DC

Giger W & Blumer M (1974) Polycyclic aromatic hydrocarbons in the environment; isolation and characterization by chromatography, visible, ultraviolet and mass spectrometry. Anal. Chem. 46: 1663–1671

Gold MH, Wariishi H & Valli K (1989) Extracellular peroxidases involved in lignin degradation by the white rot basidiomycete *Phanerochaete chrysosporium*. In: Whitaker JR &

Sonnet PE (Eds) Biocatalysis in Agricultural Biotechnology (pp 127–140). American Chemical Society, Washington, DC

Grbic-Galic D & Vogel TM (1987) Transformation of toluene and benzene by mixed methanogenic cultures. Appl. Environ. Microbiol. 53: 254–260

Grosser RJ, Warshawsky D & Vestal JR (1991) Indigenous and enhanced mineralization of pyrene, benzo[a]pyrene and carbazole in soils. Appl. Environ. Microbiol. 57:3462–3469

Grund E, Denecke B & Eichenlaub R (1992) Naphthalene degradation via salicylate and gentisate by *Rhodococcus* sp. strain B4. Appl. Environ. Microbiol. 58: 1874–1877

Gschwend PM & Hites RA (1981) Fluxes of polycyclic aromatic hydrocarbons to marine and lacustrine sediments in the northeastern United States. Geochim. Cosmochim. Acta 45: 2359–2367

Guerin WF & Jones GE (1988) Mineralization of phenanthrene by a *Mycobacterium* sp. Appl. Environ. Microbiol. 54: 937–944

Guerin WF & Jones GE (1989) Estuarine ecology of phenanthrene-degrading bacteria. Estuarine Coastal Shelf Sci. 29: 115–130

Haemmerli SD, Leisola MSA, Sanglard D & Fiechter A (1986) Oxidation of benzo[a]pyrene by extracellular ligninases of *Phanerochaete chrysosporium*: veratryl alcohol and stability of ligninase. J. Biol. Chem. 261: 6900–6903

Haigler BE & Gibson DT (1990a) Purification and properties of NADH-ferredoxin NAP reductase, a component of naphthalene dioxygenase from *Pseudomonas* sp. strain NCIB 9816. J. Bacteriol. 172: 457–464

Haigler BE & Gibson DT (1990b) Purification and properties of ferredoxin NAP, a component of naphthalene dioxygenase from *Pseudomonas* sp. strain NCIB 9816. J. Bacteriol. 172: 465–468

Hammel KE (1989) Organopollutant degradation by ligninolytic fungi. Enzyme Microb. Technol. 11: 776–777

Hammel KE, Kalyanaraman B & Kirk TK (1986) Oxidation of polycyclic aromatic hydrocarbons and dibenzo[*p*]dioxins by *Phanerochaete chrysosporiumm*. J. Biol. Chem. 261: 16948–16952

Hammel KE, Green B & Gai WZ (1991) Ring fission of anthracene by a eukaryote. Proc. Natl. Acad. Sci. 88: 10605–10608

Hammel KE, Gai ZG, Green B & Moen MA (1992) Oxidative degradation of phenanthrene by the ligninolytic fungus *Phanerochaete chrysosporium*. Appl. Environ. Microbiol. 58: 1831–1838

Heitkamp MA & Cerniglia CE (1987) The effects of chemical structure and exposure on the microbial degradation of polycyclic aromatic hydrocarbons in freshwater and estuarine ecosystems. Environ. Toxicol. Chem. 6: 535–546

Heitkamp MA & Cerniglia CE (1988) Mineralization of polycyclic aromatic hydrocarbons by a bacterium isolated from sediment below an oil field. Appl. Environ. Microbiol. 54: 1612–1614

Heitkamp MA & Cerniglia CE (1989) Polycyclic aromatic hydrocarbon degradation by a *Mycobacterium* sp. in microcosms containing sediment and water from a pristine ecosystem. Appl. Environ. Microbiol. 55: 1968–1973

Heitkamp MA, Freeman JP & Cerniglia CE (1987) Naphthalene biodegradation in environmental microcosms: estimates of degradation rates and characterization of metabolites. Appl. Environ. Microbiol. 53: 129–136

Heitkamp MA, Franklin W & Cerniglia CE (1988) Microbial metabolism of polycyclic aromatic hydrocarbons: isolation and characterization of a pyrene-degrading bacterium. Appl. Environ. Microbiol. 54: 2556–2565

Heitkamp MA, Freeman JP, Miller DW & Cerniglia CE (1991) Biodegradation of 1-nitropyrene. Arch. Microbiol. 156: 223–230

Herbes, SE & Schwall LR (1978) Microbial transformation of polycyclic aromatic hydrocarbons in pristine and petroleum contaminated sediments. Appl. Environ. Microbiol. 35:306–316.

Hites RA, LaFlamme RE & Farrington JW (1977) Sedimentary polycyclic aromatic hydrocarbons: the historical record. Science 198: 829–831

Hites RA, LaFlamme RE & Windsor JG (1980) Polycyclic aromatic hydrocarbons in marine/aquatic sediments: their ubiquity. In: Petrakis L & Weiss FT (Eds) Petroleum in the Marine Environment (pp 289–311). Advances in Chemistry Series, American Chemical Society, Washington, DC

Holland HL, Khan SH, Richards D & Riemland E (1986) Biotransformation of polycyclic aromatic compounds by fungi. Xenobiotica 16: 733–741

Jacob J, Karcher W, Belliardo JJ & Wagstaffe PJ (1986) Polycyclic aromatic hydrocarbons of environmental and occupational importance. Fresenius, Z. Anal. Chem. 323: 1–10

Jeffrey AM, Yeh HJC, Jerina DM, Patel RT, Davey JF & Gibson DT (1975) Initial reactions in the oxidation of naphthalene by *Pseudomonas putida*. Biochemistry 14: 575–584

Jerina DM, Selander H, Yagi H, Wells MC, Davey JF, Mahadevan V & Gibson DT (1976) Dihydrodiols from anthracene and phenanthrene. J. Am. Chem. Soc. 98: 5988–5996

Johnson AC & Larsen D (1985) The distribution of polycyclic aromatic hydrocarbons in the surficial sedimentsss of Penobscot Bay (Maine, USA) in relation to possible sources and to other sites worldwide. Mar. Environ. Res. 15: 1–16

Jones KC, Stratford A, Waterhouse KS & Vogt NB (1989) Organic contaminants in Welsh soils: polynuclear aromatic hydrocarbons. Environ. Sci. Technol. 13: 540–550

Keck J, Sims RC, Coover M, Park K & Symons B (1989) Evidence for cooxidation of polynuclear aromatic hydrocarbons in soil. Wat. Res. 23: 1467–1476

Keith LH & Telliard WA (1979) Priority pollutants. I. A perspective view. Environ. Sci. Technol. 13: 416–423

Kelley I & Cerniglia CE (1991) The metabolism of fluoranthene by a species of *Mycobacterium*. J. Ind. Microbiol. 7: 19–26

Kelley I, Freeman JP & Cerniglia CE (1991a) Identification of metabolites from the degradation of naphthalene by a *Mycobacterium* sp. Biodegradation 1: 283–290

Kelley I, Freeman JP, Evans FE & Cerniglia CE (1991b) Identification of a carboxylic acid metabolite from the catabolism of fluoranthene by a *Mycobacterium* sp. Appl. Environ. Microbiol. 57: 636–641

Keuth S & Rehm H-J (1991) Biodegradation of phenanthrene by *Arthrobacter polychromogenes* isolated from a contaminated soil. Appl. Microbiol. Biotechnol. 34: 804–808

Kiyohara H & Nagao K (1978) The catabolism of phenanthrene and naphthalene by bacteria. J. Gen. Microbiol. 105: 69–75

Kiyohara H, Nagao K & Nomi R (1976) Degradation of phenanthrene through *o*-phthalate by an *Aeromonas* sp. Agric. Biol. Chem. 40: 1075–1082

Kiyohara H, Nagao K, Kouno K & Yano K (1982) Phenanthrene degrading phenotype of *Alcaligenes faecalis* AFK2. Appl. Environ. Microbiol. 43: 458–461

Kiyohara H, Takizawa N, Date H, Torigoe S & Yano K (1990) Characterization of a phenanthrene degradation plasmid from *Alcaligenes faecalis* AFK2. J. Ferment. Bioeng. 69: 54–56

Kuhm AE, Stolz A & Knackmuss HJ (1991) Metabolism of naphthalene by the biphenyl-degrading bacterium *Pseudomonas paucimobilis* Q1. Biodegradation 2: 115–120

LaFlamme RE & Hites RA (1978) The global distribution of polycyclic aromatic hydrocarbons in recent sediment. Geochim. Cosmochim. Acta 42: 289–303

Lijinsky W (1991) The formation and occurrence of polynuclear aromatic hydrocarbons associated with food. Mutat. Res. 259: 251–261

Lin WS & Kapoor M (1979) Induction of aryl hydrocarbon hydroxylase in *Neurospora crassa* by benzo[a]pyrene. Curr. Microbiol. 3: 177–181

Lindquist B & Warshawsky D (1985) Identification of the 11,12-dihydroxybenzo[a]pyrene as a major metabolite produced by the green alga, *Selenastrum capricornutum*. Biochem. Biophys. Res. Commun. 130: 71–75

Mahaffey WR, Gibson DT & Cerniglia CE (1988) Bacterial oxidation of chemical carcinogens: formation of polycyclic aromatic acids from benz[a]anthracene. Appl. Environ. Microbiol. 54: 2415–2423

Manilal VB & Alexander M (1991) Factors affecting the microbial degradation of phenanthrene in soil. Appl. Microbiol. Biotechnol. 35: 401–405

McMillan DC, Fu PP & Cerniglia CE (1987) Stereoselective fungal metabolism of 7,12-dimethylbenz[a]anthracene: identification and enantiomeric resolution of a K-region dihydrodiol. Appl. Environ. Microbiol. 53: 2560–2566

McMillan DC, Fu PP, Freeman JP, Miller DW & Cerniglia CE (1988) Microbial metabolism and detoxification of 7,12-dimethylbenz[a]anthracene. J. Ind. Microbiol. 3: 211–225

Means JC, Ward SG, Hassett JJ & Banwart WL (1980) Sorption of polynuclear aromatic hydrocarbons by sediments and soils. Environ. Sci. Technol. 14: 1524–1528

Mihelcic JR & Luthy RG (1987) Degradation of polycyclic aromatic hydrocarbon compounds under various redox conditions in soil-water systems. Appl. Environ. Microbiol. 53: 1182–1187

Mihelcic JR & Luthy RG (1988) Microbial degradation of acenaphthene and naphthalene under denitrification conditions in soil-water systems. Appl. Environ. Microbiol. 54: 1188–1198

Miller EC & Miller JA (1981) Searches for ultimate chemical carcinogens and their reactions with cellular macromolecules. Cancer 47: 2327–2345

Morgan P, Lewis ST & Watkinson RJ (1991) Comparison of abilities of white-rot fungi to mineralize selected xenobiotic compounds. Appl. Microbiol. Biotechnol. 34: 693–696

Mueller JG, Chapman PJ & Pritchard PH (1989) Action of a fluoranthene-utilizing bacterial community on polycyclic aromatic hydrocarbon components of creosote. Appl. Environ. Microbiol. 55: 3085–3090

Mueller JG, Chapman PJ, Blattmann BO & Pritchard PH (1990) Isolation and characterization of a fluoranthene-utilizing strain of *Pseudomonas paucimobilis*. Appl. Environ. Microbiol. 56: 1079–1086

Narro ML, Cerniglia CE, Van Baalen C & Gibson DT (1992a) Evidence of NIH shift in naphthalene oxidation by the marine cyanobacterium, *Oscillatoria* species strain JCM. Appl. Environ. Microbiol. 58: 1360–1363

Narro ML, Cerniglia CE, Van Baalen C & Gibson DT (1992b) Metabolism of phenanthrene by the marine cyanobacterium *Agmenellum quadruplicatum*, strain PR-6. Appl. Environ. Microbiol. 58: 1351–1359

Park KS, Sims RC, Dupont RR, Doucette WJ & Matthews JE (1990) Fate of PAH compounds in two soil types: influence of volatilization, abiotic loss and biological activity. Environ. Toxicol. Chem. 9: 187–195

Patel TR & Gibson DT (1974) Purification and properties of (+)-*cis*-naphthalene dihydrodiol dehydrogenase of *Pseudomonas putida*. J. Bacteriol. 119: 879–888

Pothuluri JV, Freeman JP, Evans FE & Cerniglia CE (1990) Fungal transformation of fluoranthene. Appl. Environ. Microbiol. 56: 2974–2983

Pothuluri JV, Heflich RH, Fu PP & Cerniglia CE (1992a) Fungal metabolism and detoxification of fluoranthene. Appl. Environ. Microbiol. 58: 937–941

Pothuluri JV, Freeman JP, Evans FE & Cerniglia CE (1992b) Fungal metabolism of acenaphthene by *Cunninghamella elegans*. Appl. Environ. Microbiol. 58: 3654–3659

Ryu BH, Oh YK & Bin JH (1989) Biodegradation of naphthalene by *Acinetobacter calcoaceticus* R-88. J. Kor. Agric. Chem. Soc. 32: 315–320

Sanglard D, Leisola MSA & Fiechter A (1986) Role of extracellular ligninases in biodegradation of benzo[a]pyrene by *Phanerochaete chrysosporium*. Enzyme Microb. Technol. 8: 209–212

Savino A & Lollini MN (1977) Identification of some fermentation products of phenanthrene in microorganisms of the genus *Arthrobacter*. Boll. Soc. Ital. Biol. Sper. 53: 916–921

Schell MA (1983) Cloning and expression in *Escherichia coli* of the naphthalene degradative genes from plasmid NAH7. J. Bacteriol. 153: 822–829

Schocken MJ & Gibson DT (1984) Bacterial oxidation of the polycyclic aromatic hydrocarbons acenaphthene and acenaphthylene. Appl. Environ. Microbiol. 48:10–16

Serdar CM & Gibson DT (1989) Studies of nucleotide sequence homology between naphthalene-utilizing strains of bacteria. Biochem. Biophys. Res. Commun. 164: 772–779

Shiaris MP (1989) Phenanthrene mineralization along a natural salinity gradient in an urban estuary, Boston Harbor, Massachusetts. Microb. Ecol. 18: 135–146

Simon MJ, Osslund TD, Saunders R, Ensley BD, Suen W-C, Cruden DL, Gibson DT & Zylstra GJ (1992) Nucleotide sequences encoding the genes for naphthalene dioxygenase in *Pseudomonas putida* G7 and *Pseudomonas* sp. NCIB 9816-4. Gene (in press)

Sims JL, Sims RC & Matthews JE (1990) Approach to bioremediation of contaminated soil. Haz. Waste Haz. Mater. 7: 117–149

Smith RV & Rosazza JP (1974) Microbial models of mammalian metabolism. Aromatic hydroxylation. Arch. Biochem. Biophys. 161: 551–558

Sutherland JB (1992) Detoxification of polycyclic aromatic hydrocarbons by fungi. J. Ind. Microbiol. 9: 53–62

Sutherland JB, Freeman JP, Selby AL, Fu PP, Miller DW & Cerniglia CE (1990) Stereoselective formation of a K-region dihydrodiol from phenanthrene by *Streptomyces flavovirens*. Arch. Microbiol. 154: 260–266

Sutherland JB, Selby AL, Freeman JP, Evans FE & Cerniglia CE (1991) Metabolism of phenanthrene by *Phanerochaete chrysosporium*. Appl. Environ. Microbiol. 57: 3310–3316

Sutherland JB, Selby AL, Freeman JP, Fu PP, Miller DW & Cerniglia CE (1992) Identification of xyloside conjugates formed from anthracene by *Rhizoctonia solani*. Mycol. Res. 96: 509–517

Tagger S, Truffaut N & Le Petit J (1990) Preliminary study on relationships among strains forming a bacterial community selected on naphthalene from marine sediment. Can. J. Microbiol. 36: 676–681

Tattersfield F (1927) The decomposition of naphthalene in the soil and the effect upon its insecticidal action. Ann. Appl. Biol. 15: 57–67

Tausson WO (1927) Napthalin als Kohlenstoffquelle für Bakterien. Planta 4: 214–256

Thakker DR, Yagi H, Levin W, Wood AW, Conney AH & Jerina DM (1985) Polycyclic aromatic hydrocarbons: Metabolic activation to ultimate carcinogens. In: Anders MW (Ed) Bioactivation of Foreign Compounds (pp 177–242). Academic Press, Orlando

Treccani V, Walker N & Wiltshire GH (1954) The metabolism of naphthalene by soil bacteria. J. Gen. Microbiol. 11: 341–348

Trower MK, Sariaslani FS & Kitson FG (1988) Xenobiotic oxidation by cytochrome P-450-enriched extracts of *Streptomyces griseus*. Biochem. Biophys. Res. Commun. 3: 1417–1422

Utkin IB, Yakimov MM, Matveeva LN, Kozlyak EI, Rogozhin IS, Solomon ZG & Bezborodov AM (1990) Catabolism of naphthalene and salicylate by *Pseudomonas fluorescens*. Folia Microbiol. 35: 557–560

Vogel TM & Grbic-Galic D (1986) Incorporation of oxygen from water into toluene and benzene during anaerobic fermentative transformation. Appl. Environ. Microbiol. 52: 200–202

Walter U, Beyer M, Klein J & Rehm H-J (1991) Degradation of pyrene by *Rhodococcus* sp. UW1. Appl. Microbiol. Biotechnol. 34: 671–676

Wang X, Yu X & Bartha R (1990) Effect of bioremediation on polycyclic aromatic hydrocarbon residues in soil. Environ. Sci. Technol. 24: 1086–1089

Warshawsky D, Radike M, Jayasimhulu K & Cody T (1988) Metabolism of benzo[a]pyrene by a dioxygenase enzyme system of the freshwater green alga *Selenastrum capricornutum*. Biochem. Biophys. Res. Commun. 152: 540–544

Warshawsky D, Keenan TM, Reilman R, Cody TE & Radike MJ (1990) Conjugation of benzo[a]pyrene metabolites by the freshwater green alga *Selenastrum capricornutum*. Chem.-Biol. Interact. 73: 93–105

Weissenfels WD, Beyer M & Klein J (1990) Degradation of phenanthrene, fluorene and fluoranthene by pure bacterial cultures. Appl. Microbiol. Biotechnol. 32: 479–484

Weissenfels WD, Beyer M, Klein J & Rehm HJ (1991) Microbial metabolism of fluoranthene: isolation and identification of ring fission products. Appl. Microbiol. Biotechnol. 34: 528–535

West PA, Okpokwasili GC, Brayton PR, Grimes DJ & Colwell RR (1984) Numerical taxonomy of phenanthrene-degrading bacteria isolated from the Chesapeake Bay. Appl. Environ. Microbiol. 48: 988–993

Wild SR, Berrow ML & Jones KC (1991a) The persistence of polynuclear aromatic hydrocarbons (PAHs) in sewage sludge amended agricultural soils. Environ. Pollut. 72: 141–157

Wild SR, Obbard JP, Munn CI, Berrow ML & Jones KC (1991b) The long-term persistence of polynuclear aromatic hydrocarbons (PAHs) in an agricultural soil amended with metal-contaminated sewage sludges. Sci. Total Environ. 101: 235–253

Williams PA (1981) Genetics of biodegradation. In: Leisinger T, Hutter R, Cook AM & Nuesch J (Eds) Microbial Degradation of Xenobiotics and Recalcitrant Compounds (pp 97–107). Academic Press, New York

Wiseman A & Woods LFJ (1979) Benzo[a]pyrene metabolites formed by the action of yeast cytochrome P-450/P-448. J. Chem. Technol. Biotechnol. 29: 320–324

Yen KM & Gunsalus IC (1982) Plasmid gene organization: naphthalene/salicylate oxidation. Proc. Natl. Acad. Sci. U.S.A. 79: 874–878

Yen KM & Serdar CM (1988) Genetics of naphthalene catabolism in Pseudomonads. CRC Crit. Rev. Microbiol. 15: 247–267

Zylstra GJ & Gibson DT (1991) Aromatic hydrocarbon degradation: a molecular approach. In: Setlow JK (Ed) Genetic Engineering: Principles and Methods, Vol 13 (pp 183–203). Plenum Press, New York

Zylstra GJ, Chauhan S & Gibson DT (1990) Degradation of chlorinated biphenyls by *Escherichia coli* containing cloned genes of the *Pseudomonas putida* F1 toluene catabolic pathway. In: Proceedings of the Sixteenth Annual Hazardous Waste Research Symposium: Remedial Action, Treatment, and Disposal of Hazardous Waste (pp 290–302). (EPA/600/9–90/037)

Zylstra GJ, Cuskey SM & Olsen RH (1991) Construction of plasmids for use in risk assessment research. In: Levin MA, Seidler RJ & Rogul M (Eds) Microbial Ecology: Principles, Methods, and Applications (pp 363–370). McGraw-Hill, New York

Biodegradation 3: 369–385, 1992.

Biological degradation of explosives and chemical agents

John E. Walker & David L. Kaplan
US Army Natick Research, Development and Engineering Center, Natick, Massachusetts 01760-5020, USA

Key words: organo-nitro compounds, chemical agents, biotransformation, detoxification

Abstract

Hazardous energetic organo-nitro compounds are found as contaminants in many environments. A series of nitro aromatics, nitrate esters and nitro amines, all characteristic of this class, has been studied for their susceptibility to biological transformation. Biotransformation pathways for each of these compounds have been identified and are summarized. Implications for these findings in light of current contamination issues is discussed. The detoxification of organophosphate chemical agents focuses on the investigation of organophosphate degrading enzymes from bacteria. Certain of these enzymes, active both in solution and when immobilized onto a solid surface, are very successful in hydrolyzing and detoxifying various organophosphate chemical agents. The relationship of this research to the critical concerns of the agricultural industry regarding detoxification of organophosphorous pesticides is discussed.

Introduction

The scope of environmental contamination from hazardous wastes is overwhelming, with recent estimates for EPA Superfund sites at over 32,000 in the US alone. The cost for remediation of these sites is variably estimated at over a trillion dollars, and even at this cost most sites would not be returned to a 'pristine' condition (Abelson 1992). Most technologies currently considered for remediation are expensive and often do not permanently alleviate the pollution hazard. Bioremediation has recently received renewed consideration as a treatment option due to its potential effectiveness without adding new hazardous burdens to the environment, its comparative low cost, and its adaptability and tailorability to specific treatment scenarios and environments (EPA 1990; Thayer 1991; Fox 1992). Organic energetic compounds are found as contaminants in many environments at sites where these compounds were produced, processed and disposed. Many of these organic energetic compounds are used in the production of explosives, as well as in the synthesis of fungicides, insecticides, herbicides, pharmaceuticals, and dyes. The toxicity and mutagenicity of many of these compounds, along with their relative recalcitrance in the environment lead to concerns regarding environmental fate and impact on human and environmental health. Chemical agents also represent environmental hazards during manufacture, processing, storage, and disposal, particularly with increased emphasis on environmentally-compatible destruction of stockpiles. For both organic energetic compounds and chemical agents, a biodegradation approach may offer a useful option to safe and economic disposal.

Many questions must be answered before a biological approach to the degradation of these compounds can and should be considered. At the basic level, can the target compound(s) be shown to be susceptible to biological degradation such that nontoxic products are produced? Can the organism(s) responsible for the biodegradation, and perhaps the enzymes and genetics involved be identified and characterized? Can genetic enhancement, construction of new genetic pathways, or genetic recruitment be used to improve or tailor the natural process? Can the range and optimum environmental conditions under which the identified organisms or consortia perform be identified? Can laboratory-scale or microcosm studies be conducted to demonstrate the process? Can controlled field studies be successfully run to evaluate performance and to optimize conditions? The intent of this paper is to summarize some of our previous studies on the biodegradation of hazardous organo-nitro explosives/propellants and chemical agents (Kaplan 1991, 1992). These studies only deal with some of the questions raised above, with many questions remaining to be answered before a final assessment of the suitability of biodegradation for a given chemical or set of chemicals at a given site under a given set of conditions can be made.

It will be important to define bioremediation/biodegradation terminology for the purpose of this paper. Biodegradation includes biotransformation and mineralization. Biotransformation indicates some degree of modification of the target chemical, such as a change in functional groups, but not resulting in mineralization. Mineralization is defined as biodegradation of the target chemical where carbon dioxide, methane, water and biomass are the final products. In studies on bioremediation these terms are critical, since two goals for this process are the reduction in hazardous properties of the target compound (biotransformation), and ideally, the complete return of the compounds into the normal geochemical carbon and nitrogen cycles (mineralization).

Explosives and propellants

Organo-nitro explosives and propellants can be divided into at least three categories, (1) nitroaromatics (e.g., 2,4,6-trinitrotoluene, 2,4-dinitrotoluene), (2) nitramines (e.g., nitroguanidine, hexahydro-1,3,5-trinitro-1,3,5-triazine, octahydro-1,3,5,7-trinitro-1,3,5,7-tetrazocine), and (3) nitrate esters (e.g., nitrocellulose, nitroglyerin, ammonium nitrates). General pathways for the biodegradation of explosives and propellants can be summarized as follows (Kaplan 1991):

(1) nitroaromatics:
$\gtrdot$C-NO_2 → $\gtrdot$C-NO → $\gtrdot$C-NHOH → $\gtrdot$C-NH_2
$\gtrdot$C-NO_2 → $\gtrdot$C-NH_2 → $\gtrdot$C=O → ring cleavage → mineralization
(2) nitramines:
>N-NO_2 → >N-NO → >N-NHOH → >N-NH_2 → mineralization
(3) nitrate esters:
>C-O-NO_2 → >C-OH → mineralization

2,4,6-Trinitrotoluene (TNT)

TNT can be biotransformed, mineralized or conjugated into higher molecular weight complex products. We have previously shown that a reductive pathway exists for the biotransformation of TNT by bacteria (Fig. 1) (McCormick et al. 1976; Carpenter et al. 1978; Kaplan & Kaplan 1982a-e, 1985c; Kaplan et al. 1985; Greene et al. 1985). This pathway has been observed in a number of systems including aqueous, sewage, soil and compost. Nitroaromatic compounds such as TNT are usually considered resistant to oxidation by oxygenase enzymes due to the presence of the electron withdrawing nitro groups on the ring. Recent studies with fungal systems (*Phanerochaete chrysosporium*) have provided the first substantive evidence for mineralization of the aromatic ring of TNT although the details of the mechanism and pathway remain to be elucidated (Fernando & Aust 1990, 1991; Lebron et al. 1992).

2,4-Dinitrotoluene (2,4-DNT)

Until recently, most biodegradation pathways elucidated for 2,4-DNT involved biotransformation but not ring cleavage or substantial mineralization. We have previously elucidated a biotransformation pathway for 2,4-DNT which involves the enzymatic

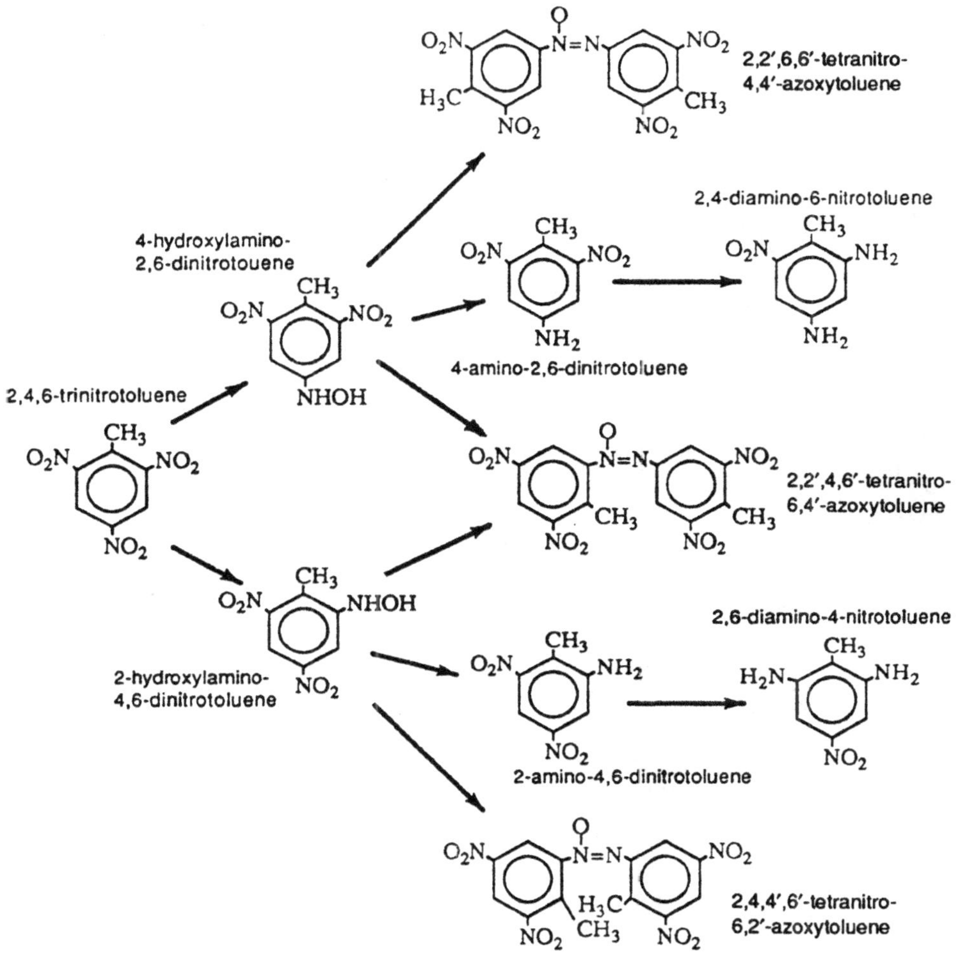

Fig. 1. Reductive biodegradation pathway for TNT.

reduction of both nitro groups to the corresponding amino groups (Fig. 2) (McCormick et al. 1978). This process occurs under aerobic conditions and supplmental carbon is required. The 4-nitro group was preferentially reduced in comparison to the 2-nitro group, similar to the preference noted for TNT. Recently, a reaction mechanism has been elucidated for the mineralization of 2,4-DNT by a reduction-oxidation system involving peroxidase enzymes and the white rot fungus *P. chrysosporium* (Valli et al. 1991).

Nitroglycerin (glycerol trinitrate)

We have shown that glycerol trinitrate is biotransformed through a series of successive denitration steps, including glycerol dinitrate and glycerol mononitrate isomers, with each succesive step slower than the previous step (Fig. 3) (Wendt et al. 1978). Glycerol is mineralized by biological systems. Glycidol and glycidol nitrate, formed by the chemical desensitization of glycerol trinitrate are also mineralized (Kaplan et al. 1982a). The biotransformation pathway of these compounds proceeds from glycidyl nitrate to 1-mononitrate to glycerol, with slower rates at each succeeding step.

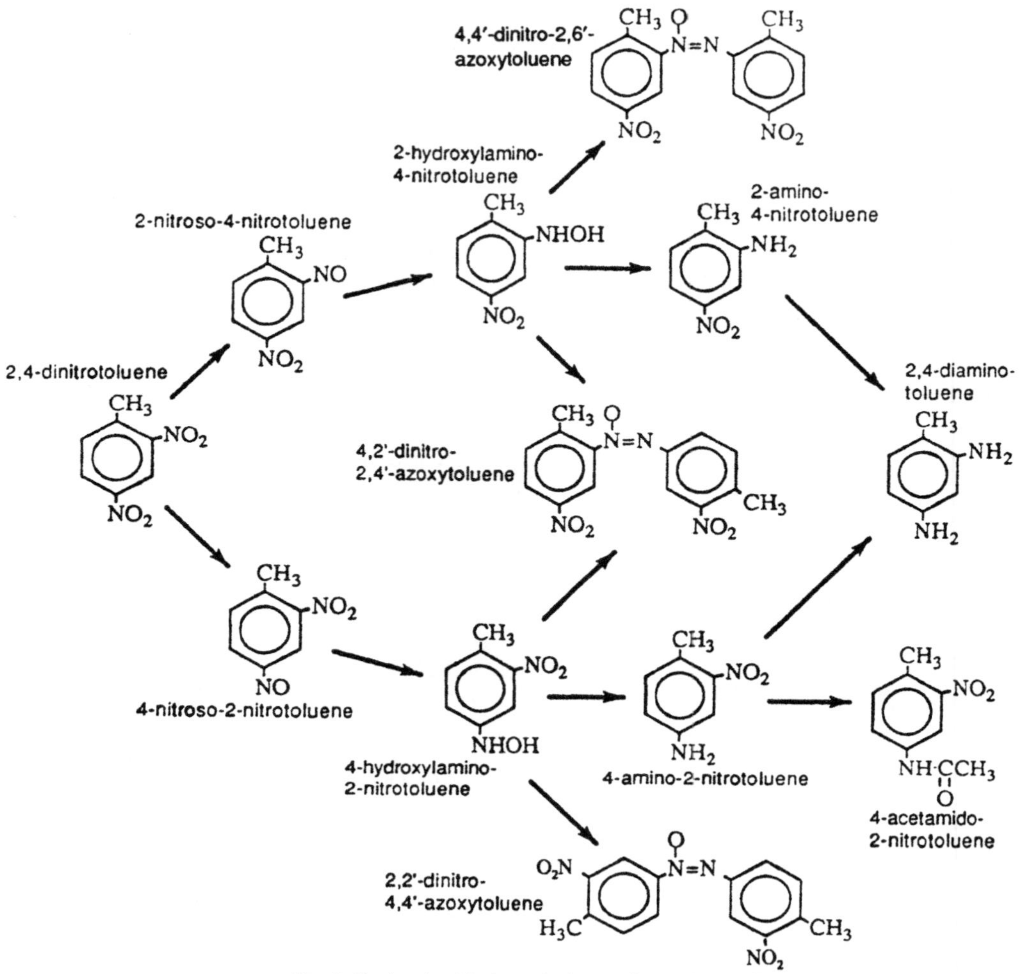

Fig. 2. Reductive biodegradation pathway for 2,4-DNT.

The steps from glycidol to glycerol and glycidyl nitrate to glycerol 1-mononitrate occur spontaneously in aqueous solutions, however it appears that the rates are accelerated by microbial activity, perhaps due to secondary effects.

Nitrocellulose

Nitrocellulose with a high degree of substitution (usually 11.1 to 14.5% nitrogen for cellulose dinitrate to trinitrate) is not subject to direct microbial attack and is generally regarded as persistent in the environment (Riley et al. 1984). In studies where the growth of organisms on nitrocellulose has been observed, growth may be due to contaminants in the preparation, growth on regions of incomplete or low degree of substitution, or possibly due to the effects of secondary metabolites on the chain chemistry (e.g., alkaline hydrolysis of the nitro-groups). We have shown that chemical pretreatment of nitrocellulose by alkaline hydrolysis was necessary to generate a modified denitrated polymer that could be attacked by microorganisms in an anerobic denitrification system with supplemental carbon (Wendt & Kaplan 1976a, 1976b; Kaplan et al. 1987). Importantly, the concentrations of nitrate and nitrite formed in process remained below drinking water limits due to the use of the denitrification process.

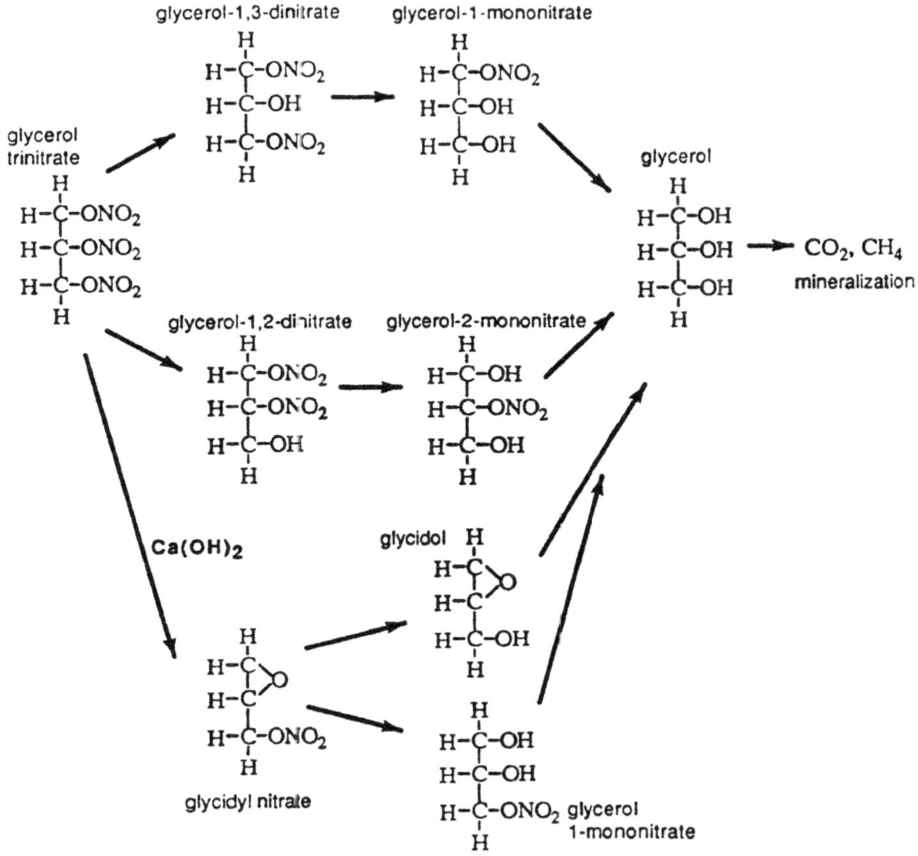

Fig. 3. Biomineralization pathway for glycerol trinitrate.

Propylene-, diethylene- and triethylene- glycol dinitrates and trimethylolethane trinitrate

We have shown that these nitrate esters are microbially transformed in a series of successive denitration steps to the correponding glycols: propylene glycol, diethylene glycol, triethylene glycol, and trimethylolethane glycol (Fig. 4) (Cornell et al. 1981), leading to mineralization apparently through a combination of biological and nonbiological processes. The relative rates of mineralization are propylene glycol > diethylene glycol > triethylene glycol > trimethylolethane glycol (Kaplan et al. 1982c).

Hydroxylammonium-, trimethylammonium-, isopropylammonium-, triethanolammonium- nitrates

We have shown that under aerobic conditions the trimethylammonium-, isopropylammonium-, and the triethanolammonium- nitrates were mineralized (Fig. 5) (Kaplan et al. 1984). Under anaerobic denitrification conditions trimethylammonium nitrate and triethanolammonium nitrate were mineralized and isopropylammonium nitrate was incompletely biodegraded. These studies were conducted with ^{14}C-labeled compounds in aerobic and anaerobic batch studies where the nitrate esters were available individually as the sole source of carbon and nitrogen and in continuous flow denitrification systems. In these studies no accumulation of intermediates was observed. The trimethyl-, isopropyl-

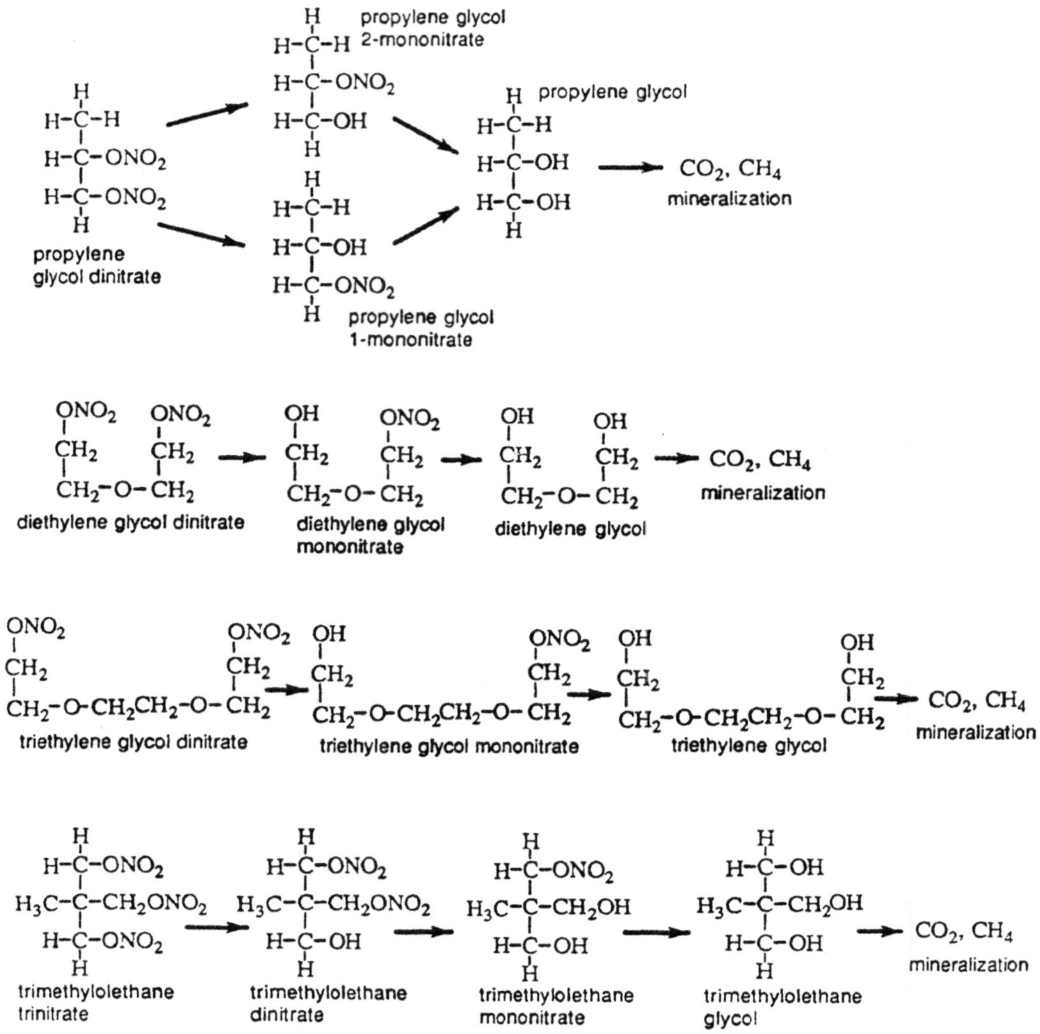

Fig. 4. Biomineralization pathways for a series of glycol nitrate propellants.

and triethanol- ammonium nitrates were mineralized in soil when studied at concentrations from 50 to 5000 mg/l, and hydroxylammonium nitrate was labile under slightly acidic and alkaline conditions. Nitrosamines such as N-nitrosodimethylamine were observed during analysis of culture samples from biodegradation studies and biodegradation studies indicated that this compound is biodegradable and is presumed to be mineralized (Kaplan & Kaplan 1986).

Triaminoguanidine nitrate

We have shown that triaminoguanidine nitrate is biodegraded under aerobic and anaerobic conditions with a requirement for supplemental carbon (Kaplan & Kaplan 1985a). No intermediates were identified during this process. Guanidine, hydrazine, urea, carbohydrazide (1,3-diaminourea), cyanamide, and cyanoguanidine were sought but not detected in culture samples.

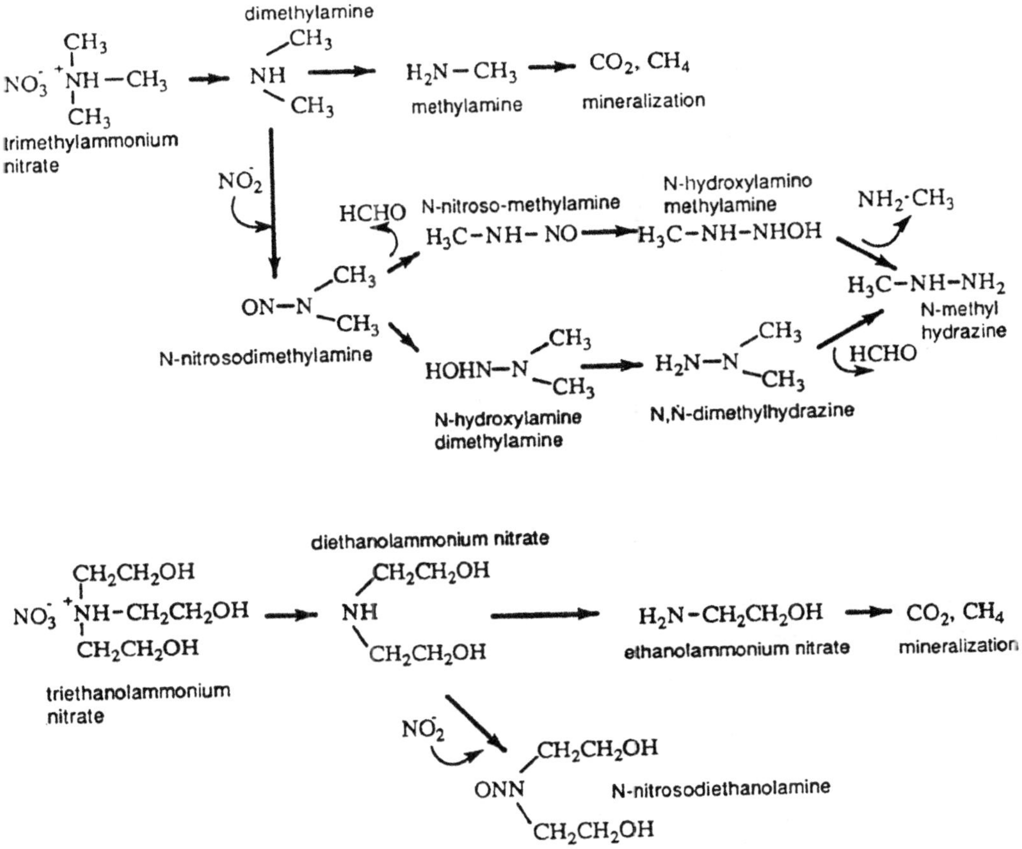

Fig. 5. Biomineralization pathways for ammonium nitrate propellants.

Hexahydro-1,3,5-trinitro-1,3,5-triazine (RDX) and hexahydro-1-N-acetyl-3,5-dinitro-1,3,5-triazine (AcRDX)

RDX and the related N-acetylated derivative, AcRDX, were studied in aqueous systems and found to be mineralized by mixed populations of microorganisms under anaerobic conditions when supplemental carbon was provided (Fig. 6) (McCormick et al. 1981, 1985a, 1985b). Intermediates in this pathway included mono-, di-, and tri-nitroso-compounds formed during the sequential reduction of the nitro groups on the parent compound. Traces of hydrazine, 1,1-dimethylhydrazine, and 1,2-dimethylhydrazine were also detected in the samples after exposure to the mixed microbial populations. No biotransformation of RDX was observed under aerobic conditions. The AcRDX was also biotransformed only under anaerobic conditions with the formation of the analogous series of mono- and di-nitroso compounds and in this case, no hydrazine or dimethylhydrazine.

Octahydro-1-3,5,7-trinitro-1,3,5,7-tetrazocine (HMX) and octahydro-1-N-acetyl-3,5,7-trinitro-1,3,5,7-tetrazocine (AcHMX)

We have reported that HMX and AcHMX were mineralized under anaerobic conditions at 50 mg/l, although rates were slower than for RDX (McCormick et al. 1985a). Biotransformation was incomplete and mono- and di-nitroso intermediates were identified (Fig. 7). In a similar fashion to RDX, supplemental carbon was necessary for the biotransformation process to function. The kinetics of

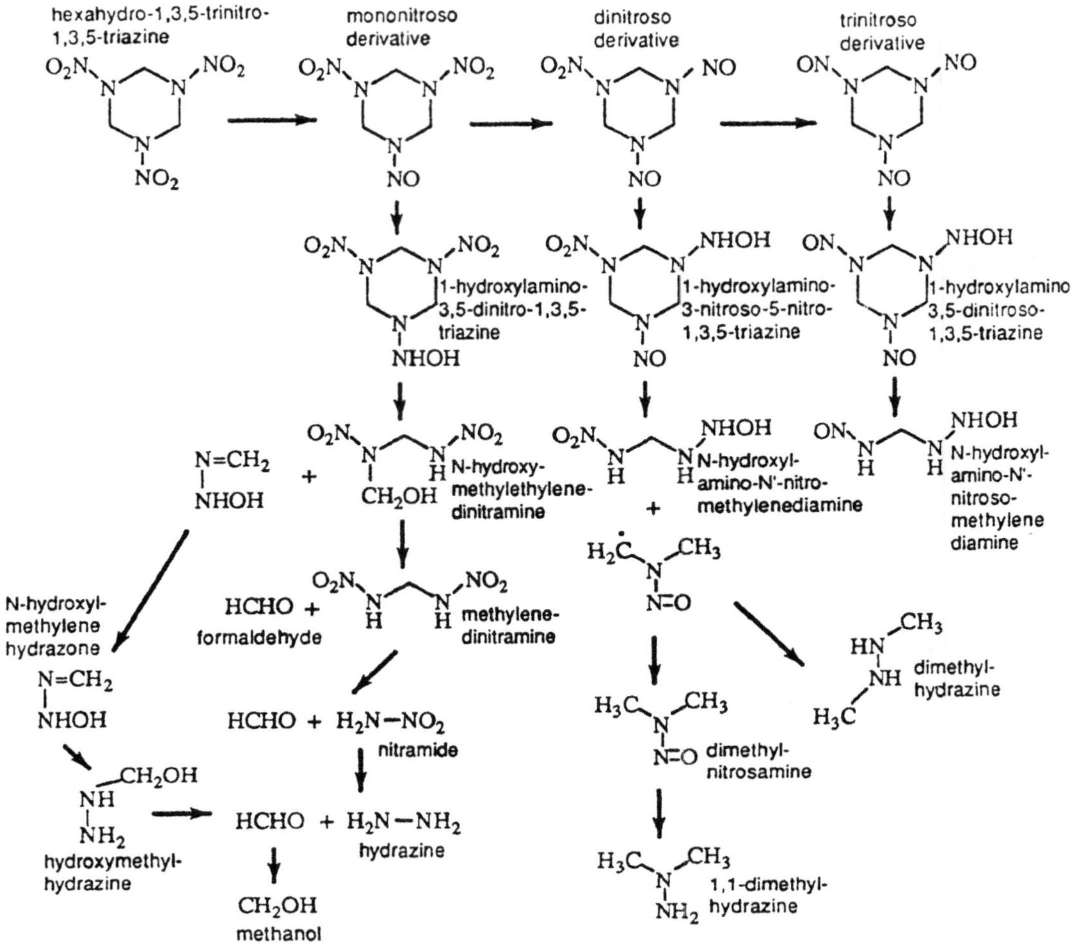

Fig. 6. Anaerobic biomineralization pathway for RDX.

biotransformation of the AcHMX were slower than for HMX, and two isomers of dinitroso-AcHMX were detected. No hydrazine, 1,1-dimethylhydrazine, or 1,2-dimethylhydrazine were detected during the biotransformation of HMX or AcHMX. HMX and AcHMX degraded at a slower rate in continuous culture under anaerobic denitrification conditions than in the batch studies and none of the intermediates were detected. No biotransformation of HMX or AcHMX occured under aerobic conditions.

Nitroguanidine

We have shown that in aqueous systems, nitroguanidine was biotransformed under anaerobic conditions in the presence of cometabolic carbon (Fig. 8) (Kaplan et al. 1982b, 1985b). Biotransformation of nitroguanidine to nitrosoguanidine was observed with subsequent nonbiologically-mediated formation of the intermediates cyanamide, nitrosamide, cyanoguanidine, melamine, and guanidine. Melamine is the trimer-cyanamide product which can also react with ammonia to form guanidine. The nitrosamide is transitory and decomposes to nitrogen gas and water. The cometabolic degradation of nitroguanidine at 150 mg/l in soil was also demon-

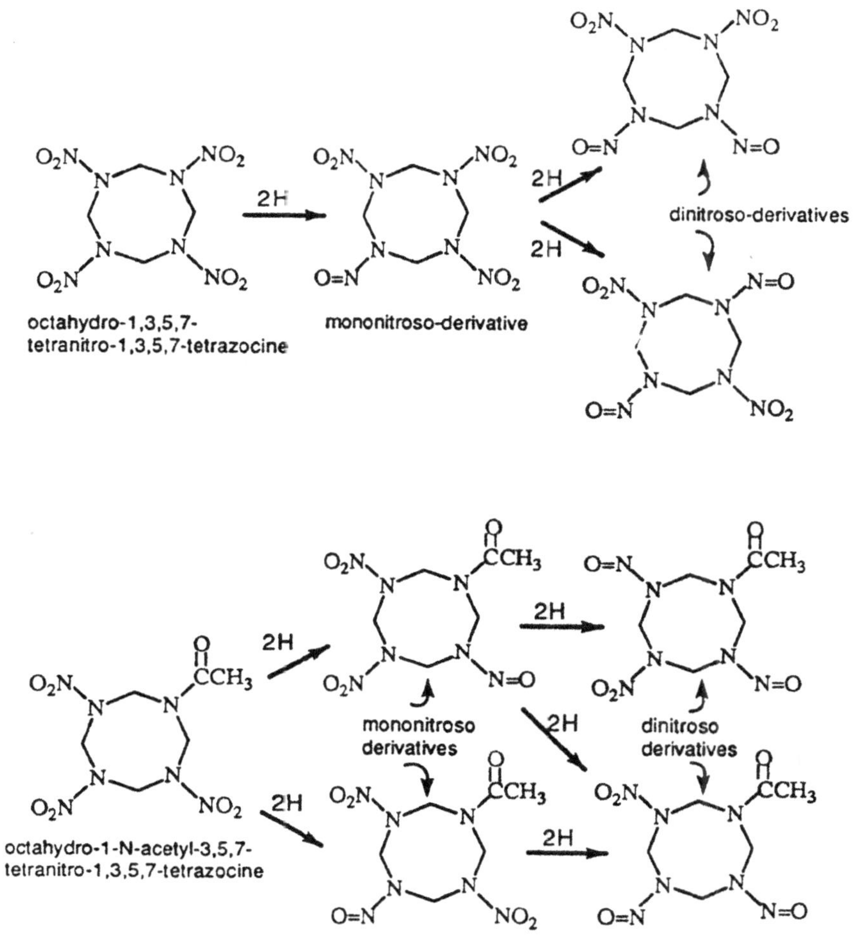

Fig. 7. Anaerobic biotransformation of HMX and acetylated-HMX.

strated (Kaplan and Kaplan 1985). Traces of nitrosoguanidine (< 100 μg/l) were detected in the leachates from the soil, and ammonia was the primary nitrogen species formed. Supplemental carbon, added in the form of glucose, was required for this transformation to occur. The concentration of glucose required for complete mineralization of the nitroguanidine ranged between 1.0 and 0.5%. The ammonia concentration detected in leachates from the soil columns correlated directly with the degradation kinetics of nitroguanidine. Nitrate and nitrite concentrations remained at or below the mg/l range throughout the study. Mass balance studies show an 85% conversion of the nitroguanidine-nitrogen to ammonium-nitrogen in these soil studies.

Chemical agents

Chemical agents are found as enviromental contaminants, particularly in soils, due to past production practices. In addition, international treaties will require the destruction of large quantities of these compounds. Environmentally compatible and cost effective approaches will be required to deal with these problems and a biological approach is one alternative being actively explored. Chemical agents can be divided into the organophosphate

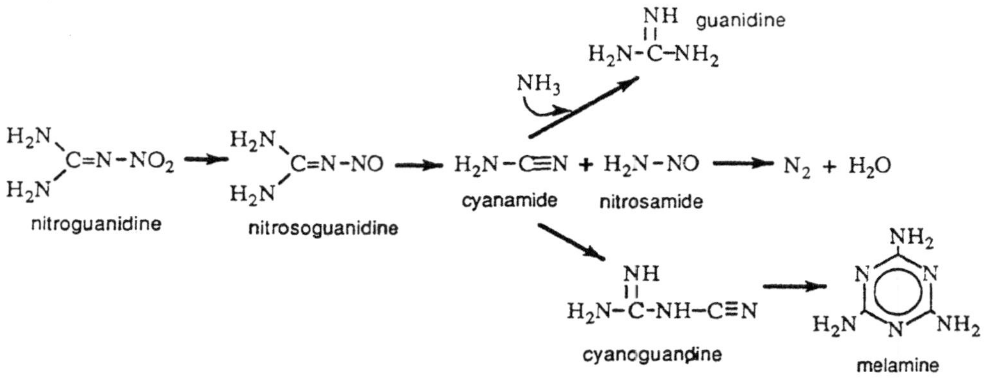

Fig. 8. Biotransformation pathway for nitroguanidine in aqueous systems.

neurotoxins [sarin, VX (malathion-type), paraoxon and parathion-type (P-F) bonds)], and mustards [HT (N-mustard) and HD (S-mustard)]. The mustards can be defined as organic compounds containing at least two chloroethyl groups ($-CH_2CH_2Cl$) linked to either a thioether or amine. The toxicity of mustards is due to alkylating effects on DNA and two chloroethyl groups are required for the toxicity. The toxicity of the organophosphates derives from their interference with neurotransmitter biochemistry such as inhibition of acetylcholinesterase activity.

Organophosphorous hydrolases (OPH)

Organophosphorous hydrolases (OPH) [EC 3.1.8.1] previously termed organophosphorous acid anhydrases or OPA anhydrases, DFPases, somanases, sarinases, paroxonases, parathion hydrolase,and phosphotriesterase have been identified in bacteria (mesophilic, thermophilic and halophilic), protozoa, invertebrates and mammals. Mazur (1946) first described a P-F splitting enzyme that he termed DFPase. Over the following decade a number of papers appeared on the general topic of the DFPases and the developments of this period have been summarized by Mounter (1963). In 1965, as an unexpected result of bathing squid giant axons in diisopropylfluorophosphate (DFP) solutions, squid nerve and indeed cephalopod nerve in general was found to have a high level of DFPase by Hoskin (1966) and coworkers. The term 'squid-type DFPase' was introduced by Garden et al. (1975) to denote that the Hoskin enzyme hydrolyzed DFP more rapidly than it hydrolyzed tabun (ethyl N,N-dimethyl-phosphoramidocyanidate) or sarin (isopropylmethylphosphonofluoridate). The squid enzyme was shown to have a molecular weight close to 26 kDa. Hoskin (1982) was able to demonstrate that the squid type DFPase has a degree of stereospecificity with respect to the phosphorous chiral center of Soman (1,2,2-trimethylpropyl methylphosphonofluoridate), that is, the enzyme preferentially hydrolyzed the toxic enantiometers of soman.

OPHs are especially plentiful in mammalian kidney and liver, and are capable of hydrolyzing DFP, tabun and sarin. Hoskin (1984) stated that there appears to be two distinctly different classes of OPH. The first class is of the 'Mazur' type OPH. They hydrolyze Soman 10 to 20 times faster than DFP, are stimulated three to five fold by manganous ion , are inhibited by Mipafox (N,N-diisopropylphosphodiamidofluoridate), and are inactivated by ammonium sulfate. Mazur type OPH have molecular weights in the range of 50 to 100 kda. This class includes OPH from mammalian tissues as well as from bacteria. The second class is called 'Squid' type OPH. It is a much narrower class than the Mazur type and includes not only squid but cephalopods in general. Unlike the Mazur type, they hydrolyze DFP five times faster than soman,

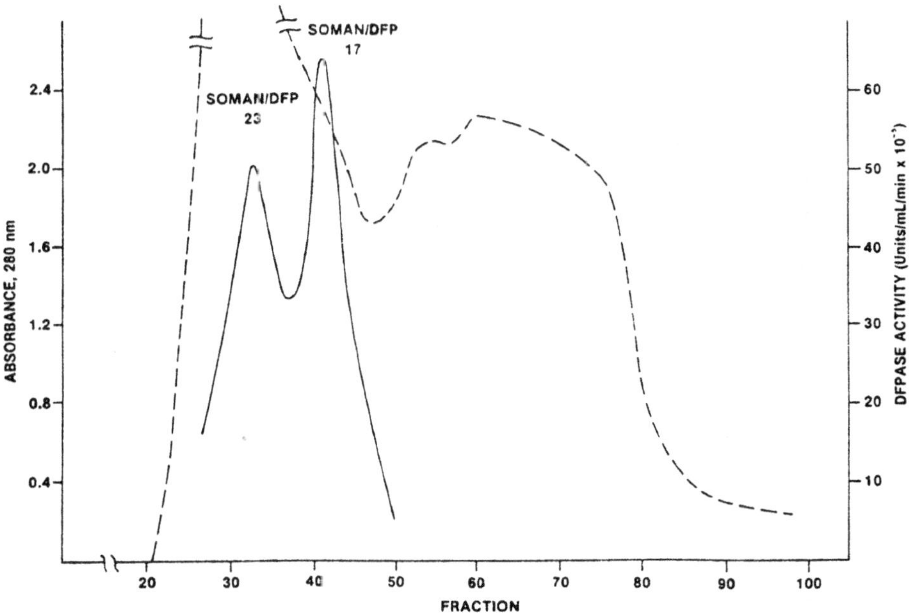

Fig. 9. Chromatogram of *E. coli* extract on g-150 column. (---) protein; (——) activity.

are indifferent to Mipafox, and tolerate ammonium sulfate precipitation as a first purification step.

Escherichia coli OPH

Zech & Wigand (1975) were the first investigators to report that a cell extract of *E. coli* exhibited two chromatographically separable forms that could hydrolyze DFP. Walker et al. (1986) showed that these fractions had a DFP/Soman activity ratio of 23 and 17, respectively. Using gel filtration (Fig. 9) and ion exhange chromatography on DEAE

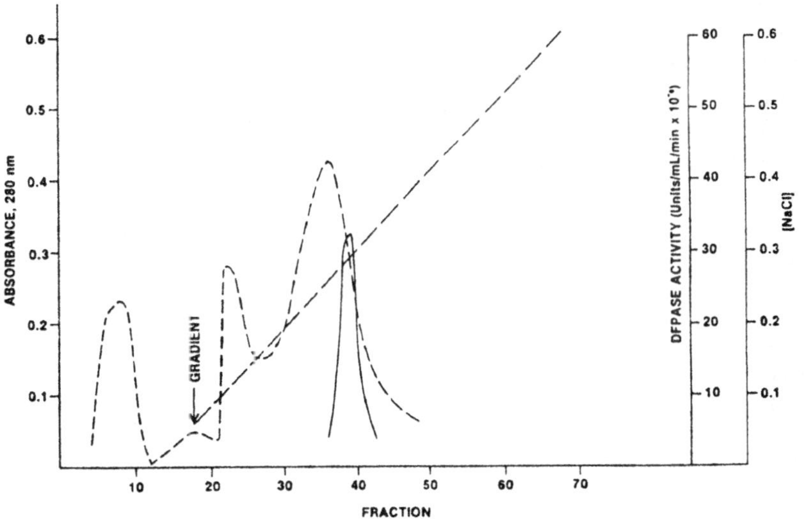

Fig. 10. A-50 chromatogram of *E. coli* fraction-1. (---) protein; (——) activity.

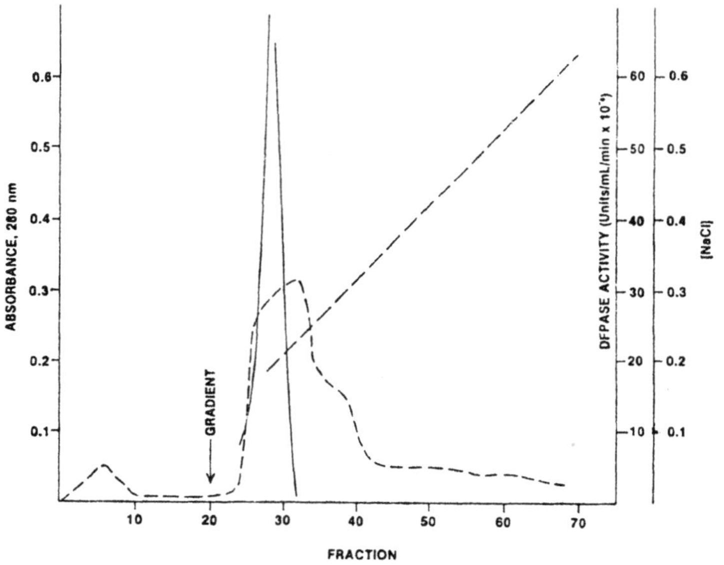

Fig. 11. A-50 chromatogram of *E. coli* fraction-2. (---) protein; (——) activity.

A-50 (Pharmacia) (Figs. 10 and 11), two active fractions were isolated.

SDS gel electrophoresis of the DEAE separated OPHs gave monomer molecular weights of 50 and 40 kDa respectively (Fig. 12). Recently Hoskin et al. (1993) reported that *E. coli* OPHs react differentially in their ability to detoxify the toxic and nontoxic pairs of soman isomers. One form hydrolyzes and detoxifies all the isomers equally, while the other preferentially hydrolyzes the nontoxic pair initially (Table 1). Akkara & Kaplan (1991) have immobilized the *E. coli* enzyme to cotton fabric and showed that it is reactive against DFP in solution (Fig. 13). These authors have shown that the immobilized *E. coli* enzyme can react against soman in the vapor state (Fig. 14).

Table 1. Enzymatic degradation of soman by *E. coli* fractions determined by two methods.

Fraction	Hydrolysis by F-electrode (%)	Detoxication by AChE inhibition (%)
G-150-I	50	48
G-150-II	50	42
A-50-(G-150-I)	50	49
A-50-(G-150-II)	50	18

Bacillus stearothermophilus *OPH*

Walker et al. (1988) isolated an enzyme from a strain of *B. stearothermophilus* (JD-100) that possesses no hydrolyzing activity against DFP but hydrolyzes and detoxifies both the toxic and non toxic isomers of soman at the same rate (Table 2). The enzyme has a molecular weight close to 80 kDa as determined by HPLC on a calibrated TSK-400 (Bio-Rad) column and by SDS gel electrophoresis, and has monomer molecular weights of 52, 40, and 36 kDa (Fig. 15). The characteristics of the enzyme are given in Table 3. The enzyme has been immobilized by covalent attachment onto cotton and hydrolyzes soman in solution (Fig. 16). On-going work is proceeding to test the immobilized thermophilic enzyme against soman vapor. The enzyme appears to retain its activity in short term heat stability tests (Table 4).

Tetrahymena thermophilia *OPH*

Landis et al. (1987) reported that *T. thermophilia* synthesizes an intracellular OPH capable of degrading DFP. Gallo et al. (1991) isolated two enzymes from *T. thermophilia* bryant which has high

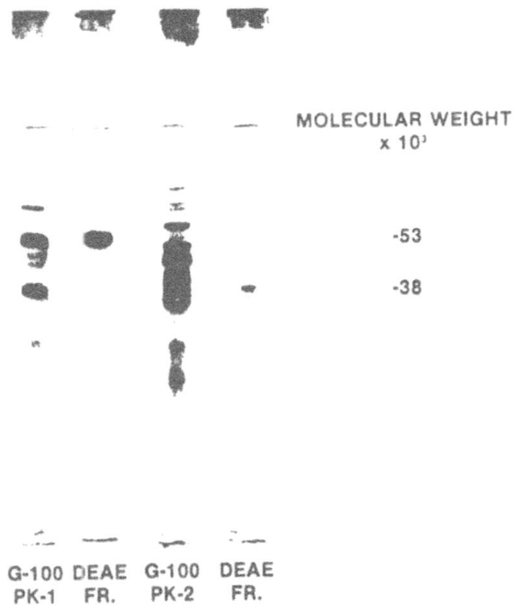

Fig. 12. SDS gel electrophoresis of A-50 fractions of *E. coli*.

OPH activity when compared to both the OPH from *B. stearothermophilus* and *E. coli* (Table 5). The enzyme could be separated into two forms by hydrophobic chromatography on phenyl sepharose (Pharmacia). Both chromatographically separated forms hydrolyzed and detoxified all isomers of soman (Table 6). SDS gel electrophoresis of the enzyme gives a molecular weight of 45 kDa.

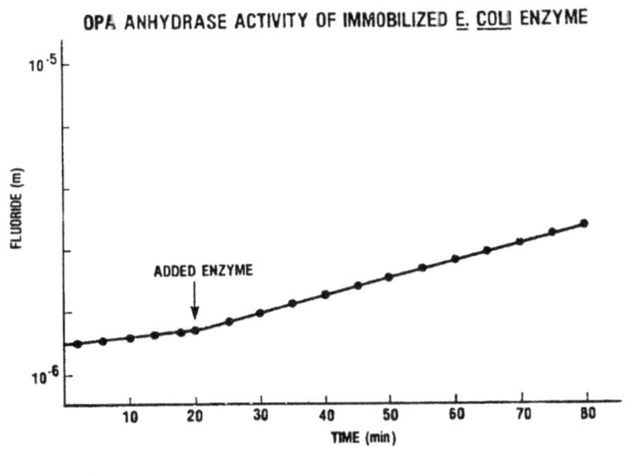

Fig. 13. Activity of immobilized *E. coli* OPH against soman in solution.

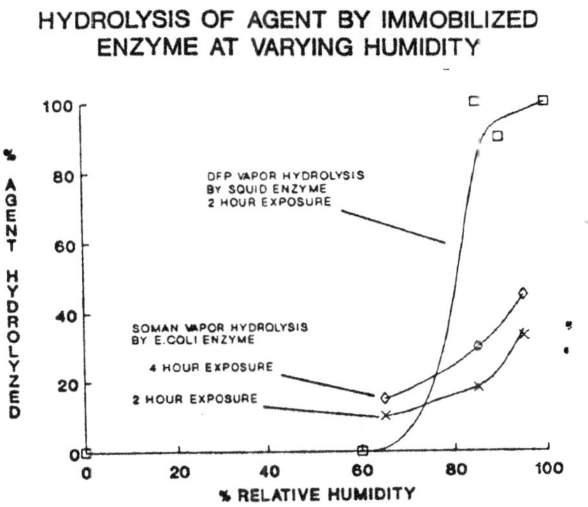

Fig. 14. Activity of immobilized *E. coli* OPH against soman vapor.

Squid OPH

The squid OPH has received the most attention as a potential enzymatic decon material against G-type agents. Extensive stability studies by the US Navy under contract with IIT research Institute (Rajan et al. 1989) have shown that the enzyme retains its activity as a lyophilate after two years storage at 4 °C. A phosphate buffered solution of the enzyme (100 mg/ml) can decontaminate 200–250 mg of soman/square centimeter of material surface. The enzyme retains >96% of its initial activity after each decontamination experiment. Walker et al. (1990) have studied the secondary and tertiary structure of the enzyme (Table 7). The OPH from squid has a calcium ion requirement for expression of activity. Circular di-

Table 2. Relative rates of hydrolysis of soman by immobilized DFPases from various sources.

Enzyme source	Racemic	Rp (AChE inhib)	Sp
Hog Kidney	100	8.6	91.4
E. coli	100	0.0	100.0
Squid	100	44.1	55.9
JD-100	100	38.4	61.1

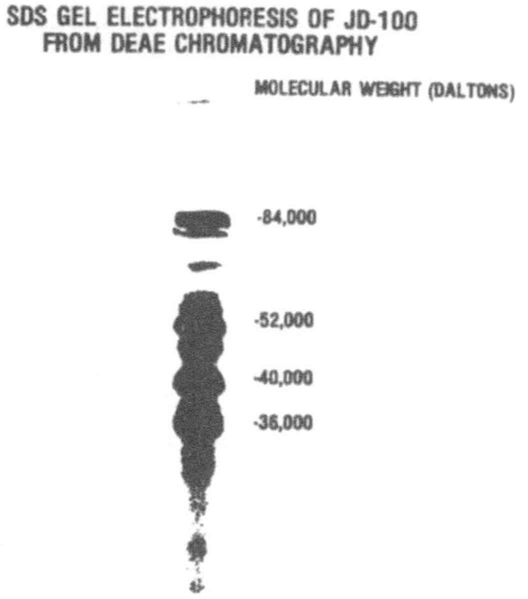

Fig. 15. SDS gel electrophoresis of *B. stearothermophilia* OPH.

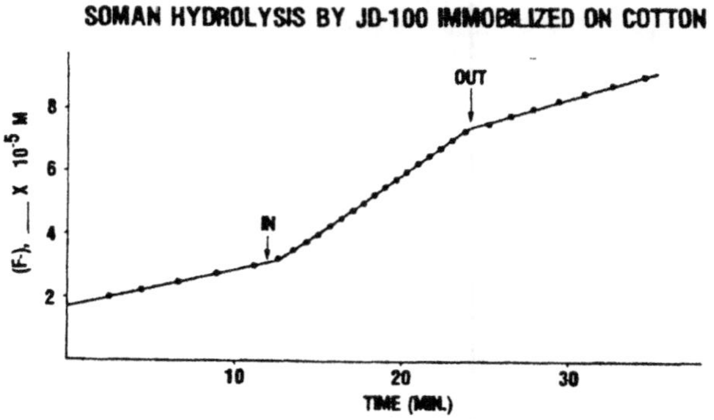

Fig. 16. Activity of immobilized *B. stearothermophilia* OPH against soman in solution.

chroism spectra as well as fluorescence life time studies of tryptophan residues, indicate that the enzyme undergoes a structural change in the presence of calcium. On-going studies are focusing on the sequence of amino acids in the active site region and the binding sites for calcium.

Biodegradation of mustard

A number of possible routes for the biodegradation of mustard have been considered, including dehalogenation, reduction and oxidation. Since mustard (dichlorodiethyl sulfide) is a strong DNA alkylating agent, it was expected to be lethal to bacteria and therefore isolation of microorganisms capable of degrading mustard has been unsuccessful. Chemostats were recently used to select for mustard degraders in a mineral salts medium with thiodiglycol as a carbon source (Harvey et al. 1991). Post selection, the thiodiglycol could be used as a sole carbon source at 10 mM. *Pseudomonas picketii* and *Alcaligenes xylosoxidans* sp. have been putatively identified as the responsible organisms. The organisms can grow on thiodiglycol, thiodiglycol sulfoxide and thioether compounds, producing carbon dioxide and biomass. The mustard is in equilibrium with the thiodiglycol and sulfonium ion intermediate in the mineral salts medium.

Table 3. Characterization of JD-100.

Molecular weight:	80,000 Daltons
Oligomeric enzyme:	Subunits; 52,000 40,000 and 36,000 Daltons
Isoelectric point:	4.0
Detoxifies all isomers of Soman	
Not inhibited by Mipafox	
Manganous ion requirement	

Discussion

In general, many of the explosives/propellants and

Table 4. Stability of JD-100.

	Activity (Units/mL)	% Loss of activity
Initial		
Solution	0.60	–
Freeze dried	0.60	–
23 °C for 2 weeks		
Solution	0.18	60%
Freeze dried	0.54	10%
55 °C for 2 weeks		
Solution	0.45	25%
Freeze dried	0.66	0.0%
4 °C for 2 weeks		
Solution	0.51	15%
Freeze dried	0.60	0.0%

Table 5. Specific activity of partially purified OPA anhydrase from 3 diverse microbial strains.[a]

Microbe type	Species	Strain	Specific activity[b]
Protozoan	*Tetrahymena thermophilia*	Brynat	0.73
Bacterium	*Baccilus stearothermophilus*	I5D	0.13
Bacterium	*Escherichia coli*	ATCC 25922	0.06

[a] Partially purified OPA anhydrase resulting from Sephadex G150 and then Phenyl-Sepharose CL-4B columns chromatography.
[b] uM PNMPP hydrolyzed/min*mg^{-1} soluble protein.

chemical agents are amenable to biodegradation. The activity of biological systems on these compounds can lead to products that represent reduced toxicity hazards when compared with the parent compound, and in many cases can lead to mineralization to return the carbon, hydrogen and oxygen to natural geochemical cycles. This can be the scenario in many instances provided the appropriate conditions can be maintained. In many cases, the reaction kinetics and the products formed in these reactions depend on the conditions under which the target compound is studied. Pathways outlined in this chapter for explosives/propellants and chemical agents form only a starting point, to indicate that a potential exists to treat the compound in a biological system.

It would be easy to generalize the pathways and reactions described. However caution must be used in the careful assessement of each compound and the conditions under which it is to be studied. In most cases, biological approaches to the degradation or biotransformation of these types of compounds can offer significant benefits such as the low process costs and the formation of less toxic compounds. Often, chemical pretreatments in combination with the biological system, as in the case of the alkaline pretreatment of nitrocellulose or glycerol trinitrate and the hydrolysis of mustard are useful and perhaps in some cases essential steps, to accelerate the follow-on biological process.

Table 6. Enzymatic hydrolysis and detoxification of soman by two Phenyl-Sepharose CL-4B OPA anhydrase fractions from *Tetrahymena thermophilia.*

Activity peak	% Soman hydrolysis[a]	% Soman detoxification[b]
Fraction F-1	50	39
Fraction F-2	50	59

[a] Soman hydrolysis was determined by measuring the release of fluoride ions by a fluoride specific electrode.
[b] Soman detoxification was determined by measuring the degree of inhibition of acetylcholinesterase by soman (Conducted at IIT).

Table 7. Physico-chemical properties of squid OPA anhydrase.

Narrow distribution; Squid nerve, saliva, hepatopancreas
Molecular weight; 26,000 Daltons
Isoelectric point; 5.5
Soman/DFP = 0.25
Km = 2.5×10^{-3} M
Hydrolyses all isomers of soman
Ca^{+2} requiring, not Ca^{+2} stimulated. Ca^{+2} may be involved in intramolecular bonding
Mipafox indifferent
Em 280 = 38,500
Disulfide bridges not essential for enzyme activity
2 of the 6 Trp residues are exposed to the microenvironment

Acknowledgements

Special thanks go to Joe Akkara, John Cornell, Ben Gallo, Neil McCormick and Diane M. Steeves for their contributions.

References

Abelson PH (1992) Remediatioon of hazardous waste sites. Science 255: 901.

Akkara JA & Kaplan DL (1991) Soman detoxifying enzyme from *E. coli*. Proceedings 1990 Scientific Conference on Chemical Defense Research,Aberdeen Proving Ground, Maryland. Report #SP-034, pp 1007–1012

Carpenter DF, McCormick NG, Cornell JH & Kaplan AM (1978) Microbial transformation of ^{14}C-labeled 2,4,6-trinitrotoluene in an activated sludge system. Appl. Environ. Microbiol. 35: 949–954.

Cornell JH, Wendt TM, McCormick NG, Kaplan DL & Kaplan AM (1981) Biodegradation of nitrate esters used as military propellants. Technical Report 81-029, US Army Natick Research, Development and Engineering Center, Natick, MA

Environmental Protection Agency (1990) Bioremediation of hazardous wastes. Report 600/9-90/041, EPA Office of Research and Development, Washington DC 1-58

Fernando T & Aust SD (1991) Biodegradation of munition waste, TNT (2,4,6-trinitrotoluene), and RDX (hexahydro 1,3,5-trinitro-1,3,5-triazine) by *Phanerochaete chrysosporium*. In: Tedder DW & Pohland FG (Eds) ACS SymEmerging Technologies in Hazardous Waste Management II, Series 468 (pp 214–232)

Fernando T, Bumpus JA & Aust SD (1990) Biodegradation of TNT (2,4,6-Trinitrotoluene) by *Phanerochaete chrysosporium*. Appl. Environ. Microbiol. 56: 1666–1671

Fox JL (1992) Contemplating large-scale use of engineered microbes. American Society for Microbiology News 58: 191–196

Gallo BG, Walker JE & Hoskin FCG (1991) OPA anhydrase synthesis by *Tetrahymena thermophilia*. Proceedings 1990 Scientific Conference on Chemical Defense Research, Aberdeen Proving Ground, Maryland. Report#SP-034, pp 1085–1092

Garden JM, Hause SK, Hoskin FCG & Roush AH (1975) Comparison of DFP hydrolyzing enzyme purified from head ganglia and hepatopancreas of squid (*Loligo pealei*) by means of isoelectric focusing. Comp. Bioch. Physiol. 52C: 95–98

Greene B, Kaplan DL & Kaplan AM (1985) Degradation of pink water compounds in soil - TNT, RDX, HMX, Technical Report 85-046, US Army Natick Research, Development and Engineering Center, Natick, MA

Harvey S, DeFrank JJ, Kamely D, Valdes JJ & Chakrabarty AM (1991) Microbial degradation of agent orange and mustard related compounds. In: Kamely D, Chakrabarty AM & Kornguthm SE (Eds) Biotechnology: Bridging Research and Applications (pp 2–220). Kluwer Academic Publishers, Boston

Hoskin FCG & Roush AH (1982) Hydrolysis of nerve gas by squid type diisopropylphosphofluoridate hydrolyzing enzyme on agarose resin. Science 215: 1255–1257

Hoskin FCG, Rosenberg P & Brzin M (1966) Re-examination of the effect of DFP on electrical and cholinesterase activity of squid giant axon. Proc. Natl. Acad. Sci. USA 55: 1231–1235

Hoskin FCG, Kirkish MA & Steinmann KE (1984) Two enzymes for the detoxification of organophoshorous compounds; sources, similarities and significance. Fundam. Appl. Toxicol. 4: S165–S172

Hoskin FCG, Walker JE, Steeves DM & Gallo BJ (1993) Stereoselectivity of soman detoxication by organophosphorus acid anhydrases from *Escheria coli*. Chemico-Biological Interactions (in press)

Kaplan DL (1991) Biotransformation of hazardous organo-nitro compounds. In: Kamely D, Chakrabarty A & Omenn G (Eds) Biotechnology and Biodegradation, Houston (pp 155-181). Gulf Publishing, Houston

Kaplan DL (1992) Biological degradation of explosives and chemical agents. Curr. Opinion in Biotechnol: 3: 253–260

Kaplan DL & Kaplan AM (1982a) 2,4,6-trinitrotoluene-surfactant complexes: decomposition, mutagenicity and soil leaching studies. Environ. Sci. Technol. 16: 566–571

Kaplan DL & Kaplan AM (1982b) Composting industrial wastes – biochemical considerations. Biocycle 3: 42–44

Kaplan DL & Kaplan AM (1982c) Mutagenicity of 2,4,6-trinitrotoluene-surfactant complexes. Bulletin Environ. Contam. Toxicol. 28: 33–38

Kaplan DL & Kaplan AM (1982d) Separation of mixtures of 2,4,6-trinitrotoluene reduction products with liquid chromatography. Anal. Chem. Acta 136: 425–428

Kaplan DL & Kaplan AM (1982e) Thermophilic biotransformation of 2,4,6-trinitrotoluene under similated composting conditions. Appl. Environ. Microbiol. 44: 757–760

Kaplan DL & Kaplan AM (1985a) Bioconversion of nitramine propellant wastewaters – triaminoguanidine nitrate. Technical Report 85-045, US Army Natick Research, Development and Engineering Center, Natick, MA

Kaplan DL & Kaplan AM (1985b) Degradation of nitroguanidine in soils. Technical Report 85-047, US Army Natick Research, Development and Engineering Center, Natick, MA

Kaplan DL & Kaplan AM (1985c) Reactivity of TNT and TNT-microbial reduction products with soil components. Technical Report 83-041, US Army Natick Research, Development and Engineering Center, Natick, MA

Kaplan DL & Kaplan AM (1986) Biodegradation of N-nitrosodimethylamine in aqueous and soil systems. Appl. Environ. Microbiol. 50: 1077–1086

Kaplan DL & Cornell JH & Kaplan AM (1982a) Biodegradation of glycidol and glycidol nitrate. Appl. Environ. Microbiol. 43: 144–150

Kaplan DL, Cornell JH & Kaplan AM (1982b) Decomposition of nitroguanidine. Environ. Sci. Technol. 16: 488–492

Kaplan DL, Emerson D, LeDoux R, Mayer J & Kaplan AM (1985) Effects of environmental factors on the transformation of 2,4,6-trinitrotoluene in soils. Technical Report 85-052, US Army Natick Research, Development and Engineering Center, Natick, MA

Kaplan DL, Riley PA, Emerson DJ & Kaplan AM (1984) Degradation of ammonium nitrate propellants in aqueous and soil systems. Environ. Sci. Technol. 18: 694–699

Kaplan DL, Riley PA, Pierce J & Kaplan AM (1987) Denitrification of high nitrate loads – efficiencies of alternate carbon sources. International Biodeterioration 23: 233–248

Kaplan DL, Walsh JT & Kaplan AM (1982c) Gas chromatographic analysis of glycols to determine biodegradability. Environ. Sci. Technol. 16: 723–725

Landis WG, Anderson RA, Durst DH, James J, Chester NA, Haley MV, Johnson DW & Tauber RM (1987) The organofluorophosphate hydrolases of *Tetrahymena thermophilia* and *Rangia cuneata*. Proceedings of 1986 Scientific Conference on Chemical Defense Research, Aberdeen Proving Ground, Maryland. Report#SP-008, pp 903–909

Lebron CA, Karr LA, Fernando T & Aust SD (1992) Biodegradation of 2,4,6-trinitrotoluene by white-rot fungus. US Patent 5,085,998.

Mazur A (1946) An enzyme in animal tissues capable of hydrolyzing the phosphorous-fluorine bond of alkyl fluorophosphates. J. Biol. Chem. 164: 271–289

McCormick NG, Feeherry FE & Levinson HS (1976) Microbial transformation of 2,4,6-trinitrotoluene and other nitroaromatic compounds. Appl. Environ. Microbiol. 31: 949–958

McCormick NG, Cornell JH & Kaplan AM (1978) Identification of biotransformation products for 2,4-dinitrotoluene. Appl. Environ. Microbiol. 35: 945–948

McCormick NG, Cornell JH & Kaplan AM (1981) Biodegradation of hexahydro-1,3,5-trinitro-1,3,5-triazine. Appl. Environ. Microbiol. 42: 817–823

McCormick NG, Cornell JH & Kaplan AM (1985a) The anaerobic biotransformation of RDX, HMX, and their acetylated derivatives. Technical Report 85-007, US Army Natick Research, Development and Engineering Center, Natick, MA

McCormick NG, Cornell JH & Kaplan AM (1985b) The fate of hexahydro-1,3,5-trinitro-1,3,5-triazine (RDX) and related compounds in anaerobic denitrifying continuous culture systems using simulated waste water. Technical Report 85-008, US Army Natick Research, Development and Engineering Center, Natick, MA

Mounter LA (1963) Metabolism of organophosphorous anticholinesterase agents. In: Handbuch der Experimentallen Pharmakologies, Ch 10, Springer-Verlag, Berlin

Rajan KS, Mainer S, Steinmann KE & Hoskin FCG (1989) Stability of squid-type DFPase and its recovery in CW-agent decontamination application. Proceedings 1988 Scientific Conference on Chemical Defense Research, Aberdeen Proving Ground, Maryland. Report #SP-013, pp 119–128

Riley PA, Kaplan DL & Kaplan AM (1984) Stability of nitrocellulose to biological degradation. Technical Report 85-004, US Army Natick Research, Development and Engineering Center, Natick, MA

Thayer AM (1991) Bioremediation: innovative technology for cleaning up hazardous waste. Chem. Eng. News, August 26: 23–44

Valli K, Brock BJ, Joshi DK & Gold MH (1992) Degradation of 2,4-dinitrotoluene by the lignin-degrading fungus *Phanerochaete chrysosporium*. Appl. Environ. Microbiol. 58: 221–228

Walker JE, Robbins FM, Smith WJ & Gallo BJ (1986) Enzymes from microbial sources with activities against G-agents. Proceedings Army Science Conference, West Point, New York, Vol 5, pp 251–256

Walker JE, Gallo BJ, Gowenlock D, Starcher B & Hoskin FCG (1988) Soman hydrolyzing and detoxifying properties of an enzyme from a thermophilic bacterium. Proceedings Army Science Conference, Fort Monroe, Virginia, Vol 3, pp 441–446

Walker JE, Connolly JJ, Steeves DM & Hoskin FCG (1990) Molecular topography of squid OPA anhydrase as revealed by spectroscopic studies. Proceedings of 1989 Scientific Conference on Chemical Defense Research, Aberdeen Proving Ground, Maryland, Report #SP-024, pp 645–651

Wendt TM & Kaplan AM (1976a) A chemical-biological treatment process for cellulose nitrate disposal. J. Water Pollut. Control Fed. 48: 660–668

Wendt TM & Kaplan AM (1976b) Process for treating waste water containing cellulose nitrate particles, US Patent 3,939,068

Wendt TM, Cornell JH & Kaplan AM (1978) Microbial degradation of glycerol nitrate. Appl. Environ. Microbiol. 36: 693–699

Zech R & Wigand KD (1975) Organophosphate-detoxicating enzymes in *E. coli*; gel filtration and isoelectric focusing of DFPase, paraoxonase and unspecific phosphohydrolases. Experimentia 15: 157–158

Biodegradation 3: 387–398, 1992.

Microbial degradation of pollutants at high salt concentrations

Aharon Oren,[1] Peter Gurevich,[1] Malkit Azachi[2] & Yigal Henis[2]
[1] *Division of Microbial and Molecular Ecology, The Alexander Silberman Institute of Life Sciences and The Moshe Shilo Center for Marine Biogeochemistry, The Hebrew University of Jerusalem, Jerusalem 91904, Israel;* [2] *Department of Plant Pathology and Microbiology, and the Otto Warburg Center for Biotechnology in Agriculture, The Faculty of Agriculture, The Hebrew University of Jerusalem, Rehovot 76100, Israel*

Key words: halophilic, archaeobacteria, eubacteria, pollutants, hydrocarbons, formaldehyde, nitroaromatic compounds

Abstract

Though our knowledge on microbial degradation of organic pollutants at high salt concentrations is still limited, the list of toxic compounds shown to be degraded or transformed in media of high salinity is growing. Compounds transformed aerobically include saturated and aromatic hydrocarbons (by certain archaeobacteria), certain aromatic compounds, organophosphorus compounds, and formaldehyde (by halotolerant eubacteria). Anaerobic microbial transformations of toxic compounds occurring at high salt concentrations include reduction of nitroaromatic compounds, and possibly transformation of chlorinated aromatic compounds.

1. Introduction

The association of hypersaline lakes with unpleasant and harmful chemical compounds has been known for thousands of years, in any case as long as written documentation on these lakes exists:

> 'Then the Lord rained on Sodom and Gomor'rah brimstone and fire from the Lord out of heaven.'
> (Genesis 19:24)

Solutions to pollution problems in hypersaline water bodies can be achieved in different ways. As we will discuss below, not all biological degradation processes known to function in freshwater environments have been shown to be operative also in the presence of high salt concentrations. Therefore, from a microbiological point of view the easiest way to stimulate the degradation of pollutants is dilution to a sufficiently low salinity. In the case of the Dead Sea it was already suggested more than two thousand years ago that when the biology of hypersaline lakes does not perform as desired, dilution with fresh water can work wonders:

> 'And he said to me, 'This water flows toward the eastern region and goes down into the Arabah; and when it enters the stagnant waters of the sea, the water will become fresh. And wherever the river goes every living creature which swarms will live, and there will be very many fish, for this water goes there, that the waters of the sea may become fresh; so everything will live where the river goes.'
> (Ezekiel 47:8–9)

However, dilution with sufficient quantities of fresh water, if these are available at all, cannot always be expected to be feasible from the point of view of the engineer or the economist.

Another approach, and one that poses much greater challenges to the microbiologist, is to make microorganisms do the job in the presence of the high salt concentrations found in the polluted environment, without prior dilution to lower the salinity.

To our knowledge no systematic survey has ever been made on the potential of bacteria to transform and/or degrade polluting chemicals at high salt concentrations. One possible reason is that preliminary attempts to demonstrate the degradation of otherwise relatively easily biodegradable pollutants in hypersaline lakes often yielded disappointing results. For example, no significant breakdown of long-chain straight hydrocarbons such as hexadecane could be demonstrated in Great Salt Lake brines at salinities exceeding 20% (Ward & Brock 1978). The ability of halophilic anaerobic microorganisms to degrade different organic substrates has been reviewed (Oren 1988; Oren 1990a), the main conclusion being that only a few easily degradable substrates such as simple sugars and amino acids can be fermented.

The ability of halophilic or halotolerant bacteria to degrade or transform different organic pollutants in the presence of high salt concentrations – when salt stress is superimposed on pollution stress, is of both basic and applied importance. Applied aspects include answers to the questions: How well can naturally occurring bacteria degrade pollutants in hypersaline lakes (the Great Salt Lake, Utah, or the Dead Sea, to give two major examples)? Can specialized halophilic or halotolerant bacteria be introduced to deal with the degradation of specific pollution events in these lakes? Can halophilic or halotolerant bacteria be used in the biological treatment of highly saline industrial waste effluents that contain toxic materials?

The first section of this chapter reviews the presently available information on microbial degradation of pollutants at high salt concentrations, under aerobic as well as under anaerobic conditions, and the nature of the bacteria involved in these processes. The discussion is limited to naturally occurring bacteria. Table 1 summarizes the range of toxic compounds that have been shown to be degraded by halophilic or halotolerant microorganisms. Another possible approach, not discussed here, would be the introduction of genes derived from nonhalophilic microorganisms coding for the ability to transform selected pollutants into halophilic or halotolerant bacteria, using techniques of genetic engineering. To our knowledge this approach has not yet been used.

In the second part of the chapter two processes, currently studied in our laboratories will be discussed: aerobic transformation of formaldehyde at high salt concentrations by halotolerant eubacteria, and anaerobic transformation of nitroaromatic compounds by obligately anaerobic halophilic eubacteria.

Table 1. Toxic compounds shown to be degraded or transformed at high salt concentrations.

Compound degraded or transformed	Halophilic organisms involved	Salt concentrations	Reference
Saturated hydrocarbons Pristane Aromatic hydrocarbons	*Halobacterium* sp.	15–31%	Bertrand et al. 1990
Hexadecane	Unidentified bacteria in the Great Salt Lake	< 20%	Ward & Brock 1978
Benzoate and other aromatic compounds (?)	*Pseudomonas halodurans*	1.8–15.5%	Rosenberg 1983
Organophosphorus compounds	Unidentified moderately halophilic bacterium	2–24% (?)	DeFrank & Cheng 1991
Pretreated lignitic material	Unidentified anaerobes and methanogens	8–12%	Oren 1990e
Chlorophenols, chlorophenoxyphenols	Unidentified	9–12% or higher (?)	Boone et al. 1989
Formaldehyde	Unidentified moderately halophilic bacterium	1–20%	This work
Nitroaromatic compounds	*Haloanaerobium, Sporohalobacter*	13–16%, possibly higher	Oren et al. 1991, This work

2. Halophilic bacteria – an overview

The ability to grow at high salt concentrations – even up to saturating concentrations of NaCl, is widespread in the prokaryotic world. Halophilic or halotolerant bacteria are found both in the kingdom of the archaeobacteria (Archaea), and in the eubacteria. Halophilic bacteria can be found in many of the major subgroups within the eubacteria: the Proteobacteria, the Gram-positive branch, the cyanobacteria, and others.

Two major physiological groups of aerobic heterotrophic bacteria are generally found in hypersaline environments:

1. Halophilic archaeobacteria – generally red-pigmented cells, most of which require salt concentrations of at least 15–20% for growth, and often grow at salt concentrations of up to saturation. This group is represented by the genera *Halobacterium, Haloferax, Haloarcula,* and *Halococcus*. Exposure to lower salt concentrations causes lysis of the cells (with the exception of *Halococcus* cells, which possess a rigid cell wall). The halophilic archaeobacteria are heterotrophs, using amino acids, and to a lesser extent carbohydrates, as carbon and energy source. Certain strains can obtain additional energy from light, using the bacteriorhodopsin proton pump. The halophilic archaeobacteria maintain an osmotic balance of their cytoplasm with the hypersaline environment by accumulating high concentrations of salts (mainly K^+ and Cl^-, to a lesser extent Na^+). This mechanism of osmoregulation requires special adaptations of the intracellular enzymatic machinery, which has to be operative in the presence of high salt concentrations, which in some species may exceed 5 M.

2. Halophilic or halotolerant eubacteria – belonging to a great variety of taxonomic and physiological groups within the eubacterial kingdom. Metabolic diversity in the salt-tolerant eubacteria is much greater than in the archaeobacteria. In contrast to the halophilic archaeobacteria, which maintain high intracellular salt concentrations, their intracellular salt concentration is low. Halophilic aerobic eubacteria maintain an osmotic balance of their cytoplasm with the external medium by accumulating high concentrations of variety of organic osmotic solutes (Trüper et al. 1991).

Though many known halotolerant eubacteria are able to grow in the laboratory at salt concentrations comparable to those preferred or tolerated by the halophilic archaeobacteria, and can be isolated from sites dominated by halophilic archaeobacteria, experiments have shown that hypersaline water bodies with salt concentrations exceeding 25% are dominated by archaeobacteria. On the basis of experiments with specific inhibitors (bile salts, causing lysis of non-coccoid halophilic archaeobacteria, and antibiotics selectively inhibiting protein synthesis in either group) it was concluded that in saltern ponds with salinities below 250 g dissolved salts per liter all heterotrophic activity can be attributed to eubacteria, while at higher salinities the archaeobacteria take over (Oren 1990b; Oren 1990c; Oren 1991a). Similarly, in the Dead Sea (total dissolved salts about 340 g per liter) specific inhibitors of archaeobacterial activity abolished all heterotrophic activity (Oren 1990c; Oren 1991a).

Relatively little is known about anaerobic degradation processes at high salt concentrations (Oren 1988; Oren 1990a). Only during the past ten years have investigators isolated fermentative bacteria which are both obligately anaerobic and halophilic from environments such as the bottom sediments of the Dead Sea and the Great Salt Lake. A new family (the Haloanaerobiaceae) was created to accomodate the three new genera described (*Haloanaerobium, Halobacteroides,* and *Sporohalobacter*) (Oren 1991b). *Haloanaerobium praevalens,* an organism present in high numbers in sediments of the Great Salt Lake, is able to grow at salt concentrations from 3 to as high as 25%, with an optimum at 13%; *Halobacteroides* and *Sporohalobacter* species described to date grow in a more restricted range of salt concentrations of 5 to 22% (Oren 1991b). All of these species ferment simple substrates (sugars, certain amino acids) to products including acetate, butyrate, hydrogen and carbon dioxide. To what extent representatives of the Haloanaerobiaceae are widespread and contribute to anaerobic degradation processes at high salt concentrations is unknown.

The anaerobic halophilic eubacteria of the family Haloanaerobiaceae present an exception to the general finding that halophilic or halotolerant eubacteria have low intracellular salt concentrations, and balance their intracellular environment by accumulating high concentrations of organic osmotic solutes. In the Haloanaerobiaceae no organic osmotic solutes could be detected, and intracellular salt concentrations were reported to be high (Oren 1986; Rengpipat et al. 1988). This means that intracellular enzymes should be adapted to the presence of high salt concentrations, and this may, as discussed above, restrict the metabolic potential of such organisms.

Further metabolism of the primary fermentation products (acetate, hydrogen, and others) at high salt concentrations is also limited. The upper limit of salinities at which dissimilatory sulfate reduction has been established to occur is about 20%. Methanogenesis does occur at higher salinities, but is based on the transformation of a few specialized substrates only (notably methylated amines) (Oren 1990d). No halophilic or halotolerant methanogenic bacteria are known that derive their energy from the acetate or from hydrogen + carbon dioxide, the main substrates for methanogenesis in freshwater environments.

3. Microbial degradation or organic pollutants at high salt concentrations – a literature survey

3.1. Degradation under aerobic conditions

3.1.1. Processes performed by archaeobacteria

Little is known to what extent the assumption of the apparently limited metabolic potential of the halophilic archaeobacteria is due to the restricted range of compounds that have been tested as potential substrates for members of the *Halobacterium – Haloferax – Haloarcula* group. Only recently some indications were obtained that the halophilic archaeobacteria may have a greater potential in the degradation of pollutants than previously assumed. Using enrichment cultures with the saturated C-20 hydrocarbon eicosane as single carbon and energy source, a number of red extremely halophilic archaeobacterial strains able to degrade eicosane were isolated from a salt marsh in the south of France with a salinity of 310 g per liter (Bertrand et al. 1990). One of these isolates (designated strain EH4) was found to be able to degrade a range of different hydrocarbons, both even and odd carbon number saturated hydrocarbons (tetradecane, hexadecane, eicosane, heneicosane), saturated isoprenoid alkanes (pristane) and aromatic hydrocarbons (acenaphtene, phenantrene, anthracene and 9-methyl anthracene). After 30 days of incubation at 32 °C with 0.5 g per liter of hydrocarbon, and at a salinity of 310 g per liter, between 48 and 88% of the added amount of straight-chain hydrocarbons was degraded. Aromatic hydrocarbons were biodegraded to a lesser extent, but nevertheless significantly (between 19 and 24%). Palmitic acid was also degraded by the strain. NaCl concentrations of at least 2.5–3 M were required for growth and eicosane degradation. No information is available as yet on the geographic distribution of these or similar strains in hypersaline environments. The only field study known to us on the degradation of hydrocarbons at high salt concentrations was performed in the Great Salt Lake, Utah (Ward & Brock 1978). Enrichment cultures with mineral oil as substrate yielded growth at salinities of up to 17.2%, while no growth was obtained above 20% salt. No measurable amounts of $^{14}CO_2$ were evolved during 5 days incubation of brine samples of salinities of 20.4% and higher supplemented with [1-^{14}C]hexadecane – though upon longer incubation some signs of degradation were observed. The published information does not identify the bacteria involved. However, the relatively low salinities at which the process did occur suggest that eubacteria rather than archaeobacteria were primarily involved. Marine eubacteria degrading petroleum components show an optimum salinity of 0.4 M (close to that of seawater), but significant degradation was observed at 2 M NaCl; the inhibitory effect of salinity was larger for the biodegradation of aromatic and polar fractions than of the saturated fraction (Mille et al. 1991).

In view of the finding of the ability of hydrocarbon degradation in certain halophilic archaeobacteria (Bertrand et al. 1990) a reexamination of the metabolic potential of existing and newly isolated strains of the *Halobacterium – Haloferax – Haloarcula* group may be worthwile, as the group may yield additional surprises.

3.1.2. Processes performed by eubacteria

The halophilic or halotolerant aerobic eubacteria are more promising than the halophilic archaeobacteria as a source of organisms degrading organic pollutants. First, eubacteria in general have a much greater metabolic versatility than the archaeobacteria. Eubacterial genera such as *Pseudomonas* are known for their wide range of substances they degrade, including toxic organic compounds. Second, halophilic eubacteria may be more useful than halophilic archaeobacteria in the degradation of unusual and toxic compounds because of their mode of haloadaptation. In contrast to the halophilic archaeobacteria, which maintain high intracellular salt concentrations, their intracellular salt concentration is low. Halophilic aerobic eubacteria maintain an osmotic balace of their cytoplasm with the external medium by accumulating high concentrations of a variety of organic osmotic solutes (Trüper et al. 1991). Thus, the intracellular enzymes responsible for the degradation processes may be 'conventional' enzymes of the types encountered in non-halophilic bacteria, and there is no need for special salt-tolerant (or even salt-requiring) enzymes such as found in the archaeobacterial halophiles.

In view of the above it is surprising that no systematic study of the potential of halophilic or halotolerant eubacteria to degrade unusual compounds has ever been done. Studies in which halophilic eubacteria were characterized, especially by means of numerical taxonomy, which involves the examination of a large number of properties, hardly ever mention the ability or lack of ability to degrade aromatic compounds or other unusual and/or toxic chemicals. The only substrate that was regularly tested in the course of studies published on the characterization of halophilic or halotolerant eubacteria by numerical taxonomy is benzoate. Only a small percentage of the isolates examined were found to grow on benzoate (Del Moral et al. 1988; Garcia et al. 1987; Ventosa et al. 1982). In addition, out of the 58 strains examined, one single strain of *Vibrio costicola*, growing at salt concentrations between 0.5 and 20%, was reported to degrade quinate (1,3,4,5-tetrahydrocyclohexanecarboxylic acid) (Garcia et al. 1987). Degradation of benzoate was investigated in somewhat more detail in *Pseudomonas halodurans*, an isolate growing at salt concentrations between 1.8 and 15.5% (or maybe higher – it was stated that 22% NaCl, and 'NaCl concentrations greater than 10-fold that of seawater' are tolerated). The stain was found to cleave aromatic rings by *ortho* cleavage (Rosenberg 1983).

The range of substrates that can be transformed and/or degraded at high salt concentrations by aerobic halophilic or halotolerant eubacteria may be much greater than commonly assumed, and may include certain pollutants, even those considered to be among the more recalcitrant compounds. This was illustrated by the recent isolation of a halotolerant eubacterium degrading organophosphorus compounds (DeFrank & Cheng 1991). The isolate – designated JD6.5 – is a moderately halophilic eubacterium, and was isolated from Grantsville Warm Springs, just south of the Great Salt Lake, Utah, a spring with a salt content of 24%, a temperature of 24° C, and a pH of 6.0. No data were published on the range of salt concentrations tolerated by strain JD6.5, but it was described as being obligatory halophilic, requiring at least 2% NaCl for growth. It was routinely grown in a medium containing 5% NaCl and 1% $MgSO_4 \cdot 7H_2O$ as major salts. The strain is a Gram-negative short rod, and was tentatively identified as an *Alteromonas* sp. The organism was found to possess high levels of enzymatic activity against several highly toxic organophosphorus pesticides and chemical warfare agents, acting as cholinesterase inhibitors. Compounds degraded by strain JD6.5 include DFP (diisopropylfluorophosphate), NPMPP (*p*-nitrophenylmethyl-(phenyl)phosphinate), MPEPP (*p*-nitrophenylethyl(phenyl)phosphinate), and Paraoxon (diethyl *p*-nitrophenyl phosphate). The predominant enzyme responsible for the degrading activity (designated organophosphorus acid anhydrase 2) was purified to homogeneity and characterized. In assays using diisopropylfluorophosphate as substrate optimal activity was obtained at a pH of 8.5 at 50 °C. Assays were performed in the presence of 0.5 M NaCl, but no information was given on the salt requirement or tolerance of the enzyme. The natural substrate and function of organophosphorus acid anhydrases is not known, and at present it is impossible to speculate what role it plays in the normal metabolism of strain JD6.5.

3.2. Degradation under anaerobic conditions

To a certain extent it can be expected that those compounds that can be degraded aerobically by halophilic aerobic archaeobacteria can also be oxidized anaerobically using alternative electron acceptors. Many halophilic archaeobacteria are denitrifiers (Mancinelli & Hochstein 1986), and some can also use other terminal electron acceptors, such as dimethylsulfoxide or trimethylamine N-oxide (Oren & Trüper 1990). Possible exceptions are those cases in which molecular oxygen is required in the biochemical reactions involved in the transformation process, as can for example be expected to be the case in the degradation of hydrocarbons by certain representatives of the extremely halophilic archaeobacteria (Bertrand et al. 1990). As discussed above, the range of compounds known to be degraded by this group is very limited anyhow, and therefore its potential use in degradation of pollutants at high salt concentrations under anaerobic conditions is doubtful.

Few attempts have been made to exploit the properties of anaerobic halophilic bacteria in the biotransformation of unusual chemicals. One preliminary study dealt with the possibility of anaerobic biotransformation of chemically pretreated lignitic material with the production of methane (Oren 1990e). This study was initiated in view of the development of biological gasification processes for the utilization of Texas lignite after chemical pretreatment (alkaline hydrolysis, leading to breakdown to simple water-soluble aromatic compounds). One of the methods proposed for the biogasification of the pretreated lignite makes use of underground salt caverns in Texas as cheaply available bioreactors. In the course of time the salt concentration in these reactors is expected to increase as a result of dissolution of salt from the underground salt domes. Therefore a study was made on the feasibility of methanogenic biodegradation of the complex mixture of chemicals obtained at high salt concentrations (Oren 1990e; Wise 1987). It was shown that chemically pretreated lignite, inoculated with anaerobic hypersaline sediment samples as a source of bacteria, can give rise to a limited amount of methane and hydrogen evolution in an anaerobic fermentation at high salt concentrations. The optimal salt concentration for the process was found to be between 8 and 12%, and above 16% salt the reaction rates were negligible. The optimal temperature was between 35 and 42 °C. Methane yields were low – between 0.006 and 0.01 ml of methane per ml of lignite preparation containing 2.4 g total solids per liter. No methane could be shown to be formed from hydrogen + carbon dioxide or acetate at these salt concentrations. It was postulated that the methane evolved was derived from $-O\text{-}CH_3$ groups that are abundant in the alkaline hydrolysis products formed during the lignite pretreatment. The nature of the microorganisms responsible for the formation of the methanogenic precursor (probably methanol) was not clarified.

The possible occurrence of reductive dechlorination of chlorophenols and chlorophenoxyphenols under anaerobic hypersaline conditions was reported in hypersaline groundwaters near Alkali Lake, a hypersaline alkaline lake in the desert of south central Oregon, U.S.A. (Boone et al. 1989). No clear data were given with regards to the actual salinity of the groundwater; it was reported that the lake has a salinity of 90 to 120 gram per liter total dissolved salts, but this value seems much too low in view of the statement that 'The lakes and groundwater contain large amounts of dissolved cations sodium (5.1 M) and potassium (0.23 M), with carbonate and bicarbonate (1.6 M, total concentration), chloride (1.3 M), and sulphate (0.48 M) being the major anions'. Between 1969 and 1971 more than 25,000 55-gallon drums containing chlorophenols, chlorophenoxyphenols, and other herbicide manufacturing wastes were deposited in the valley near the lake. In 1976, the drums were placed in shallow trenches, crushed, and covered with gravel. The contents contaminated the groundwater. In a study performed in 1982 chlorophenols (2,4-dichlorophenol and 2,4,6-trichlorophenol) and chlorophenoxyphenols were readily detectable in the upper, oxic groundwater at a level concentration about 20 mg/kg of soil, but in the anoxic zone the concentration of the compounds dropped more rapidly than their diffusion and dispersion would predict. All data obtained led to the conclusion that the chlorinated compounds are probably biologically degraded in the anoxic subsurface. The partial pressure of hydrogen in samples from this anoxic zone was found to be high, and this was suggested to be related to the rapid reductive transformation of chlorinated phenolic compounds (Boone et al. 1989). No information was obtained on the nature of the microorganism(s) involved in the process.

4. Aerobic transformation of formaldehyde at high salt concentrations by a halotolerant eubacterium

Formaldehyde is a carbonyl compound with a high polarity and a very high reactivity. As a nucleophilic agent it reacts with amines (it readily combines with amino groups in proteins), amides, sulfides, purines, and a variety of other compounds, with the formation of C- or N-methyloyl groups and methylene bridges. Formaldehyde is widely used in medicine, agriculture, and industry. Among its applications we may mention its use as a high effective disinfectant for the killing of bacteria, fungi and viruses, as a fixing fluid and preservative in histological technique, and its use in the transformation of toxins to toxoids. Formaldehyde is known to have mutagenic activity to both eukaryotic and prokaryotic organisms.

In spite of its toxicity at higher concentrations, formaldehyde is a common intermediate in metabolic pathways of bacteria. Table 2 gives a – not necessarily exhaustive – survey of some biochemical reactions by which formaldehyde may enter the carbon and/or energy metabolism of bacteria. Certain reactions are limited to specialized groups of bacteria. For example, formaldehyde can be used as sole carbon and energy source in methylotrophic bacteria – a small group of organisms that can grow on methane, methanol and other one-carbon compounds. In these organisms formaldehyde is an intermediate in the oxidation of methane or methanol to carbon dioxide (Attwood & Quayle 1984). To our knowledge no halophilic or halotolerant representatives of this group have been described.

Many bacteria which do not use formaldehyde as sole carbon and energy source do incorporate and/or transform formaldehyde. Possible mechanisms may either be based on reaction with the coenzyme tetrahydrofolate to yield methylene-tetrahydrofolate, or may involve oxidation to formate by the enzyme formaldehyde dehydrogenase (Attwood & Quayle 1984). The latter reaction may serve as a detoxification mechanism, enabling bacteria to tol-

Table 2. Some pathways for the microbial utilization of formaldehyde.

Pathway	Organisms performing the reaction
A. Assimilatory pathways	
Ribulose monophosphate pathway	Methylotrophic bacteria
Reaction with xylulose 5-phosphate	Methylotrophic bacteria
Serine pathway	Methylotrophic bacteria
Formation of methylene-tetrahydrofolate	Probably widespread
B. Dissimilatory pathways	
Oxidation to CO_2	Methylotrophic bacteria
Oxidation to formate by formaldehyde dehydrogenase	Possibly widespread
Reduction or disproportionation, via methylene-tetrahydromethanopterin	Methanogenic bacteria

Information given in this table was derived from Attwood & Quayle (1984), Escalante-Semerena et al. (1984), and Kaulfers & Marquardt (1991).

erate high formaldehyde concentrations. This phenomenon has been investigated in some depth in clinical isolates of formaldehyde-resistant members of the Enterobacteriaceae (Kaulfers & Marquardt 1991). Resistant strains of *Escherichia coli* and *Serratia marcescens* can be grown in media containing 200 ppm formaldehyde and higher, one strain even up to 700 ppm (Kaulfers & Brandt 1987; Kaulfers & Laufs 1985). In these strains formaldehyde resistance appeared to be plasmid-bound, as shown by conjugation, transformation, and plasmid-curing experiments, and the property can be transferred from *Serratia* to *E. coli.* A 62 MDa conjugative plasmid determining formaldehyde resistance was isolated from formaldehyde-resistant *E. coli* strains. Resistant strains could be cured of the plasmid by treatment with acridine orange, and as a result they became formaldehyde-sensitive (Kaulfers & Brandt 1987). Clinical formaldehyde-resistant isolates of different Enterobacteriaceae were shown to possess a NAD- and glutathione-dependent formaldehyde dehydrogenase activity, which was absent in formaldehyde-sensitive strains.

In the course of a search for formaldehyde-transforming halophilic or halotolerant bacteria soil was collected at a storage site for formaldehyde near a chemical plant that uses formaldehyde in the production of glue (S. Sarig, Bar-Ilan University, Ramat-Gan, pers. comm.). Soil samples were used to inoculate enrichment cultures containing a mineral medium with 10% NaCl, other salts, and supplemented with 100 ppm glycine betaine and 20 or 50 ppm formaldehyde. After incubation for 4 weeks at 30° C in a shaker bacterial growth was obtained. Upon repeated transfers in the same medium, in all cases accompanied by disappearance of the added 20 or 50 ppm formaldehyde within 3 days, a consortium of bacteria was obtained that could be grown in media containing 100 ppm glycine betaine and 100 ppm formaldehyde. This consortium was found to consist of at least four different bacteria, which were obtained in pure cultures. The isolate (designated MA-C) that proved to be the most formaldehyde-resistant (growing well in media supplemented with 100 ppm formaldehyde), and the most active in the transformation of formaldehyde, was selected for further study.

Isolate MA-C is an as yet unidentified rod-shaped Gram-negative, motile eubacterium. It is highly halotolerant, and grows well at NaCl concentrations from 0 to 20% and possibly higher. Mineral media with single carbon sources (several sugars, acetate, succinate, citrate, glutamate) support excellent growth. Under anaerobic conditions acid is produced from a variety of sugars. A complete account on the strain and its formaldehyde metabolism will be published elsewhere; the relevant results are summarized below.

We performed assays for formaldehyde dehydrogenase in cell-free extracts of strain MA-C, using the procedure described by Kaulfers & Marquardt (1991). A high NAD- and glutathione-dependent activity was measured. NAD could not be replaced by NADP, and no activity could be measured in the absence of glutathione. The enzyme proved constitutive, as was the case in the Enterobacteriaceae (Kaulfers & Marquardt 1991), but activities measured were somewhat higher in cells previously exposed to 20 or 50 ppm formaldehyde, compared with cells that had not been in contact with formaldehyde for many subsequent transfers. Activities measured were between 650 and 850 nmol NAD reduced/min·mg protein at room temperature (in cells grown in the presence of 20 ppm formaldehyde, either in medium containing 0.5% yeast extract, 0.5% tryptone, or in medium containing 0.5% Na-succinate and 0.05% yeast extract, in addition to 10% NaCl, other salts). In cell extracts prepared from cells grown in similar media, but in the absence of formaldehyde activities were between 280 and 480 nmol/min·mg protein. In formaldehyde-resistant Enterobacteriaceae, rates between 14 and 25 units/mg protein were reported (Kaulfers & Marquardt 1991); as no definition of the units used was given, comparison with our results is not possible. The formaldehyde dehydrogenase activity of cell-free extracts of strain MA-C is sensitive to salt: when 2.5, 5.0 or 7.5% NaCl were present in the assay mixture, the rates obtained were respectively 63, 41, and 31% of the control without salt. Being a eubacterium, strain MA-C probably has a low intracellular salt concentration, and it can be expected that the intracellular formaldehyde dehydrogenase is not exposed to high salt concentrations such as found in the outside medium. It remains to be determined whether in strain MA-C the formaldehyde dehydrogenase activity is plasmid-coded, or even possibly due to a plasmid identical or similar to those responsible for formaldehyde resistance in *E. coli* or *Serratia* (Kaulfers & Brandt 1987; Kaulfers & Laufs 1985).

To examine the fate of the formaldehyde metabolized by strain MA-C, we used radioactively labeled formaldehyde as a tracer, and tested whether label from formaldehyde is incorporated into the cells, oxidized to CO_2, or to non-volatile dissolved products. ^{14}C-labeled formaldehyde was added to cultures in the presence of 25 ppm unlabeled formaldehyde, and the amount of label incorporated into the cells, and that remaining in the culture supernatant before and after acidification, was determined. No significant incorporation of radioactivity into the cells was observed, and the radioactivity of the culture supernatant slowly declined (within days), probably as formaldehyde is transformed to volatile products.

To assay for the accumulation of formate in the supernatant of formaldehyde-metabolizing cultures of strain MA-C we used an enzymatic assay for formate, based on the reduction of NAD by formate dehydrogenase from *Pseudomonas oxalaticus*. No significant amounts of formate could be demonstrated in the supernatant fluid of cultures fed with formaldehyde. Formate, added to cultures at concentrations of 3.7 or 7.4 mM (250 or 500 mg/l) sodium formate) disappeared from the medium within one day, while no significant change in formate concentration was observed in sterile control experiments, showing that strain MA-C is able to further metabolize formate. Experiments with ^{14}C-labeled formate showed that only a minor part of the added formate (less than 5%) was incorporated by the cells. A NAD-dependent formate dehydrogenase activity was detected in cell extracts of cells grown in the presence of 7.4 mM formate (activity between 14 and 34 nmol NAD reduced/mg protein·min at 45 °C). In extracts of cells not exposed to formate, the activity was only 10–25% of that of formate-grown cells. NAD could not be replaced by NADP. The optimum temperature of the formate dehydrogenase was 45 °C, and the activity of the enzyme was greatly inhibited by NaCl.

5. Anaerobic transformation of nitroaromatic compounds by obligately anaerobic halophilic bacteria

Nitrosubstituted aromatic compounds such as nitrobenzene and nitrophenols are widely used in the manufacture of azo dyes, explosives, pharmaceuticals, and especially pesticides: insecticides (e.g. parathion), herbicides, and fungicides. During chemical or biological breakdown of these pesticides in soil and water nitroaromatic compounds are released into the environment.

Though nitrophenols and other nitrosubstituted aromatic compounds are generally considered to be highly resistant to microbial degradation, a number of aerobic bacteria have been shown to degrade nitrophenols. In this process the nitro group is released as nitrite, followed by oxidation of the aromatic ring. Also under anaerobic conditions degradation of nitroaromatic compounds occurs, and in this case the initial step in the process is probably a reduction of the nitro group to an amino group. A number of bacteria have been shown to reduce nitrosubstituted aromatic compounds to their corresponding amino derivatives. These include obligately anaerobic *Clostridium* and *Eubacterium* species isolated from the human intestinal tract (Rafii et al. 1991), and unidentified bacteria present in the rumen of sheep (Gurevich et al. 1993). A few reports have been published on anaerobic bacterial consortia degrading nitrophenols to methane (Berry et al. 1987; Boyd et al. 1983). During methanogenic fermentation of nitrophenols in sewage sludge, the nitro group was again presumed to undergo reduction to form a substituent amino group. The resulting intermediate aminophenol subsequently underwent further mineralization (Boyd et al. 1983).

In the course of our studies on the metabolic abilities of obligately anaerobic halophilic eubacteria belonging to the family Haloanaerobiaceae (Oren 1991b) it was found that two species tested, *Haloanaerobium praevalens* (isolated from the Great Salt Lake), and *Sporohalobacter marismortui* (isolated from sediments on the shore of the Dead Sea) rapidly reduced *p*-nitrophenol to *p*-aminophenol (Oren et al. 1991). The experiments were performed in the presence of 13–14% NaCl (13.5–15.8 g/liter total dissolved salts). Other nitrosubstituted aromatic compounds transformed by both strains included nitrobenzene, *o*-nitrophenol, *m*-nitrophenol, *o*-nitroaniline, *m*-nitroaniline, *p*-nitroaniline, 2,4-dinitrophenol, and 2,4-dinitroaniline. Most of these compounds, when added to growing cultures at concentrations of 50 to 100 ppm, were completely transformed within 24 h, but at the highest concentrations growth rates were somewhat lowered.

The transformation of *p*-nitrophenol by *H. praevalens* was investigated in more detail. Concentrations of up to 125 ppm were tolerated, but at concentrations above 25 ppm growth was slowed down significantly. However, at all concentrations tested, the *p*-nitrophenol added had disappeared completely within 24 h, and was quantitatively converted to *p*-aminophenol. No degradation of aromatic amines could be demonstrated by the organisms tested, and it was concluded that the only transformation that took place was a reduction of the nitro groups to amino groups. Nitrite was not detected in *p*-nitrophenol-transforming cultures. Incubation of growing cultures of *H. praevalens* in the presence of ^{14}C-labeled *p*-nitrophenol showed that the compound was neither incorporated by the cells, nor degraded to acid-volatile compounds.

The obligately anaerobic bacteria of the family Haloanaerobiaceae were not previously known for their metabolic versatility, and the range of substrates used was reported to be restricted to simple sugars and amino acids (Oren 1988; Oren 1990a; Oren 1991b). It is not clear what enzymatic activity is involved in the reduction of nitroaromatic compounds in these organisms. In fact, the nature of similar activities in non-halophilic anaerobic microorganisms is still to be determined. It is highly improbable that the halophilic strains examined have ever been exposed to significant concentrations of nitrophenols and other nitroaromatic compounds in their natural habitat (the Great Salt Lake, the Dead Sea), and therefore the function of the activity remains enigmatic. It may be imagined that the reaction serves for the detoxification of toxic nitro compounds. Another possibility is that the nitro groups are used as an electron sink during fermentation processes, allowing the formation of more oxidized end products and a higher energy yield. No microorganisms are yet known able to further transform aminophenols under either aerobic or anaerobic conditions at high salt concentrations. In this context it may be stated that our knowledge on microbial transformations of aminophenols in freshwater environments is likewise extremely limited.

The discovery of anaerobic reduction of nitroaromatic compounds by members of the Haloanaerobiaceae indicates that the variety of organic compounds that can be transformed and/or degraded in hypersaline environments may be much greater than previously assumed. The results also suggest that anaerobic halophilic bacteria show a certain promise in biotechnological processes, and practical use may be found for their newly discovered properties.

6. Conclusions

The above survey shows that our knowledge on the degradation of pollutants at high salt concentrations is still extremely limited. This may simply be due to the fact that the potential of halophilic bacteria for biodegradation of toxic compounds has never been systematically investigated. Our lack of knowledge by no means implies that halophilic bacteria do not have a potential to degrade toxic compounds. The opposite may be true – a world of unusual microorganisms is waiting to be tested for useful properties. The finding of features such as hydrocarbon degradation in certain halophilic archaeobacteria (Bertrand et al. 1990), degradation of organophosphorus compounds (DeFrank & Cheng 1991), transformation of formaldehyde in halotolerant eubacteria (described here), and reductive transformation of nitroaromatic compounds (Oren et al. 1991) and possibly chlorophenols and chlorophenoxyphenols (Boone et al. 1989) by obligately anaerobic halophilic eubacteria suggests that a reexamination of the metabolic potential of extant and newly isolated strains of halophilic or halotolerant bacteria may yield additional surprises.

Acknowledgements

We thank Shlomo Sarig (Bar-Ilan University, Ramat-Gan, Israel), for the use of unpublished data on the formaldehyde-transforming strain MA-C, and Carol A. Baker (Monsanto Co.) for useful suggestions and thorough reading of the manuscript. Our studies on the transformation of formaldehyde and nitroaromatic compounds are supported by a grant from Monsanto Co., St. Louis, MO.

Note added in proof

The use of hydrocarbons by members of the Halobacteriaceae (Bertrand et al. 1991) was discovered independently by a Russian group (Kulichevskaya IS, Milekhina EI, Borzenkov IA, Zvyagintseva IS & Belyaev SS (1991) Oxidation of petroleum hydrocarbons by extremely halophilic archaebacteria. Mikrobiologiya (English translation) 60: 596–601).

References

Attwood MM & Quayle JR (1984) Formaldehyde as a central intermediary metabolite of methylotrophic metabolism. In: Crawford RL & Hanson RS (Eds) Microbial Growth on C_1 Compounds (pp 315–323). American Society for Microbiology, Washington, DC

Berry DF, Francis AJ & Bollag J-M (1987) Microbial metabolism of homocyclic and heterocyclic aromatic compounds under anaerobic conditions. Microbiol. Rev. 51: 43–59

Bertrand JC, Almallah M, Acquaviva M & Mille G (1990) Biodegradation of hydrocarbons by an extremely halophilic archaebacterium. Lett. Appl. Microbiol. 11: 260–263

Boone DR, Johnson RL, Chen DC, Mathrani IM & Mah RA (1989) Methanogenesis and reductive dechlorinations in an alkaline, hypersaline sediments and groundwater. In: Da Costa MS, Duarte JC & Williams RAD (Eds) Microbiology of Extreme Environments and its Potential for Biotechnology (pp 205–215). Elsevier Applied Science, London

Boyd SA, Shelton DR, Berry D & Tiedje JM (1983) Anaerobic biodegradation of phenolic compounds in digested sludge. Appl. Environ. Microbiol. 46: 50–54

DeFrank J & Cheng T-C (1991) Purification and properties of an organophosphorus acid anhydrase from a halophilic bacterium. J. Bacteriol. 173: 1938–1943

Del Moral A, Prado B, Quesada E, Garcia T, Ferrer R & Ramos-Cormenzana A (1988) Numerical taxonomy of moderately halophilic Gram-negative rods from an inland saltern. J. Gen. Microbiol. 134: 733–741

Escalante-Semerena JC, Leigh JA, Rinehart KL, Jr. & Wolfe RS (1984) Formaldehyde activation factor, tetrahydromethanopterin, a coenzyme of methanogenesis. Proc. Natl. Acad. Sci. USA 81: 1976–1980

Garcia MT, Ventosa A, Ruiz-Berraquero F & Kocur M (1987) Taxonomic study and amended description of *Vibrio costicola*. Int. J. Syst. Bacteriol. 37: 352–360.

Gurevich P, Oren A, Sarig S & Henis Y (1993) Reduction of aromatic nitro compounds in anaerobic ecosystems. Water Sci. Technol. (in press)

Kaulfers P-M & Brandt D (1987) Isolation of a conjugative plasmid in *Escherichia coli* determining formaldehyde resistance. FEMS Microbiol. Lett. 43: 161–163

Kaulfers P-M & Laufs R (1985) Übertragbare Formaldehydresistenz bei *Serratia marcescens*. Zbl. Bakt. Hyg., I. Abt. Orig. B 181: 309–319

Kaulfers P-M & Marquardt A (1991) Demonstration of formaldehyde dehydrogenase activity in formaldehyde-resistant Enterobacteriaceae. FEMS Microbiol. Lett. 79: 335–338

Mancinelli S & Hochstein LI (1986) The occurrence of denitrification in extremely halophilic bacteria. FEMS Microbiol. Lett. 35: 55–58

Mille G, Almallah M, Bianchi M, van Wambeke F & Bertrand JC (1991) Effect of salinity on petroleum degradation. Fresenius J. Anal. Chem. 339: 788–791

Oren A (1986) Intracellular salt concentrations of the anaerobic halophilic eubacteria *Haloanaerobium praevalens* and *Halobacteroides halobius*. Can. J. Microbiol. 32: 4–9

— (1988) Anaerobic degradation of organic compounds at high salt concentrations. A. v. Leeuwenhoek 54: 267–277

— (1990a) Anaerobic degradation of organic compounds in hypersaline environments: possibilities and limitations. In: Wise DL (Ed) Bioprocessing and Biotreatment of Coal (pp 155–175). Marcel Dekker, New York

— (1990b) Estimation of the contribution of halobacteria to the bacterial biomass and activity in solar salterns by the use of bile salts. FEMS Microbiol. Ecol. 73: 41–48

— (1990c) The use of protein synthesis inhibitors in the estimation of the contribution of halophilic archaebacteria to bacterial activity in hypersaline environments. FEMS Microbiol. Ecol. 73: 187–192

— (1990d) Formation and breakdown of glycine betaine and

trimethylamine in hypersaline environments. A. v. Leeuwenhoek 58: 291–298

— (1990e) Microbial formation of methane from pretreated lignite at high salt concentrations. In: Wise DL (ed) Bioprocessing and Biotreatment of Coal (pp 449–463). Marcel Dekker, New York

— (1991a) Estimation of the contribution of archaebacteria and eubacteria to the bacterial biomass and activity in hypersaline ecosystems: novel approaches. In: Rodriguez-Valera F (Ed) General and Applied Aspects of Halophilic Microorganisms (pp 25–31). Plenum Press, New York

— (1991b) The genera *Haloanaerobium, Halobacteroides,* and *Sporohalobacter.* In: Balows A, Trüper HG, Dworkin M, Harder W & Schleifer K-H (Eds) The Prokaryotes. A Handbook on the Biology of Bacteria: Ecophysiology, Isolation, Identification, Applications 2nd ed, Vol II (pp 1893–1900). Springer-Verlag, New York

Oren A & Trüper HG (1990) Anaerobic growth of halophilic archaeobacteria by reduction of dimethylsulfoxide and trimethylamine *N*-oxide. FEMS Microbiol. Lett. 70: 33–36

Oren A, Gurevich P & Henis Y (1991) Reduction of nitrosubstituted aromatic compounds by the halophilic anaerobic eubacteria *Haloanaerobium praevalens* and *Sporohalobacter marismortui.* Appl. Environ. Microbiol. 57: 3367–3370

Rafii F, Franklin W, Heflich RH & Cerniglia CE (1991) Reduction of nitroaromatic compounds by anaerobic bacteria isolated from the human gastrointestinal tract. Appl. Environ. Microbiol. 57: 962–968

Rengpipat S, Lowe SE & Zeikus JG (1988) Effect of extreme salt concentrations on the physiology and biochemistry of *Halobacteroides acetoethylicus.* J. Bacteriol. 170: 3065–3071

Rosenberg A (1983) *Pseudomonas halodurans* sp. nov., a halotolerant bacterium. Arch. Microbiol. 136: 117–123

Tindall BJ (1991) The family Halobacteriaceae. In: Balows A, Trüper HG, Dworkin M, Harder W & Schleifer K-H (Eds) The Prokaryotes. A Handbook on the Biology of Bacteria: Ecophysiology, Isolation, Identification, Applications, 2nd ed, Vol I (pp 768–808). Springer-Verlag, New York

Trüper HG, Severin J, Wohlfahrt A, Müller A & Galinski EA (1991) Halophily, taxonomy, phylogeny and nomenclature. In: Rodriguez-Valera F (Ed) General and Applied Aspects of Halophilic Microorganisms (pp 3–7). Plenum Press, New York

Ventosa A, Quesada E, Rodriguez-Valera F, Ruiz-Berraquero F & Ramos-Cormenzana A (1982) Numerical taxonomy of moderately halophilic gram-negative rods. J. Gen. Microbiol. 128: 1959–1968

Ward DM & Brock TD (1978) Hydrocarbon degradation in hypersaline environments. Appl. Environ. Microbiol. 35: 353–359

Wise DL (1987) Meeting report – first international workshop on biogasification and biorefining of Texas lignite. Res. Conserv. 15: 229–247

List of contributors

A. Abeliovich, Environmental Microbiology Unit, The Jacob Blaustein Institute for Desert Research, Ben-Gurion University, Israel

E. Adler, Marine and Coastal Environment Division, Office of the Environment, Jerusalem, Israel

O. Ardon, The Otto Warburg Center for Biotechnology in Agriculture, Faculty of Agriculture, P.O. Box 12, Rehovot, 76100, Israel

R. M. Atlas, Dept. of Biology, University of Louisville, Louisville, KY 40292, USA

M. Azachi, Dept. of Plant Pathology and Microbiology, and the Otto Warburg Center for Biotechnology in Agriculture, The Faculty of Agriculture, The Hebrew University of Jerusalem, Rehovot 76100, Israel

T. Barkay, Microbial Ecology and Biotechnology Branch, U.S. Environmental Research Laboratory, Gulf Breeze, FL 32561, USA (Address for correspondence: US EPA, 1 Sabine Island Drive, Gulf Breeze, FL 32561, USA)

E. A. Bayer, Dept. of Biophysics, The Weizmann Institute of Science, Rehovot, Israel

S. Belkin, Environmental Microbiology, Ben Gurion University of the Negev, The Jacob Blaustein Institute for Desert Research, Sede Boker Campus 84990, Israel

T. N. P. Bosma, Dept. of Microbiology, Wageningen Agricultural University, Hesselink van Suchtelenweg 4, 6703 CT Wageningen, The Netherlands

A. W. Boyle, State University of New York, College of Environmental Science & Forestry, Syracuse, NY 13210, USA

P. Broda, Dept. of Biochemistry & Applied Molecular Biology, University of Manchester Institute of Science and Technology, Manchester M60 1QD, UK

W. P. de Bruin, Dept. of Microbiology, Wageningen Agricultural University, Hesselink van Suchtelenweg 4, 6703 CT Wageningen, The Netherlands

C. E. Cerniglia, Microbiology Division, National Center for Toxicological Research, Jefferson, AR 72079, USA

A. M. Chakrabarty, Dept. of Microbiology & Immunology (M/C 790), University of Illinois College of Medicine, 835 South Wolcott Avenue, Chicago, IL 60612, USA

D. Daubaras, Dept. of Microbiology & Immunology (M/C 790), University of Illinois College of Medicine, 835 South Wolcott Avenue, Chicago, IL 60612, USA

N. P. Finkelstein, I.M.I., Haifa, Israel

O. Gat, Dept. of Food Engineering and Biotechnology, The Technion, Technion City, Haifa 32000, Israel

J. A. Glaser, Environmental Protection Agency, Risk Reduction Engineering Laboratory, Cincinnati, OH 45268, USA

B. Gorodecki, The Otto Warburg Center for Biotechnology in Agriculture, Faculty of Agriculture, P.O. Box 12, Rehovot, 76100, Israel

P. Gurevich, Division of Microbial and Molecular Ecology, The Alexander Silberman Institute of Life Sciences and The Moshe Shilo Center for Marine Biogeochemistry, The Hebrew University of Jerusalem, Jerusalem 91904, Israel

Y. Hadar, The Otto Warburg Center for Biotechnology in Agriculture, Faculty of Agriculture, P.O. Box 12, Rehovot, 76100, Israel

H. Harms, Dept. of Microbiology, Wageningen Agricultural University, Hesselink van Suchtelenweg 4, 6703 CT Wageningen, The Netherlands

J. P. Hassett, State University of New York, College of Environmental Science & Forestry, Syracuse, NY 13210, USA

Y. Henis, Dept. of Plant Pathology and Microbiology, and the Otto Warburg Center for Biotechnology in Agriculture, The Faculty of Agriculture, The Hebrew University of Jerusalem, Rehovot 76100, Israel

C. Holliger, Dept. of Microbiology, Wageningen Agricultural University, Hesselink van Suchtelenweg 4, 6703 CT Wageningen, The Netherlands (Present address: EAWAG/ETH, CH-6047 Kastanienbaum, Switzerland)

J. Horn, Center for Environmental Diagnostics and Bioremediation, University of West Florida, Pensacola, FL 32514, USA

D. L. Kaplan, US Army Natick Research, Development and Engineering Center, Natick, MA 01760-5020, USA

Z. Kerem, The Otto Warburg Center for Biotechnology in Agriculture, Faculty of Agriculture, P.O. Box 12, Rehovot, 76100, Israel

A. Khasin, Dept. of Food Engineering and Biotechnology, The Technion, Technion City, Haifa 32000, Israel

F. V. Kremer, Environmental Protection Agency, Center for Environmental Research Information, Cincinnati, OH 45268, USA

J. G. Kuenen, Kluyver Laboratory for Biotechnology, Delft University of Technology, Julianalaan 67, 2628 BC Delft, The Netherlands

A. Kushmaro, Dept. of Molecular Microbiology and Biotechnology, The George S. Wise Faculty of Life Sciences, Tel-Aviv University, Tel-Aviv, Israel

R. Lamed, Dept. of Molecular Microbiology and Biotechnology, George S. Wise Faculty of Life Sciences, Tel Aviv University, Ramat Aviv, Israel

R. Legmann, Dept. of Molecular Microbiology and Biotechnology, The George S. Wise Faculty of Life Sciences, Tel-Aviv University, Tel-Aviv, Israel

J. R. van der Meer, Dept. of Microbiology, Wageningen Agricultural University, Hesselink van Suchtelenweg 4, 6703 CT Wageningen, The Netherlands (Present address: EAWAG/ETH, CH-8600 Dübendorf, Switzerland)

D. Minz, Dept. of Molecular Microbiology and Biotechnology, Tel-Aviv University, Tel-Aviv, Israel

J. G. Mueller, SBP Technologies, Inc., EPA Environmental Research Laboratory, Gulf Breeze, FL 32561-5299, USA

J. P. Nakas, State University of New York, College of Environmental Science & Forestry, Syracuse, NY 13210, USA

A. Oren, Division of Microbial and Molecular Ecology, The Alexander Silberman Institute of Life Sciences and The Moshe Shilo Center for Marine Biogeochemistry, The Hebrew University of Jerusalem, Jerusalem 91904, Israel

P. H. Pritchard, Environmental Protection Agency, Environmental Research Laboratory, Gulf Breeze, FL 32561-5299, USA

H. H. M. Rijnaarts, Dept. of Microbiology, Wageningen Agricultural University, Hesselink van Suchtelenweg 4, 6703 CT Wageningen, The Netherlands

L. A. Robertson, Kluyver Laboratory for Biotechnology, Delft University of Technology, Julianalaan 67, 2628 BC Delft, The Netherlands

J. C. Rogers, Environmental Protection Agency, Environmental Research Laboratory, Athens, GA 30613, USA

E. Z. Ron, Dept. of Molecular Microbiology and Biotechnology, Tel-Aviv University, Tel-Aviv, Israel

E. Rosenberg, Dept. of Molecular Microbiology and Biotechnology, Tel-Aviv University, Tel-Aviv, Israel

E. Saouter, Center for Environmental Diagnostics and Bioremediation, University of West Florida, Pensacola, FL 32514, USA

G. Schraa, Dept. of Microbiology, Wageningen Agricultural University, Hesselink van Suchtelenweg 4, 6703 CT Wageningen, The Netherlands

Z. Schwartz, Dept. of Food Engineering and Biotechnology, The Technion, Technion City, Haifa 32000, Israel

Y. Shoham, Dept. of Food Engineering and Biotechnology, The Technion, Technion City, Haifa 32000, Israel

C. J. Silvin, State University of New York, College of Environmental Science & Forestry, Syracuse, NY 13210, USA

S. W. Tanenbaum, State University of New York, College of Environmental Science & Forestry, Syracuse, NY 13210, USA

R. Taube, Dept. of Molecular Microbiology and Biotechnology, The George S. Wise Faculty of Life Sciences, Tel-Aviv University, Tel-Aviv, Israel

M. Tros, Dept. of Microbiology, Wageningen Agricultural University, Hesselink van Suchtelenweg 4, 6703 CT Wageningen, The Netherlands

R. Turner, Environmental Science Division, Oak Ridge National Laboratory, Oak Ridge, TN 37831, USA

J. E. Walker, US Army Natick Research, Development and Engineering Center, Natick, MA 01760-5020, USA

A. J. B. Zehnder, Dept. of Microbiology, Wageningen Agricultural University, Hesselink van Suchtelenweg 4, 6703 CT Wageningen, The Netherlands (Present address: EAWAG/ETH, CH-8600 Dübendorf, Switzerland)

Z. Zosim, Dept. of Molecular Microbiology and Biotechnology, Tel Aviv University, Ramat Aviv, Israel

Index*

A

Acetivibrio cellulycicus 59
Achromobacter 165
acid rain 115, 118, 119
Acidianus 119
Acinetobacter 165, 178, 179, 181, 215, 220, 230
Actinomadura sp. 105
adhesion
  bacteria to cellulose 55
  bacteria to hydrocarbons 185, 213–216, 221
  fertilizer F-1 to hydrocarbons 220
  halogenated pollutants 141, 142, 154–156
  hydrocarbons to sediments 227
Aeromonas 41, 230
agriculture
  cellulytic waste 49–52, 61
  fertilizer pollution 38
  halogenated organic compounds 175–177
  herbicides 245, 269, 271
  lignocellulose 65–79, 95–110
  nitrogen pollution 133
  pesticides 141, 142, 144, 175–177, 245, 254, 271
Agrobacter 165
Albedo 61, 62
Alcaligenes
  A. eutrophus 2, 23, 24, 39, 41–43, 152, 154, 165, 183, 230
  A. faecalis 132, 230
  A. xylosoxidans 258
algae 42, 116, 119, 136–137, 142, 227, 229, 231, 235
alginic acid 7–10, 42
Alteromonas sp. 267
ammonia (see nitrogen)
anthracene 228, 231, 234, 235
aroclor (see PCBs)
arsenate 23, 24, 39, 40
Arthrobacter 39–41, 132, 164–165, 177–179, 182, 184, 230
Aspergillus niger 60, 61
asphaltene 201
avidin-biotin complex 62

B

Bacilli
  B. cereus 230
  B. stearothermophilus 83, 87–92, 256–259
  B. subtilus 15, 39
Bacteroides cellulosolvens 59, 60
Beggiatoa 117
Beijerinckia 230
benzene 144
benzo (e) pyrene 204, 228, 230, 231, 234, 236, 237
bile salts 265
biobleaching 83–92, 97
biodegradation
  brominated alkanes 175–187
  cellulose 47–63, 97, 98
  chemical agents 253–258
  chlorinated compounds 141–157, 161–172
  evolution of pathways 1–6
  explosives and propellants 245–253, 259
  high salt concentrations 263–273
  lignin 65–79
  lignocellulose 98–110
  nitrogen compounds 131–138
  petroleum 191–211, 213–225, 264
  polychlorinated biphenyls 161–172
  polycyclic aromatics 203–204, 227–238
  sulphur compounds 115–128
biofilm 42, 116, 122, 215
biogas 96
biometer flask 193, 194
bioremediation
  brominated alkanes 175–187
  cadmium 37–45
  cellulose 50–63, 97
  chlorinated compounds 141–157, 161–172
  definition 1
  explosives and chemical agents 245–259
  genetically engineered microorganisms 1–6
  lignocellulose 98–101
  mercury 28–34
  nitrogen pollution 131–138
  petroleum 191–225
  polychlorinated biphenyls 161–172
  sulphur 115–128
biphenyl 230
Brevibacterium 165
brominated compounds (see halogenated)

C

Cadmium 23, 24, 28–33
caffeine 66
carbon dioxide 193, 194, 209, 235, 258, 265, 266, 268
carcinogens 227, 228
catechols 2–7, 97
cellulases 49–63, 83, 97–101, 105

* The page numbers in this index refer to the page numbers in square brackets.

cellulose 47–63, 65–79, 83–85, 87, 97–101
cellulosome 54–63
chemical agents 245, 253–259
  organophosphates 254–257
  mustards 254, 258, 259
chemostat 5, 53, 121, 258
chlorinated compounds (see halogenated)
chlorobenzenes 141
Chlorobium 120
chloroform 144
Chromatium 117, 120
chromium 21
Citrobacter 39, 40, 42
Clostridia 271
  C. cellobioparum 59
  C. thermocellum 52–63, 98
coal 122, 123, 127, 128, 228
cobalt 23, 39
coffee pulp 66
Commonas testosteroni 161, 166, 169, 170
copper 23, 24, 42, 181
corrosion 119, 123
Corynebacterium 165, 178, 180, 181, 230
Creosote 228
Cunninghamella elegans 231, 233, 235
Cyanobacteria 42, 231, 235, 265
Cystic fibrosis 6–10

D

DDT 79, 141
Dead Sea 263–265, 272
dehalogenation 141, 143, 147, 161, 163–165, 171–172, 175, 177, 179–184, 187, 258, 268
delignification 83–92, 97–99
denitrification 121, 124, 125, 131–132, 137–138, 143, 249, 268
desorption 141, 142, 185, 213–215
Desulfovibrio desulfuricans 28
detoxification 227, 229
dioxins 2, 85
DNA
  cloning 107, 154
  colony hybridization 14–17
  fingerprinting 27
  polymerase chain reaction (PCR) 13, 16, 17, 105
  probes 14, 75, 104
  sequences 6, 7–10, 17, 30, 31, 101, 104, 105, 152–153
  transformation 95, 104
  transposon 153

E

Emulsan 215
emulsification 175, 185–187, 198, 199, 213–215, 217, 219
Enterobacter 165
Erwinia amylovora 15
Escherichia coli 14, 15, 17, 18, 24, 121, 152, 256–259, 270, 271
ethanol 50, 181
ethylene 134
evolution 1–6, 141, 148–153
explosives and propellants 245–253, 259, 271

F

fertilizers
  cadmium pollution 38
  nitrogen pollution 136
  oleophilic 191–225
fimbriae 215
Flavobacterium 41, 165, 230
fluorenes 204, 228, 231
formaldehyde 269–272
fungi
  cellulose-decomposing 52, 53, 65
  lignocellulose-decomposing 65–79, 85, 86, 95–110
  nitrifying 132
  polycyclic aromatic hydrocarbon-degrading 227, 229, 231–235

G

gas chromatography 203, 213, 224
genetically engineered microorganisms (GEMs) 1–6, 13–20
  cellulose-degradation 62
  containment 17–20
  lignin-degradation 101, 102
  monitoring 13–17
  PCB degradation 171
Great Salt Lake 264–266, 272
greenhouse effect 115, 118

H

Haloanaerobium 264–266, 272
Halobacterium 264–267
halogenated compounds 2–6
  biodegradation 141–157
  brominated alkanes 175–187
  chlorinated aromatics 83, 85, 141–157
  chlorocatechol 2–6
  chloroform 144
  dioxins 2, 85
  fluorine derivatives 233
  oxidation by nitrifiers 134
  perchloroethylene 144–146
  polychlorinated biphenyls (PCBs) 2, 15, 161–172
  trichloroethylene (TCE) 134–135, 141, 143, 146, 147
halophiles 264–273
hemicellulase (see xylanase)
hemicellulose 47, 52, 83–92, 95, 97–100
hopane 203, 205

hydrocarbons
anthracene 228, 231, 234, 235
asphaltene 201
benzene 144
hopane 203, 205
naphthalene 16, 204, 219, 228, 230, 231, 233–236
oxidation by nitrifying bacteria 134, 135
petroleum bioremediation 191–210, 213–225, 266
phenanthrene 194, 195, 204, 228, 230, 231, 234–236
phytane 191, 201–203, 205
polycyclic aromatics 203, 204, 227–238
pristane 201, 264
toluene 15, 151, 154, 219, 230
Hyphomicrobium 121, 127, 178, 181, 182

I
iron 132, 152, 161, 233

K
Klebsiella aerogenes 39, 40, 42
Klebsiella pneumonia 39, 40
kraft process 84, 86, 97

L
laccase 75
lead 39, 42
lignin 47, 52, 65–78, 83–92, 95–110
lignin peroxidase 75, 86, 95, 100–110, 229
Listeria 15

M
malathion 254
mass ion spectrometry 203
mercury 16, 23–34, 38–40, 119, 183
metallothioniens 23, 38, 42
metals 23–34, 37–45
methane 134
Methylosinus trichosporium 181
mushroom 66, 96 (see Pleurotus ostreatus)
mustards 254, 258, 259
Mycobacterium 177, 179, 230, 236–238

N
naphthalene 15, 16, 204, 219, 228, 230, 231, 233–236
nitramines 246, 249–252
nitrate 121, 131–137, 216
nitrite 131–137
nitrification 131–137
nitroaromatics 246–248, 264, 271–273
TNT (2,4,6-trinitrotoluene), 246
DNT (2,4-dinitrotoluene) 246–247
Nitrobacter 133, 136, 137
nitrocellulose 248
nitroesters 246–250
nitroglycerin 247–250
nitrocellulose 248
nitrogen
biotransformations 131–138
denitrification 121, 124–125, 131–132, 137–138
explosives 245–253, 258–259
fertilizers 136, 191–225
nitrification 131–137
nitrogen fixation 131–132
pollution 131–132, 135
nitrogen fixation 131–132
nitroglycerin 247–248, 249
nitroguanidine 252–253
Nitrosolobus multiformis 135
Nitrosomonas 133–137, 181
Nitrospira 133, 135
Nocardia 170

O
oleophilic fertilizer (see fertilizers)
organophosphates 254–256, 264, 267, 268
Oscillatoria 117

P
PAHs (see polynuclear aromatic hydrocarbons)
paper and pulp 49, 50, 51, 66, 67, 75, 79, 83–92, 96, 97, 117, 127, 162
parathion 254, 271
pectin 61, 62
perchloroethylene (PER) 144, 146
pesticides 141, 142, 144, 175–177, 245, 254, 271
Phanerochaete chrysosporium 65, 66, 76, 86, 95, 98–110, 231, 235, 246, 247
phenanthrene 194, 195, 204, 228, 230, 231, 234–236
phosphate 132, 192, 195–198, 207, 213–216, 219, 220
phytane 191, 201–205
plasmids 3–6, 14, 18, 23, 24, 27, 38, 39, 107, 151, 183, 218, 230, 233
Pleurotus ostreatus 65–79, 96
polychlorinated biphenyls (PCBs) 2, 15, 79, 161–172
polycyclic aromatic hydrocarbons (PAHs) 203–205, 227–238
pristane 201, 264
Proteus mirabilis 39, 41
Pseudomonas 14, 41, 44, 141, 147–149, 152, 154–156, 163, 175, 178, 179, 183, 184, 230, 267
P. aeruginosa 6–9, 24, 39
P. cepacia 5, 17, 42
P. cruciviae 39
P. fluorescens 15, 178, 230
P. halodurans 264, 267
P. oxalaticus 271
P. picketti 258
P. pseudoalcaligenes 165

P. putida 2, 15, 32, 39, 42, 151, 155, 161, 164, 165, 170, 171, 230, 233
P. syringae 25
P. testosteroni 164
pyrenes 204

R
Rhizobium 15
Rhizoctonia solani 231, 235
Rhodococcus fascians 39, 154, 230
Rhodococcus rhodochrous 161, 170, 171, 230
Rhodospirilla 120
Riftia 121
Ruminococcus albus 59

S
Saccharomonospora viridis 105
Salmonella 15
sarin 254
Serratia marcescens 270, 271
soman 254–256, 257
Sphingomonas 155, 156
Spirillum 41
Sporohalobacter 264, 272
Staphylococcus aureus 23, 24, 39, 40
starch 54, 98
Streptomyces 95, 100, 105–110, 230
strontium 42
suicide vectors 18, 19
Sulfolobus 119, 121, 122
sulphate-reducing bacteria 115, 116, 119, 144
sulphur compounds 115–128, 132
acid rain 115, 118, 119
aquatic 119, 124, 125
atmospheric 118, 119
coal 122, 123, 127
emissions 117
hydrocarbons 116, 117
production 126
terrestrial 119
sulphur-oxidizing bacteria 115, 116, 120–128

T
Tetrahymena thermophilia 256, 259
Thermomonospora fusca 105
Thermomonospora mesophila 105
thermophilic microorganisms 52–63, 83, 87–92, 119, 121, 128, 254
Thiobacilli 120
T. ferrooxidans 121, 123
T. denitrificans 121
T. neapolitanus 121, 126
T. thiooxidans 119
T. thioparus 121, 127
T. versutus 121, 122
Thiomicrospira 121
Thiosphaera pantotropha 132
TNT (see nitroaromatics)
toluene 15, 151, 154, 219, 230
transposon 153
triaminoguanidine nitrate 250
trichloroethylene (TCE) 134–135, 141, 143, 146, 147
Trichoderma reesei 90, 103

U
uranium 25, 38, 42
urea 196, 197, 198, 216

V
Vibrio costicola 267
Xanthobacter autotrophicus 178, 181–182
xenobiotics 141–143, 148, 161, 235
xylanase 57, 83, 86–92, 95, 97, 105–110
xylene 154, 219

Y
Yersinia 15

Z
zinc 23, 37–40, 42
Zoogloea 41, 42

GPSR Compliance

The European Union's (EU) General Product Safety Regulation (GPSR) is a set of rules that requires consumer products to be safe and our obligations to ensure this.

If you have any concerns about our products, you can contact us on ProductSafety@springernature.com

In case Publisher is established outside the EU, the EU authorized representative is:

Springer Nature Customer Service Center GmbH
Europaplatz 3
69115 Heidelberg, Germany

Batch number: 10816681

Printed by Printforce, the Netherlands